Wildlife
Wild Places

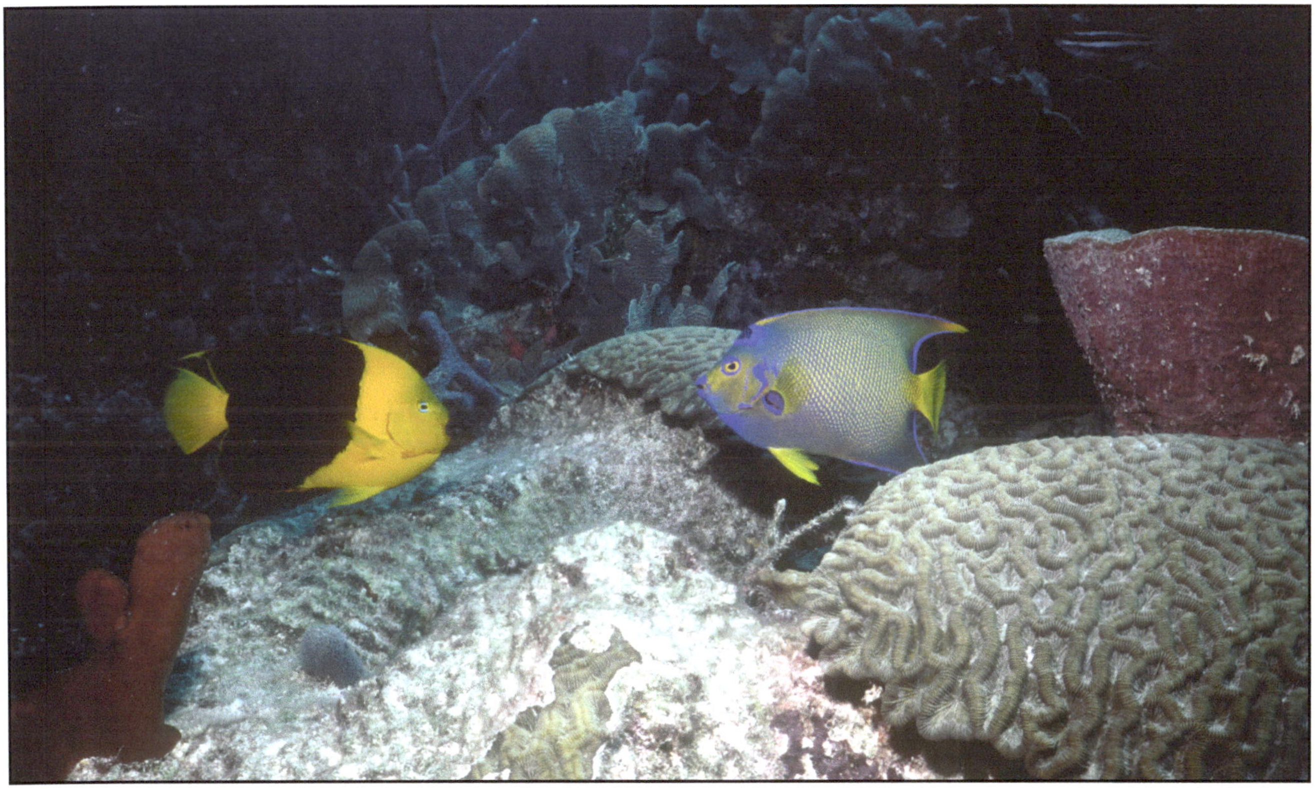

Dedication

I would like to dedicate this book to the three men who made me what I am today.

Alex Ferguson took a fourteen year old boy in 1956 and taught me all there was to know about the brand new science of scuba diving.

Arturo, a native of the Amazon forest, taught me to see, when there was nothing to see and to hear, when there was nothing to hear.

Captain Henny Tromm, owner/captain of the Isla Mia, shared his home on the sea for three years in three countries.

Good men all, sadly they are gone but not forgotten.

Jim Beveridge

Wildlife Wild Places

Text and Photographs by Jim Beveridge

Producciones de la Hamaca

Caye Caulker, BELIZE

2012

Published by *Producciones de la Hamaca*
Caye Caulker, BELIZE
<producciones-hamaca.com>

ISBN: 978-976-8142-31-3

Front Cover: (*top*) Jaguar (*Panthera onca*)

(*bottom*) Cave at Actun Tunichil Muknal National Monument

Back Cover: Christmas Tree Worms (*Spirobranchus giganteus*) on a coral head

Producciones de la Hamaca is dedicated to:
- Celebration and documentation of Belize's rich, diverse cultural heritage,
- Protection and sustainable use of Belize's remarkable natural resources,
- Inspired, creative expression of Belize's spiritual depth.

Contents

Preface

Even as a young kid, I was fascinated by wildlife!

It all started with stealing caterpillars from our neighbour's cabbages and relocating them into my mother's vegetable garden. I studied them until they became chrysalises, then, ultimately pretty white butterflies.

From caterpillars, I advanced to frog spawn, then through to tadpoles and ultimately froglets, which I then released at the edge of a "burn". (Scots for stream)

As I grew close to my teenage years, the big thing with kids in the U.K. was collecting birds' eggs. This basically meant piercing it at both ends and blowing out the liquid yolk and white inside, leaving an empty shell which was then put on display. I didn't like this, for it seemed akin to murder. I did start to watch birds, and I even got to identify several of them, such as skylarks and robins.

My interest in both frogs and my feathered friends came to a stop when I turned twelve and moved up into secondary school. This school offered swimming instruction as part of physical education.

I was a non-swimmer and had problems learning how to stay on the surface, without coughing and spluttering, but I had a good pair of lungs and had no trouble swimming underwater. I was normally the winner of fifty yard races, even though I swam the total distance underwater while my opponents swam on the surface.

I was fourteen years old when I was allowed to join the Scottish Sub-Aqua Club as a junior member. Shortly after this, I joined the Royal Life-Saving Society.

By then I could swim on the surface, as well as underwater. Two nights a week I learned scuba diving and two other nights attended life-saving classes. On Saturdays, I went to the local swimming pool and swam two miles. During my teenage years, all thoughts of wild life, and everything else went out of my mind. However, I did excel at swimming.

I was awarded the Award of Merit by the Royal Life Saving Society in 1961. Shortly after my twentieth birthday in 1962, I sat the exam and became the youngest First Class Diver in Britain.

I revived my interest in wildlife, but all the wildlife I loved was marine, so I began underwater photography. It wasn't until the 1990s that I became once more interested in amphibians and birds, and expanded my photography to include land creatures.

In March of 1965, I left Scotland and went to Canada. I was lucky in that I hit Toronto during a big construction boom. During the summer months I worked as much overtime as possible, sometimes over eighty hours a week. I lived cheaply, and then around October when the chilly winds started to blow I moved into my camper and headed south, sometimes alone and sometimes with a friend.

The National Parks of the U.S. were my winter residence for two years, but gradually I went a little further south each year until southwest Mexico was a regular destination. Strange as it may seem, I actually saved money with this kind of lifestyle because I worked lots of overtime at time and a half and even double time. By 1969 I had enough saved that I could travel around the world for a year.

As I will describe, much to my surprise, I ended up in Belize, the small nation snuggled between Guatemala in the west and south, Mexico to the north, and the Caribbean Sea to the east.

Formally called British Honduras, Belize received its independence on September 21st, 1981. According to the 2010 census, the population of Belize is approximately 312,000. Belize has an area of only 8,805 square miles, of which a little over two percent is arable. The north section of Belize is low lying with sugar cane as the main crop. The south of Belize has the Maya Mountains and the Mountain Pine Ridge, a large area of rugged pine forests.

Off the eastern shoreline of Belize is the Caribbean Sea and the Belize Barrier Reef, a continuous coral reef, approximately 175 miles long within Belizean waters, and extending north into Mexico's territory to a point just south of Cozumel. The Belize Barrier Reef is second only to Australia's Great Barrier Reef in length. The Barrier Reef is alive and vibrant over most of its length except for a little storm and tourism damage when close to resort areas. Beyond the Barrier Reef, Belize is also blessed with three atolls that also offer superb diving.

Wildlife abounds in Belize. In fact, roughly forty percent of Belize's total area is protected. This includes Cockscomb Basin Wildlife Sanctuary, better known as the "Jaguar Preserve", Crooked Tree Wildlife Sanctuary, which is superb for birding, and the Rio Bravo Management and Conservation Area approximately a quarter of a million acres.

In this book, I will be sharing with you photographs and stories from the last 42 years that highlight the life forms of unique eco-systems and their inhabitants.

Chapter 1
Roads Least Travelled
Baja California

In the Spring of 1969 I left Toronto on a trip with two friends. For me, it was the start of a trip around the world. For my friends it was an extended trip to Mexico. My companions were a German couple Maria and Reinhard, whom I had befriended several years previously in Toronto.

Our mode of transport was a 1964 Ford window van that had previously been a Bell Telephone repair truck, but I converted it into a camper. It took us four weeks to reach the Mexican border at San Diego and over two months to drive the length of Baja California. In the late 1960s, Baja California was truly a magical and wonderful place – a peninsula located at the northwest coast of Mexico, and running south with the Pacific Ocean on the west coast and the Sea of Cortez to the east.

Only the extreme north and the extreme south were populated. The remaining territory in between was uninhabited – a desert wilderness full of flowering cacti and strange bizarre plants unique to Baja called "boojum trees." The boojum is not a tree, neither is it a cactus. It is an alien, other-world-looking plant that grows over thirty feet high with no branches excepting rigid leafy stalks protruding at right angles from the lower trunk and a spray of white flowery twigs at the top. As far as I know, the boojam grows nowhere else on earth.

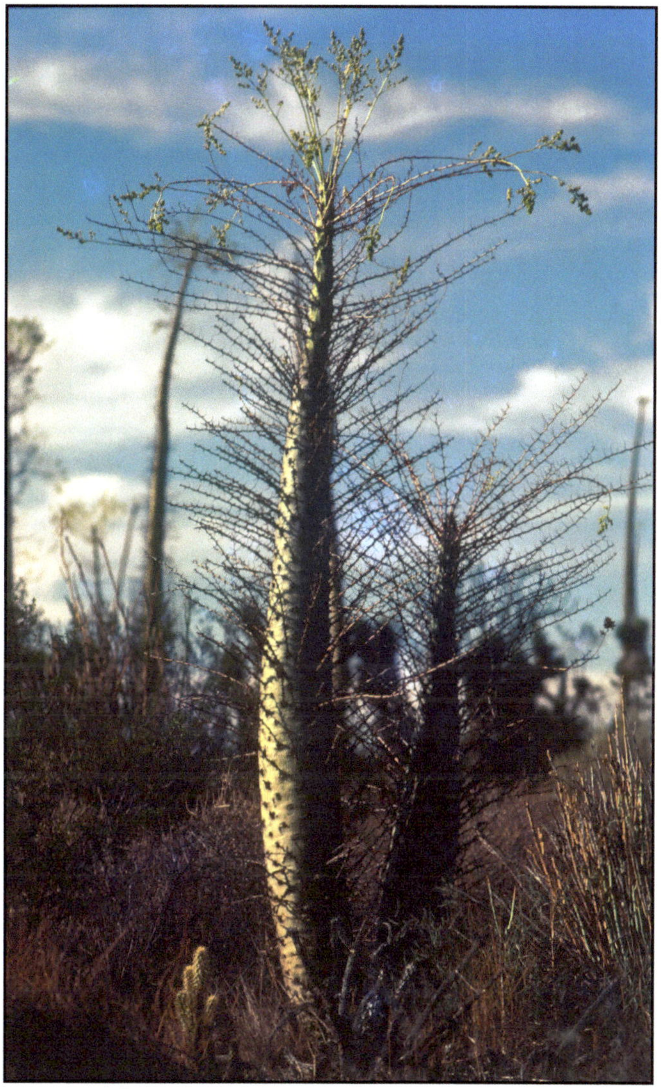

Boojum Tree (*Fouquieria columnaris*)

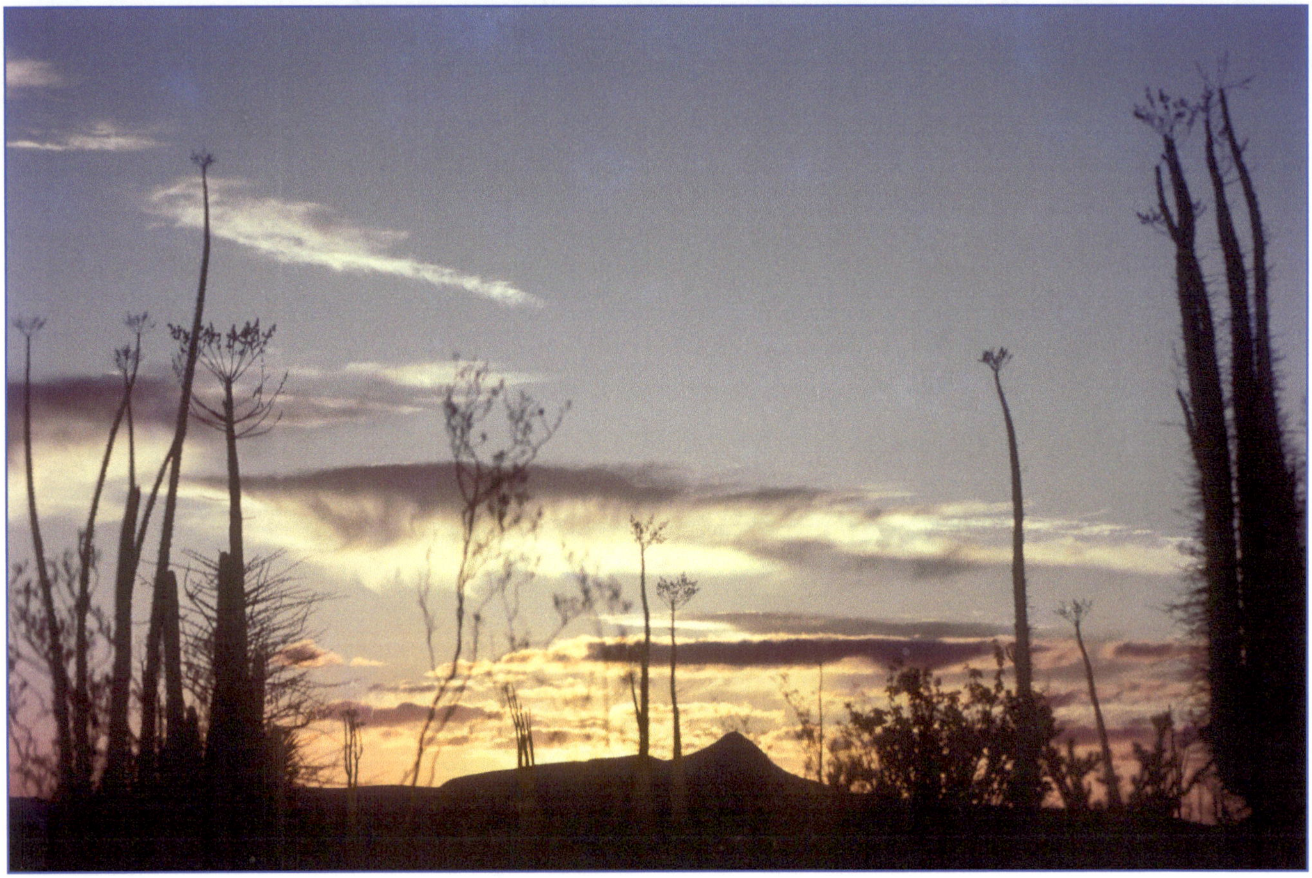

A rising sun illuminates boojum trees

The length of Baja is slightly more than 700 miles as the crow flies, but our route easily doubled that distance by zigzagging us at an inebriated snail's pace.

Our snail was, of course, my five-year-old Ford window van camper capable of sleeping three, and equipped with four spare wheels and four additional five-gallon fuel tanks. We knew that we wouldn't see a gas station until we arrived at La Paz at the southern tip of the peninsula. We also carried enough dried and canned foods, plus bottled water to last three persons in a desert for over a month, as well as my camera and snorkeling equipment, plus one spear gun with which I hoped to supplement our meagre food supply.

Back then Baja had no roads, only sandy tracks seven feet wide at the most. Shear rock faces challenged any vehicle to make it to the top and down the opposite side without incident or accident. The trip through Baja started relatively easy for the first 50 miles, then we left civilization behind us and replaced it with beautiful desert scenery. Then the road degraded to a narrow sandy track; to this we paid no mind for the surrounding area was covered with beautiful flowering cacti. We drove until dusk, then pulled off the track and made our first meal as the sun began to sink in the west, painting our immediate eastern desert with a myriad of shades of yellows and reds of the flowering cacti.

The next morning I went for a walk in the desert before dawn. I was amazed at the diversity of the cacti, for it seemed that every size, shape and flower was represented in an amazingly small area. I was still walking west when growing shadows told me that the sun was now rising.

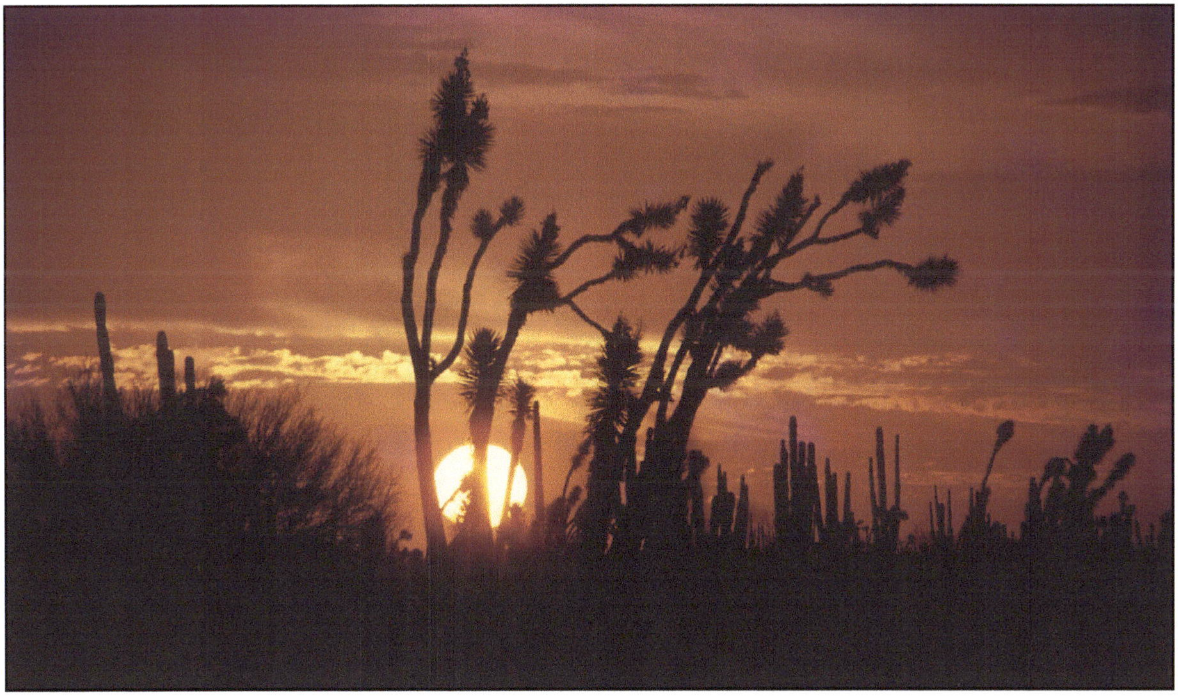

Sunset in Baja

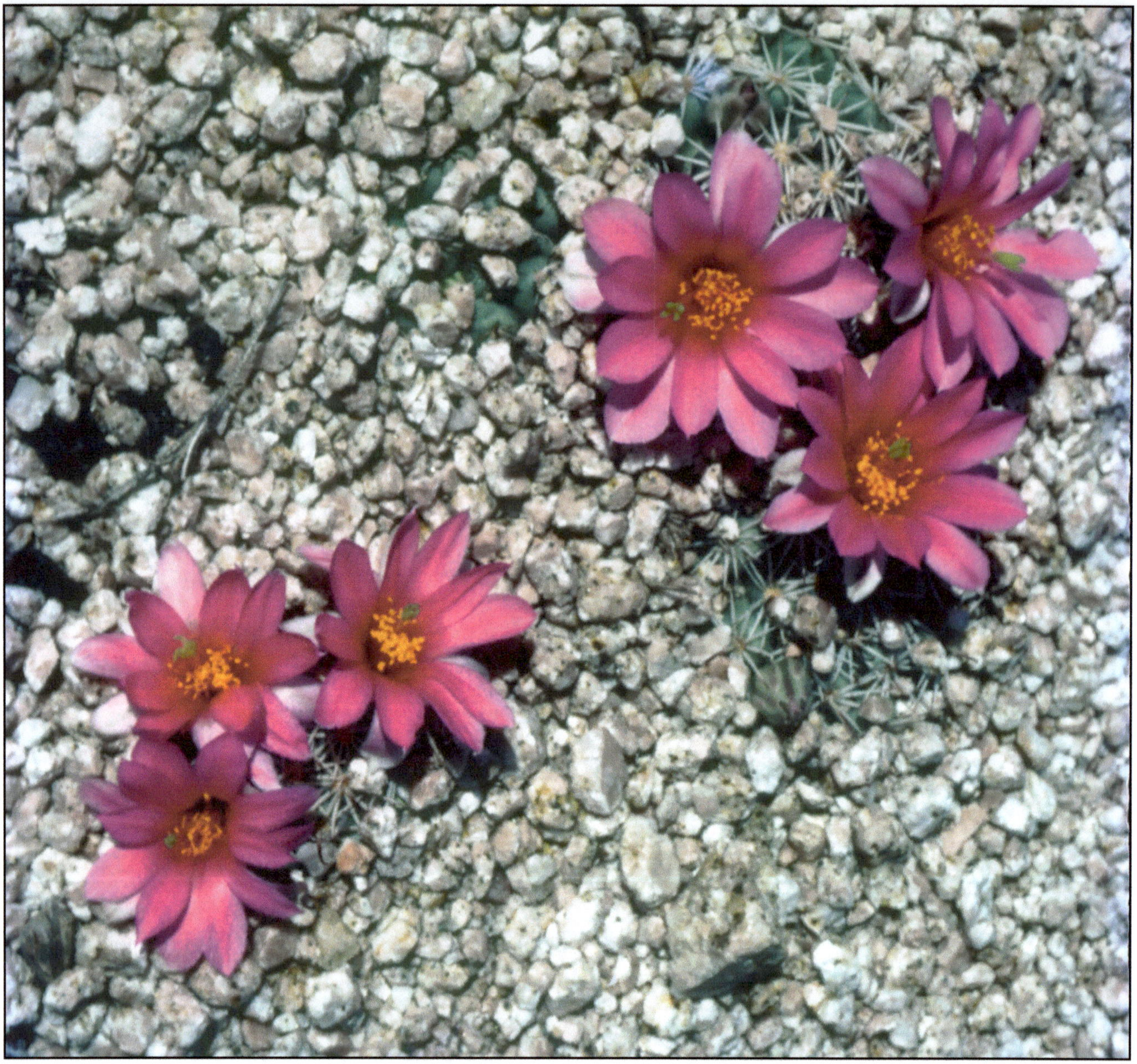

Cacti blossoms in Baja

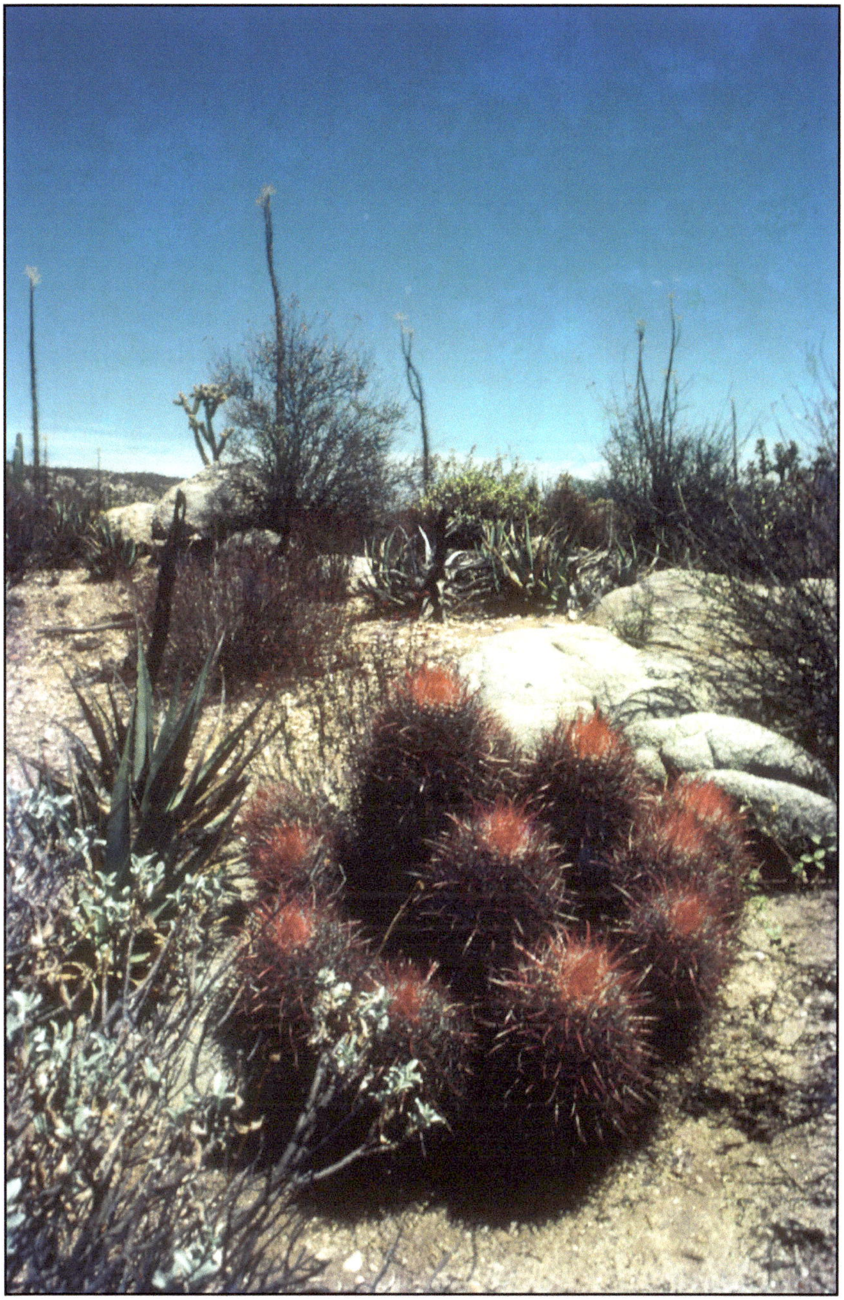

Cacti blossoms in Baja

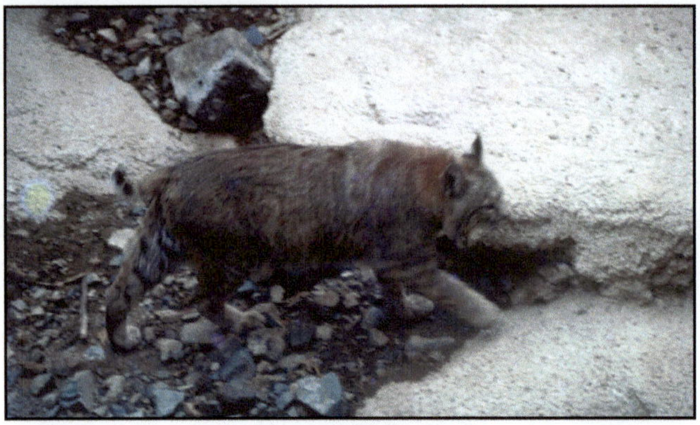

Bay Lynx (*Lynx rufus*)

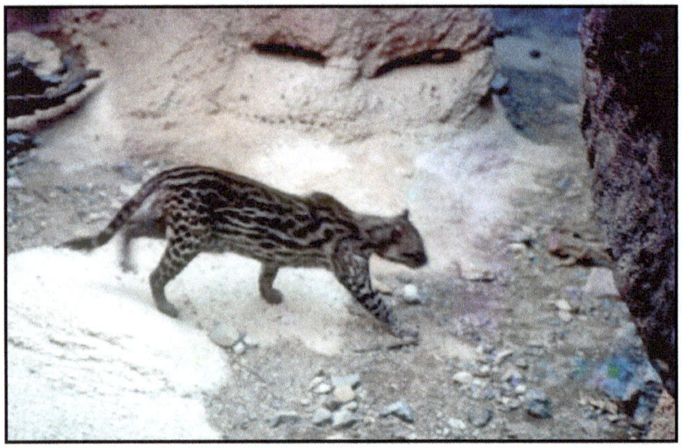

Ocelot (*Felis pardalis*)

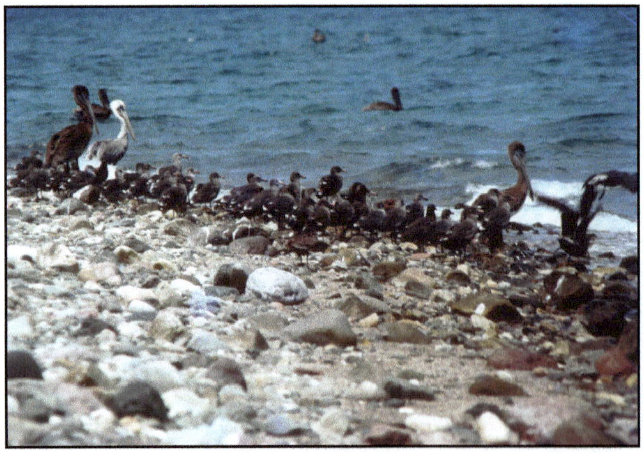

Baja sea birds

I had gone no more than 50 yards when I caught movement to my right with my peripheral vision. I turned quickly but was looking at a small group of large cacti. Suddenly a medium large animal broke away from the cacti. It ran about 20 feet, then stopped, long enough for me to take its picture. It was a lynx! The first I had ever seen. It stayed a few seconds, then broke into a slow trot and disappeared behind a rocky outcrop. Like all large cats, lynx are extremely difficult to see in the wild, never mind photograph. What a perfect start to a trip!

Fourteen years later, I made a second trip through Baja California, this time with Dorothy who would later became my wife. In approximately the same area that I had experienced the lynx so long before, we came across a second large cat—not a lynx, but an ocelot! Living in Belize I was familiar with ocelots but it took me some time before I could bring myself to believe that this was the same animal that lived in the Belizean forests. Baja was far from being a forest, rather it was a dry desert. The animal itself was recognisable in spite of its markings being quite different. I found out later that the ocelot has a huge range from southern Amazonas in Brazil to north Mexico. This covers a lot of extremely different terrain.

The lynx, like the ocelot, were both remarkable sightings and I counted myself extremely lucky not only in catching a glimpse of them but also in capturing them on film. Of course, the rarest creature in Baja back then was man, which no doubt was a significant factor in having such a close encounter. In 2011, I doubt that I would be so lucky.

There are many other creatures a sharp-eyed traveller might come across. Along the coastline of the Sea of Cortez, there are many seabirds.

In the desert, there are many reptiles one may see during the dawn and dusk period. These bury themselves during the heat of the day. Both lizards and snakes can be found then, some of them quite venomous. I was lucky coming across quite a rare creature. It was a fringe-toed lizard. This is quite a bizarre creature and seldom observed. It emerges only at dusk and dawn to feed on various insects.

The next day we came across an old man sitting by the side of the road, a two-foot-square cardboard sign was propped against his knees. It read in English, "Spare a dollar for the road cleaner-upper". There was no sign of a building, nor any other signs of habitation! The road, an eight-foot-wide indentation was certainly clean of any garbage, as was the fifty or sixty miles we had just driven south on. We were very puzzled; the old man didn't explain himself, but just acted dumb. We gave him twenty pesos, at that time a little less than two US dollars, plus a bottle of coke and a few cookies. That road-cleaner upper man was the last person we saw until we hit the southern point two months later!

After a week or so, our once sandy path became rocky, steep and shear. One particular hill took us two days to conquer. It was long, narrow, and smooth. The surface was shiny rock, almost slate like. The hill was close to a half mile long, and the road was, at several points, only a lane wide, with sheer drops on either side. On looking over the side, I noted that at least two vehicles in the past had not made the top. Their wreckage was strewn across the rocks, seventy feet below!

I asked Maria and Reinhard to climb up and station themselves at points that were particularly steep or narrow. I put the truck in first and started climbing. I reached Maria's position, but not Reinhard's. Still in first gear, it stopped; the drive wheel had no traction because the road was too smooth and sheer.

Luckily, the handbrake held. I thanked my lucky star that I had the foresight to get a brake job before leaving Toronto. I took off in first gear again with Maria and Reinhard pushing, but still no go. I had them look for four large rocks to put behind the wheels, to fortify the handbrake, then I left the driver's seat.

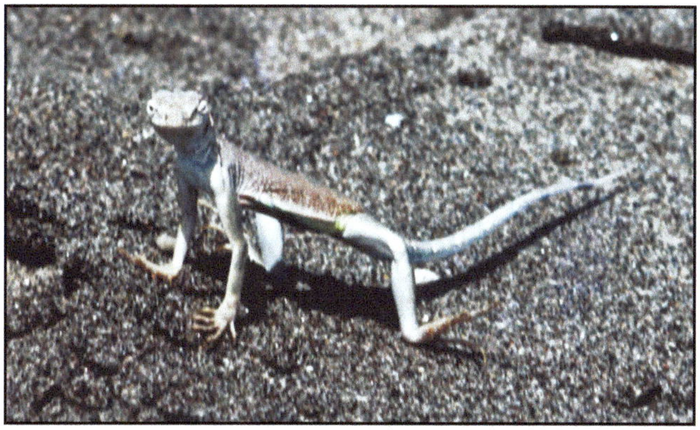

Fringe-toed Lizard (*Uma inornata)*

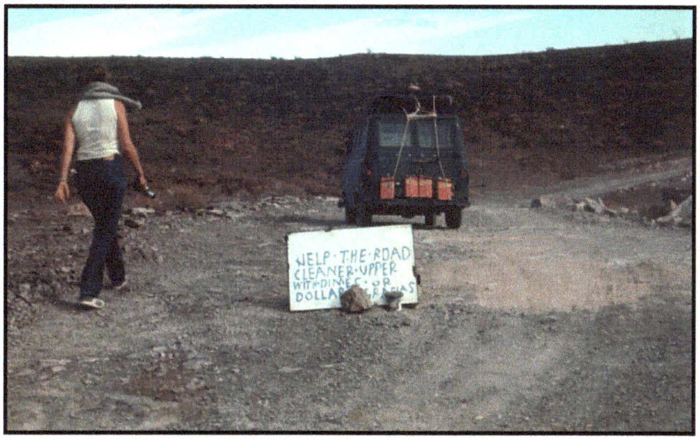

The road cleaner-upper sign

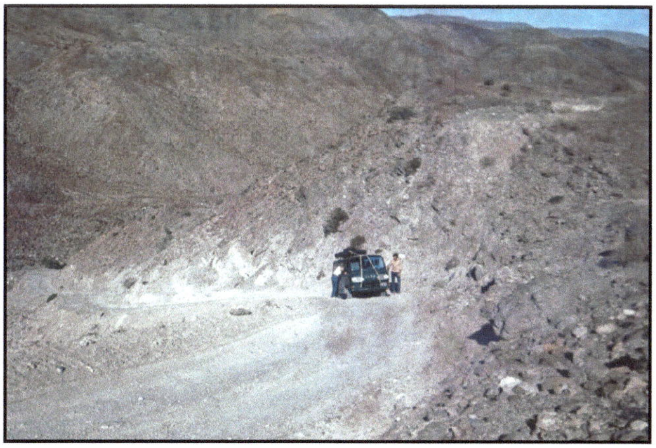

Stuck on a smooth, steep Baja road

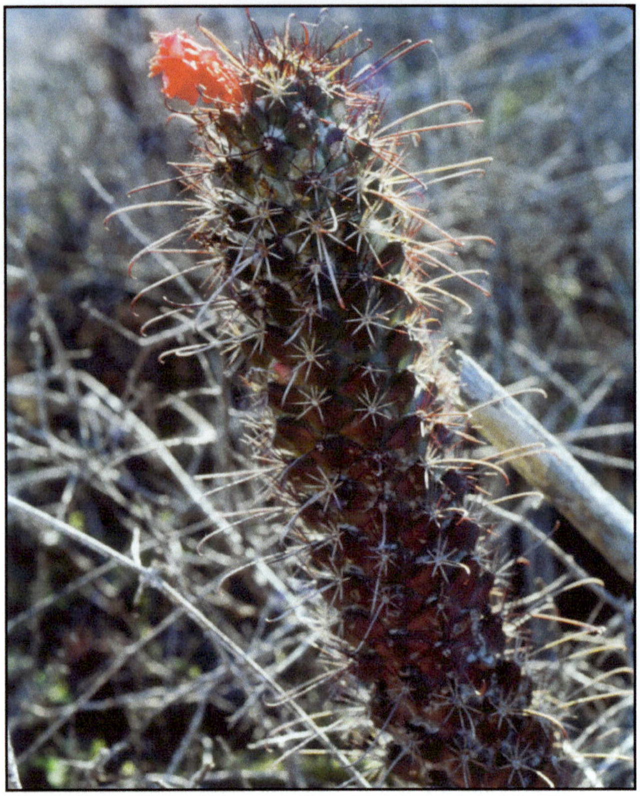

(*above and below*) Baja cacti

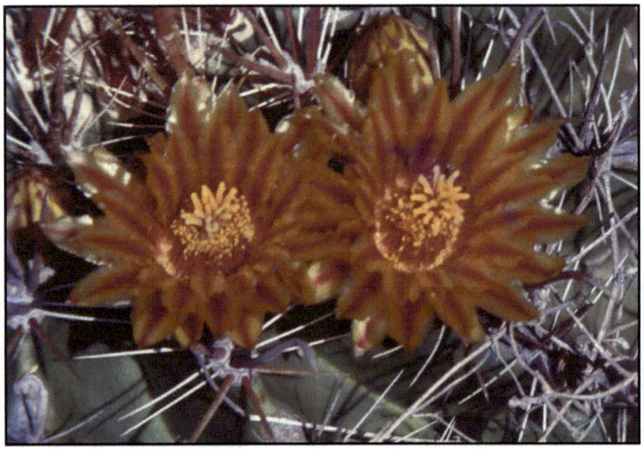

Together we removed twenty gallons of gasoline, four spare wheels, about fifty pounds of canned food, altogether between five and six hundred pounds, and placed it all along the edge of the road. Then with Reinhardt and Maria pushing and me treading carefully on the gas pedal, we took off at the grand speed of 4 mph and made it to the crest of the hill. I stopped the truck, fixed the handbrake and to be doubly sure, placed rocks on both sides of two wheels. I then walked a quarter mile down the incline to help Maria and Reinhard carry up our weighty goods.

We estimated that we had travelled approximately 45 percent of our total distance, and thought the worst was now beyond us and one more week would take us to the car ferry, which in turn would carry us across to mainland Mexico.

The road was narrow and below the level of the surrounding desert. It was quite picturesque with both large and small flowering cacti were growing alongside rock formations.

Up to this point, Reinhard and I had been the drivers, but Maria felt that she would like to drive at least one section of Baja, before we hit the mainland. No problem, I gave up the driver's seat and Maria took over, while I went into the bunk and promptly fell asleep.

Approximately three hours passed, then all hell broke loose. I was thrown out of the bunk and no sooner had I landed than I was bounced off the side door. I tried to stand and had no sooner regained a semblance of balance when there was a loud bang and a sudden stop and I flew forward between the passenger and driving seats.

As I climbed out, my initial impression was that the truck had escaped harm. Then I saw that the front right wheel pointed at right angles from the truck and the left front wheel also pointed out at right angles. I was no mechanic, but I realized that we had a problem!

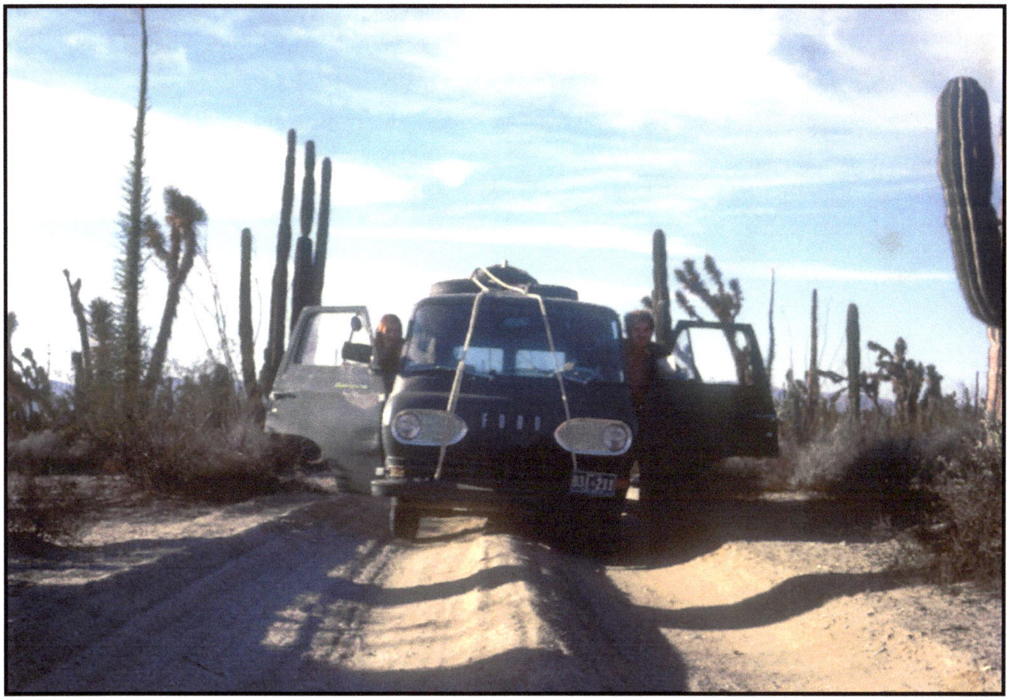

Switching drivers in a Baja desert

The front end had ploughed into the soft sand, and it was difficult to get the jack under the truck. However, eventually we managed to raise the two front wheels about six inches above the sand and pushed some flat rocks under them.

Then we could see that the tie rod, the metal bar or tube that connects the two front wheels together, was broken dead centre! Our little tower of flat rocks on which the truck was balancing was far from safe, so we substituted our four spare wheels for the rocks.

Our problem then was how to securely repair the tie rod. Without a welder it didn't seem possible and we were over four hundred miles from help, a long walk in the desert!

I remembered that before leaving Toronto, I had reservations about taking my tool-box because it weighed close to eighty pounds. In the end I decided to take it, and, like taking four spare wheels, that decision saved us.

The tools were spread out on a tarp with the two parts of the broken tie rod. Reinhard and I stared and contemplated. We were both handy with our hands. He was an engineer; I was a plumber and pipefitter.

In my set of tube wrenches, I found one that was a little larger than the outside diameter of the tie rod. Holding the wrench with vice grips, we used my hacksaw to cut off the hex end, leaving only a straight tube which slipped over the broken ends of the tie rod.

Then came the tedious work, for I only had a small wheel drill. But a day and a half later we had four holes drilled through the former tube wrench, using a concrete nail to start it off and a high speed drill bit used very slowly. Through our new coupling holes we scratched the tie rod with a concrete nail. It

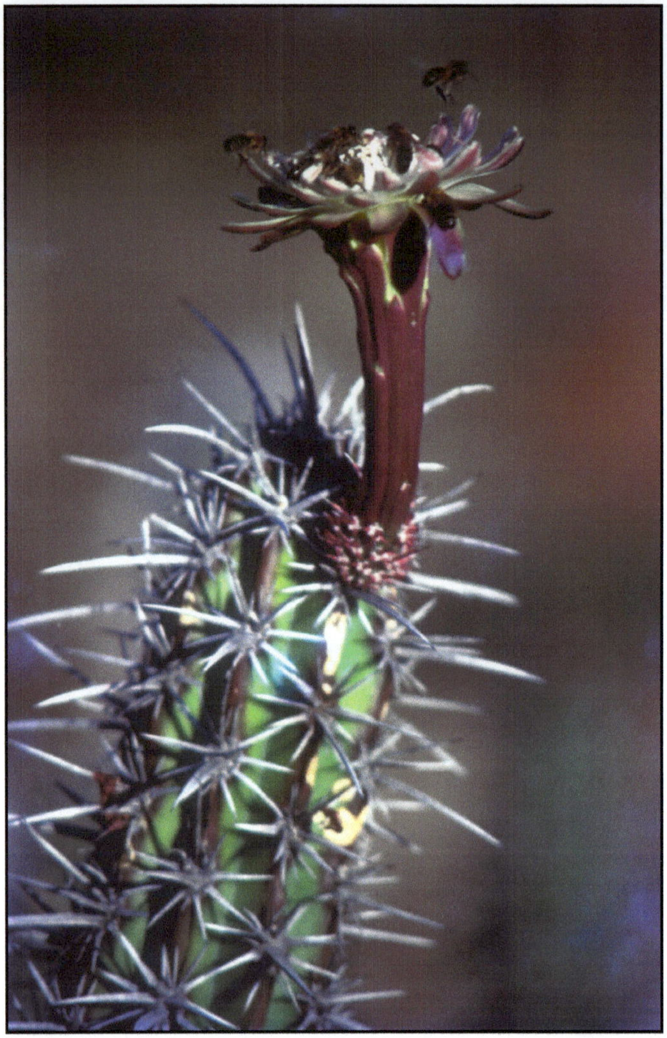

Beautiful flowering cacti

took us another day and a half to drill the four holes in the tie rod.

The moment of truth had come. Our repaired tie rod got us the remaining 400 miles to the car ferry. In mainland Mexico, our emergency repair was removed and the tie rod welded.

When we reached Acapulco, I parted ways with Maria and Reinhard, and I drove to Oaxaca, one of my favourite Mexican Towns. From there I headed to the Yucatan Peninsula for a few weeks of snorkelling and fishing at the Barrier Reef.

Camping at Laguna Milagro, I made friends with many young Americans. Most, but not all, were draft-dodgers avoiding Vietnam. It was my intention to sell my truck before heading any further, but it took ten days to sell it, during which time I met Larry who was travelling to British Honduras to visit friends who were homesteading there. Just prior to leaving the U.S., he had purchased a new Toyota Land Cruiser of which he was extremely proud. We both were ready to leave Mexico at the same time, so he invited me, now a pedestrian to ride with him!

There was a sign directing us to Belize City, but after a couple miles of single lane, broken rock, pot-holed track, we figured we had somehow missed our way and returned to the border. We were duly corrected and sent back to our original route. Larry wanted to test his new Land Cruiser. Gradually he built up speed until he was bumping along at sixty miles an hour. We had driven about 80 miles when suddenly there was a loud crack, followed by the roar of the engine. He had broken the back axle!

It was three p.m. and there were neither signs of any traffic nor of buildings. We waited for a vehicle, any vehicle, but none appeared. We pushed the land cruiser off into a little clearing. Larry slept that night in his vehicle; I slept in my hammock which I'd strung between two trees!

Next day about 11 a.m. a truck came by, Larry made a deal with the driver to tow the Toyota the remaining miles to Belize City. We were dropped off at a mechanic shop. I said goodbye to Larry and took a taxi to the bus terminal, but the bus proved to be a large truck, which got me to the Guatemalan border several hours later. After only one day and a half I had rushed out of Belize.

Now, looking back to 1969 in Belize, I recall only one emotion! I was "underwhelmed." Little did I realise that this country would be my final home!

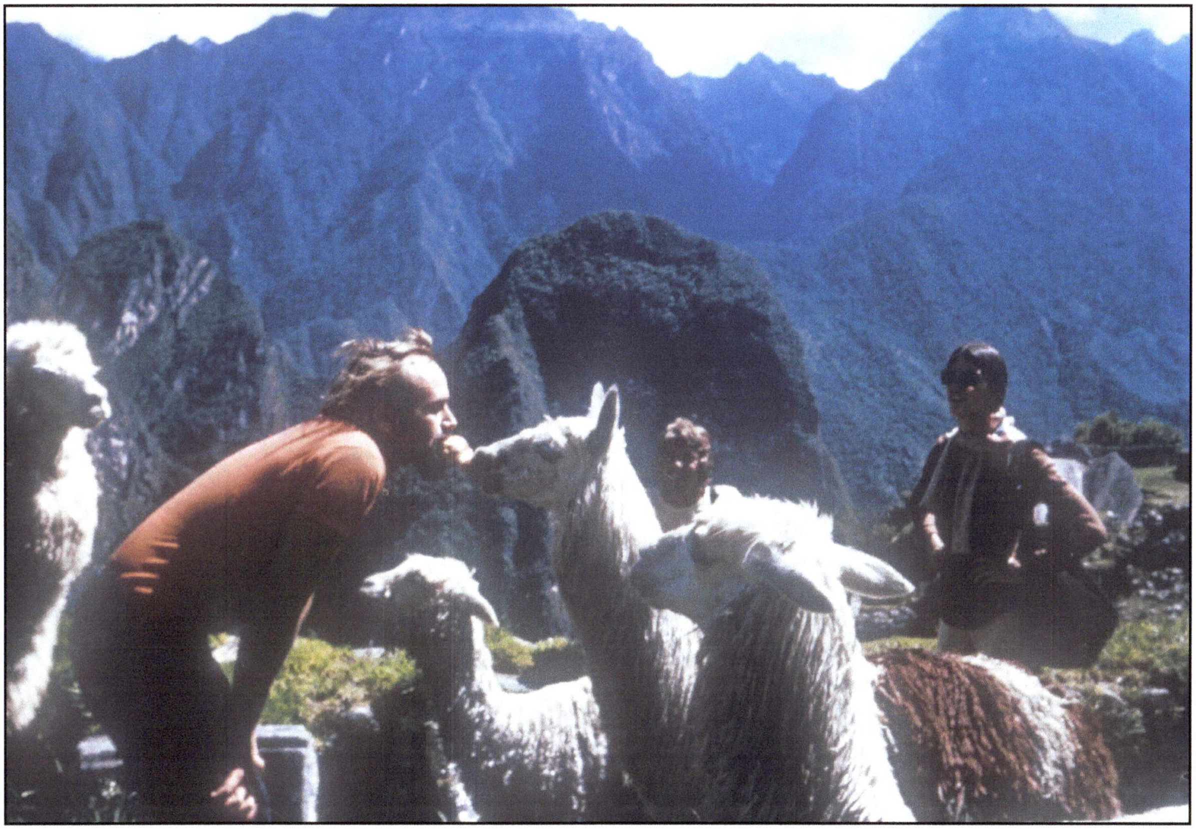

Trisha Corbet

Using a piece of bread to make friends with an alpaca at Machu Picchu

South America

On arriving in Antigua, Guatemala, I met a couple who were on the verge of heading in separate directions. The male half of the couple found it necessary to return home to the U.S., but Trish Corbet wanted to go on to South America. When they found out that I was travelling onwards to South America they invited me to coffee at a small café. Would I mind escorting her down south? I didn't mind at all.

Two days later, Trish and I flew to Bogota, and then travelled by bus and truck, first to Ecuador, then Peru. In Peru we stayed at Cusco, then Machu Picchu. When we returned to Lima, Trish was not feeling very well. A doctor confirmed that she had a kidney infection, gave her some antibiotics, and advised her to return to the U.S. as quickly as possible. Early the next day, we both flew to Bogota, and without leaving the airport, I waited until the Miami flight to see her off.

I was alone and back at my starting point in South America! I had little wish to repeat that long bus/truck trip through the mountains of South America, so I walked through the airport looking at the different destinations.

At the Satena airline desk, I noted that a plane was leaving in half an hour going to Leticia, a small town in Amazonas, Columbia. It sounded interesting, so I bought a one-way ticket. When I boarded, I noted two

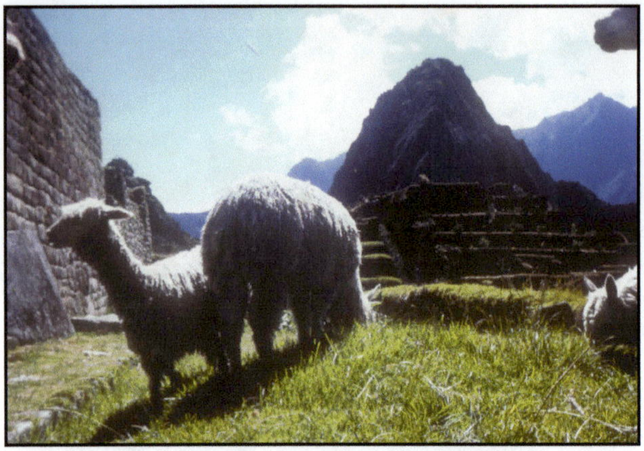

Alpaca (*Vicugna pacos*) at Machu Picchu

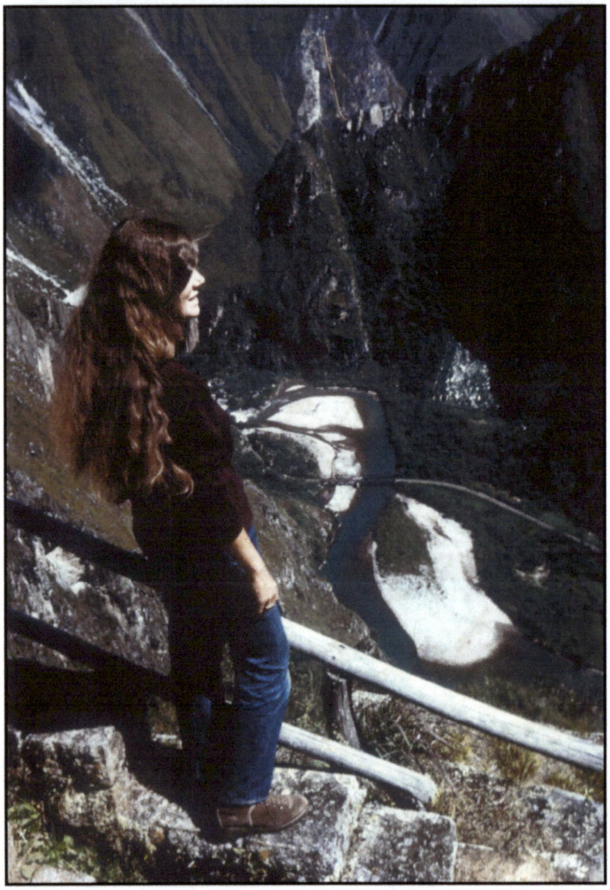

Trish at Machu Picchu

things: the plane was not a jet, but an old Hercules freight plane, and I was the only passenger! In spite of my misgivings, I arrived safely. Leticia lies on the banks of the River Solimoes, the major western tributary of the Amazon River. Across the river, which was a mile wide, was Brazil. Walk south along the bank four miles and you were in Peru.

Approximately 200 miles south of Machu Picchu in the Peruvian Andes lies an 18,300-foot mountain called Nevada Mismi. It is from here that one of the headwaters of the Amazon River start as seeping melt-water from a glacier, to ultimately become the Maranon River, which will combine with other mountain torrents, in addition to many forested waterways to become the greatest river in the world, known as the Amazon. The Amazon River system drains the flood plains and forests of Brazil, Bolivia, Columbia, Ecuador, Peru, Venezuela and the Guianas, approximately 40 percent of the South American Continent!

These waters ultimately enter the Atlantic Ocean at a point exactly on the equator. Here the mouth of the river is 210 miles wide, and such is the volume of water, that freshwater can be pulled from the ocean, when out of sight of land.

The river has over three thousand documented species of fish, many of which are dangerous to man, the most dreaded being the piranha. Actually there are numerous species of piranha and most of them only eat fish. It is the schooling piranha that are dangerous. Other dangerous fish include the bull shark, which is known to swim as far as 2,200 miles up the river, and catfish that may reach ten feet in length.

The fish that the native people are most afraid of is the Candiru. This is a tiny, five-eighth of an inch long fish, that is so attracted to urine, that it will follow traces to the source. Its dorsal spines fold back, and allow forward motion through the urinary track, but it's one way only. A hospital visit is generally necessary!

Of course, everyone has heard about the electric eel. I was unfortunate in having an encounter with an electric eel, and it was really no thanks to me that I survived the encounter!

I was walking along a tributary edge at the beginning of the dry season when the river water was receding. I happened to notice what I thought was a lungfish that was trapped away from the water. I decided to do my good deed for the day, so I attempted to lift it up!

I don't know how long I was unconscious, but when I awoke, I was lying flat on my back with my feet facing the eel, which was still lying motionless. If I was prepared to save the life of a lungfish, why not an electric eel? I found two sections of broken branches about five feet in length and dropped one end of the wood across the eel's back and with one finger touched the other end—no signs of a shock. So I pushed both branches under the eel and edged it into the water. It lay there, in one foot of water for five or six minutes, then slowly swam out.

At the time, I was totally unaware of how lucky I was. Electric eels are one of the most deadly electric fish in the world, fully capable of releasing a discharge of 550 volts at one ampere and keep it up for hours! They have been known to kill a man on contact, and knock a horse unconscious at twenty-five feet while it is crossing a river. I felt particularly lucky that I survived an encounter with an electric eel, and can only suppose that it was particularly weak, after being dehydrated by several days of equatorial sun!

An electric eel is not actually an eel. It is the only member of the genus *Electrophorus*, and is more closely related to catfish than to eels. They grow as long as six or seven feet and about 50 pounds in weight. Their habitat is in shallow flood plains, swamps, and creeks of the Amazon and Orinoco Rivers, feeding mostly on invertebrates, fish, and amphibians. Electric eels have to surface to breathe air. If you hold one under water, it will drown! On top of that, adult eels are blind!

Another large fish is the arapaima, a primitive looking fish with enlarged armoured scales and continuous caudal, rear dorsal, and anal fins. The arapaima can reach a length of over eight feet and weigh in excess of 200 pounds, from which 75 percent of the flesh can be used as food. The meat is delicious; this I can attest to myself, as I ate it quite often. It is hunted both by net and harpoon, and sadly it is at a disadvantage, for like the electric eel, it has to surface to breathe.

Apart from fish, the Amazon and its tributaries also have many other creatures dangerous to man. These include four species of Caiman, the largest of which is the black caiman at 15 feet and the smallest being the dwarf and the smooth-fronted at approximately four feet. Then there is the anaconda, which tends to lie in the shallow lagoons with only its nostrils showing.

Leticia could be described as a medium-sized village. There were two places where one could find a bed for the night, one was large and expensive looking, owned by a Greek-American; the other was small, with six rooms, plus a place where one could sling their own hammock. Being a Scot, I chose to sleep in my hammock.

A young American couple in their twenties moved into one of the rooms and argued most of the night. On their third day, the male part of the couple walked out carrying his luggage. The following day I was sitting having the strong black coffee, locally called a "tinto", in the tiny room they called "the Restaurant," when the women asked if she could join me. Gradually the story came out: they were childhood sweethearts in New York City. At the age of sixteen, the boy's family moved to Los Angeles. They were heartbroken, but wrote to each other every week. Six years later they decided to elope. They met in Miami and after some searching decided on a place where they were sure no one would ever find them: Leticia!

Later that day as we were walking along Leticia's main street, which paralleled the river, we happened to notice a truck, with a cage full of squirrel monkeys. Immediately the woman decided she wanted a monkey. I tried to tell her that she didn't want a monkey and couldn't take it back to N.Y. But she wouldn't listen! She had decided that if she couldn't have a man, a monkey was the next best thing.

That is how I met Ray, a very large Texan who had lived in South America for many years. He and his partner George were in the animal business supplying mammals, birds, and tropical fish to the "civilized" world. They were shipped to zoos, research centres and pet stores around the world. As an animal lover, I was a little dismayed at this. Ray was as adamant as I was that a monkey was not a good idea. He tried hard to dissuade her, but to no avail. She walked out carrying a young squirrel monkey.

Next morning early, as I was sipping my tinto, she came downstairs with tears in her eyes, and her cheeks were streaked and red. The monkey had kept her awake all night. Her sheets were stained with feces, and wet with urine. Even worse, it had bit her on the hand. She no longer wanted the damned monkey, so she asked if I could take it back.

When I took the little guy back to the Quarantine Centre, he was so exhausted that he just lay quietly in my arms. Ray was expecting me and handed me the five dollars she had paid. We spoke for a long time about the animals, the river and the forest. Eventually I returned to the hotel to return the money, but she was gone! The woman cleaning the room said she got a taxi to the airstrip.

The next day, I visited Ray again and met his partner George. As I turned to walk away, he asked if I would like a job. He said there would be no money, but I would be well fed and have a place of my own to sleep. I thought about it for possibly thirty seconds, and then accepted.

George and Ray dealt about ninety percent with tropical aquarium fish, which in the early nineteen seventies was very big business. On one side of the building was the living room and bedrooms for Ray and George, while the left was filled with aquaria holding multiple species of freshwater tropical fish. Going through the back door to the yard, were still more fish. These were kept in large ground-level concrete tanks, each tank being twenty feet long, by ten wide with larger fish than those in the aquaria—lungfish, arapaima, catfish, and even electric eels.

Behind the fish tanks were several cages with a variety of birds and mammals. The squirrel monkeys of the previous day were there, plus various birds of the *Psittacidae* family—Scarlet, Blue-and-gold Macaws, and Black-headed Caiques.

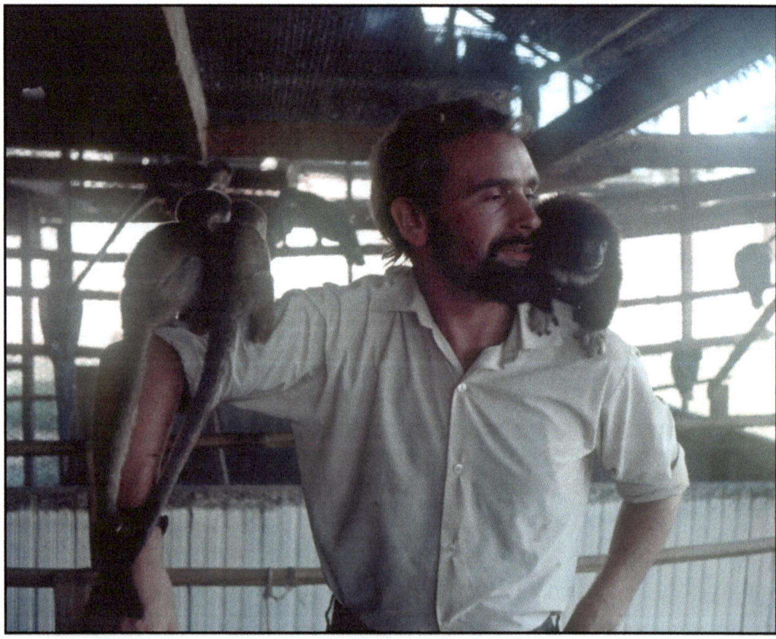

Allison Ray

Making friends with squirrel monkeys (*Saimiri sciureus*) and a saki monkey (*Pithecia monachus*) at the Quarantine Station

My accommodation was called a "balsa", which is a large raft made from balsa logs, on which a building is built. It floated on the river, but was tied securely to the shore. Access was by a gang-plank. My particular balsa had two bedrooms, plus an open area My new home was quite comfortable, and very private.

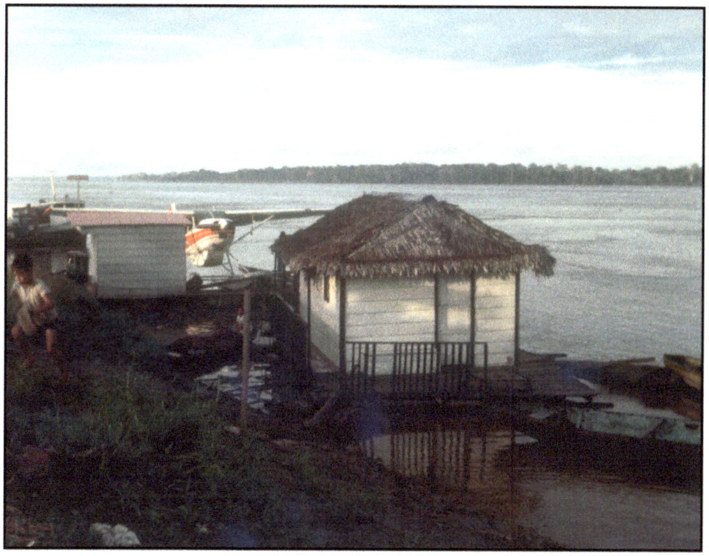

My Balsa

So far there was no mention of the type of work I was to do. They already had about six local indigenous women who cleaned up the house and cages, and fed the animals. I felt that Ray had sensed my dislike for using animals for medical research.

After two days of hanging around with nothing to do, I finally asked. "Oh!" he exclaimed, "You are going to be a boat captain and guide!" I was to pick up tropical river fish that various groups or families of Ticuna Indians had collected in their fish traps and bring them back to Leticia.

My second responsibility was to act as a guide for several scientific groups who wanted to study various creatures in the bush. My face must have shown something akin to terror, for he hastened to add, that I would have a Ticuna Indian to act as my guide. He then introduced me to the boat and to Arturo who was to be my guide.

Arturo was in his fifties, relatively small and wiry, but exuded an aura of capability. Even better, he spoke a few words in English.

The boat was a thirty-foot planked vessel, powered by an inboard diesel motor. It had no name, but could have been a stand-in for Humphrey Bogart's famous "African Queen".

Arturo's and my first trip was to collect fish from the various villages. Prior to leaving, approximately ten half oil drums, cut lengthwise along the centre, were secured on the roof of the boat. Then various items such as axes, machetes, lanterns, clothing and dried food stuff were stored inside the boat.

Eventually, with me at the wheel, we took off upstream. Five hours later I heard Arturo cry out. He pointed to a small opening on the right side and I pulled off the massive clay-coloured river into a creek at a much reduced speed. As we chug-chugged up the black waterway, I marvelled at the trees on both sides, which were massive and close to a 100 feet tall.

Suddenly we were in a large black water lagoon. I was so amazed at the sudden open area that I failed to look back to note the opening we had just exited. When I finally did so, my wake was gone, and there was no single opening, but four. I had to put my trust in Arturo, and in his good health.

Three hours later we reached our first community. There were about four or five pole and thatch buildings, from which several naked children ran towards us. Following were less excited adults, both men and women. I secured the boat to a couple of posts set in

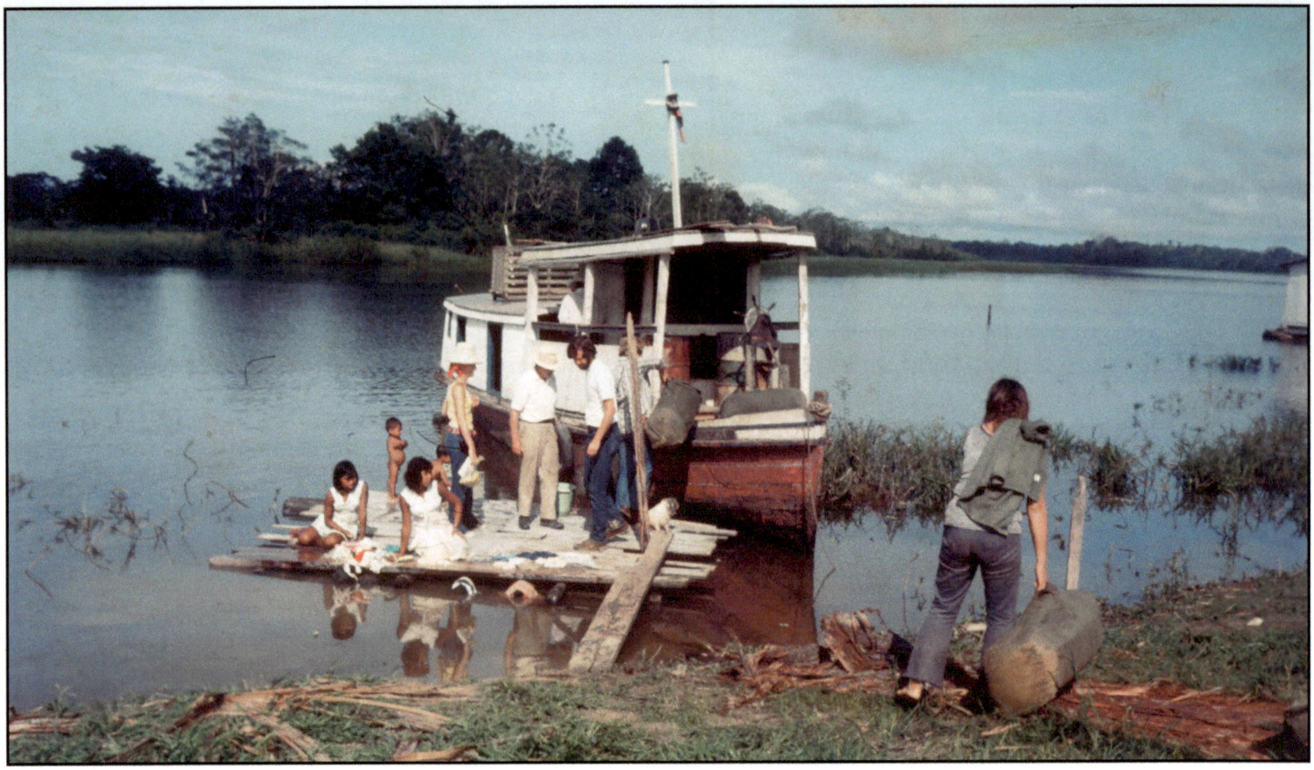

African Queen look-a-like in Amazonas

the mud and joined Arturo ashore. Arturo brought out, the goods that we had stored at the start of the trip and after their fish were inspected, deals were made both between Arturo and the men, then the men and their woman folk. What was more important—flour or a machete, fish hooks or needles and thread?

The palm-leaf-covered half-drums sitting by the waterside contained a number of small fish. Arturo untied the drums on the roof of the boat. With lots of help, the half drums with fish were lifted on the roof, wedged securely in place, and covered with palm leaves to shade the fish from the sun. Arturo showed me the fish traps that were used to capture the fish, and then we were called for refreshments. To this day, I have no idea what I was drinking. That night we slept on the boat and in the following days repeated the process three more times.

Four days later back in Leticia the fish were transported back to the station to recover from the stress they had suffered. Later they were transported to Bogota, then onwards to Miami one month later. In 1969 freshwater tropical fish were the single biggest import into Miami.

It was five months later before I experienced my second duty, catering to the scientific community. In many cities of Europe and North America, university professors and other collectors make trips to primitive places in the world to collect the specimens of their wildest imaginations. These trips were not as numerous as the tropical fish outings, only five in two years.

For Arturo and me, these trips were much more fun than a fish delivery service.

Subjects ranged from butterflies to bird watching, tree frogs to orchids, and the most popular—moths and other night flying insects. The trip I recall with fondness was a group of five or six night insect fanatics from Austria. They were anxious to find a particular moth species, thought to be in our area. Prior to their arrival, Arturo and I had re-cleared the bush in the area we reserved for such activities.

The group had brought all manner of equipment, which included a small generator, a large white screen, ultra-violet floodlights and a case of killing jars. Ray had supplied for them, six canvas-backed chairs, plus one each, for Arturo and me.

It was to be a moonless night and at about 9 pm they readied their equipment and one hour later, there were bugs of every description flying around the ultra-violet lit screen. Arturo and I had our own seats and watched with interest and amusement as the "grey beards", as Arturo called them, almost came to blows. All the learned gentlemen wanted to be the one to claim credit for a new species.

By midnight, it could have been a scene from hell! Giant beetles, moths and tree-hoppers were blackening the screen. Above, bats of numerable species were having the meal of their lives, while at ground level, huge bufo marine toads and mole crickets were gorging themselves on a feast to end all feasts. But still, the grey beards would jump high, and parry with their nets in an effort to capture that one particular moth out of thousands and secure it into their killing jar.

Arturo's stoic face would break into a grin now and again until finally he embarrassed himself by laughing. As daybreak approached, the grey beards had still not found satisfaction. However, they still had two more nights of watching bugs! I asked Arturo if he would be there watching the following night. He replied with his most stoic look, "No, I would break my face!"

Gradually Arturo and I became friends. He taught me how to walk silently through the forest, how to see, when you think there is nothing to see, and how to listen when there is seemingly nothing to hear.

One day Arturo decided that we should go see his family or friends. We had walked about two hours when when we heard some grunts and squealing from

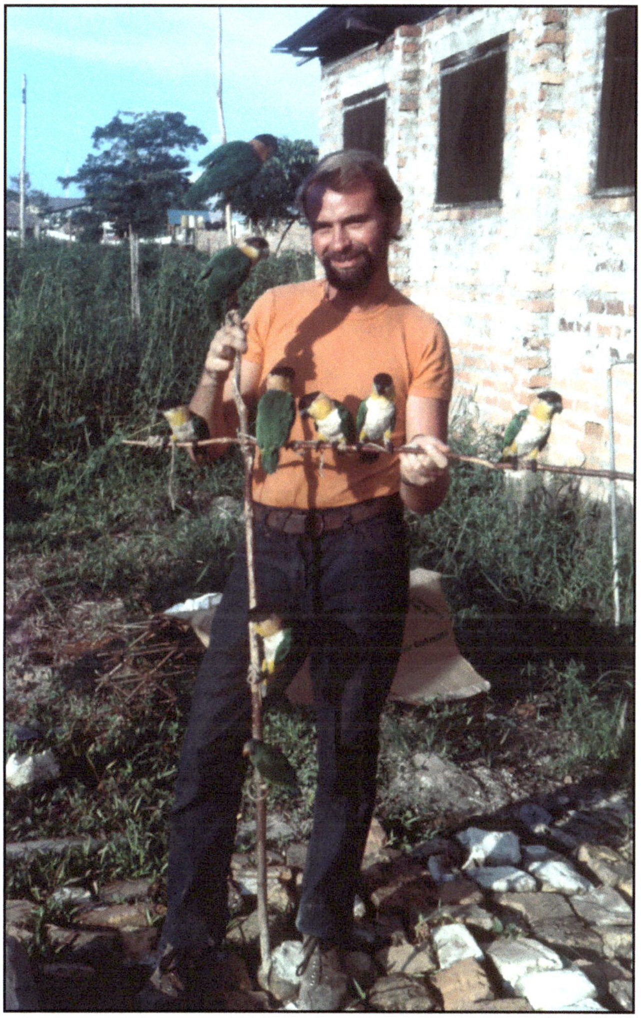

The author with Black-Headed Caiques (*Pionites melanocephalus*)

quite close by. I recognised it as peccary, but they were moving away from us, so I did not stop walking.

Abruptly, Arturo grabbed my arm and pulled me to his side. He put a finger to his lips and shook his head. "Jaguar" he mouthed. I was about to ask him what he said when a black jaguar moved out of the bush about three or four yards ahead. We both stared, for obviously, something was wrong with it.

It crouched down with watering eyes and tongue hanging down. As it shook its head, it seemed to be totally unaware of us. I remembered my camera, was hanging around my neck and took one picture. At least when my remains were found, they would know the reason of my death.

My thoughts then swung to perhaps an illness. Maybe it was rabid? Somehow it seemed worse to be killed by a rabid jaguar, than by a healthy one? Then all of a sudden the big cat gave an almighty sneeze!

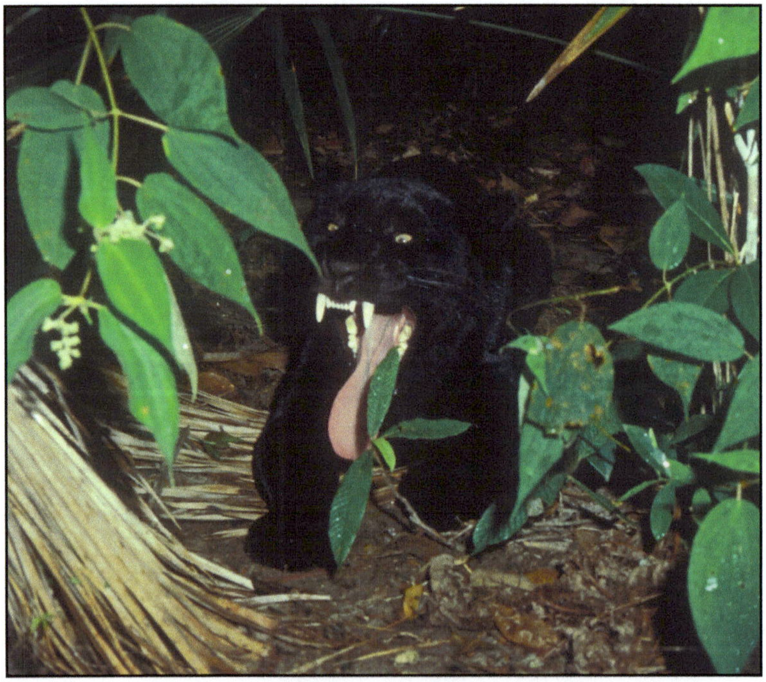

Black Jaguar (*Panthers onca*) in Amazonas

It shook its head two or three times, and rose to its feet. It looked at us with a sort of a silly apologetic glance, turned and was gone!

According to Arturo, when the peccary ran from the jaguar, they ran through a particular area of vegetation and their collective body mass caused a plant or fungi to emit fine seeds or spores which affected the following jaguar.

I was puzzled by the jaguar, for I thought melinistic jaguars were quite rare. Ray assured me that they were seen every now and again in the northern section of Amazonas. In fact, Mike Tsalikas, the Greek American who owned the large Anaconda Hotel used to have one at his place. Mike also was in the animal business in Leticia and was a part owner of Tarpon Springs Zoo in Florida.

When I was trying to identify a particular snake, a beautiful constrictor which I had bought from a forest Indian, it was Mike who identified it as an emerald tree boa. It was pretty mean, not at all like the true boa constrictor, which tames quite easily.

Buying snakes and other creatures from the Indians had become sort of a habit for me. Selling animals was the forest Indians only income, according to both Ray and Arturo. I learned that if the Indian could not sell the creature, he would eat it and if it was inedible, it would be beaten to death. But it would never be released. This was also the way animal dealers like Mike, and to a lesser degree Ray and George got their animals that were shipped all over the world.

I learned to keep a low profile when the Indians came in their canoes. I also learned to bargain when they did corner me. I rarely spent more than the equivalent of three U.S dollars. Unless the creature was too young

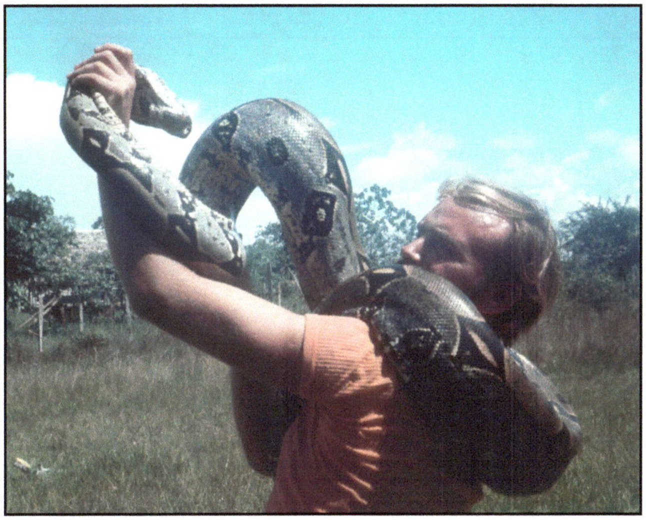

Holding a 14-foot boa constrictor (*Boa constrictor*)

to survive on its own, I released every one into the forest that same day.

The most expensive creature I ever purchased was a Hyacinth Macaw. That was the most beautiful bird that I'd ever seen. I paid 260 pesos (US $5.20). Unfortunately, I couldn't release it, as it could not yet fly. I left it with Ray, and I believe it went to a U.S. zoo after I left.

Other creatures I purchased from the Indians, and released, included several species of monkeys, various constrictor snakes, which included a boa constrictor that was fourteen feet long and weighed ninety-two pounds, and an anaconda of about the same length, but twenty pounds lighter.

Anacondas are the largest snakes in the world. In Amazonas, we often heard about giant anacondas 40 feet long, covered with green moss and as round as an oil drum. Unfortunately, no scientists have seen one that big. Anacondas are an aquatic snake, and adults seldom leave the water. Among their main prey are caiman.

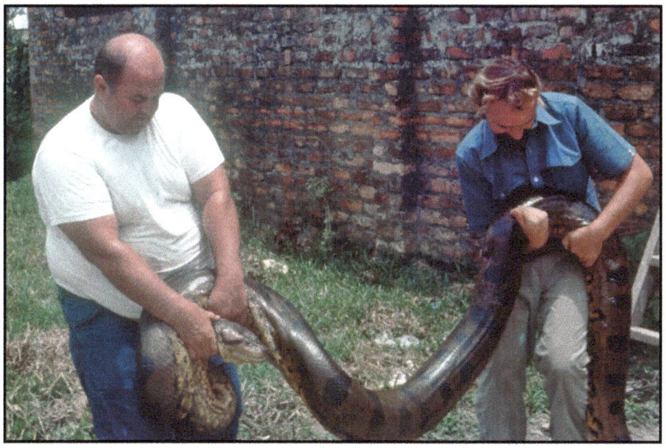

Ray and myself taking delivery of a 220-pound Anaconda (*Eunectes murinus*)

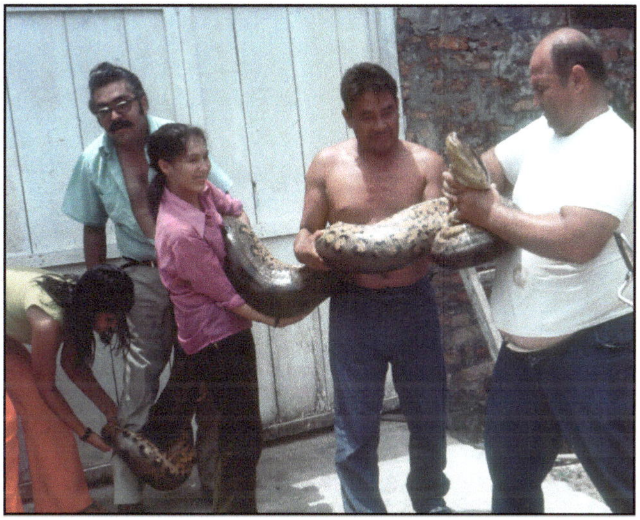

Everyone wants to get in the picture

Ray had an order from a zoo in Tokyo for a large anaconda. I was there when the Indians brought one that was 220 pounds and over twenty feet long.

Mike was one hundred percent into birds and mammals. He owned an island in the river, which was well forested. Here, he kept squirrel monkeys by the hundreds. The monkeys were for research and were captured and shipped to the U.S. labs on demand.

Mike also gave orders to the Indians for special items. Unfortunately, Mike was a businessman who was not always around when Indians brought in their merchandise. This sometimes led to a difficult situation for himself, and a comical situation for the rest of Leticia.

Mike once got a big order from a U.S. pesticide company to test their products on forest cockroaches. So Mike passed this order to the Indians. The order was filled, but unfortunately Mike was out of the country when the Indians staggered into the hotel with a 200-pound sugar sack filled with two inch long cockroaches. The Indians were upset, to say the least. They had paddled their dugout eight miles with these cockroaches. Don't even mention two weeks of night shift work catching cockroaches in the Amazon Forest.

Mike wasn't expected back that day. Indians of the Amazon expect promptness, so two hundred pounds of jumbo sized cockroaches were dumped in the hotel foyer. When Mike returned and questioned why they didn't leave the full sack, he was told that the sack was very valuable. Without the sack how they could bring more cockroaches!

Another time, some Indians managed to catch an Amazonas Pink Porpoise. They knew Mike always bought animals, so they loaded it into a twelve foot dugout canoe and paddled for two days, until they reached Leticia.

Of course, Mike was away on business when the porpoise and six men in three dugout canoes arrived. They carried it up to the Anaconda Hotel and left it in the swimming pool where Mike found it on his return.

The Amazonas Pink Porpoise was locally known as "boto". Even back in 1970, it was one of the most endangered animals in the world and still is on CITES Red List. Mike had no option but to take it upriver and release it, but he paid the Indians to prevent the Indians taking it back to their village and having a Bar-B-Q.

In the beginning, I gave myself a maximum of two years to stay in Amazonas. When the two years were up, I purchased a ticket on Satena Airline to Bogota. Satena was not what one would call a luxury airline. My seat for the flight was a 200lb sack full of dried fish, but still I counted myself lucky, for I'd spoken with many others who had sat on much less pleasant seats. Mine was at least dry!

On landing, I got a taxi to the American Express Office, where I cashed a personal cheque. My next move was to return to the airport, where I planned to purchase a ticket to Guatemala, the first stage of my return home, but alas, that was not to be!

I am originally from Scotland. A name known by all Scots is Scotland's National poet, "Rabbie Burns". One of his best known poems was called "Ode to a Mouse". The final line reads, "The best laid plans of mice and men gang aft agley, and leave us naught but grief and pain, for promised joy!"

And so it was for me. As I walked out of the American Express Office, I was involved in a collision, first from a man in front who bumped heavily on my right side, then almost immediately by another who banged into my left back. I took one step before I realised that it was orchestrated. The cash and travellers cheques I'd received in Amex were in my front pocket, but my passport was missing.

I accelerated my walk to catch up with the pickpocket, he nervously looked back, then accelerated his walk, I kept with him. He then made a right turn into a secondary street. I followed him, as I turned, he was running, I also ran and saw him disappear into a large church. I ran to the back of the church, thinking he would run right through. There I found two Columbian policemen. With my very poor Spanish I tried to explain, it took a while. The police eventually went into the church, but it was now empty!

That same day, I went to the British Embassy, where I explained my predicament. A new passport would take a few days only, for luckily I had other British I.D papers. So I booked into a hotel for two days. When I returned to the Embassy, the official did not have a happy face. No passport! It seemed that the Columbians didn't have any record of my entering the country legally. As an illegal the Embassy couldn't issue me a new passport. They suggested that I visit the D.A.S. Office, the Columbian Border Security.

The uniformed officer at the D.A. S wanted to know where I entered Columbia. I told him that I flew from Lima to Bogota, and prior to that, from Guatemala City to Bogota. This made him even more suspicious because there were two entries into Columbia, and their paperwork showed neither! He told me to come back in ten days! Luckily I had made some friends in Bogota and stayed with them.

Ten days later, I visited the D.A.S. once more. The officer still could not find a record of my entry into the country. He got a little bit nasty and demanded to see everything that was in my wallet, so I pulled out my local money, my traveller's cheques, then some odds and ends like drivers licenses, one Canadian and one British. Now my wallet was empty, or so I thought. He demanded to examine it.

After a few minutes he let out a cry of triumph! He pulled out a small white square of folded paper. He slowly and theatrically unfolded the scrap of paper. It was a receipt from American Express in Lima, for payment of two airline tickets, Lima to Bogota. This proved that I had flown into Columbia legally. He told me to visit the British Embassy the following day. He thought a little and asked, "Why two tickets?" So I explained about my friend who had been ill. Then he shook my hand and said that he was glad he didn't have to lock me up. I was also glad!

The next day I flew back to Leticia for two more months. On the 2nd of September 1972, I flew from Columbia to Guatemala City, then travelled overland to British Honduras. My plan was to fly to New Orleans and then onwards to Toronto, and home!

I overnighted in Belize City. The following morning I was walking along Albert St. on route to a travel agency and someone hailed me. I did not recognise the blonde woman immediately, but I did recognise her dog, a charcoal grey, 140-pound Irish wolfhound, named "Doncher."

Earlier in this trip, while driving south in Mexico, I had picked up Margie and Doncher as hitchhikers. My window van had plenty of room, even for an Irish Wolfhound. I dropped them both off at Palenque, never expecting to see them again. At the end of my trip here they were again. Margie was living in Belize, on a small island, Caye Caulker, and she invited me out for a few days. I didn't know it at the time, but I was going home.

Margie had a sailboat, so we sailed out to Caye Caulker, all three of us, Margie, me, and Doncher. The next day, we sailed out to the reef, which changed my preconceived notions about British Honduras. Now, forty years later I'm still on Caye Caulker!

I first came to British Honduras, as it was then called, in early 1969. I stayed two days only, but returned in February of 1972. A diver since my fifteenth year, I had made a point to bring my Nikonus Calypso Camera, the first underwater camera made from a design by Jacques Cousteau. I've been hooked ever since!

Since then, I've been taking pictures in both the sea and the rainforest and loved every minute of it. In writing this book, it took not a little thought. Where to begin?? The sea or the forest?

But in the end my decision was made in favour of the sea and why not? Life began in the sea, and my life pretty much also began in the sea.I learned to dive in 1957 in the cold stormy waters of the North Sea and even then, I've never looked back!

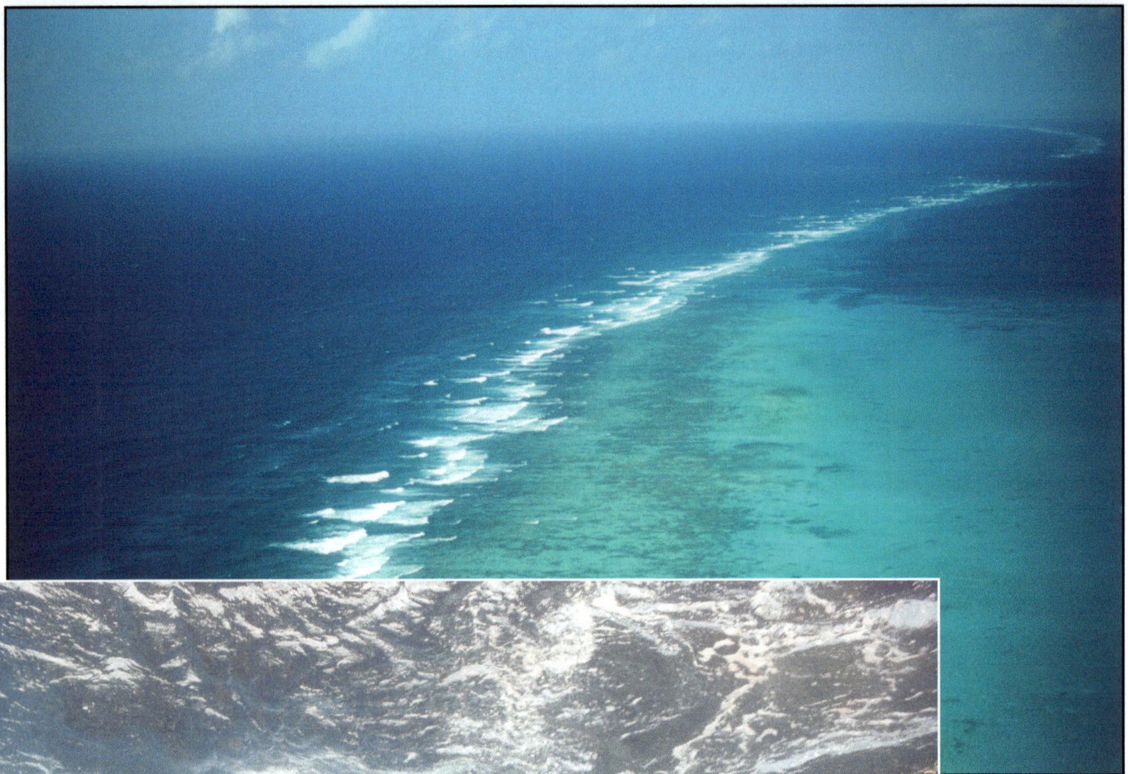

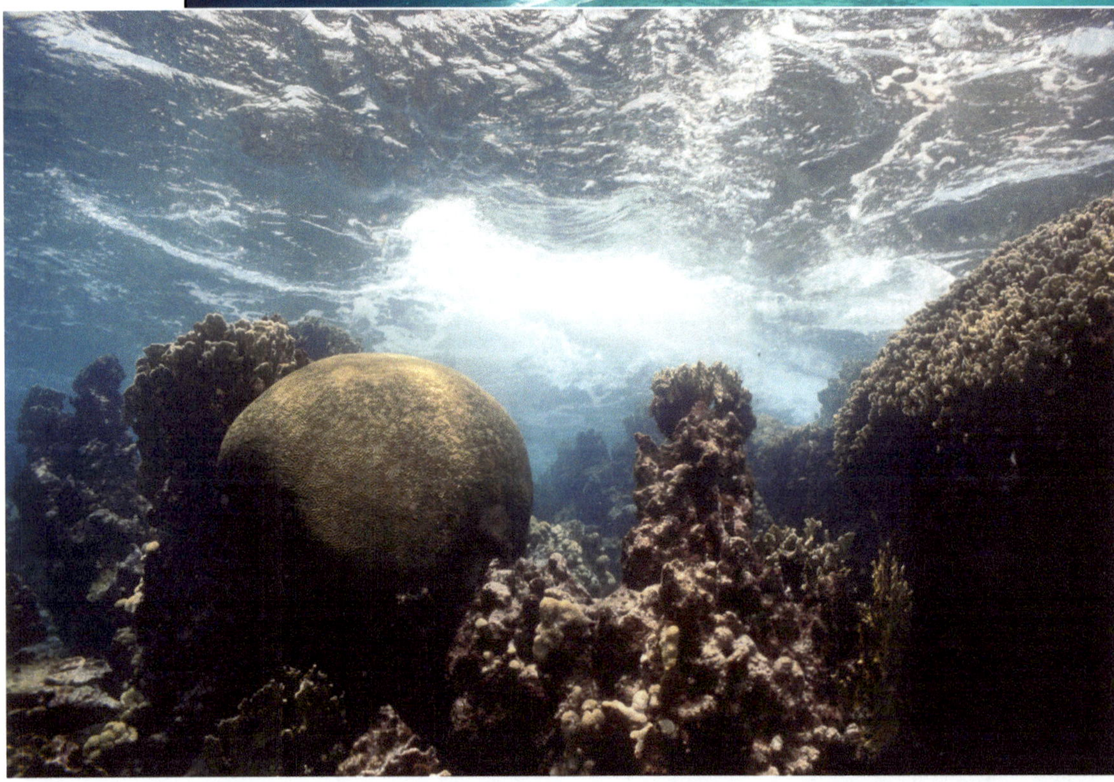

(*above*) Aerial view of the Belize Barrier Reef and (*below*) Submarine view of the Belize Barrier Reef

Chapter II The Coral Seas of Belize

Corals

Coral creatures are a symbiotic relationship between tiny colonial animals of the Phylum *Cnidaria* that create stony polyps of many shapes and sizes, and zooxanthellae, one-celled animals with chloroplasts. The coral polyps provide shelter for the zooxanthellae, while the zooxanthellae captures energy from the sun through photosynthesis. The zooxanthellae also give the colour to coral.

In order to build the stony structures, corals extract calcium carbonate from the sea water. The physically larger coral species are called reef-building corals. They include the staghorn, elkhorn, lettuce, star, brain, and pillar corals. In order to survive, all corals need a very delicate balance. They require a firm attachment to the sea floor, continuous movement of the surrounding water, fixed salinity, clarity of water and, most important of all, a stable temperature of between 70 and 85^0 F.

When corals are stressed by high temperature or acidity, the polyps expel the zooxanthellae, which bleaches the coral to a stark white. When the stress is over, some of the coral may recover by capturing the zooxanthellae again. But coral live at the upper level of their temperature range. Unfortunately, with increasing temperatures due to the present global warming crisis, a living coral reef may, in our grandchildren's lifetime, be only a memory.

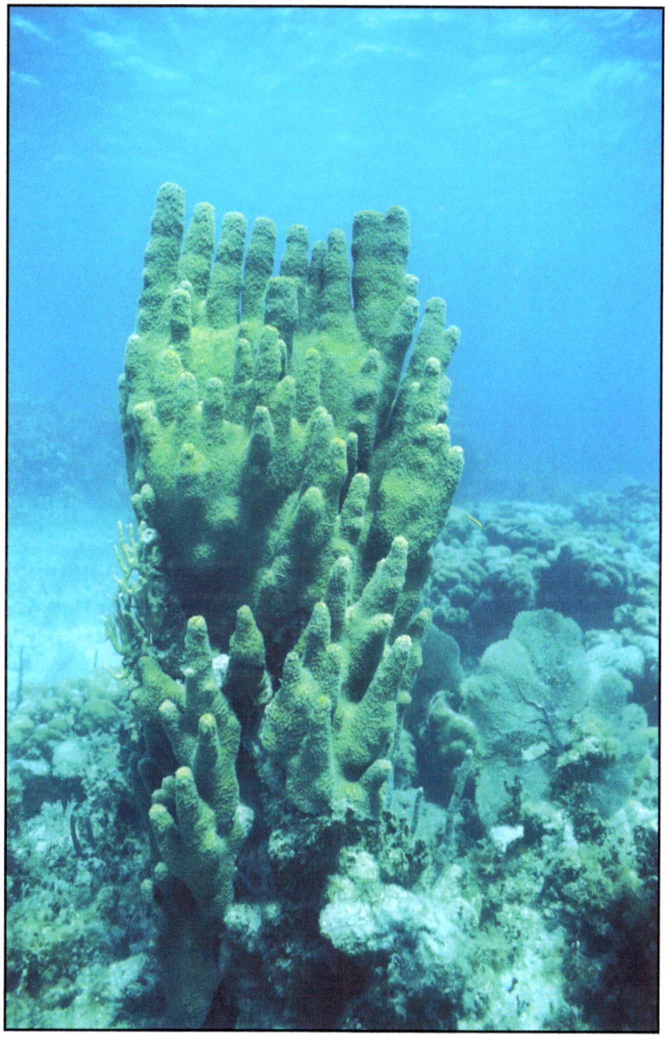

Pillar Coral (*Dendrogyra cylindrus*)

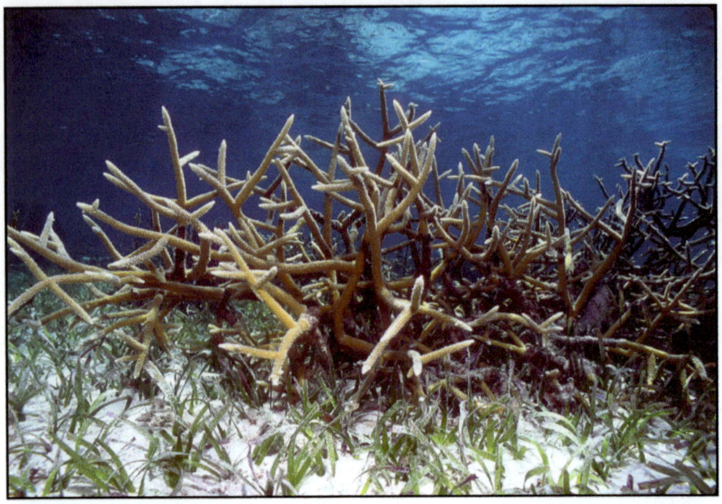

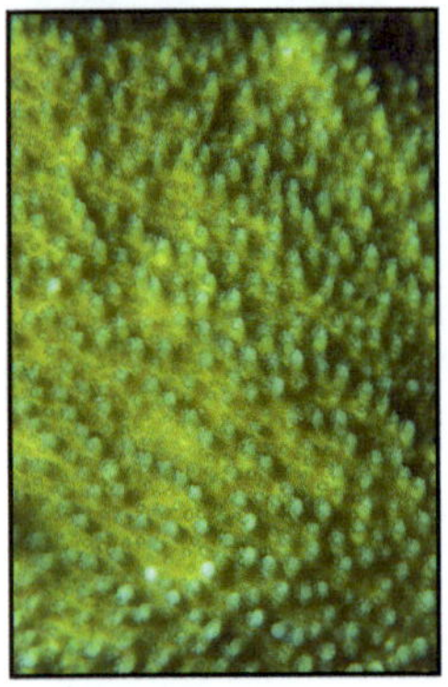

(*above and right*)
Staghorn Coral
(*Acropora cervicornis*)

(*below and left*)
Elkhorn Coral
(*Acropora palmata*)

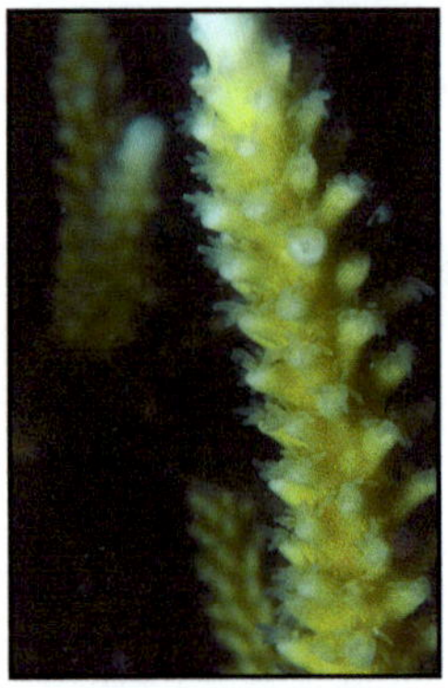

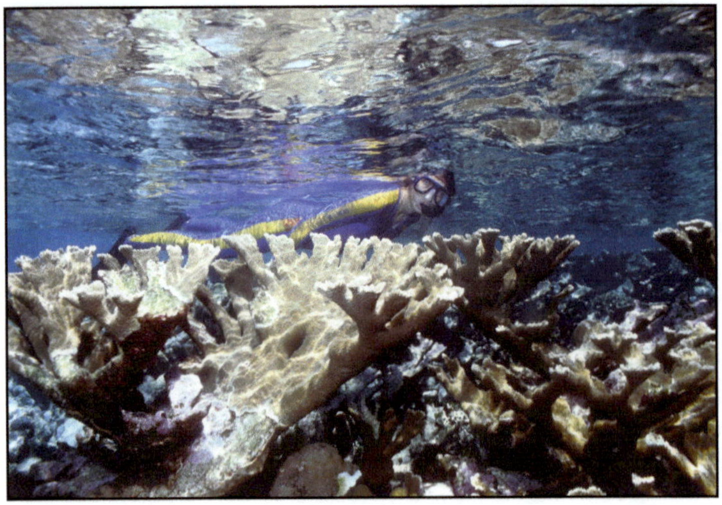

The coral animal may be a reef-building coral, or a soft coral such as a gorgonian, black coral, lace or fire corals. Both have cup-shaped bodies with one central opening which acts both as anus and a mouth. In order to feed, the tentacles surrounding this opening extend though the skeletal structure where they capture minute organisms by way of nematocysts (stinging cells). These stings, excepting fire corals, are harmless to man.

Corals reproduce in two ways. The first way is by budding. Budding is an asexual growth related to individual colonies. If one examines a coral head closely, it will be noted that new corallites may be seen erupting from between mature polyps. In this manner a coral colony grows and as it does so, it will enlarge as time passes.

The second way for a coral to reproduce is far more complicated, this growth is sexual. Some coral species have mixed sexes, producing both sperm and eggs within the same colony, so fertilization can take place within the colony, and the resulting larvae can either develop internally, or be ejected into the open sea, depending on species.

Other coral species have colonies that are totally female, while all male corals may be resident in an entirely different colony. These corals release their reproductive cells by the millions, once a year synchronized with other coral colonies of that same species, but hopefully opposite sexes. Fertilized eggs from this synchronized mass spawning drift for many days before settling on the hard sea floor in shallow transparent water. This a tiny calcium structure will slowly start growing and multiplying. Belize has approximately one hundred and fifty coral species, both hard and soft corals.

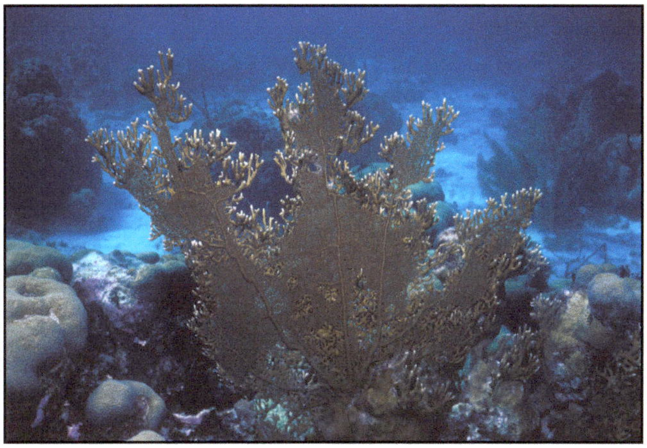

Branching Fire Coral (*Millepora alcicornis*)

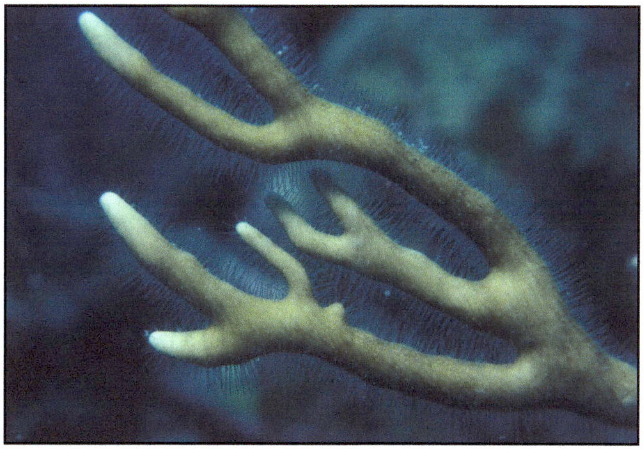

Branching Fire Coral (*Millepora alcicornis*)

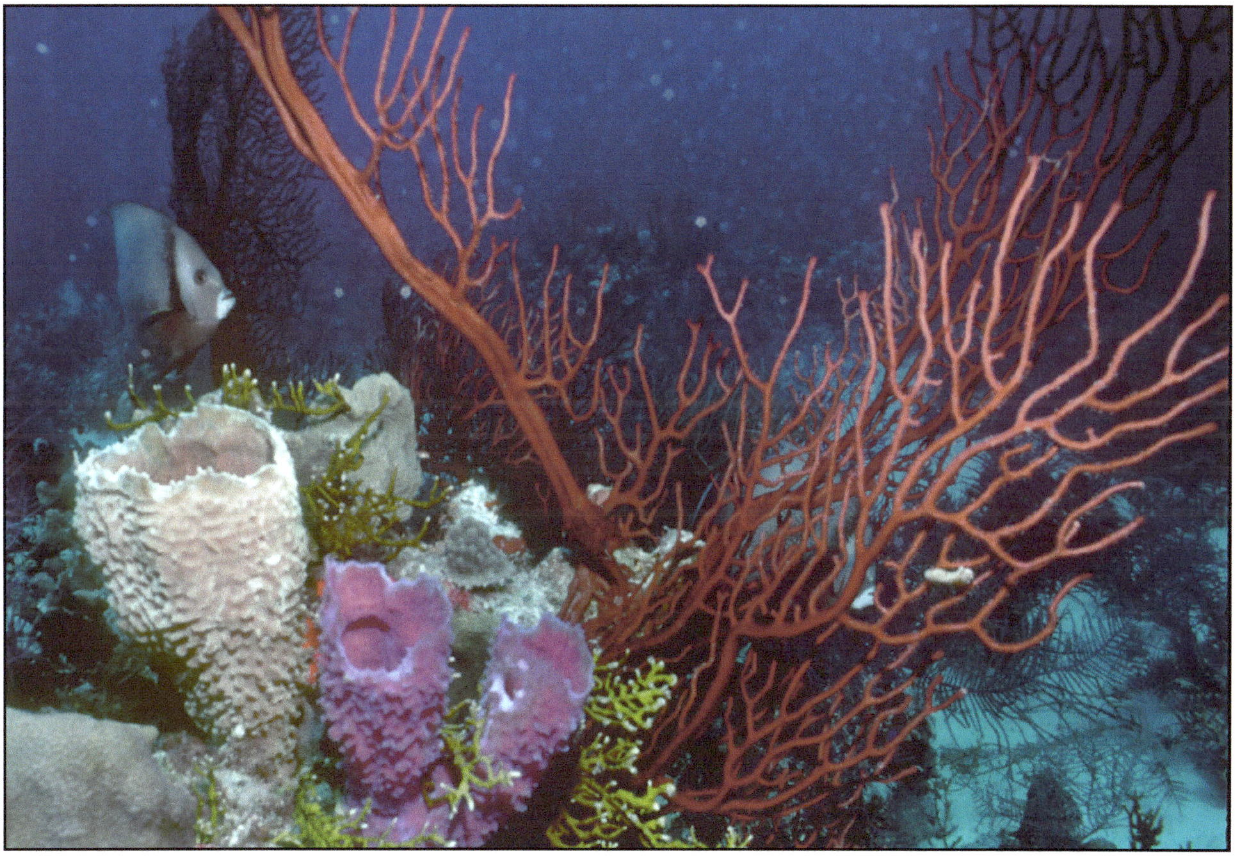

Coral garden with fire coral, sponges, and gorgonians. A gray angelfish is in the background.

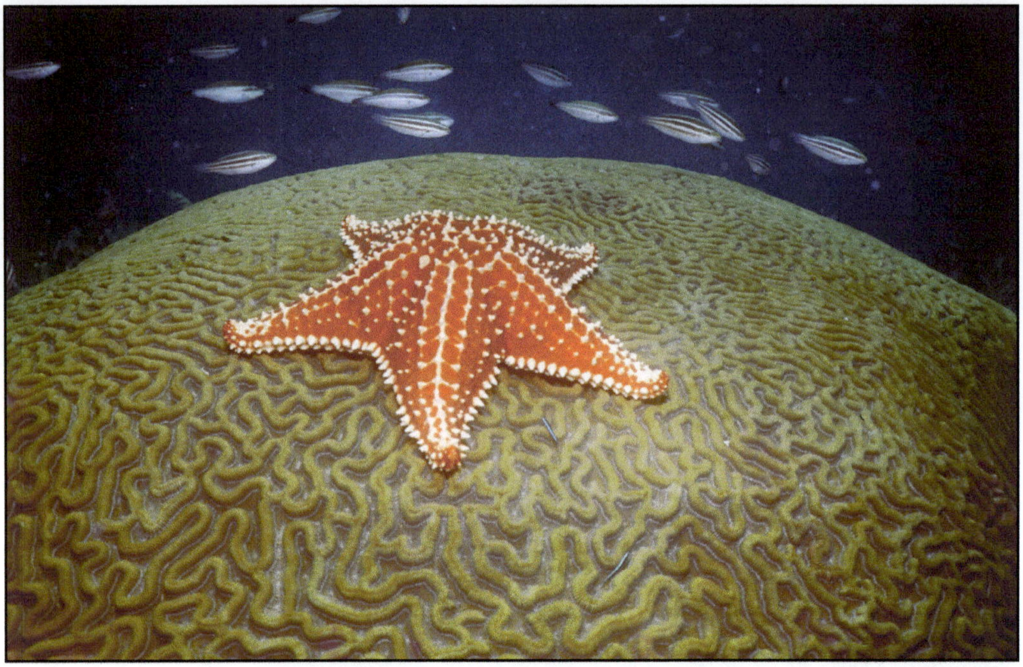

Cushion Sea Star (*Oreaster reticulatus*) on Grooved Brain Coral (*Diploria labyrinthiformis*)

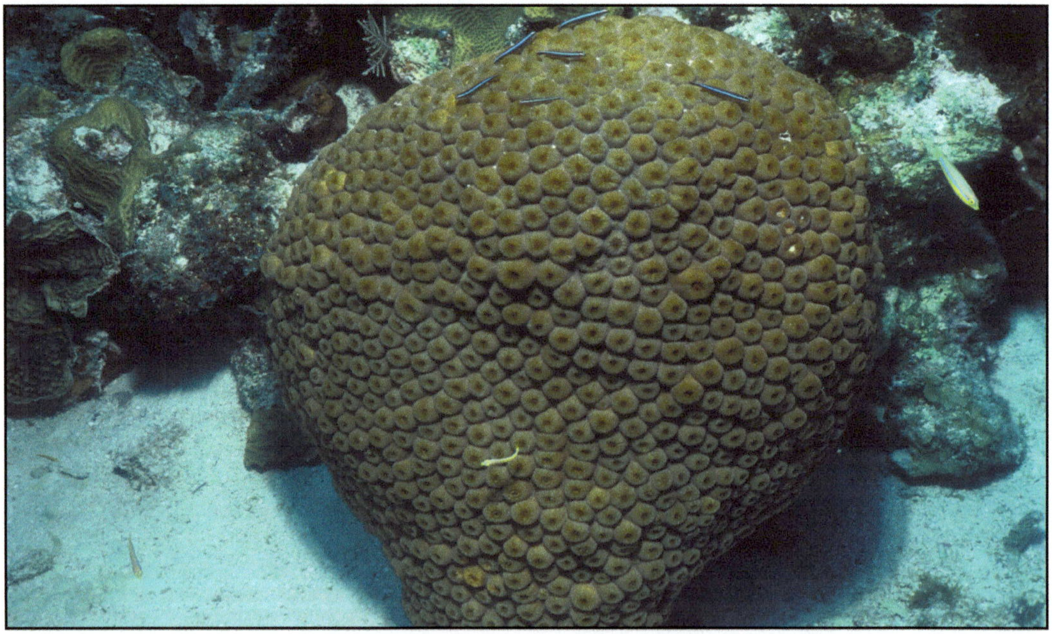

Great Star Coral (*Monastrea cavernosa*)

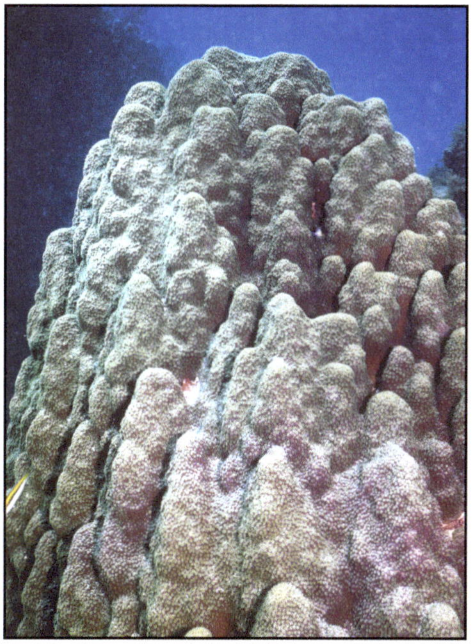

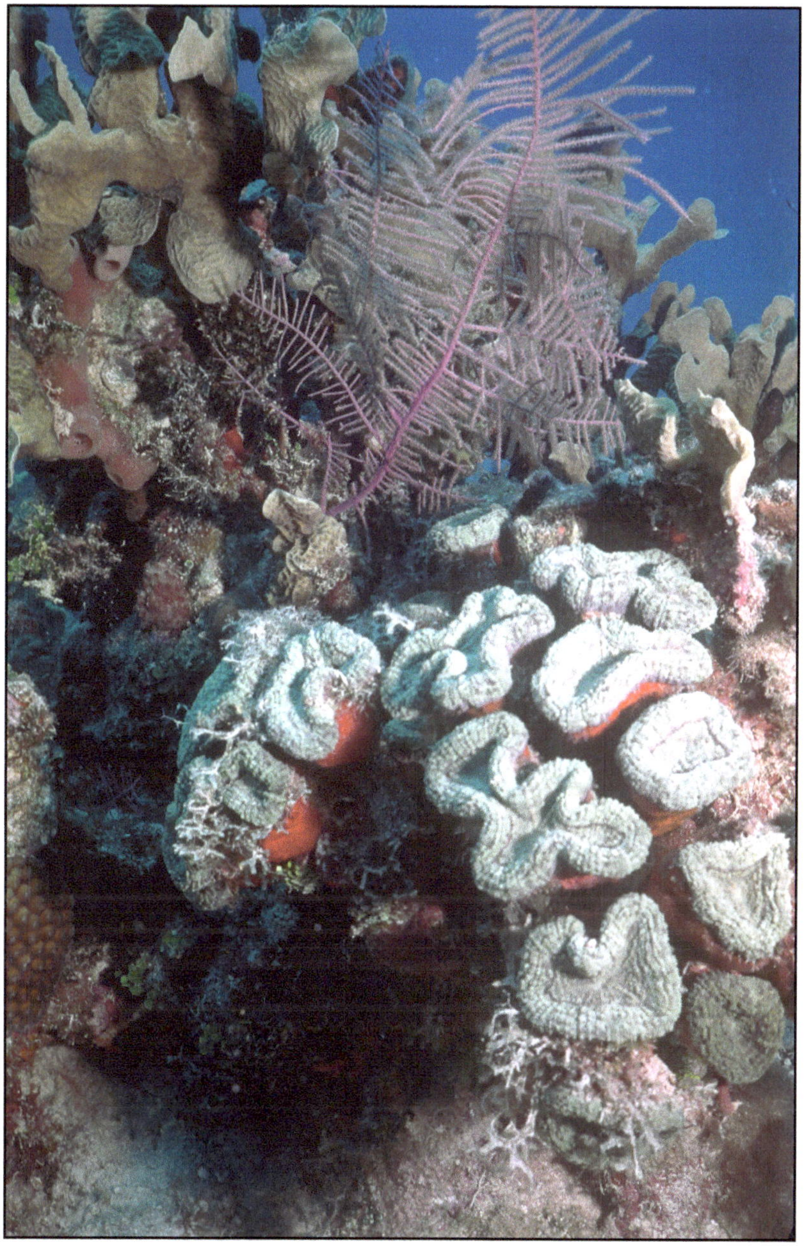

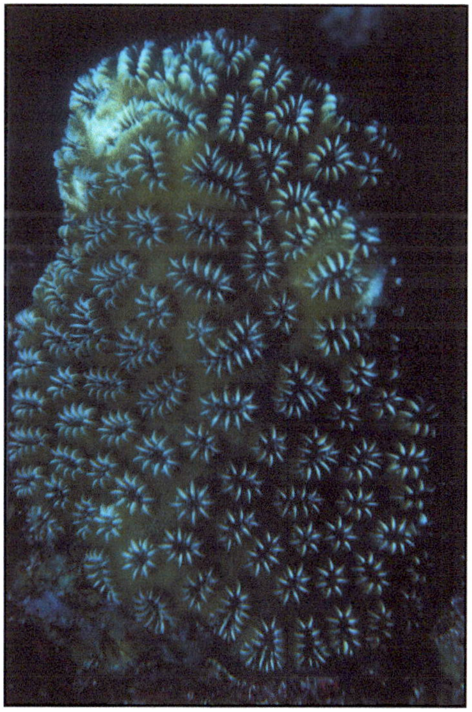

(*above*) Spiny Flower Coral (*Mussa angulosa*)

(*upper left*) Boulder Star Coral (*Montastrea annularis*)

(*lower left*) Elliptical Star Coral (*Dichocoenia stokesii*)

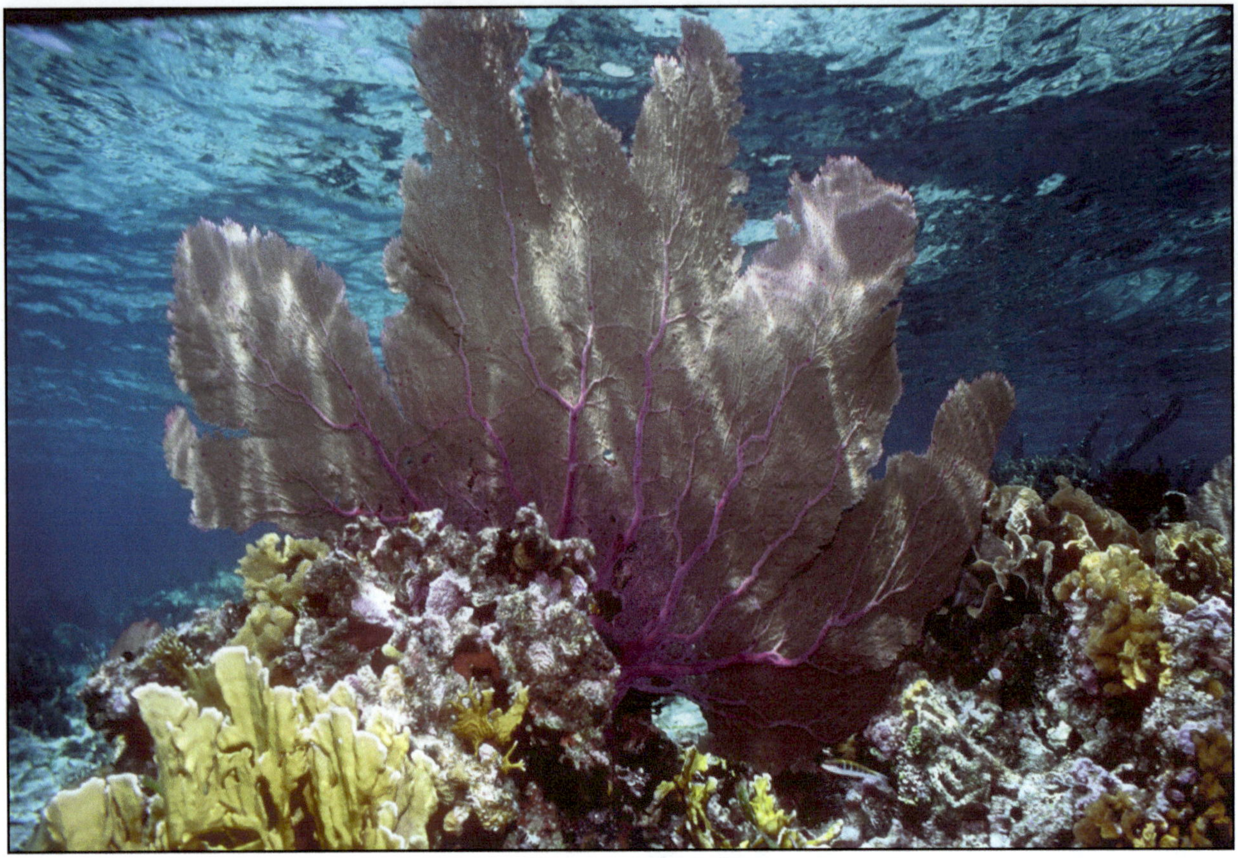

Common Sea Fan (*Gorgonia ventalina*)

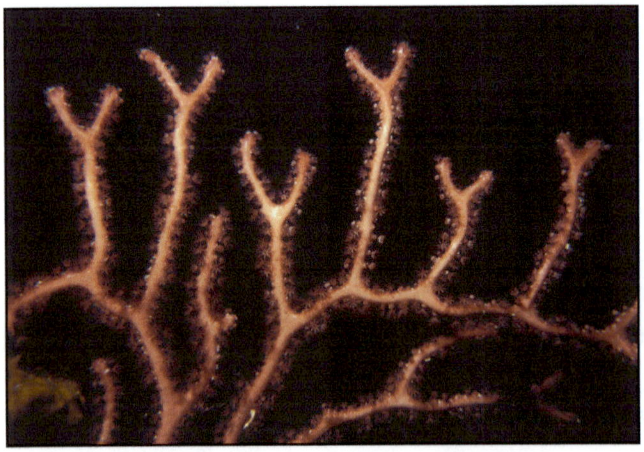

Deep Water Sea Fan (*Iciligorgia schrammi*)

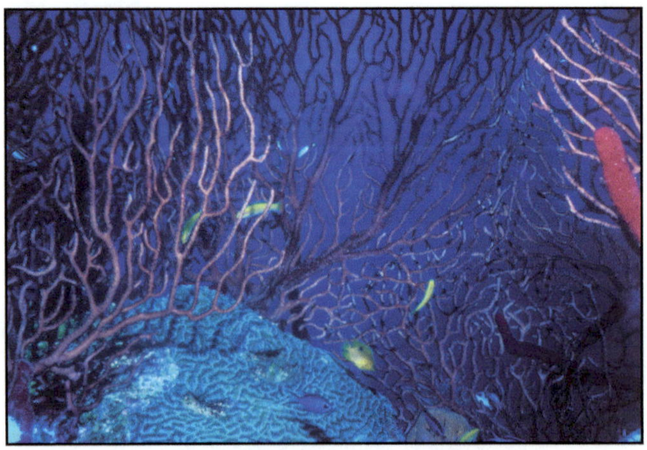

Deep Water Sea Fan (*Iciligorgia schrammi*)

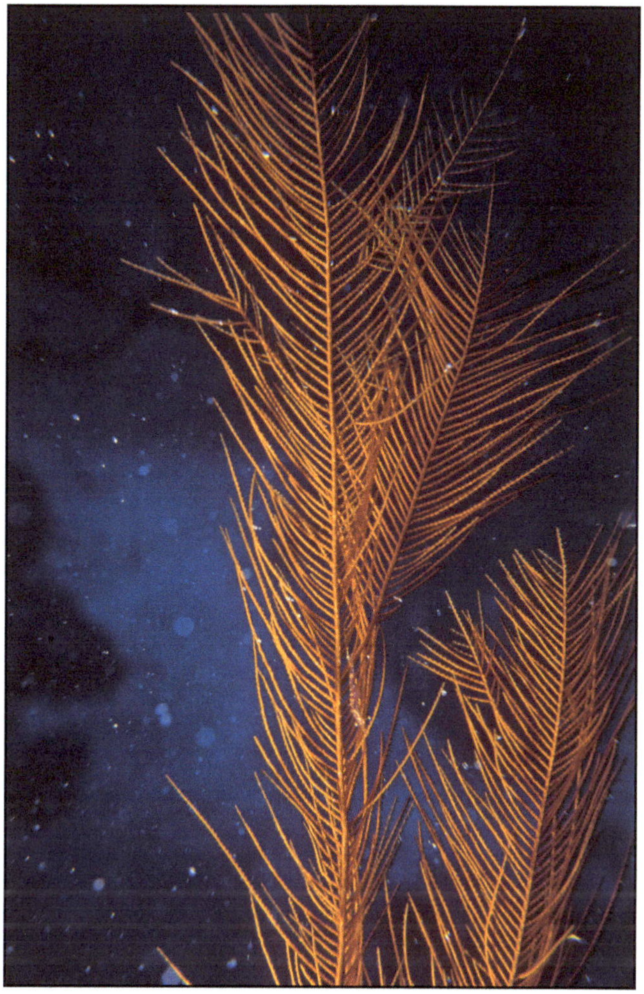

Feather Black Coral (*Antipathes pennacea*)

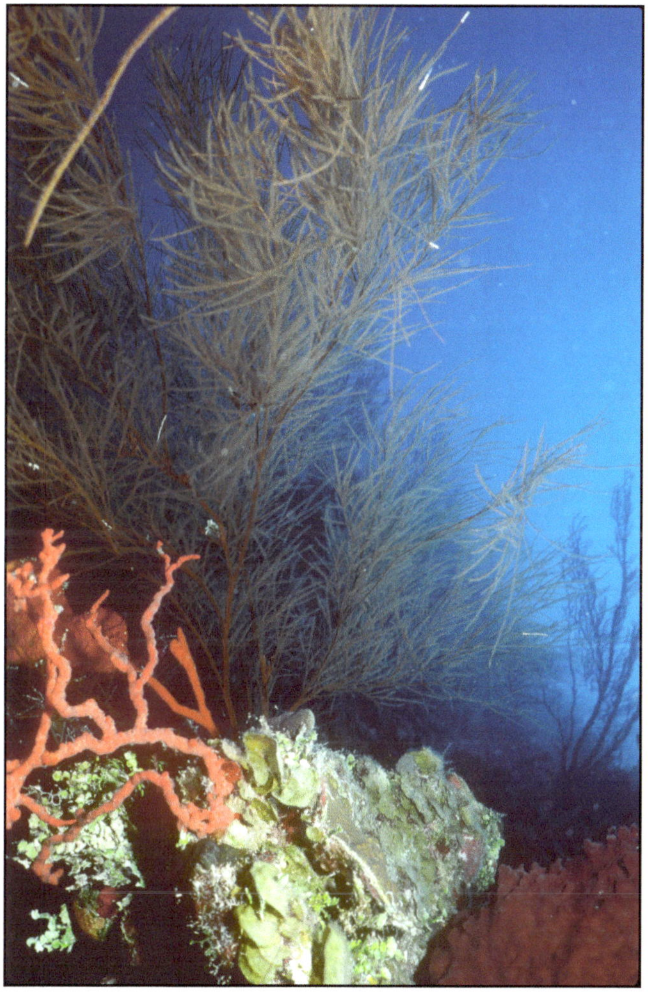

Black Coral (*Antipathes* sp.)

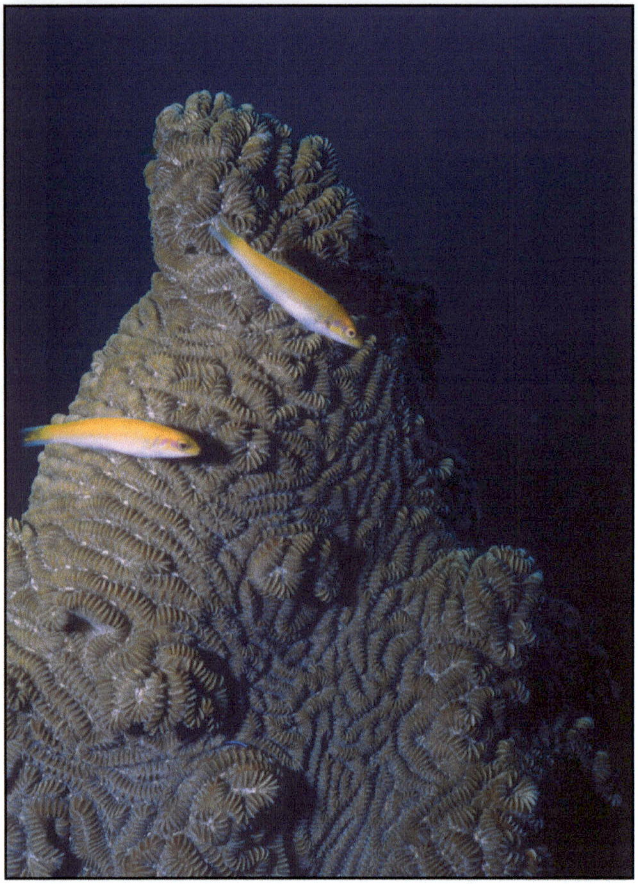

Maze Coral (*Meandrina meandrites*)

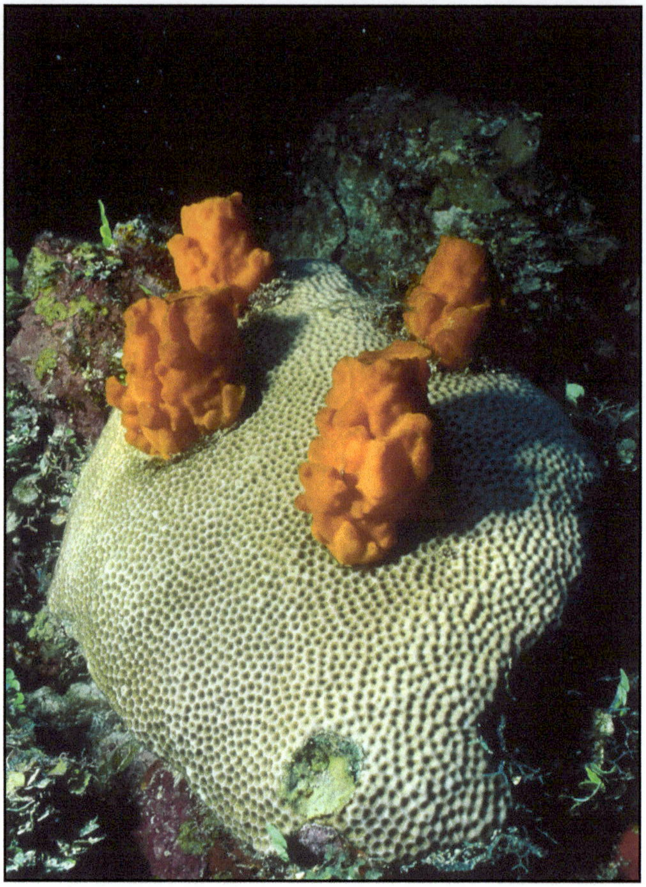

Lesser Starlet Coral (*Siderastrea radians*)

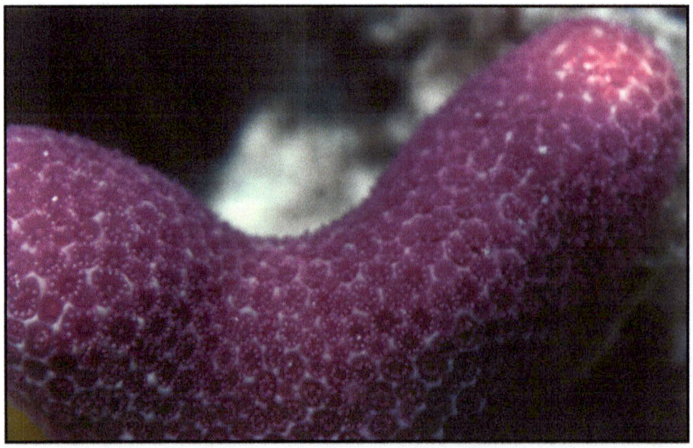

Finger Coral (*Porites porites)*

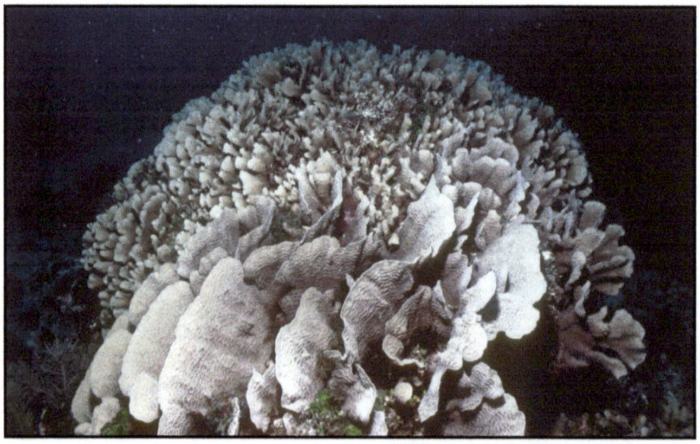

Thin Leaf Lettuce Coral (*Agaricia tenufolia*)

Anemones and Jellyfish

Also belonging to the Phylum *Cnidaria* are anemones and jellyfish. Anemones, like corals spend their adult life in one spot as stationary creatures. Whereas corals are colonial, anemones live singly.

There is one characteristic shared by all cnidarians, the presence of stinging cells. The toxins released by Caribbean corals are harmless to man, with the exception of the fire coral. Anemones also have a very mild toxin.

In contrast, jellyfish are not stationary. They wander the seas under their own power, or assisted by the vagaries of the sea and the winds. Toxins of some jellyfish species in Belize can be quite dangerous for a swimmer.

The upside-down jellyfish is normally found in grassy or muddy flats, however on occasions it will take off from the bottom and go for a swim. This jellyfish is about seven inches in diameter. Its tentacles are short and heavy. Contact with this jellyfish will result in only a slight sting.

One of the most common jellyfish in Belize is the sea thimble. Though small, rarely over three-quarters of an inch in diameter, they make an appearance in the thousands, usually in March or April. They swarm close to the surface.

Sea thimbles are not dangerous—redness from its mild sting maybe the only symptom. However, a couple of weeks later their invisible larvae in the water called "pica-pica" in Belize cause a rash. A tee-shirt or other clothing will only catch the larvae and make the situation worse.

Our only real nasty common jellyfish is the very beautiful, Portuguese man-of-war. Apart from being an extremely attractive creature, the Portuguese man-of-war luckily enough is very noticeable. The body of this jellyfish consists of a blue to purple float which is very visible above the surface. On this float is a

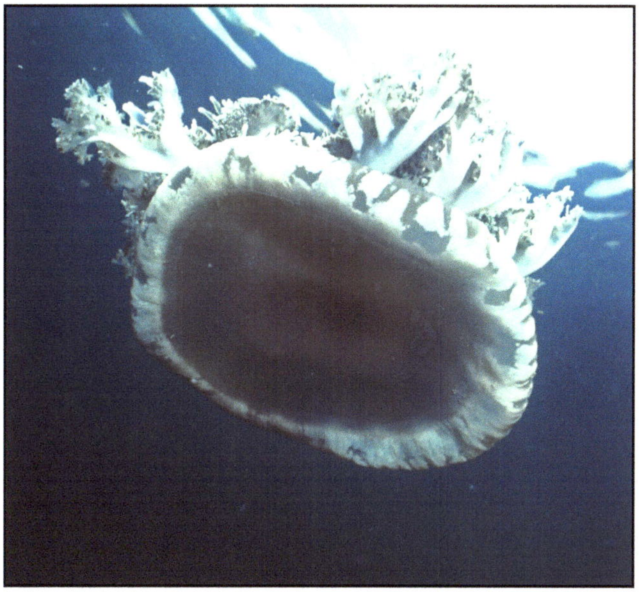

(*above and below*) Upsidedown Jellyfish (*Cassiopea frondosa*)

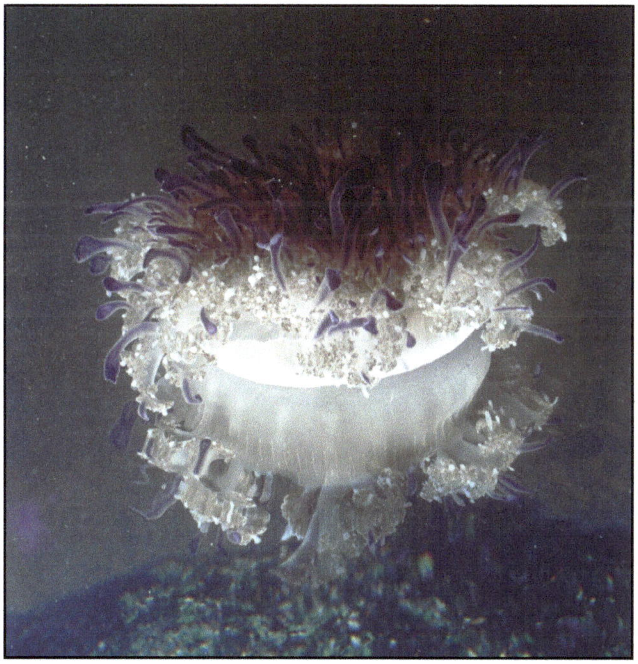

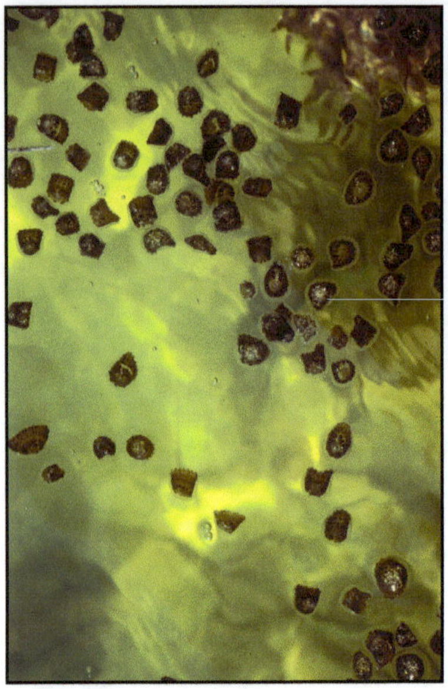

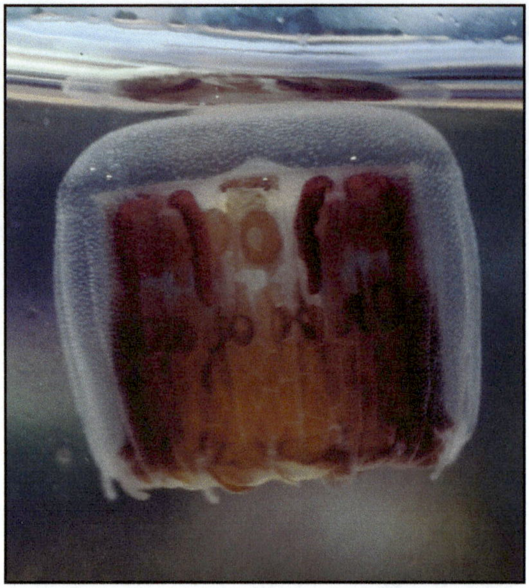

(*far left*) Sea Thimbles (*Linuche unguiculata*)

(*left*) Sea Thimble up close

(*below*) Portuguese Man-of-War (*Physalia physalis*)

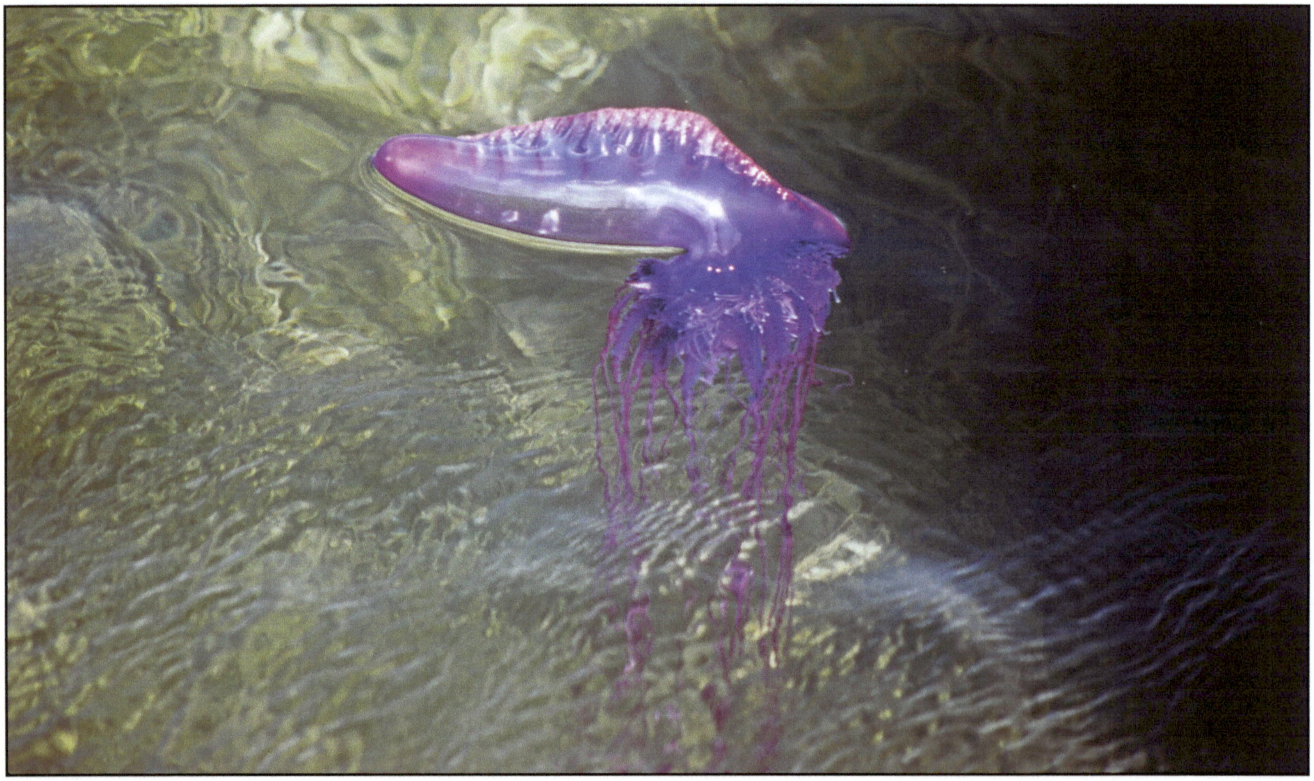

flattened sail-like extension. Not seen from the surface, are the twenty foot long tentacles that trail behind the surface float in hope of snaring a fish. These tentacles are extremely toxic to swimmers.

This creature is well-named, for it is an excellent sailor. I watched a man-of-war, when the prevailing winds looked to drive it ashore, it came about and tacked several times in an effort to change its direction and head back out to sea. Alas, the wind was too strong, and it didn't make it, but I felt obliged to help it on its way.

Blue Button

Although it might be mistaken for a jellyfish, the blue button is in a class almost of its own. It belongs to the same Phylum (*Cnidaria*) with corals and jellyfish, and the class *Hydrozoa*. The blue button belongs to the chondrophore group, along with the by-the-wind sailor (*Velella velella*).

Although blue buttons are not true jellyfish, they also are colonial animals, each one contributing something quite unique to the colony. It may be a tiny section of a small tentacle, or perhaps part of the central disc, but each has its part.

Blue buttons exist in vast colonies that are normally pelagic. They are found in both tropical Caribbean and Pacific waters. They are rarely seen, except when part of a mass stranding along tropical coasts.

Normally, blue buttons drifts on the sea surface where they play a role in the food chain as a passive drifter, feeding on either organic debris or zooplankton. On the other hand, it is food for many organisms. Like corals and jelly fish it has a single body opening which acts as both a mouth and a rectum!

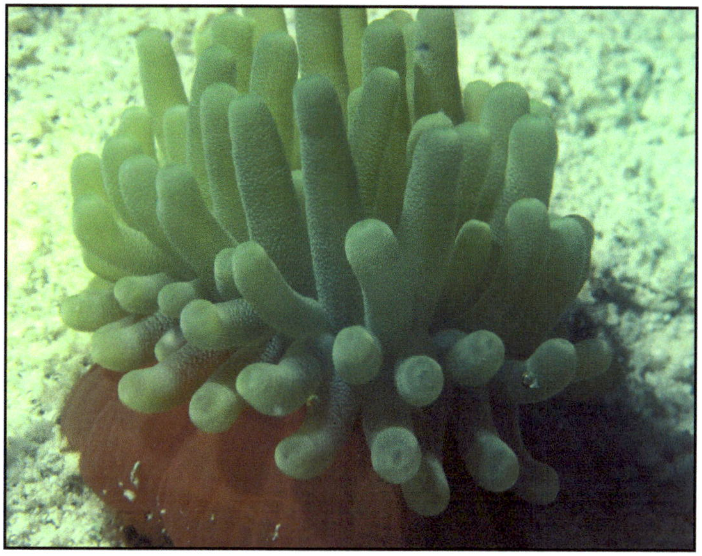

Giant Anemone (*Condylactis gigantea*)

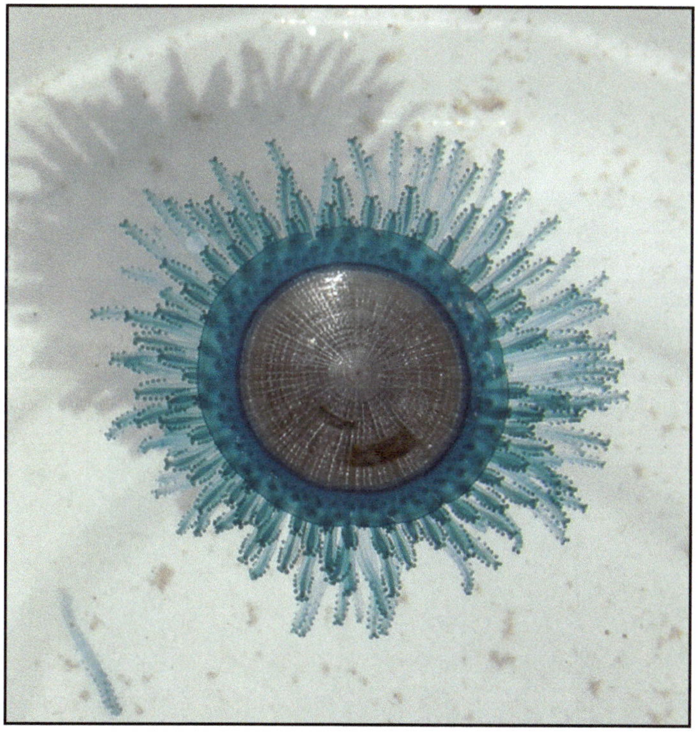

Blue Button (*Porpita porpita*)

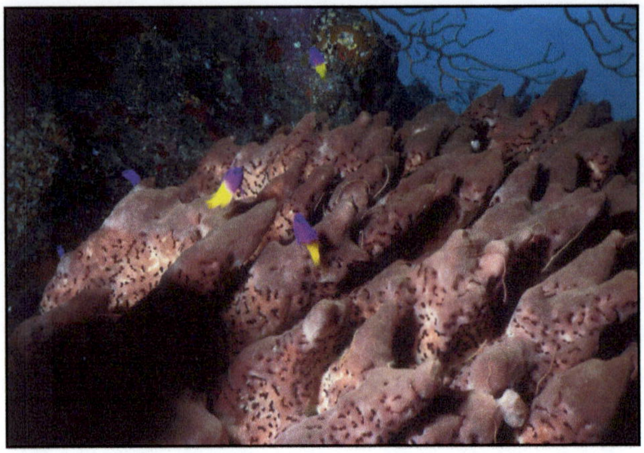

Chicken Liver Sponge (*Chondrilla nucular*)

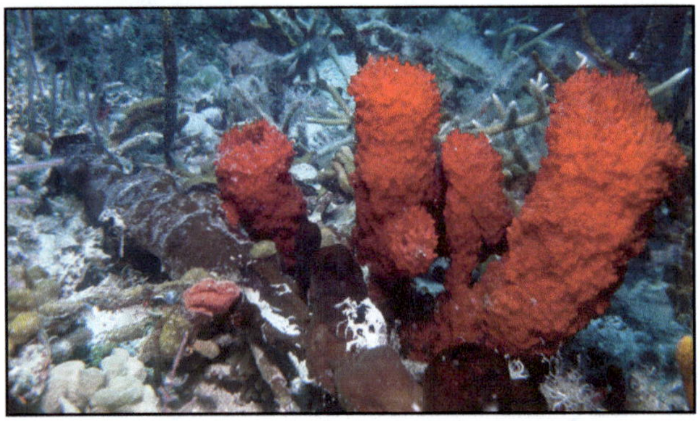

Strawberry Vase Sponge (*Mycale laxissima*)

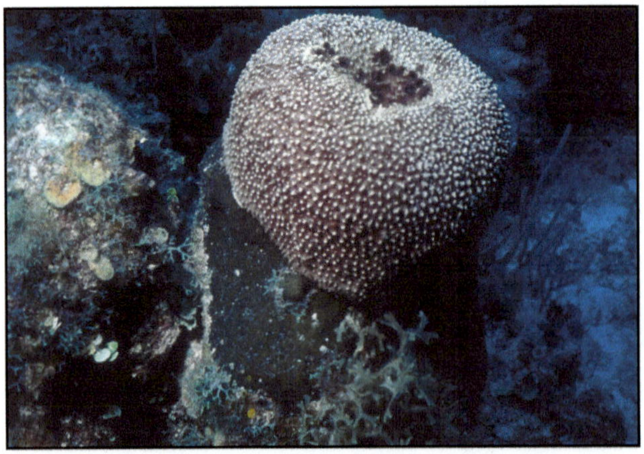

Black-Ball Sponge (*Ircinia strobilina*)

Sponges

Sponges, with the possible exception of the small colourful reef fish of the coral grottos, are very likely the most colourful and varied inhabitants of Belize's coral gardens. They can be found in an unbelievable number of colours, a great variety of sizes, and in an incredible number of shapes. Man did them an injustice. They were removed from the Kingdom of animals and now they share a kingdom with algae and slime mold.

Sponges belong to the Phylum *Porifera*. In Belize, there are sixty or so known species of sponges. Even within a single species, they can be so unlike that one can easily mistake them for several different species. Their bodies are extremely diverse, varying from huge barrel-like forms, to tubes, vases, balls, sheets, and even branch-like. There are even groups that will bore into, or encrust living corals.

A sponge has a multi-cellular body, but it has neither organs, a nervous system nor a circulatory system. Instead it is a colony of specialized individual cells. When separated, each cell can multiply and reassemble as a sponge.

Sponges are filter feeders; to consume breathe or remove waste, they must maintain a constant seawater flow. Most of the world's 500 plus species feed on bacteria or particles even smaller.

Sponges can either reproduce sexually or asexually; asexual sponges are relatively few and reproduce by budding. Most poriferans are hermaphroditic, producing sperm or eggs during different periods. Sperm is ejected into the sea in a cloud so dense, that it's not unlike a smoking chimney. Some of these sperm will be captured by a female sponge of the same species. The method of capture is via its in-current pores in exactly the same way as bacterial food is captured. Fertilization occurs and the fertilized eggs develop into larvae that are ejected into the sea where they settle and slowly develop into juvenile sponges.

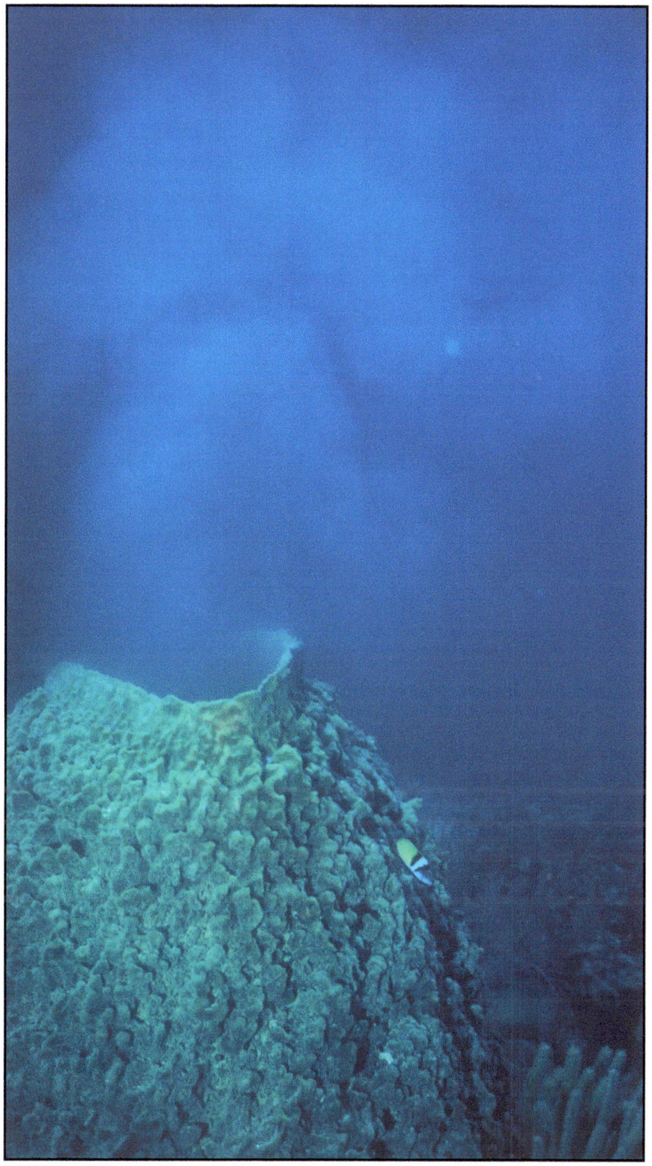

Giant Barrel Sponge spawning

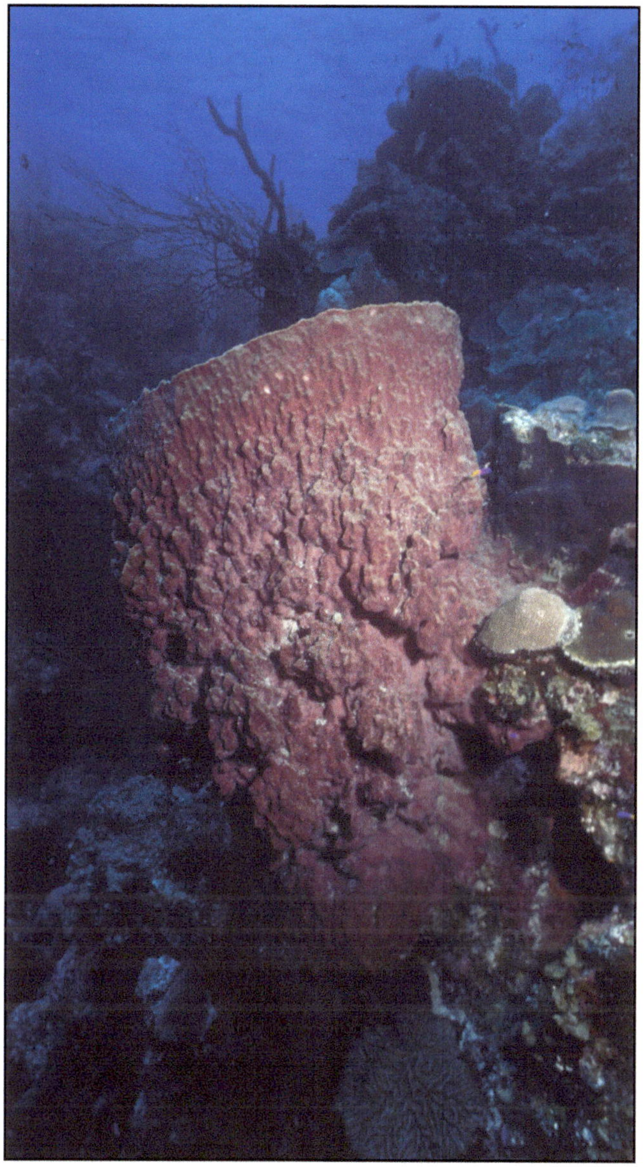

Giant Barrel Sponge (*Xestospongia muta*)

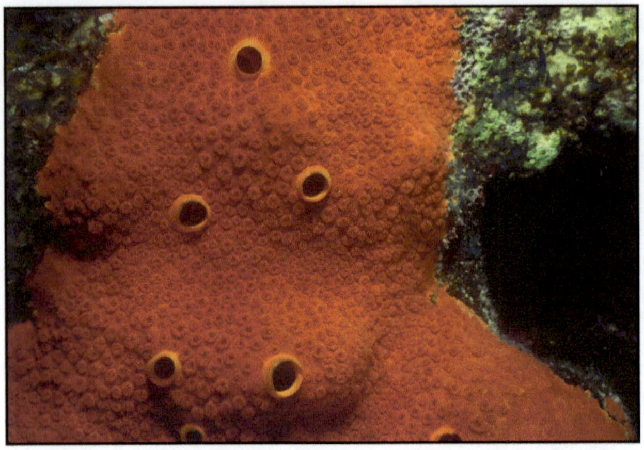

Red Boring Sponge (*Cliona delitrix*)

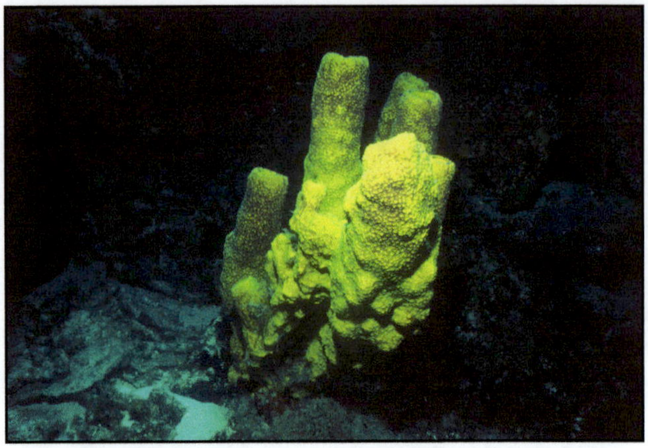

Branching Tube Sponge (*Pseudoceratina crassa*)

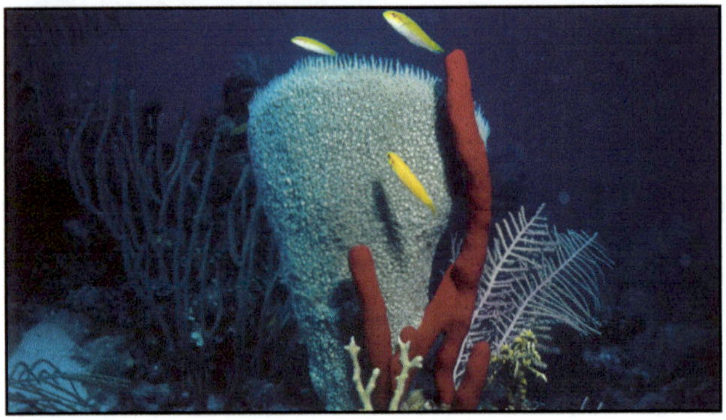

Pink Vase Sponge (*Niphates digitalis*)
and Erect Rope Sponge (*Amphimedon compressa*)

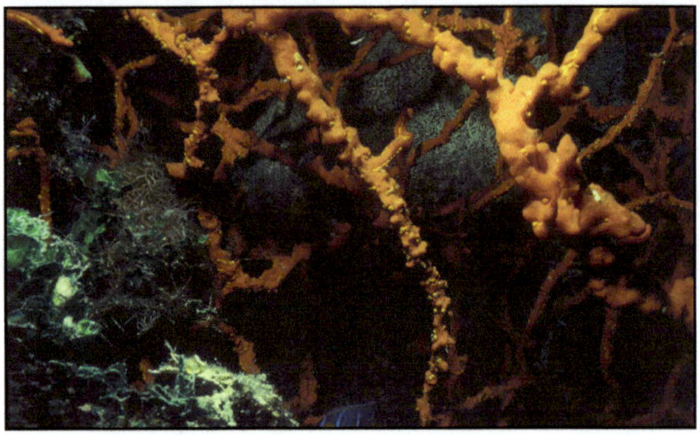

Thin Rope Sponge (*Rhaphidophlus juniperinus*)

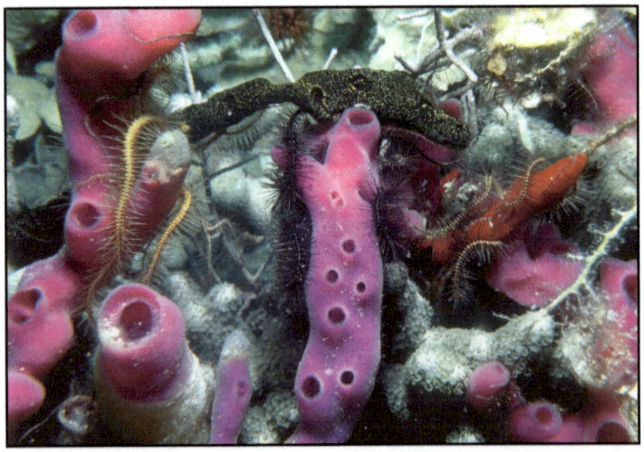

Lavender Rope Sponge (*Niphates erecta*)

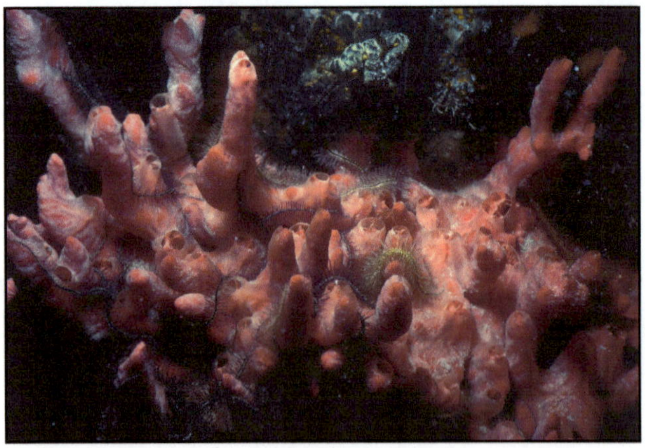

Pink Lumpy Sponge (*Monanchora unguifera*)

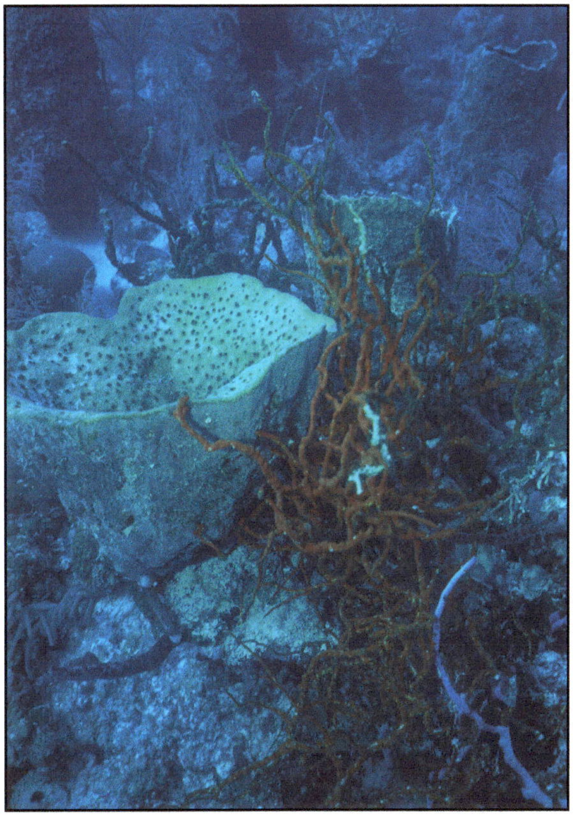

Vase and Thin Rope Sponges

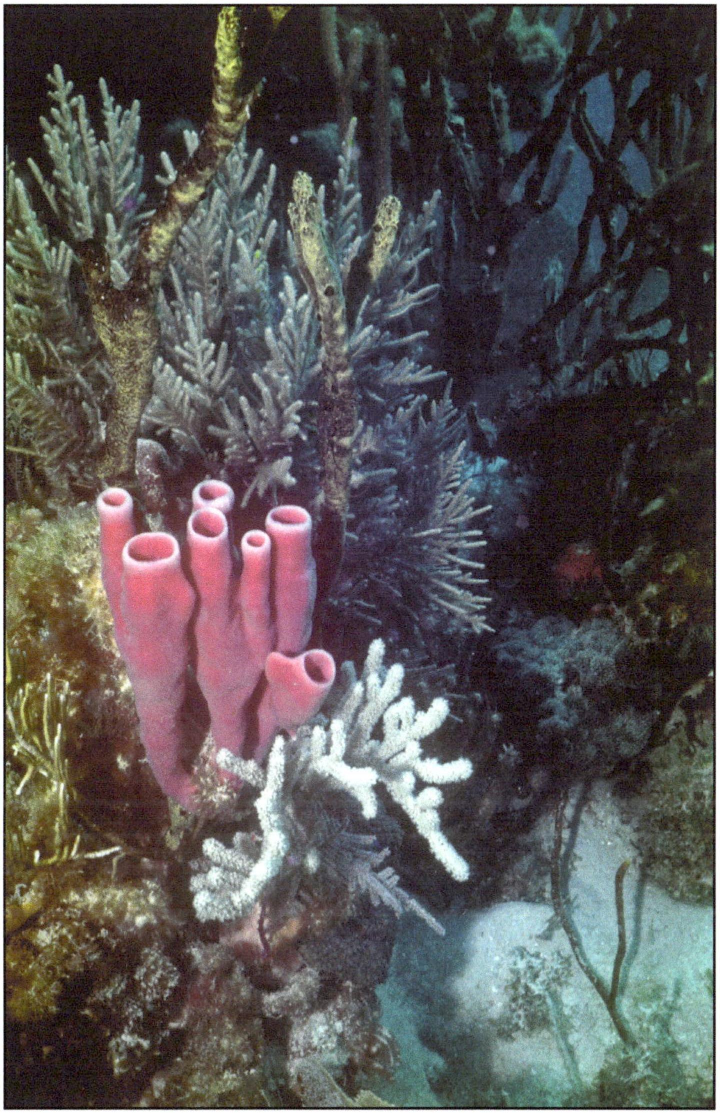

Stove-pipe Sponge (*Aplysina ascheri*)

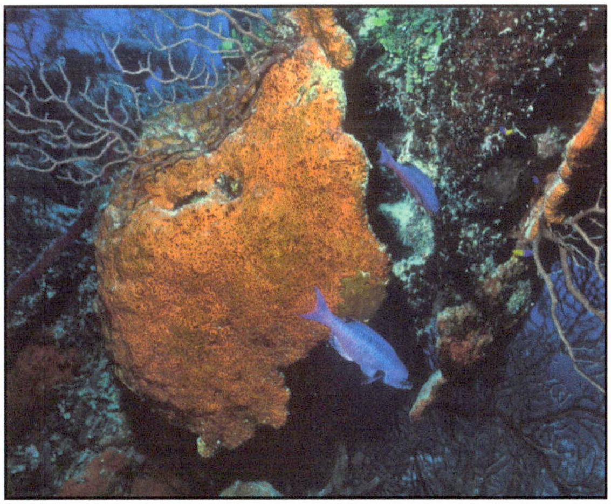

Orange Elephant Ear Sponge (*Agelas clathrodes*)

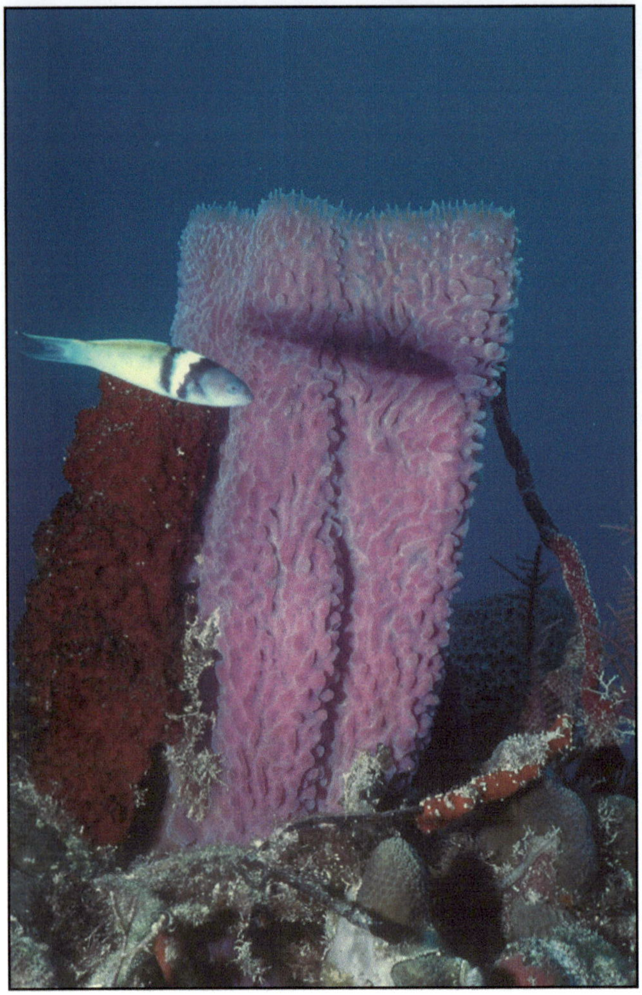

Azure Vase Sponge (*Callyspongia plicifera*)

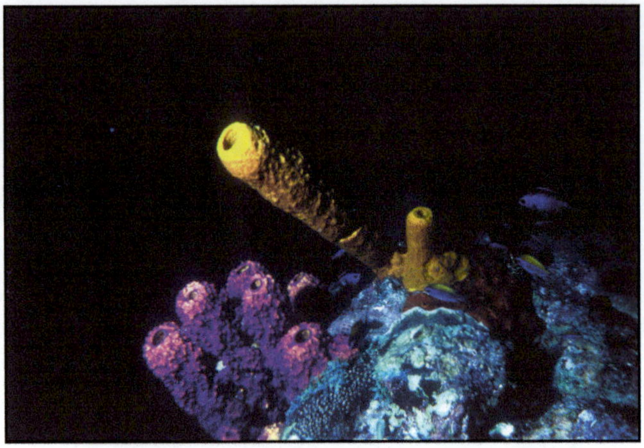

Yellow Tube Sponge (*Aplysina fistularus*) and Branching Tube Sponge (*Pseudoceratina crassa*)

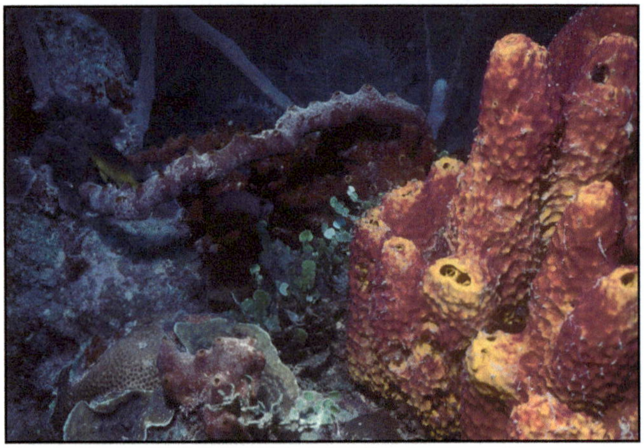

Branching Tube Sponge (*Pseudoceratina crassa*)

The small, brightly coloured strawberry vase sponge favours walls rather than the seafloor and may encrust black corals. Reproduction and brooding occurs within the sponge and when the larvae are released they are already well developed. Many will not survive, for those that land on sand or silt will suffocate. The survivors will generally adhere to the same wall, lower or adjacent to the parent sponge.

Sponges have several predators. It has been documented that the queen angelfish and the rock beauty's diet is 95% sponge. The gray and French angel fish also feed on sponges. Other sponge predators include the box fishes and the filefish. The hawksbill turtle lives on them. No fish, however, will completely devour a sponge. It is more a nibble here, a nibble there. Who knows, maybe each sponge has a different flavour.

In the 1920s, when people actually bathed with real sponges, the sponge industry in Belize was booming. The program called for numerous small cuttings of the sponge species *Hippospongia lachne*, wool or velvet sponge. Centred in Turneffe Islands, the sponge cuttings were placed on concrete discs and firmly planted in the lagoon beds. For a number of years the sponge market boomed with record growth. Then a major hurricane swept across Turneffe and wiped out most of the crop. By 1935 the industry had rebounded and another crop was sent to market in both Britain and the U.S.

Unfortunately one catastrophe followed another. In 1941 a fungal infection decimated Belize's sponge industry, hurricanes in 1942 and 1945 aggravated an already desperate situation. Several sponge growers persevered, but the killing blow struck in 1961 when Hurricane Hattie swept directly across Turneffe lagoon before destroying much of Belize City. Even without Hurricane Hattie, however, the sponge industry was finished. Synthetic sponges came into common use in the early 1960s.

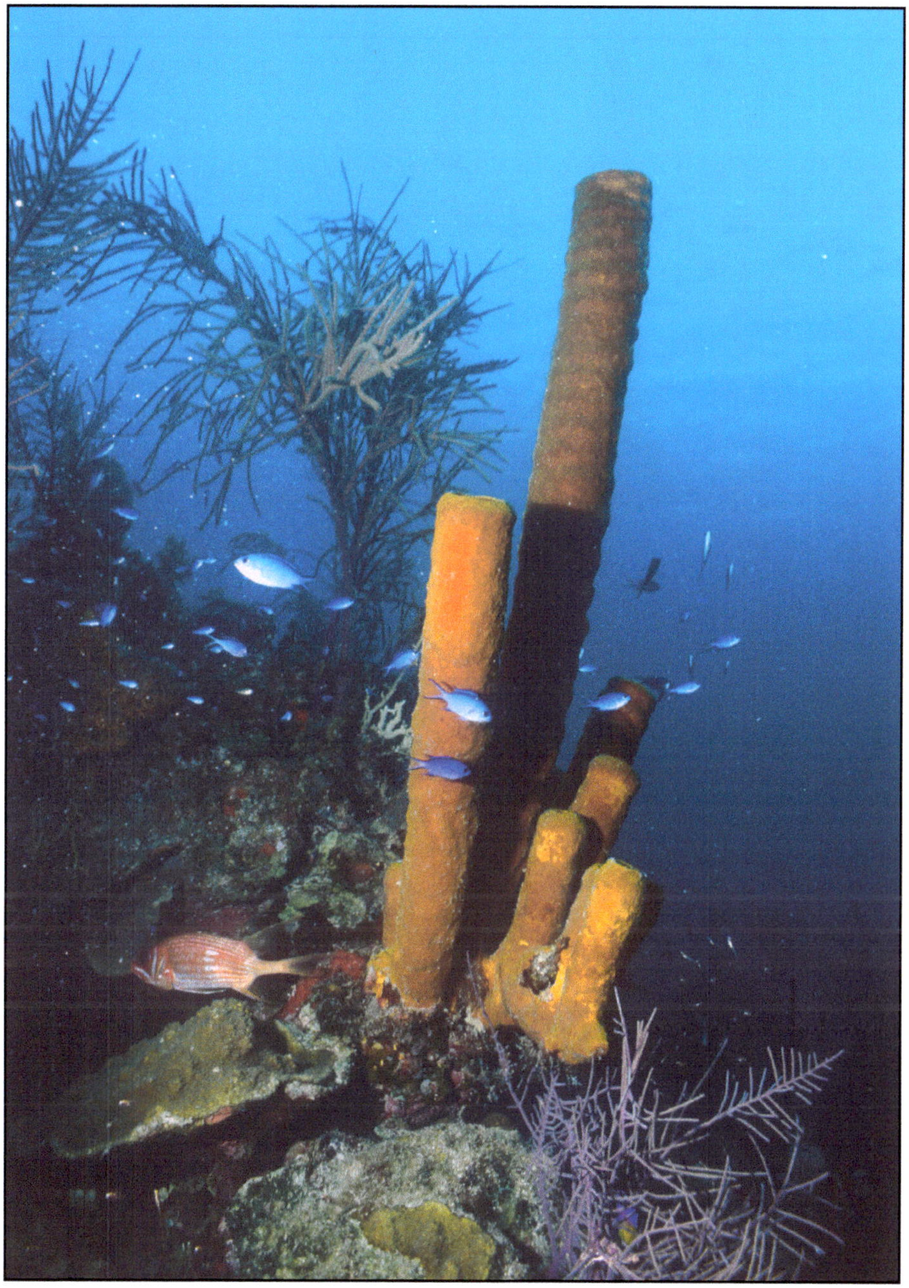

Yellow Tube Sponge (*Aplysina fistularus*)

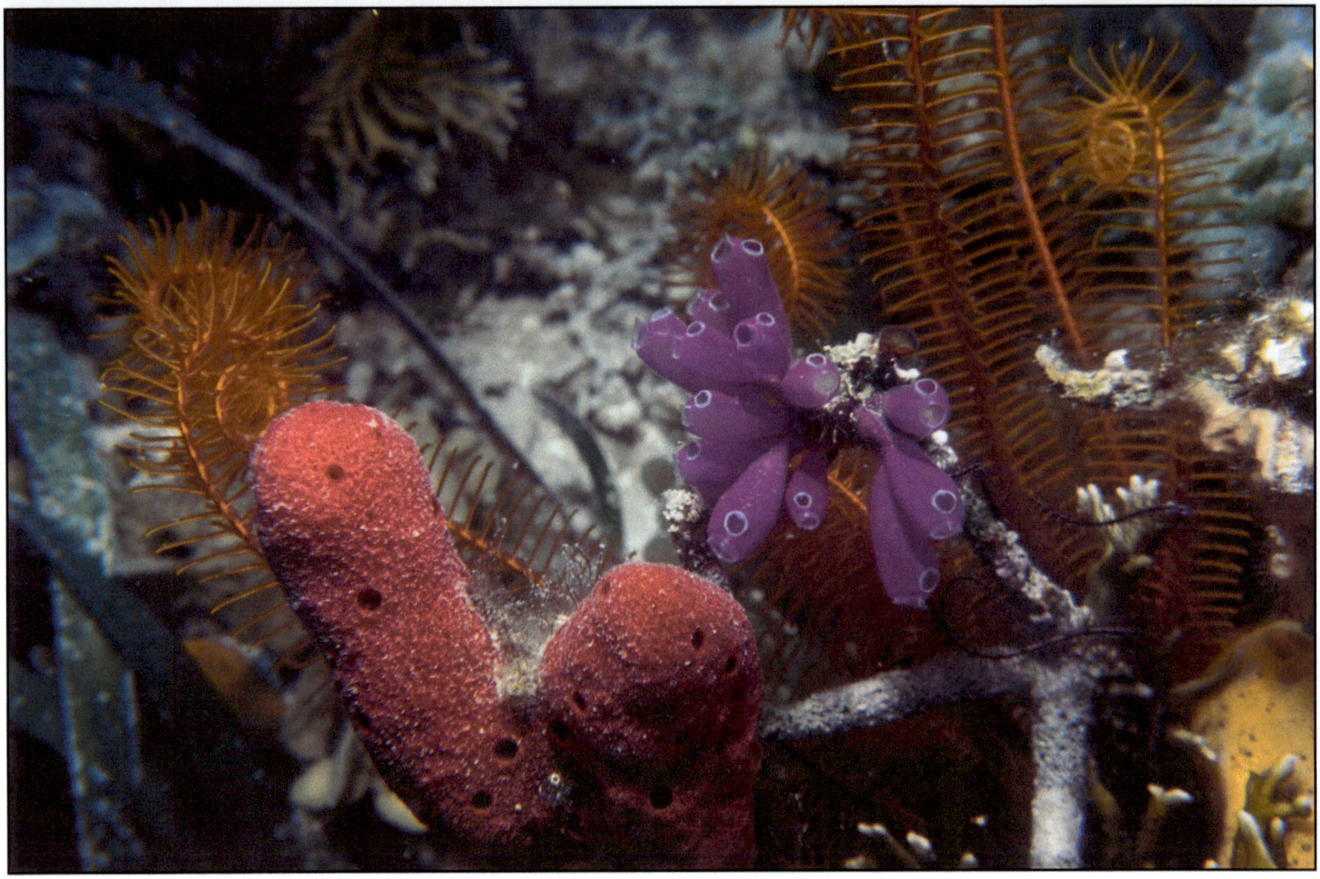

Blue Bell Tunicate (*Clavelina puerto-secensis*)
with Gold Crinoid (*Davidaster rubiginosa*)

Ascidian Tunicates

Tunicates are probably our ancestors. Very distant ancestors to be sure, but they belong to the Subphylum *Uracordata* of the Phylum *Cordata*, which means these colourful little critters have a rudimentary spine, which makes them ancestors of the fish, reptiles, birds and us, the mammals. Almost all Chordates have a backbone and as such, they are vertebrates. A Urochordate is not yet a vertebrate *per se*, but they do have a central nerve cord in the dorsal area, and a notacord which is the equivalent of the spine in vertebrates.

It has been theorised that at some time in the prehistoric past, vertebrates evolved from tunicates. The theory suggests that tunicate larvae swam up rivers, and during the process they developed muscles and sense organs eventually evolving into vertebrates.

Unlike corals and sponges, tunicates are highly evolved creatures. They have relatively intricate muscles, a nervous system and a heart that pumps their equivalent of blood. They have incurrent and excurrent siphons with which they breathe and feed. The main difference between a sponge and a tunicate is that a

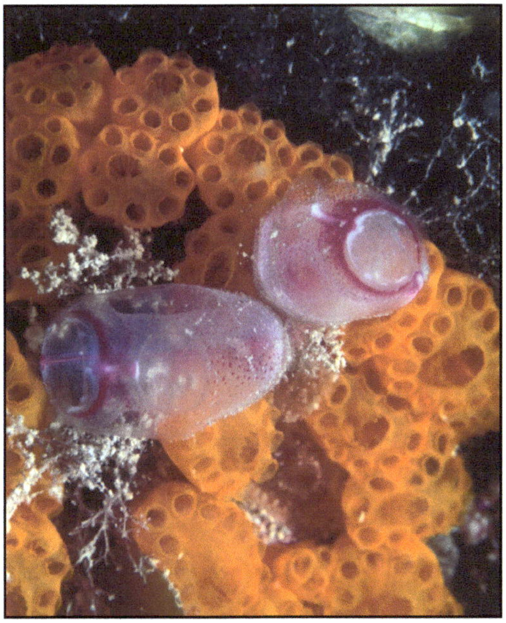

Painted Tunicate (*Clavelina picta*)
and Button Tunicate (*Distaphlia corolla*)

sponge cannot react if touched. Because they have a nervous system, tunicates can react by quickly closing their siphons.

This creature's name, "tunicate" is derived from its body covering, a cellulose coating called a tunic. This tunic is quite unique, for cellulose is a product of the plant world and a tunicate seems to be the only animal capable of producing it. Ascidian tunicates of the class *Ascidiacea* are characterized by a tougher tunic than the other classes of tunicates.

Once known as sea-squirts, ascidian tunicates are found in the underwater mangrove systems where few snorkers go, so they are seldom seen and might be mistaken for sponges. Like sponges they are attached to the stratum below. Most ascidian tunicates are very colourful and can be blue, yellow, orange, or red as well as black or white, depending on the species.

Pelagic tunicates (*Thaliacea* and *Larvacea*) are those that are unattached and swim in the open ocean by means of jet propulsion, the jet being an extension of the excurrent siphon. Most pelagic tunicates are transparent and are very seldom noticed. Pelagic tunicates reproduce by budding. The juvenile attaches to the rear of the parent tunicate, its sibling then attaches to its posterior and so on, until there is the likeness to a train. Each juvenile is a copy in miniature of the parent tunicate.

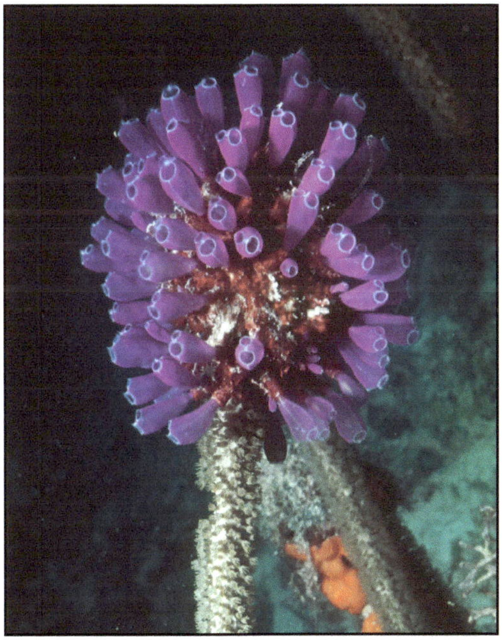

Blue Bell Tunicate (*Clavelina puerto-secensis*)

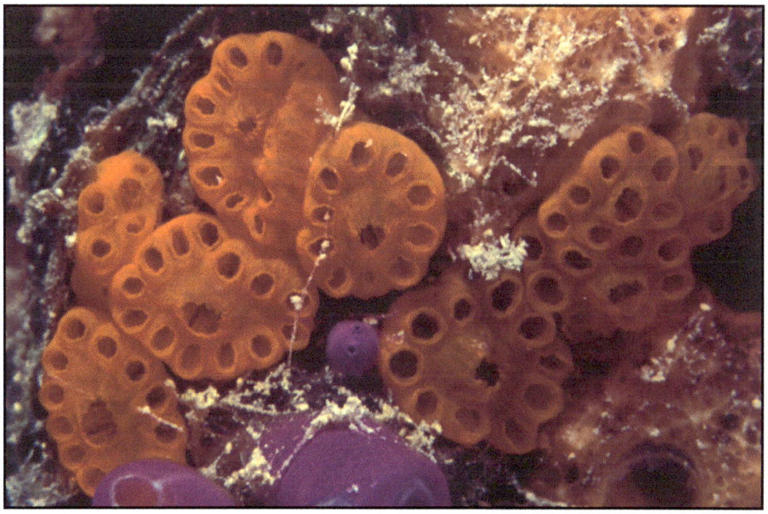

Button Tunicate (*Distaphlia corolla*)

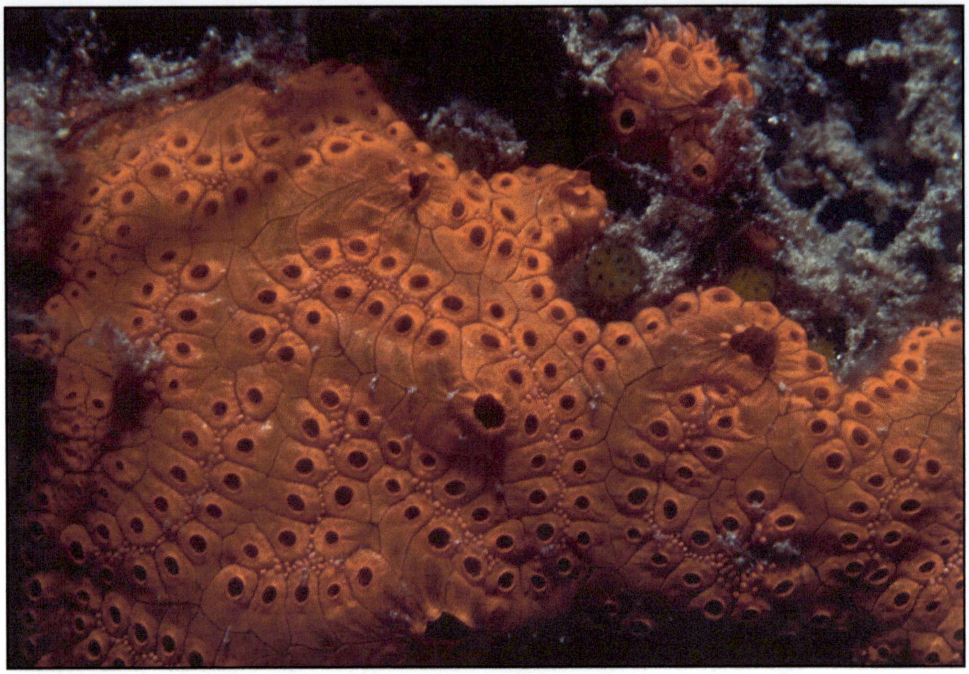

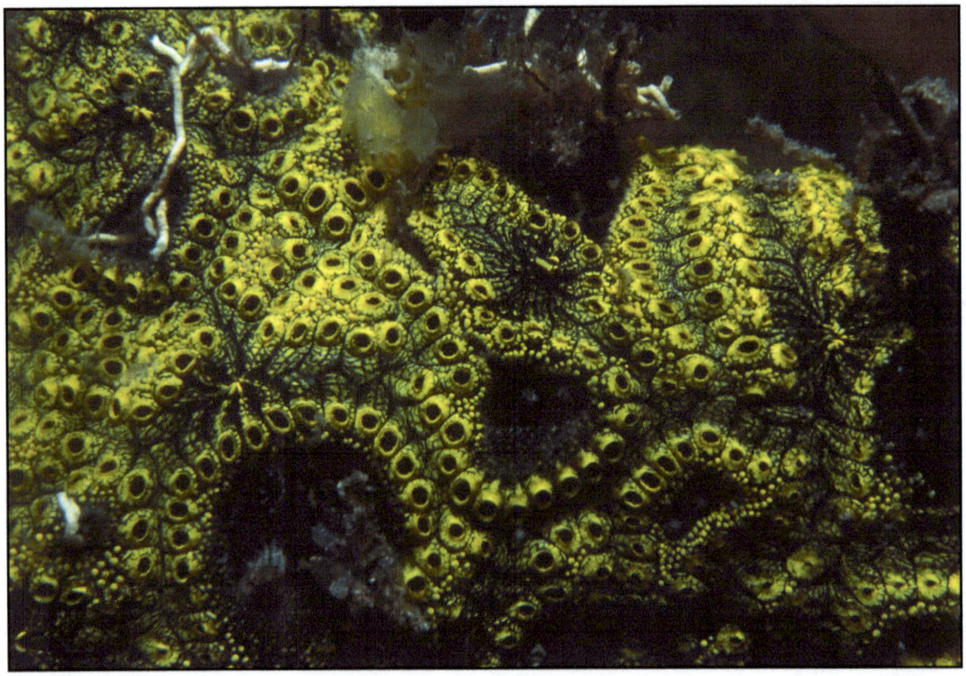

Encrusting Social Tunicates
(*Symplegma* sp.)
(*above and left*)

Segmented Worms

If someone describes a certain creature's head so covered with worms that several could be seen protruding from their host, there would no doubt be cries of disgust and feelings of revulsion. However, if this conversation took place on a dive boat, the participatants might be impatiently waiting for their dive guide to terminate his lecture on Christmas tree worms so they could jump into the sea, to see for themselves. These worms grow on giant coral heads, as illustrated on the back cover of this book.

Christmas tree worms and other segmented worms belong to the same Phylum as ordinary earth worms, but bear no outward resemblance whatsoever. Belonging to the family *Sabellidae,* their worm-like bodies are enclosed in hard stony tubes that are constructed of white sand cemented together with a sticky substance which is discharged from a special gland on the tube worm. These tubes protrude from the head of the brain and other species of coral. From the top end of these stony tubes the worms raise their heads and do their thing—breathe. For these tree-like appendages are actually gills.

At first glance, they look like a flower garden, blooms of variable bright colours and shapes. A snap of your fingers sends them back into their protective tubes. The Christmas tree worm is the most common of the segmented worms. Others include feather dusters, horseshoe worms and fan worms.

Tube worms come in a variety of colours, even within the same species. Their unique feather-like gills also work to catch prey such as plankton.

The gills are extremely sensitive to both movement and light, such as a dive light during a night dive. When disturbed, it will quickly disappear. One only needs to stay motionless for several seconds and the feathery gills will slowly emerge from their nondescript tubes.

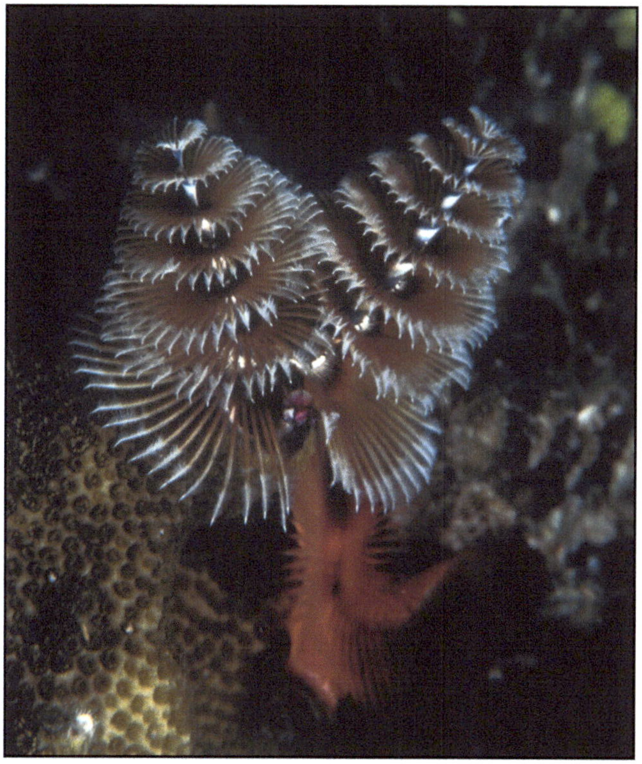

Christmas Tree Worm (*Spirobranchus giganteus*)

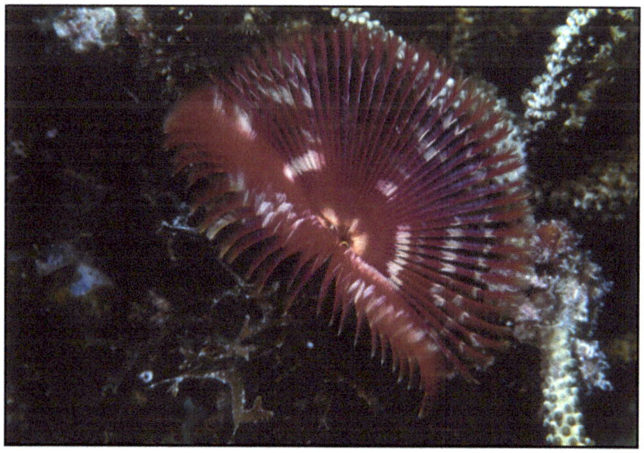

Split-crown Feather Duster Worm (*Anamobaea orstedii*)

The shape of the gills is the only clue to the identity of the hidden worm as most species have variable colours.

Another segmented worm, that the snorkeler or diver may come across is the fire worm. Of the three common fire worms in the Caribbean, the bearded fire worm is the one you would mostly encounter.

A fire worm may grow to six or seven inches long. They can be found under loose rubble, on corals, or in grass or algae. In most cases one need not look for them. They feed on a variety of marine organisms such as corals, anemones, and tunicates, and will normally be encountered, in the time it takes to complete one dive.

The fire worm is a "look, but don't touch" creature. It is an extremely colourful animal. It can afford to be, for it has no predators. The body colour varies from shades of red, orange, or green, along its sides. The body is segmented with tufts of gill filaments. If disturbed the worm will display numerous bristles which are both brittle and venomous. Contact with a fire worm causes an extremely agonizing and burning wound.

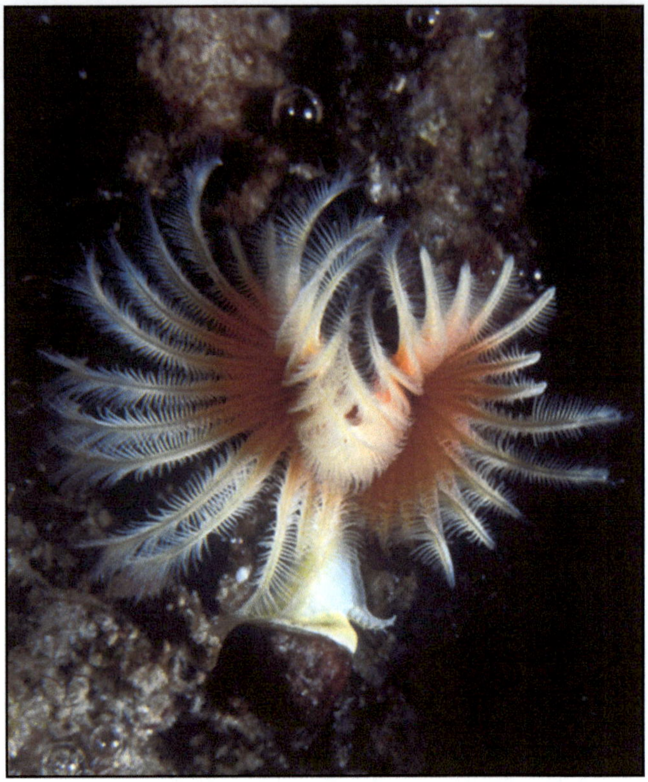

Star Horseshoe Worm (*Pomatostegus stellatus*)

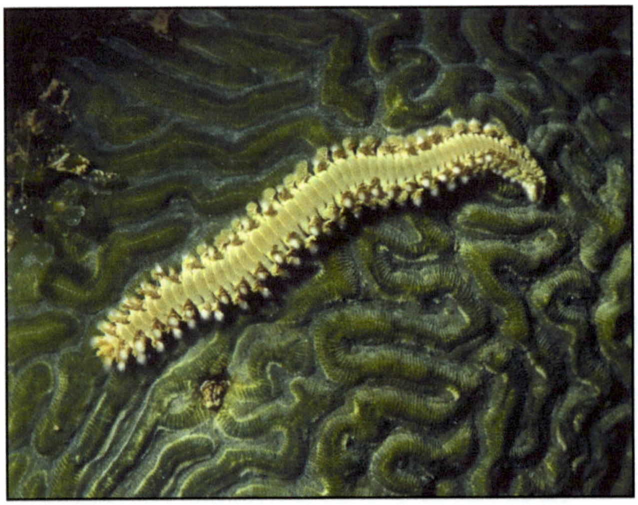

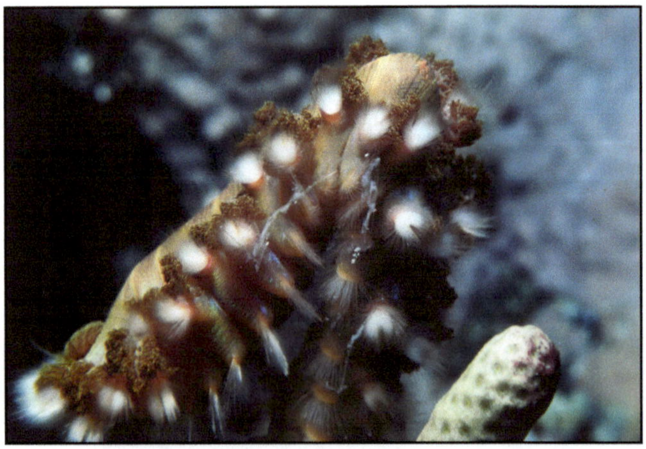

Bearded Fireworm (*Hermodice carunculata*)
(*left*) whole worm, (*above*) close-up

Arthropods, Molluscs and Echinoderms

With close to two million described creatures, arthropods are the largest Phylum in the Animal Kingdom. This includes the arthropods of the land, spiders and insects, which won't be covered in this book. But here you will find the arthropods that one might expect to see during a one-hour dive in a Caribbean garden.

The one arthropod that divers look out for is the spiny lobster. Most divers have seen a spiny lobster, if not in the sea, then perhaps on a dinner plate. The spiny lobster can grow to over two foot long, if given the chance. However, they have many predators not least of which is man.

They are found in all marine habitats within Belizean waters. One of Belize's main exports is lobster caught in the grass flats using a trap, unbaited but made in such a way that the lobsters find easy entry through a rectangular funnel, but escape impossible.

For divers in Belizean waters, it is illegal to take lobster while wearing scuba gear. Therefore, for a sports diver it is not uncommon to see lobsters as they do not see man as a predator.

During daylight hours, lobsters tend to hide in caves or under recesses in the reef. Hiding or not, their long antennae give their existence away by protruding out into the open. It is interesting that their tail will always face their favourite hole in the wall.

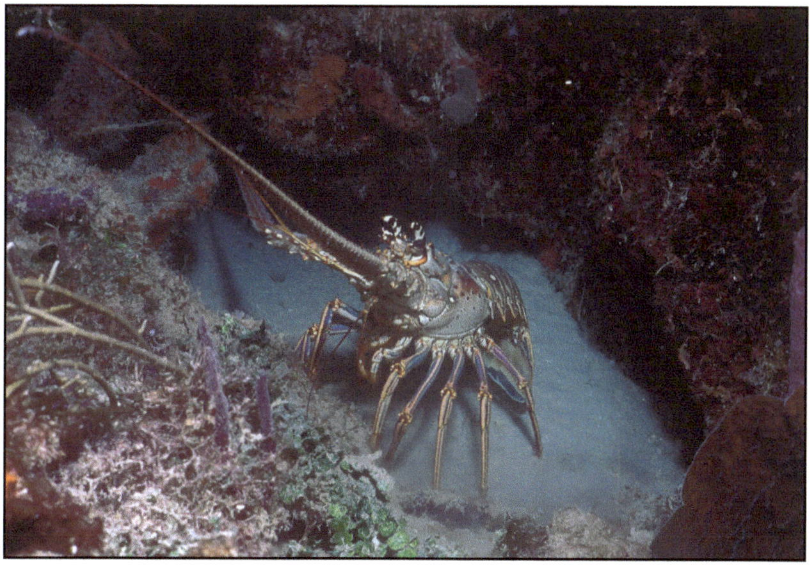

Caribbean Spiny Lobster (*Panulirus argus*)

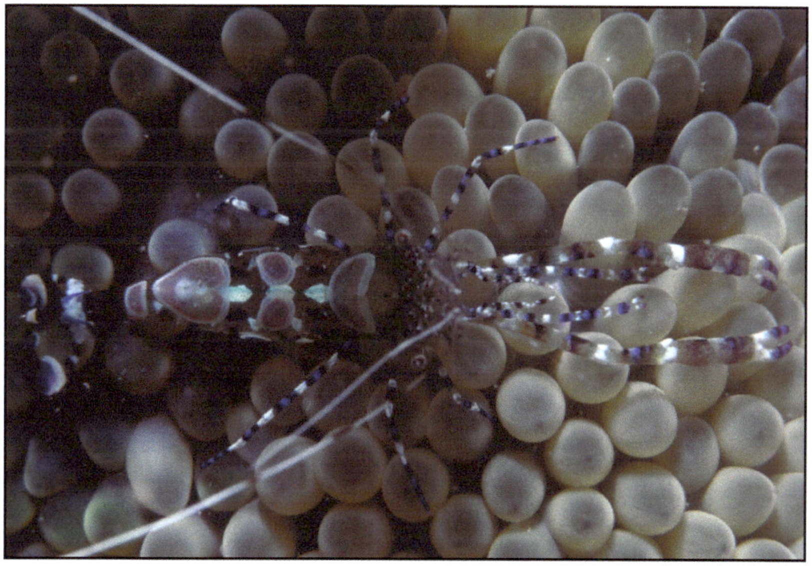

Spotted Cleaner Shrimp (*Periclimenes yucatanicus*)

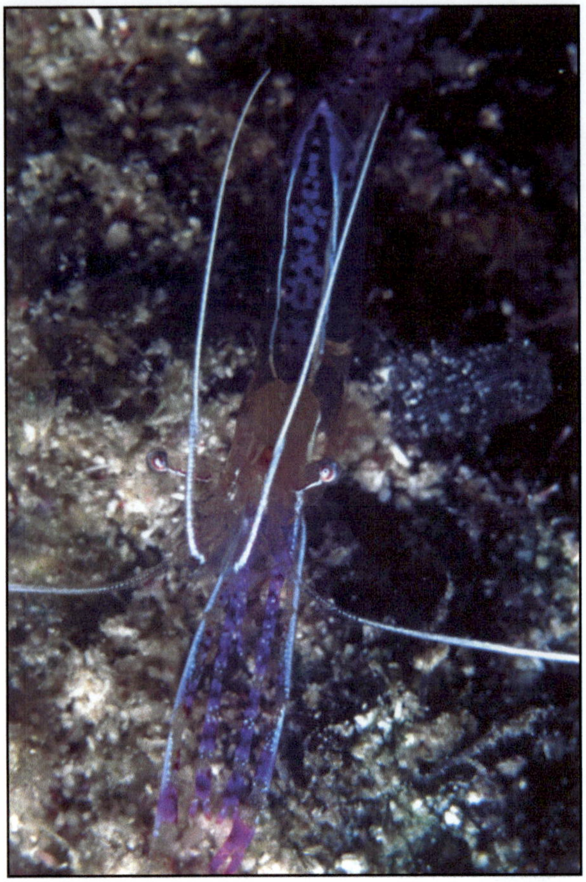

Pederson Cleaner Shrimp (*Periclemenes pedersoni*)

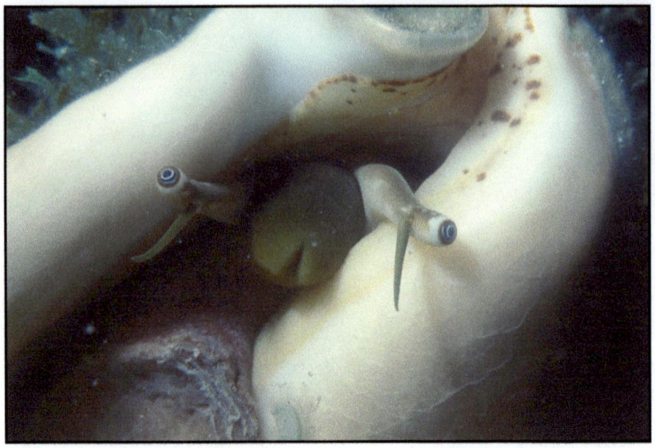

Queen Conch (*Strombus gigas*)

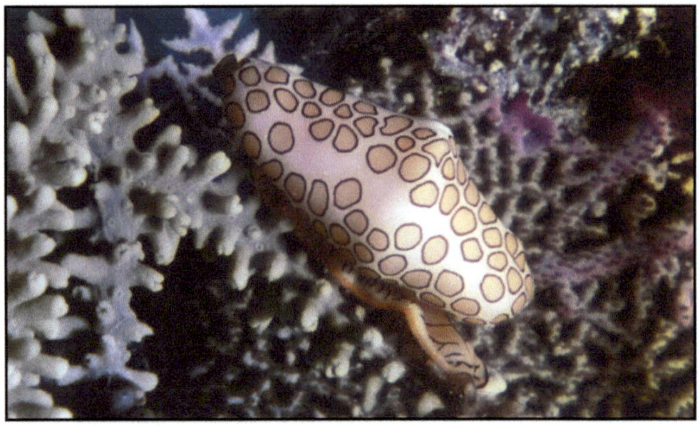

Flamingo Tongue (*Cyphoma signatum*)

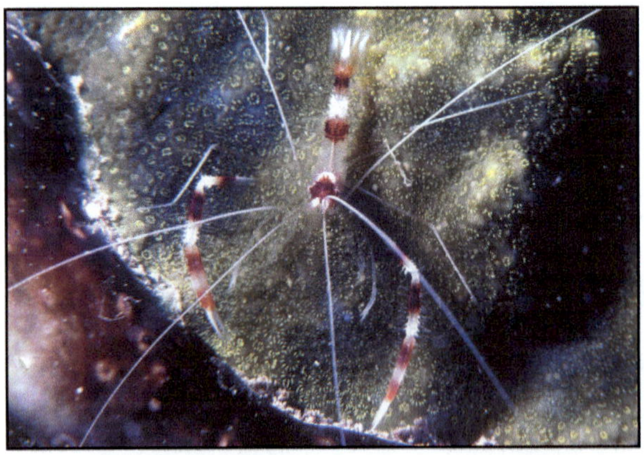

Banded Coral Shrimp (*Stenopus hispidus*)

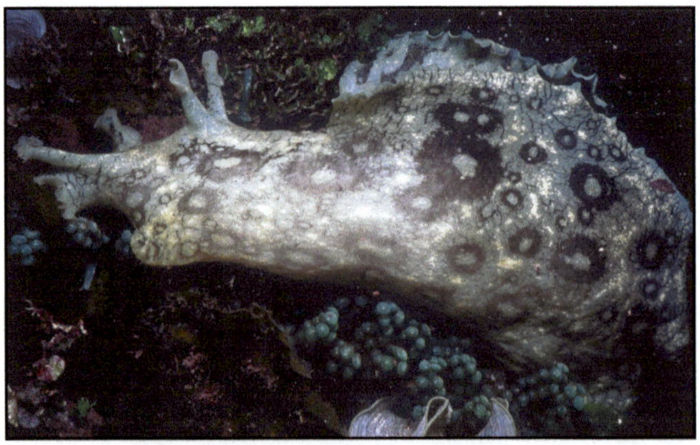

Spotted Sea Hare (*Aplysia dactylomela*)

Spiny lobsters have ten legs, two pair of antennae and three distinct body parts. A smaller cousin of the lobster is the shrimp. The shrimps of the coral gardens are not the shrimp you would expect to see at your favourite restaurant. Coral shrimps are at the most about two inches long and quite colourful. They are cleaning shrimp, and you will find them either cleaning anemones or larger fish. At night, they leave their host, to feed in the debris of the sea floor.

Other reef creatures include molluscs. The queen conch is the most recognized mollusc in Belize. The name mollusc is derived from the Latin *mollusca*, which means soft body. Like the conch, many other molluscs compensate for this soft unprotected body by secreting calcium carbonate and creating a protective outer covering, the shell.

Other molluscs manage to live without this protection. Illustrated are several molluscs, both with shells and without.

The last major group of sea creatures are the echinoderms, which include sea stars, feather stars, sea urchins, sea cucumbers, brittle stars, and basket stars.

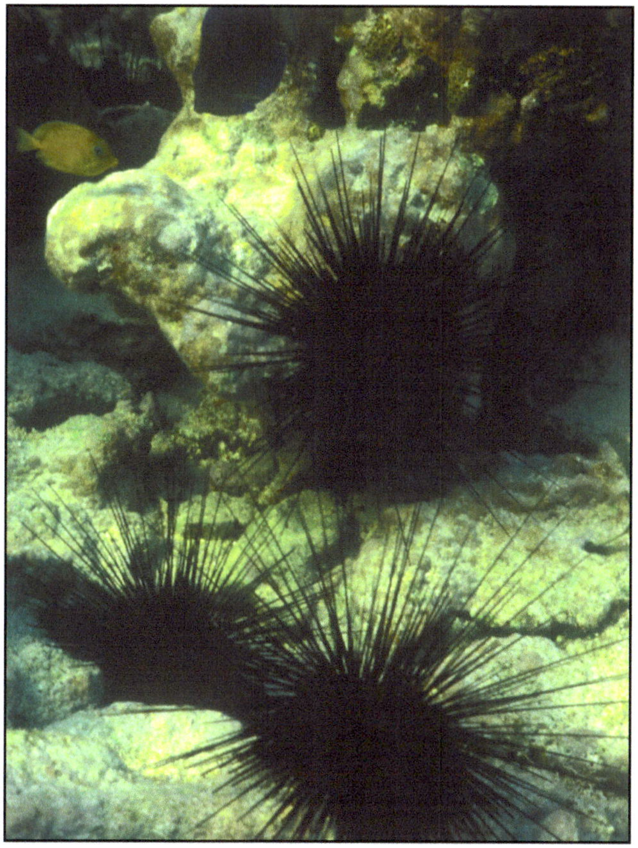

Long-spined Sea Urchin (*Diadema antillarum*)

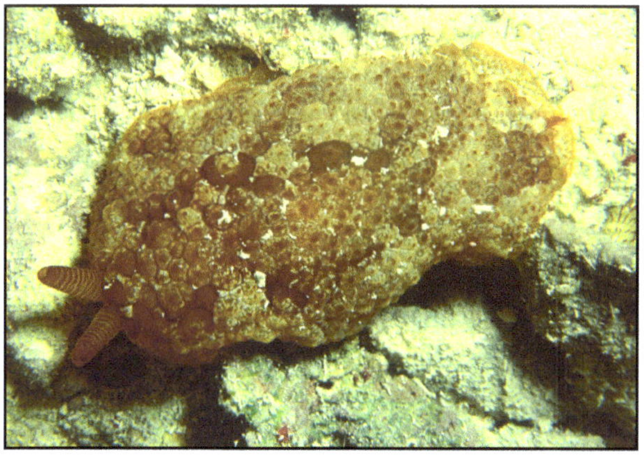

Warty Sidegill Slug (*Pleurobranchus areolatus*)

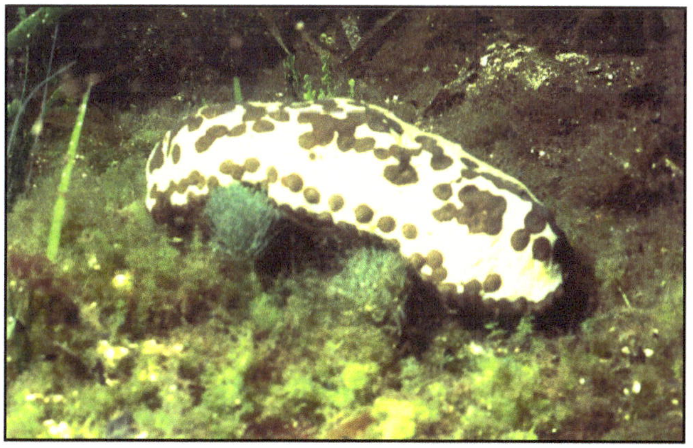

Three-rowed Sea Cucumber (*Isostichopus badionotus*)

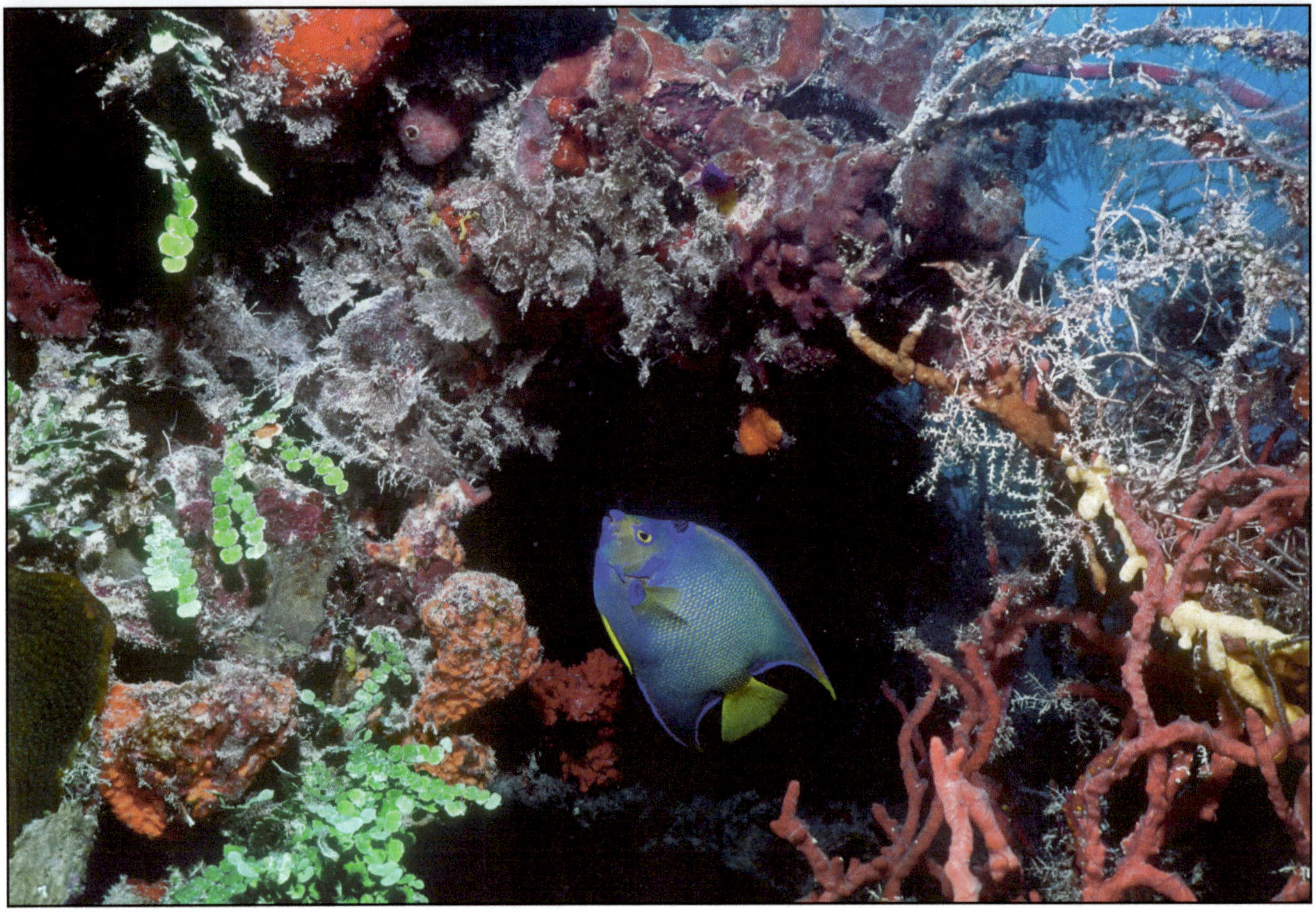

Queen Angelfish (*Holacanthus ciliaris*)

Fish of the Coral Sea

As I stand at the edge of the boat I feel a little like a medieval knight in armour, eighteen pounds of lead are strapped round my waist. A forty-pound metal tank is strapped to my back. Sweat runs into my eyes and drips off my nose.

For the hundredth time, I ask, "why am I doing this?" I lift my right foot, which is almost two feet long, and take a giant step. There is an eight foot drop, a splash, and then I remember! I let myself drift slowly down through this familiar invisible sea, and all of a sudden there is a kaleidoscope of colour. Multi-hued fish move out of my way as I slowly drift down a wall that is vibrant with the reds, yellows and purples of sponges.

I watch the bottom come up, and with a slight motion of my hand, I guide myself to a clear sandy area and look around. I now weigh two pounds, rather than the two hundred and thirty pounds I was when I plummeted into the sea. I press my buoyancy control and give the valve a brief squeeze. A few silver bubbles head for the surface. I am now weightless.

The coral sea, along with its various life forms has got to be one of the most beautiful environments on earth. To my right is a maze coral. A coral colony, many living organisms, all created in various shapes, sizes with intricate detail. Each structure is created by thousands of tiny creatures, every structure exact to its specs. Sponges of every colour abound, every shape imaginable.

The variety of animal life in any other ecosystem cannot compare with a coral reef system. Their density and complexity put coral sea life forms in a class of their own. That is a big statement, I know, but in the following pages I'm going to do my best to make you into divers.

As I slowly swim along the base of the wall, I am joined by two queen angel fish. They are predominantly blue with bright yellow tails; their fins also terminate with yellow tips. As I fin past a small cave, a six -foot long green moray eel eases out of the crevice in the sponge covered wall. His head turns in my direction; there is neither fear nor animosity in his eyes, only curiosity at this weird looking fish, which is emitting silvery looking spawn from its head.

I move on and come across a fissure in the wall. Glancing in, I see light, a cave running up forty feet, emerging behind the drop of the wall. I move in—the interior is covered with many encrusting red sponges. I ascend and the shaft widens, now the sun illuminates

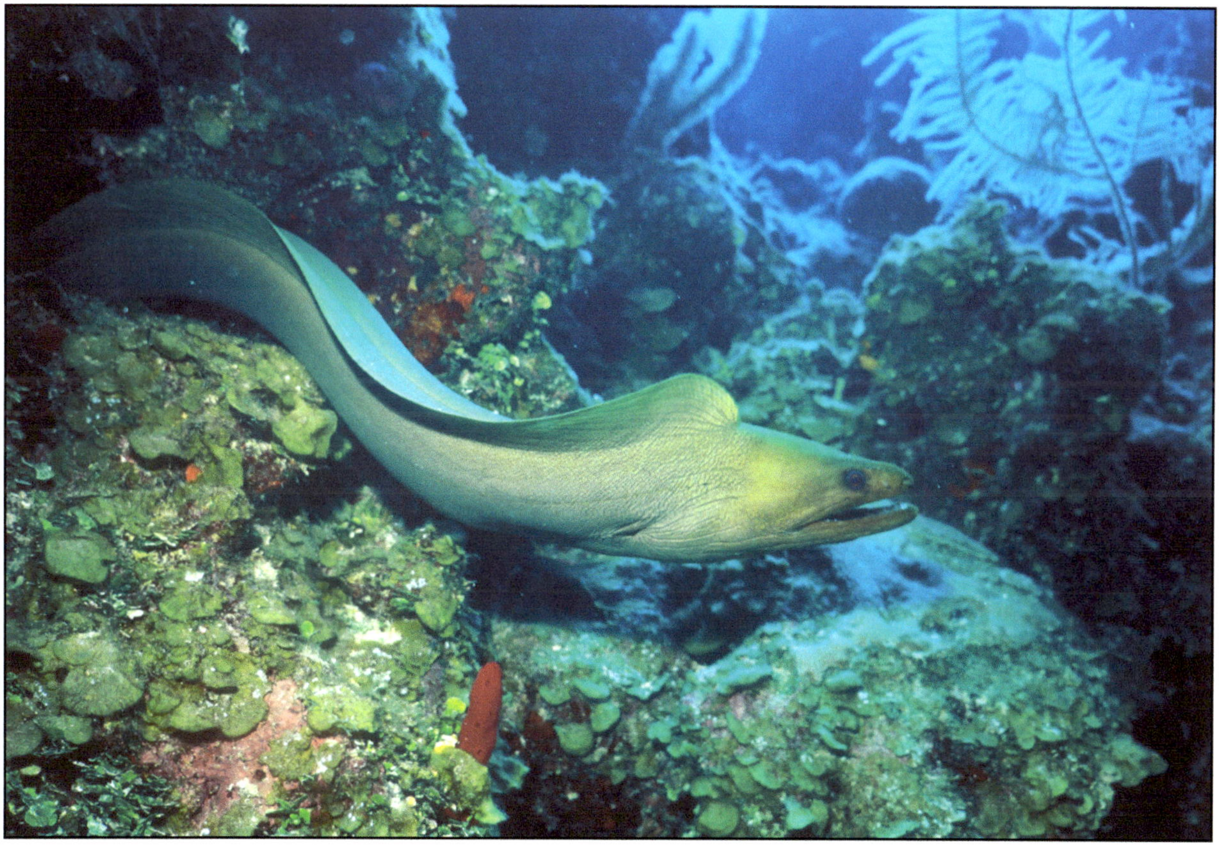

Green Moray Eel (*Gymnothorax funebris*)

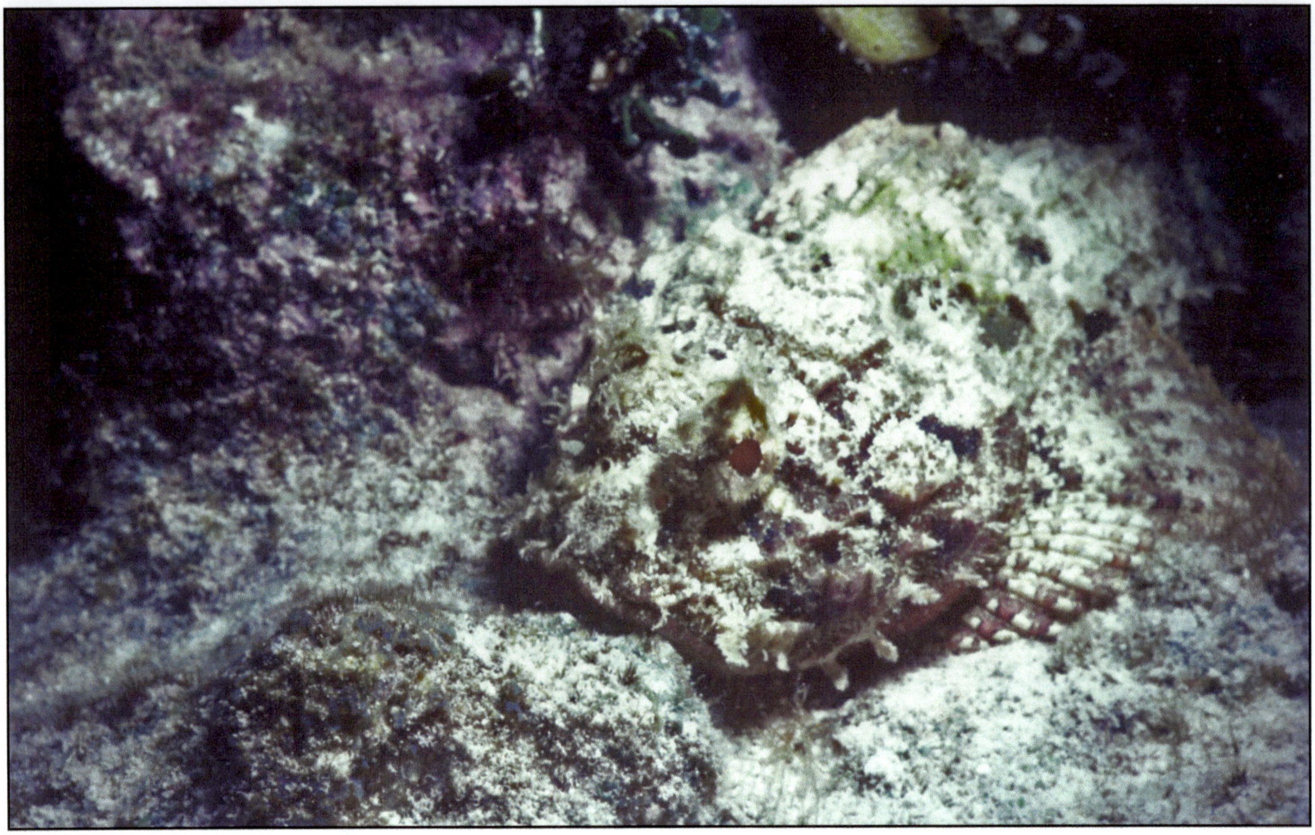

Spotted Scorpionfish (*Scorpaena plumieri*) Can you find it?

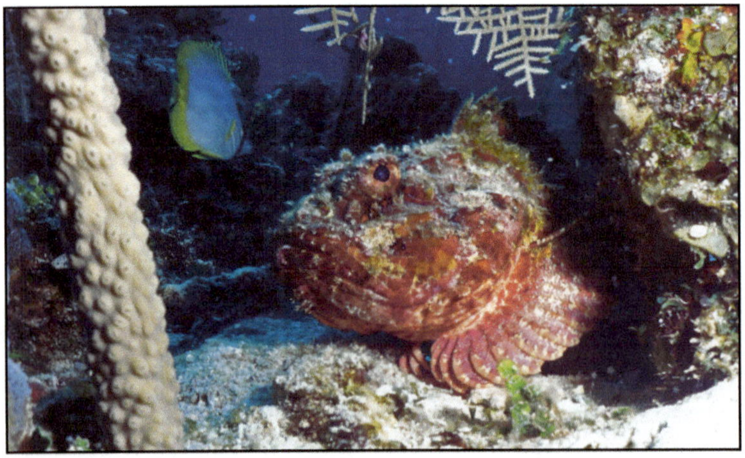

Spotted Scorpion Fish (*Scorpaena plumieri*)

the cave. As I ascend, I pass what looks like an irregular shaped piece of dead coral; I squeeze past, not wanting to scrape myself on it. It moved! I look more closely. Gradually a form takes shape. It is a large scorpionfish.

As I edge past, I examine it. A single red eye examines me. The spines of the frontal dorsal fin are venomous, and can cause extensive pain and illness. The spotted scorpionfish is a master of camouflage. Blending in with the background, they will stay unmovable for hours. Whatever the background, be it sand, rubble, or coral, a spotted scorpionfish will blend in and disappear!

I continue ascending until I suddenly swim into clear water. Above, the surface is a silvery sky. The depth is now a little less than thirty feet. I move back to the wall and look down. The coral garden which I've just left is another thirty feet below. I turn to head back to the dive boat and a movement catches my eye. Three eagle rays, gliding in a V formation, are swimming towards me. I purge my remaining air from my BC and settle down to watch their fly-by. They are not aware of me and continue to hold a steady course.

Their backs are blue-black with numerous smallish white spots. The wings are over eight feet across and their heads are well pronounced with a protruding tapering nose. Whip-like tails trail behind ten feet or more. Not seen from my angle, there will be several venomous spines at the base of their tails.

I have seen eagle rays on many occasions, but never quite as close. Eagle Rays feed on molluscs that they find by settling down in sandy areas and digging in a shuffling manner. The fly-by is over so I rise and make my way back to the boat, another dive logged.

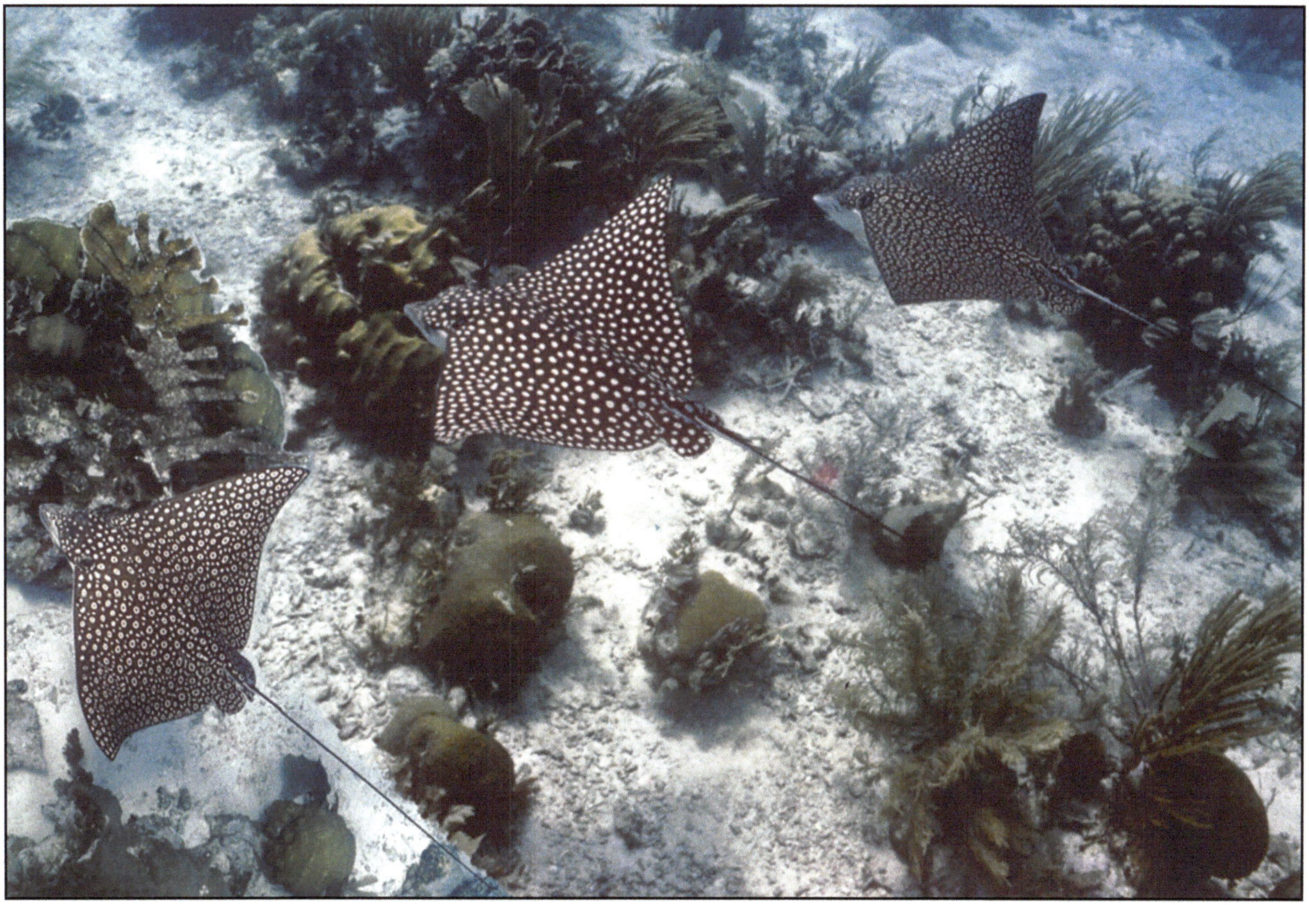

Spotted Eagle Rays (*Aetobatus narinari*)

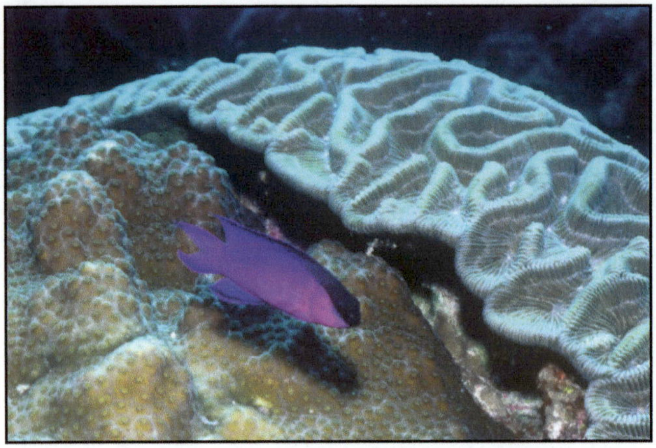

Black-capped Basslet (*Gramma melacara*)

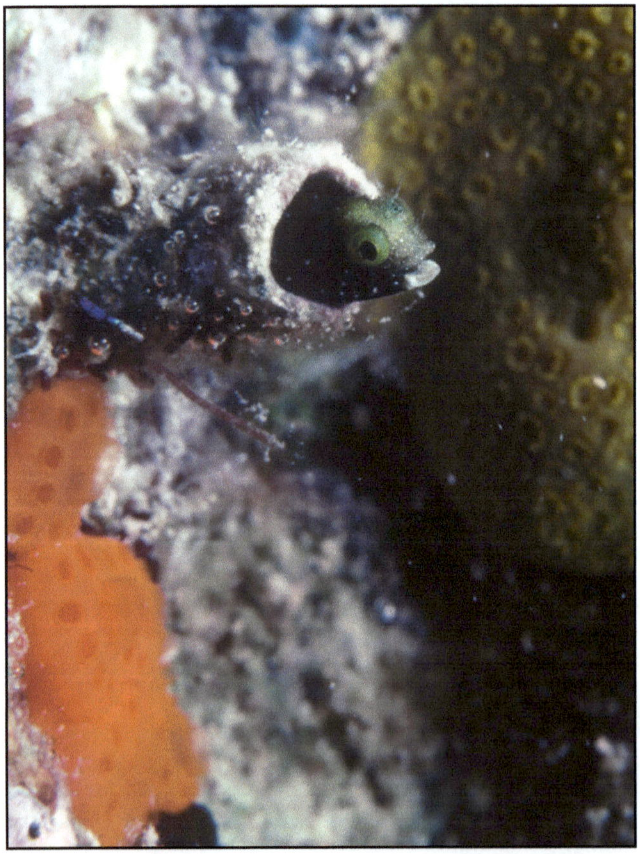

Spinyhead Blenny (*Acanthemblemaria spinosa*)

Most reef fish are small, seldom over five or six inches in length and extremely colourful. They will include angelfish, basslets, blennies, damselfish, butterflyfish, filefish, wrasses, and many other highly colourful little guys.

The reef also has many larger fish such as the parrotfish, groupers, eels, and so on. However, many of these larger fish have a symbiotic relationship with the small guys, who act as cleaning fish, removing parasitic pests from the mouth, eyes, and body of the larger fish.

But, for all its tranquillity, the coral gardens are not without their predators. Tourist divers have often marvelled at the numbers of fish going quietly about their business as peaceful as a giant aquarium. Major predators may pass by or, like the great barracuda, hover about thirty foot above the sea bed, in the same place for hours, directly above a hundred reef fish going about their daily business.

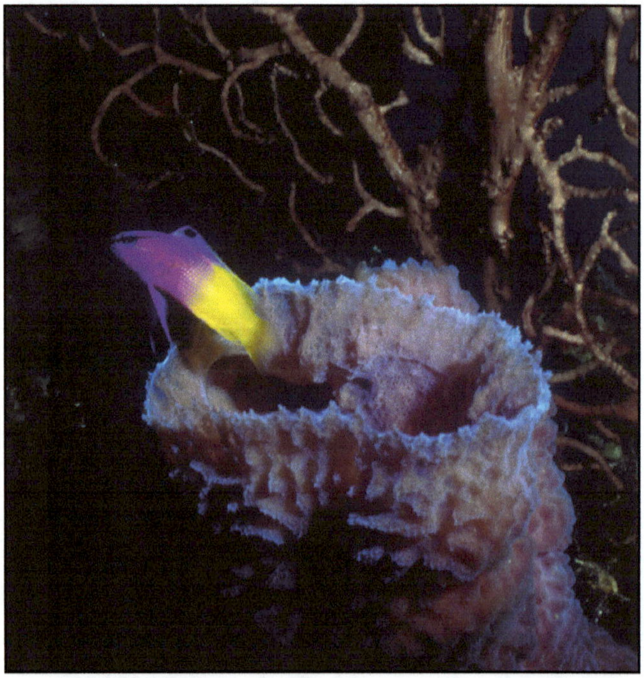

Fairy Basslet (*Gramma loreto*) and Vase Sponge

This peacefulness is only an illusion! Those reef fish are extremely aware of the dangers lurking above, and in their midst, and the slightest aggressive motion is enough to activate a vanishing act. The predators are aware of this also and wait until the time is right, and the time is right at dusk. As the sun approaches the horizon, the greater part of its light is reflected from the surface—instant dusk. Only then do the piscivores take advantage of the changing shadows and strike.

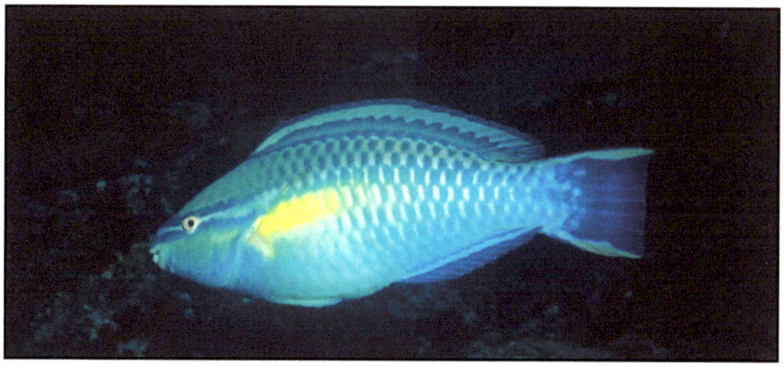

Princess Parrotfish (*Scarus taeniopteus*)

Other carnivores may feed mainly on invertebrates, such as, lobster and shrimp. Still other carnivores are planktivores that feed on swimming crustaceans and other tiny free-swimming creatures. Planktivores may range in size from the tiny inch and a half long fairy basslet, to the fifty-five foot long whale shark.

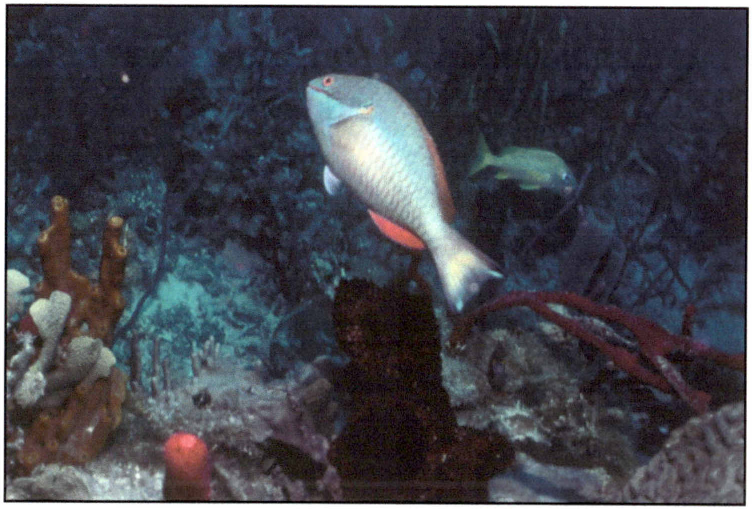

Red-band Parrotfish (*Sparisoma aurofrenatum*)

Some fish are herbivores that feed mainly on algae and the many small invertebrates that densely inhabit the algae foliage. These fish include parrotfish and surgeonfish.

These herbivores are very important. Without Belize's herbivores, the Belize Barrier Reef would be dead. In the tropics algae grow extremely fast and without the algae eaters, we would soon have an algae reef, rather than a coral reef. Helping the parrotfish and the surgeonfish are several other herbivores, especially invertebrates, the most important of which are the long-spined sea urchins. It has been estimated that a single long-spined sea urchin can keep one square metre of reef clear of algae indefinitely.

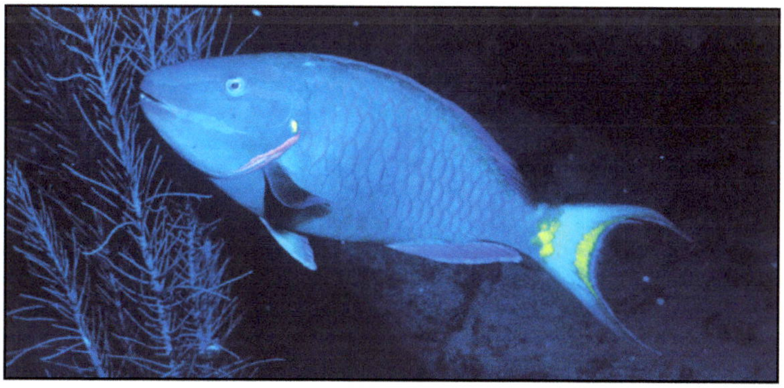

Stoplight Parrotfish (*Sparisoma viride*)

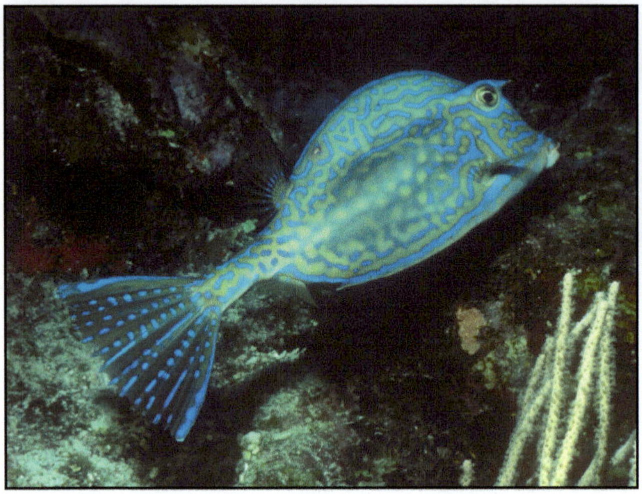

Scrawled Cowfish (*Acanthostracion quadricornis*)

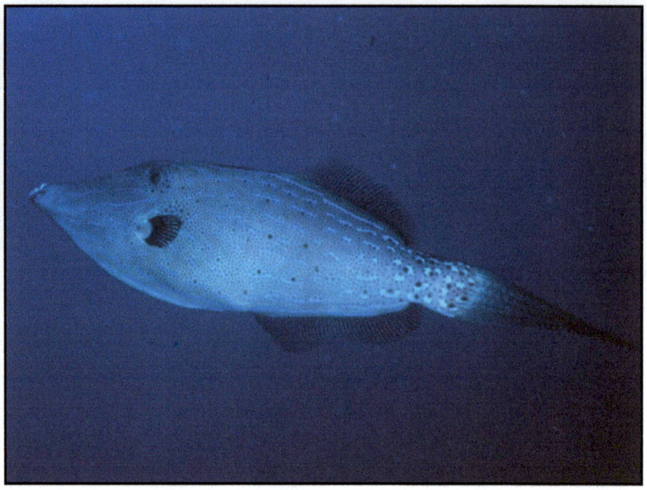

Scrawled Filefish (*Aluterus scriptus*)

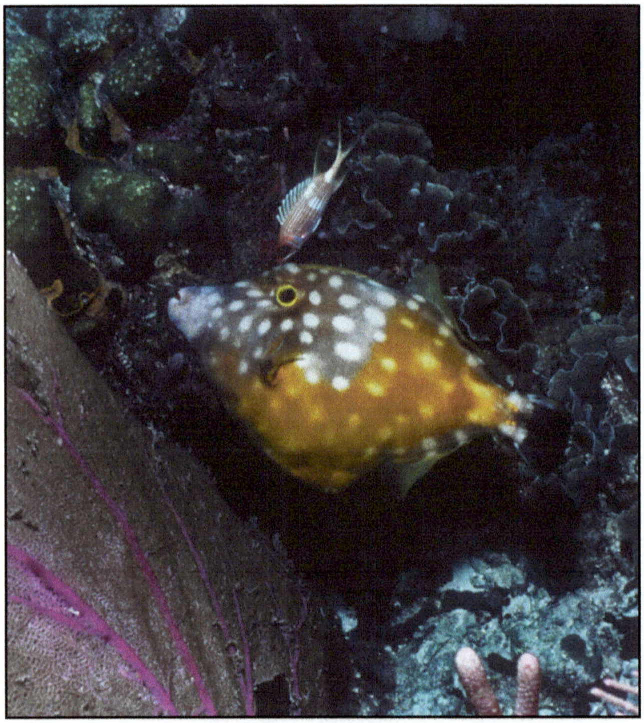

White-spotted Filefish being cleaned by a Squirrelfish (*Holocentrus adscensionis*)

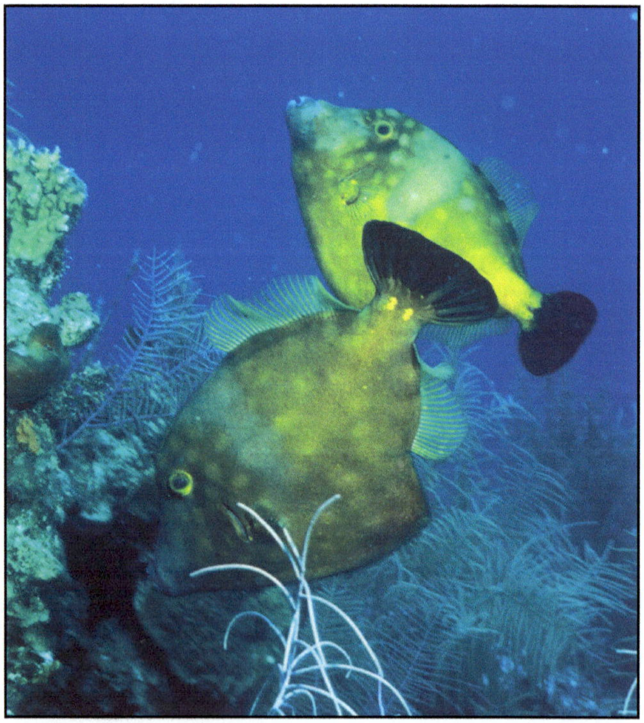

White-spotted Filefish (*Cantherines macrocerus*)

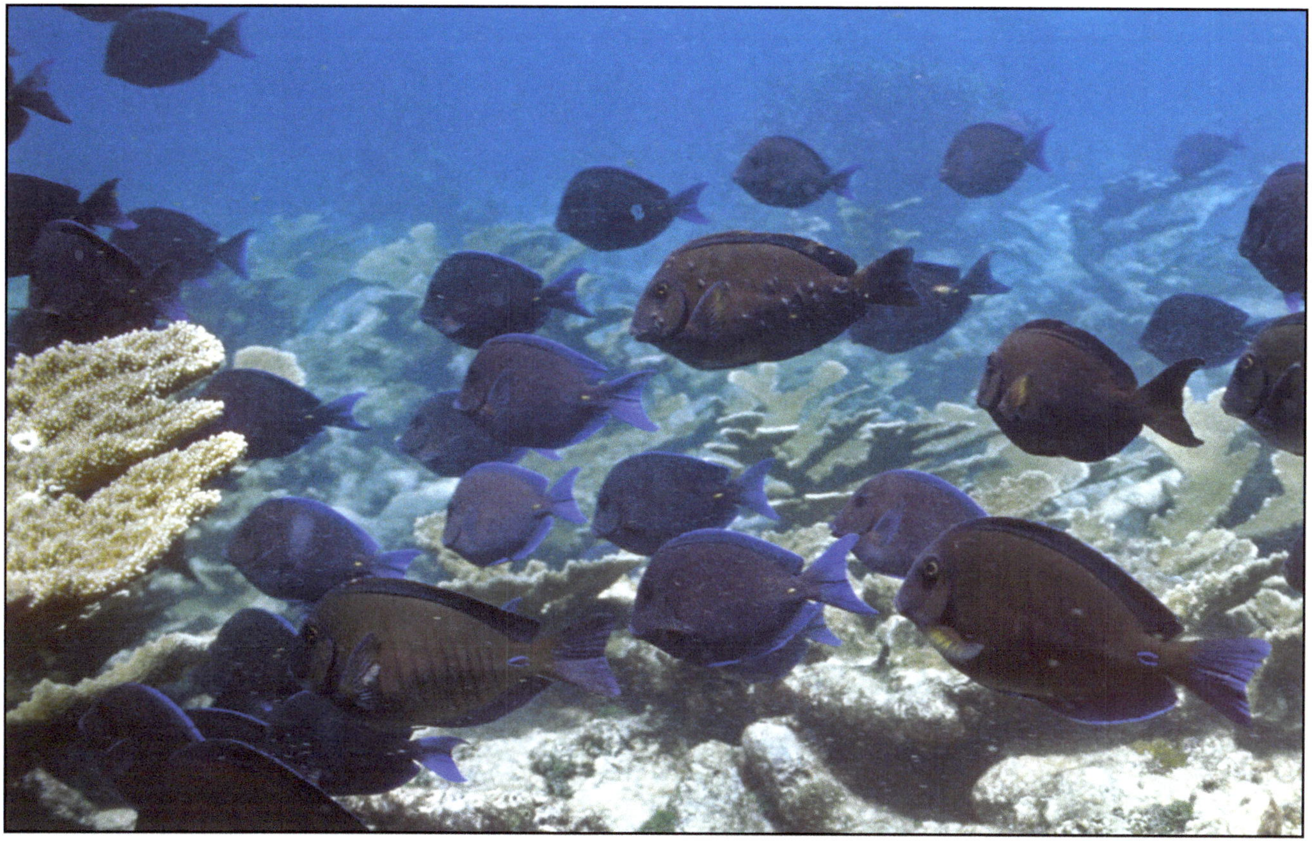

Schooling Blue Tang and Doctorfish

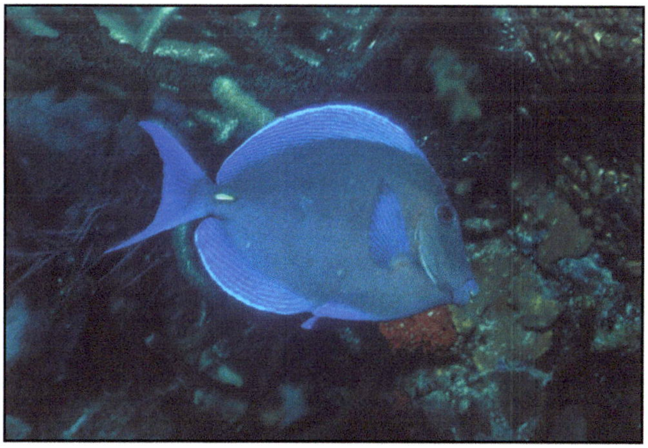

Blue Tang (*Acanthurus coeruleus*)

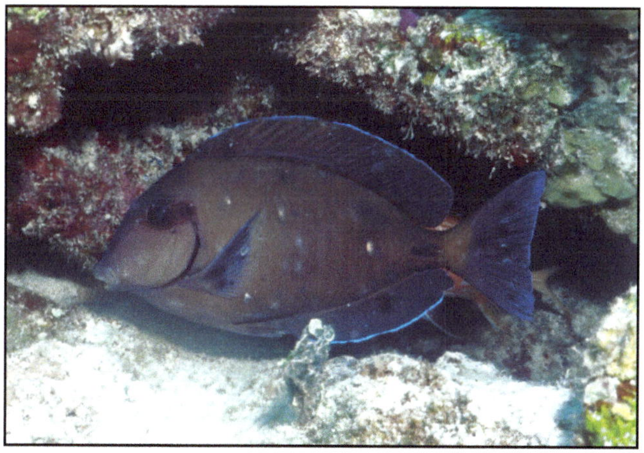

Doctorfish (*Acanthurus chirurgus*)

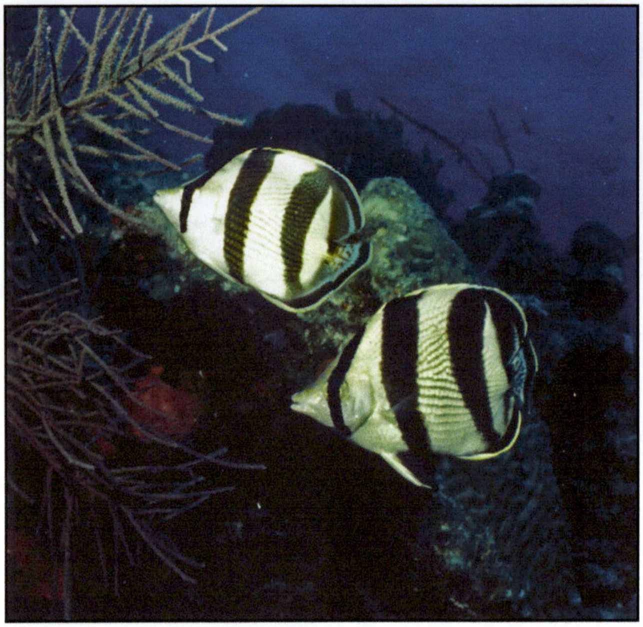

Banded Butterflyfish (*Chaetodon striatus*)

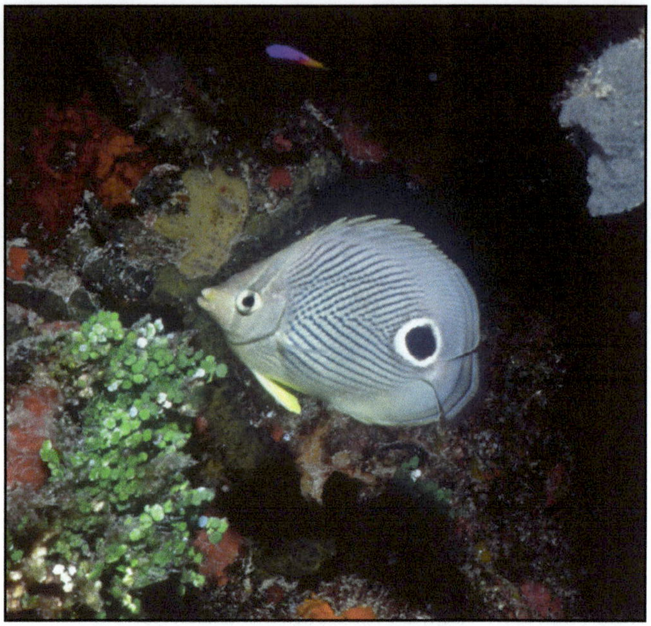

Four-eyed Butterflyfish (*Chaetodon capistratus*)

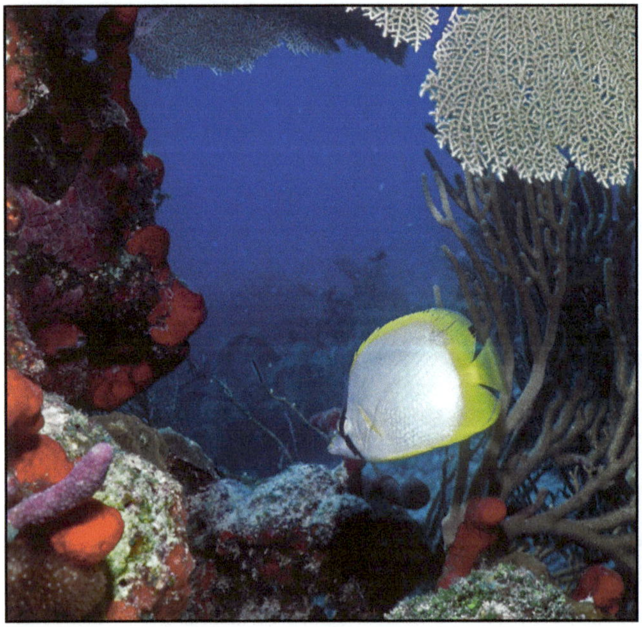

Spotfin Butterflyfish (*Chaetodon ocellatus*)

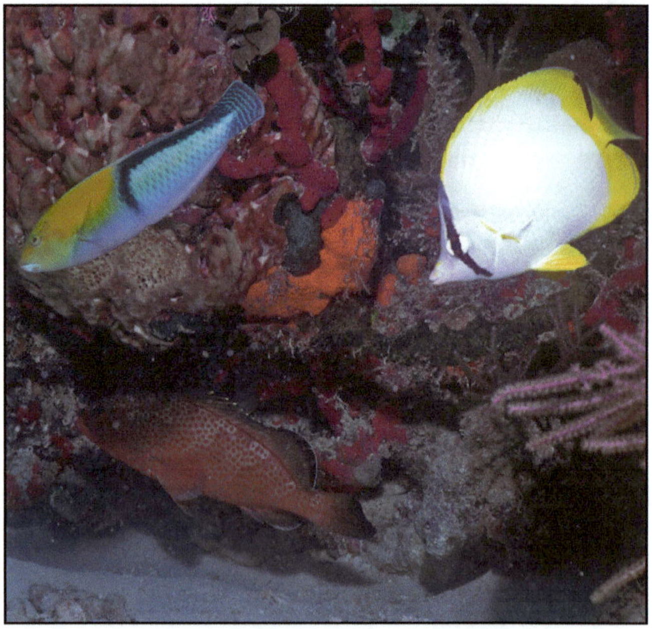

Yellowhead Wrasse (*Halichoeres garnoti*)
Spotfin Butterflyfish, and Graysby (*Cephalopholis cruentatus*)

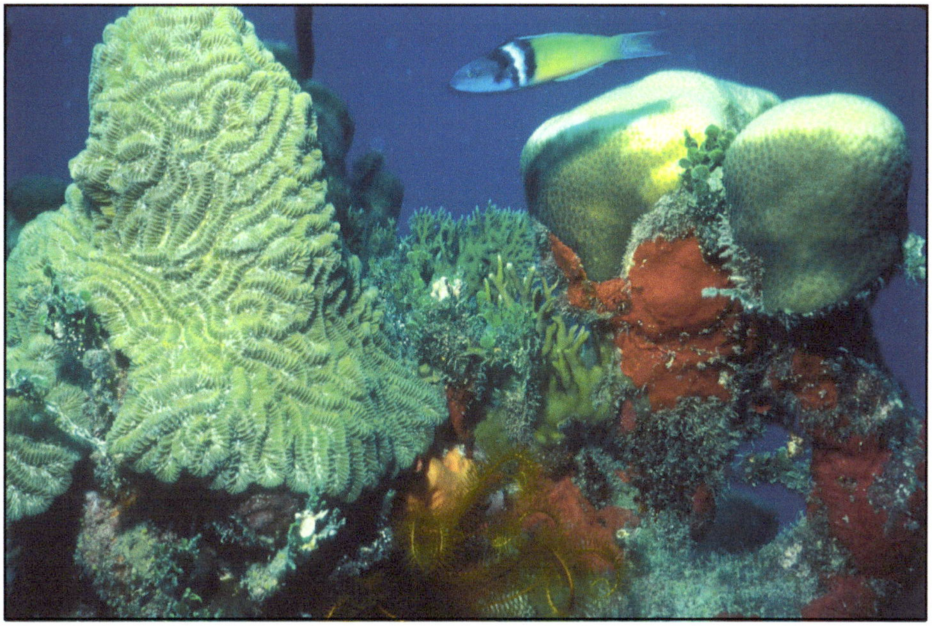

Bluehead Wrasse (*Thalassoma bifasciatum*)

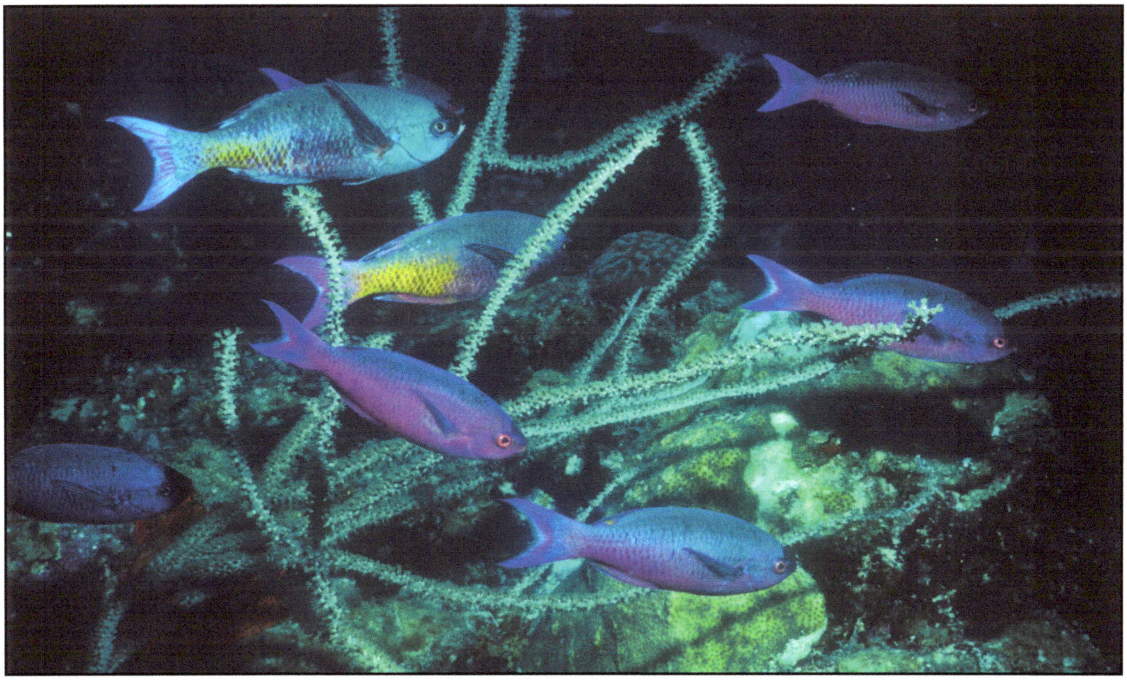

Creole Wrasse (*Clepticus parrae*)

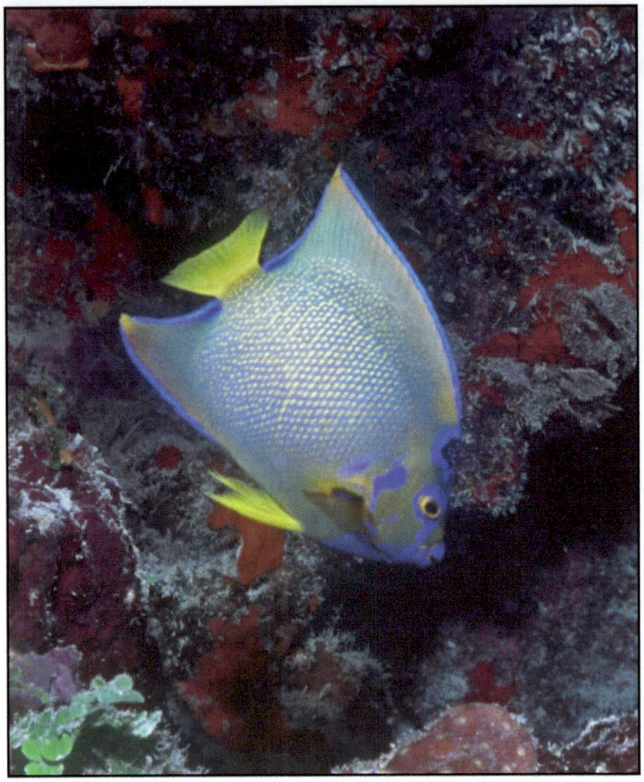

Queen Angelfish (*Holacanthus ciliaris*)

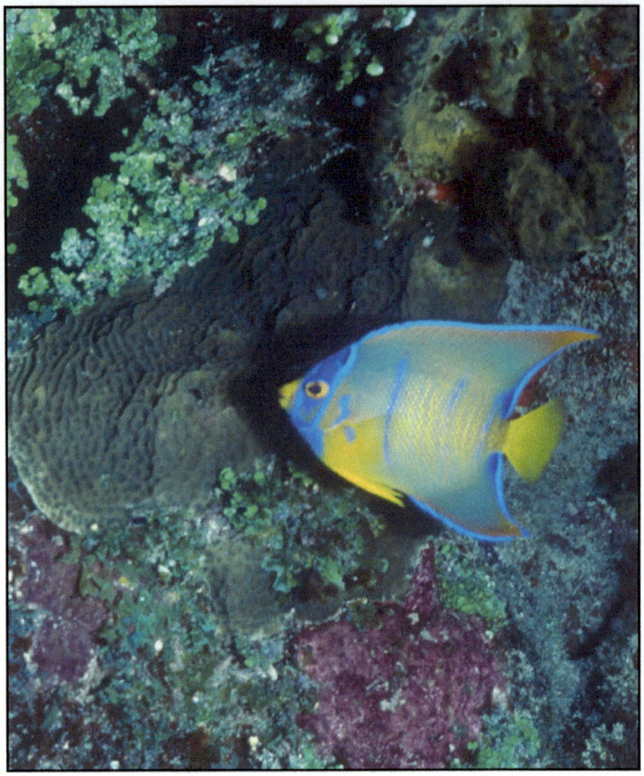

Intermediate Queen Angelfish

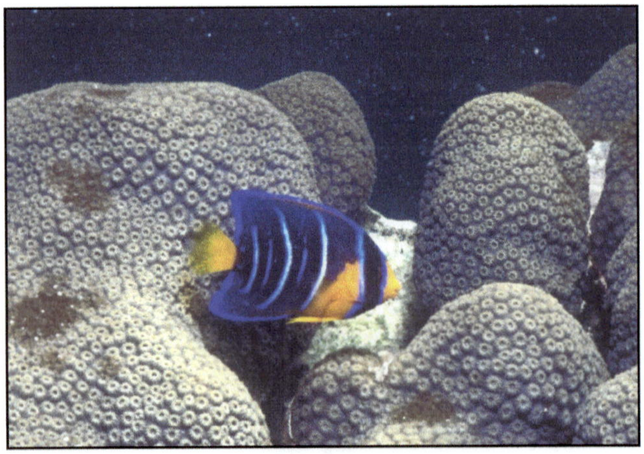

Juvenile Queen Angelfish

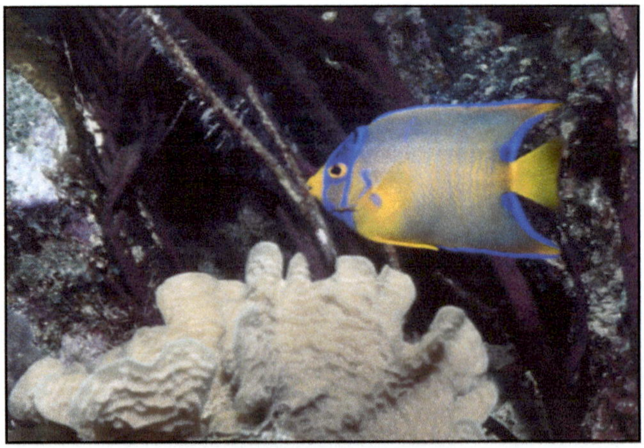

Intermediate Queen Angelfish

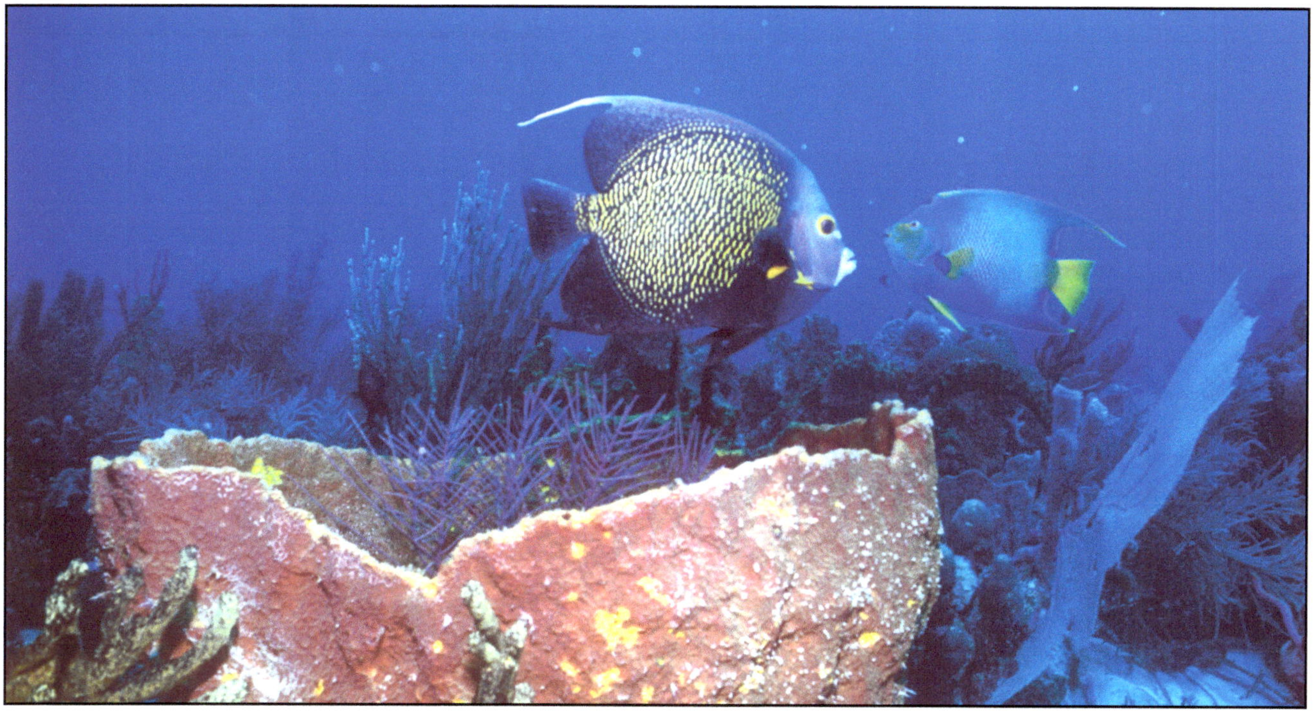

French Angelfish (*Pomacanthus paru*) and Queen Angelfish (*Holacanthus ciliaris*)

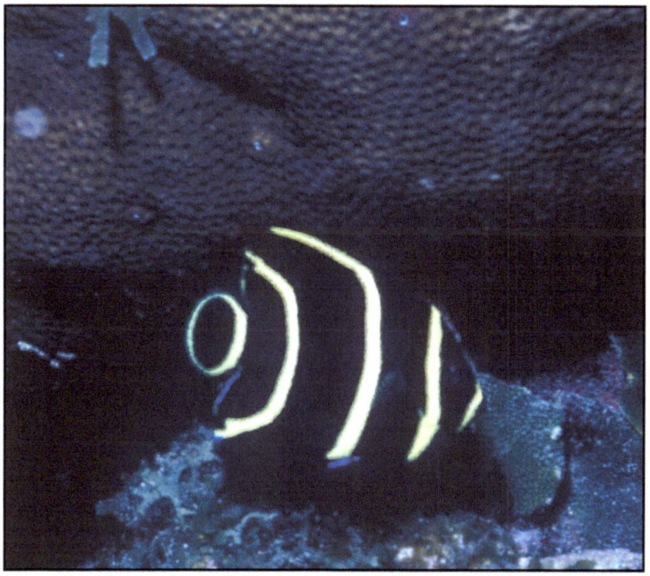

Juvenile French Angelfish

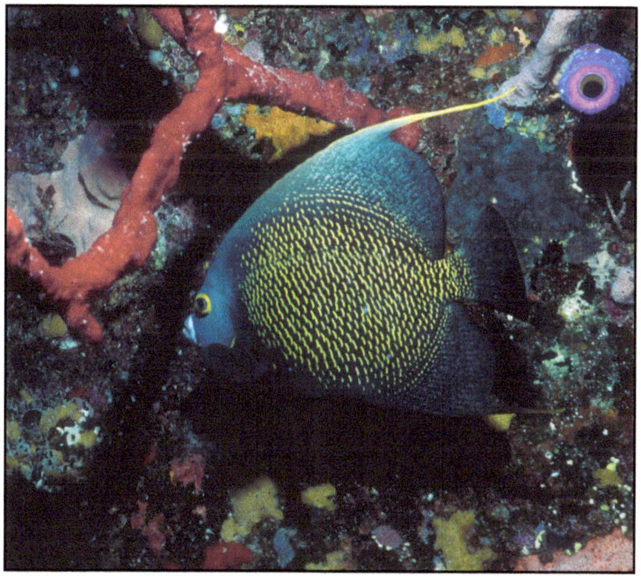

French Angelfish (*Pomacanthus paru*)

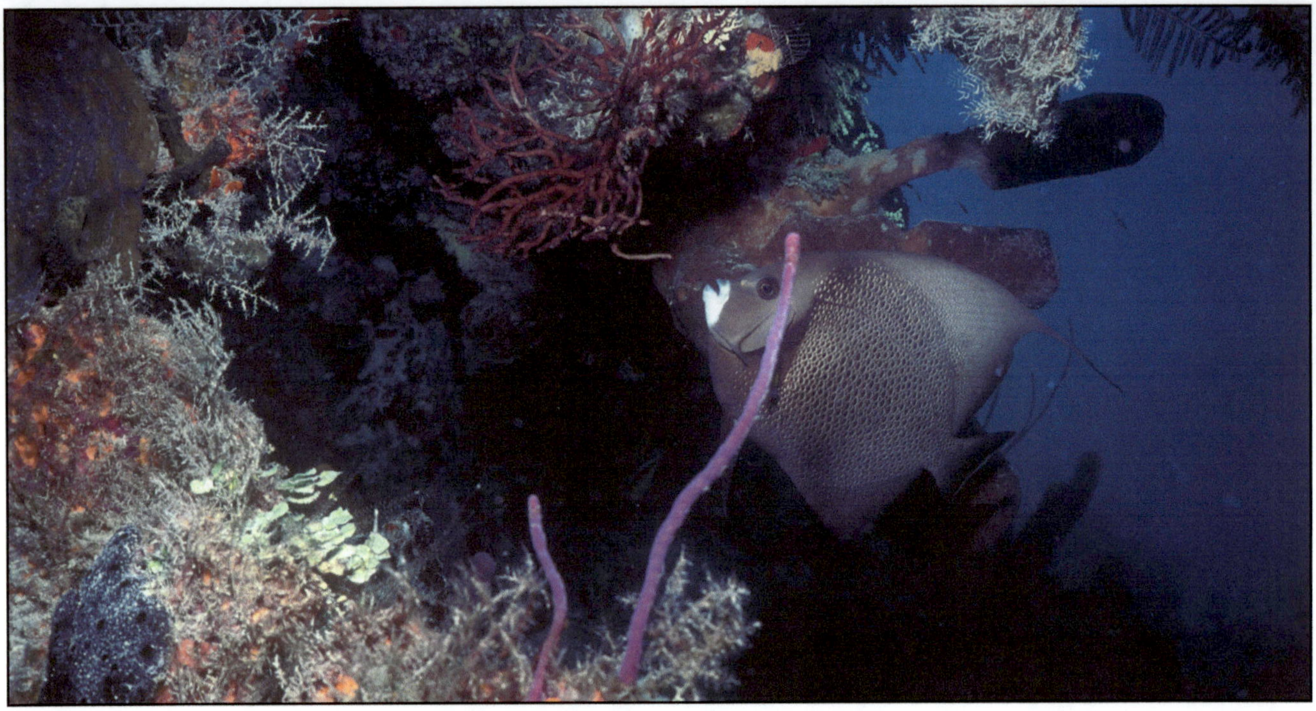

Grey Angelfish (*Promacanthus arcuatus*)

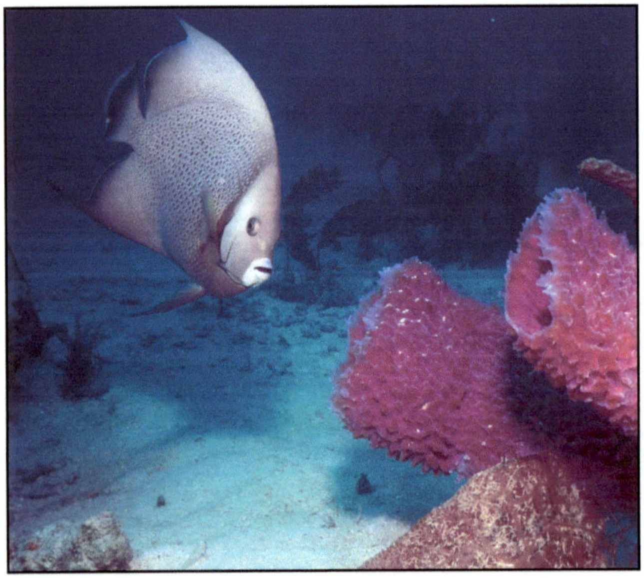

Grey Angelfish with Sponges

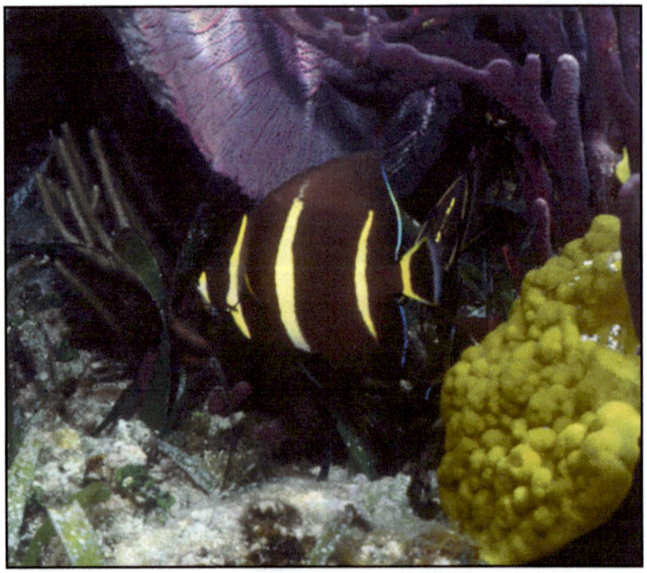

Juvenile Grey Angelfish

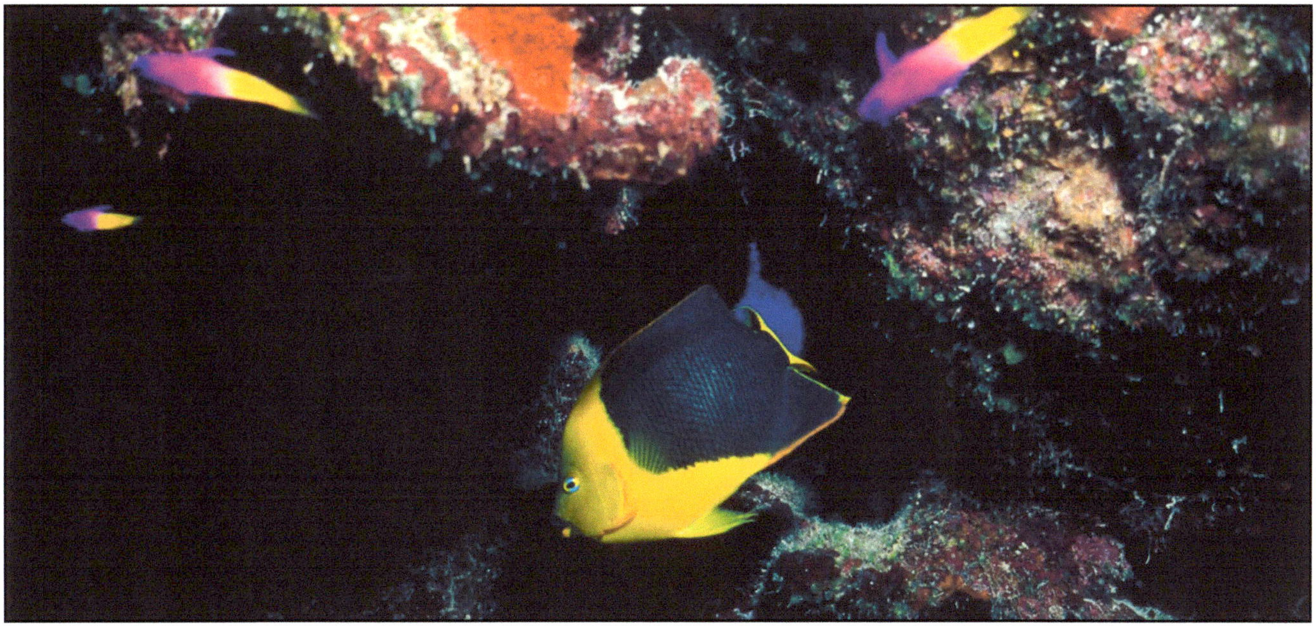

(*above and below*) Rock Beauty (*Holacanthus tricolor*)

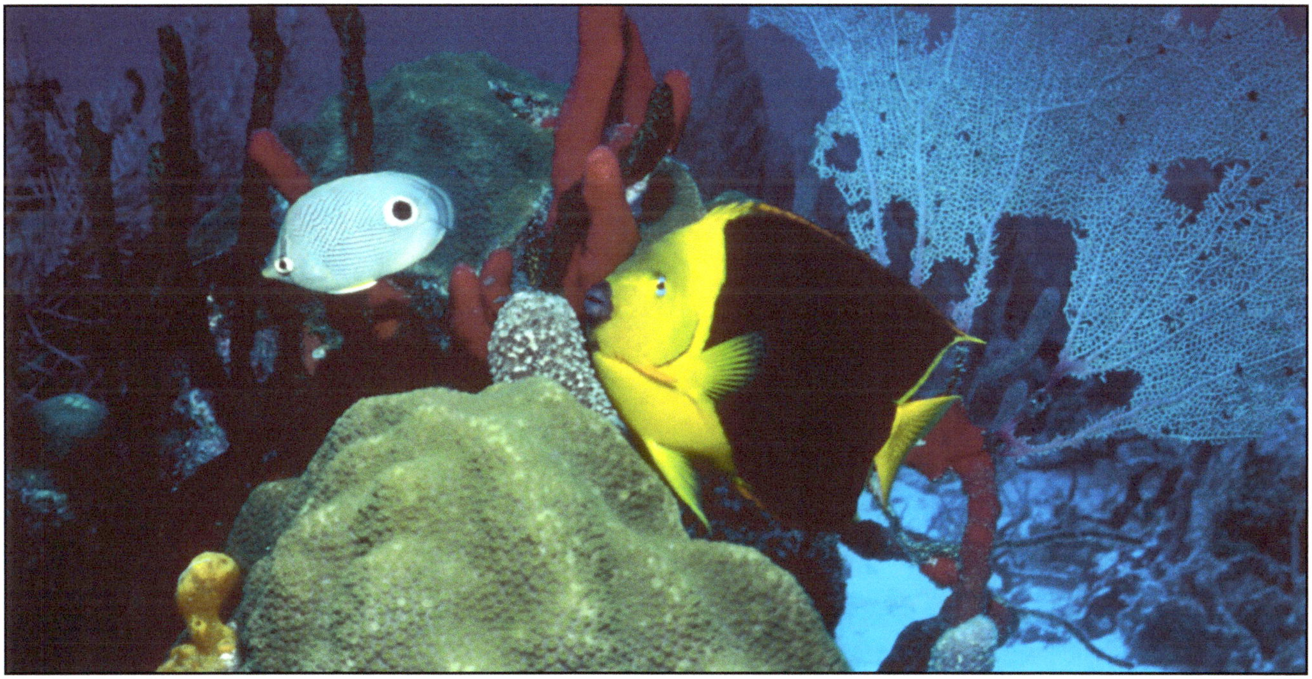

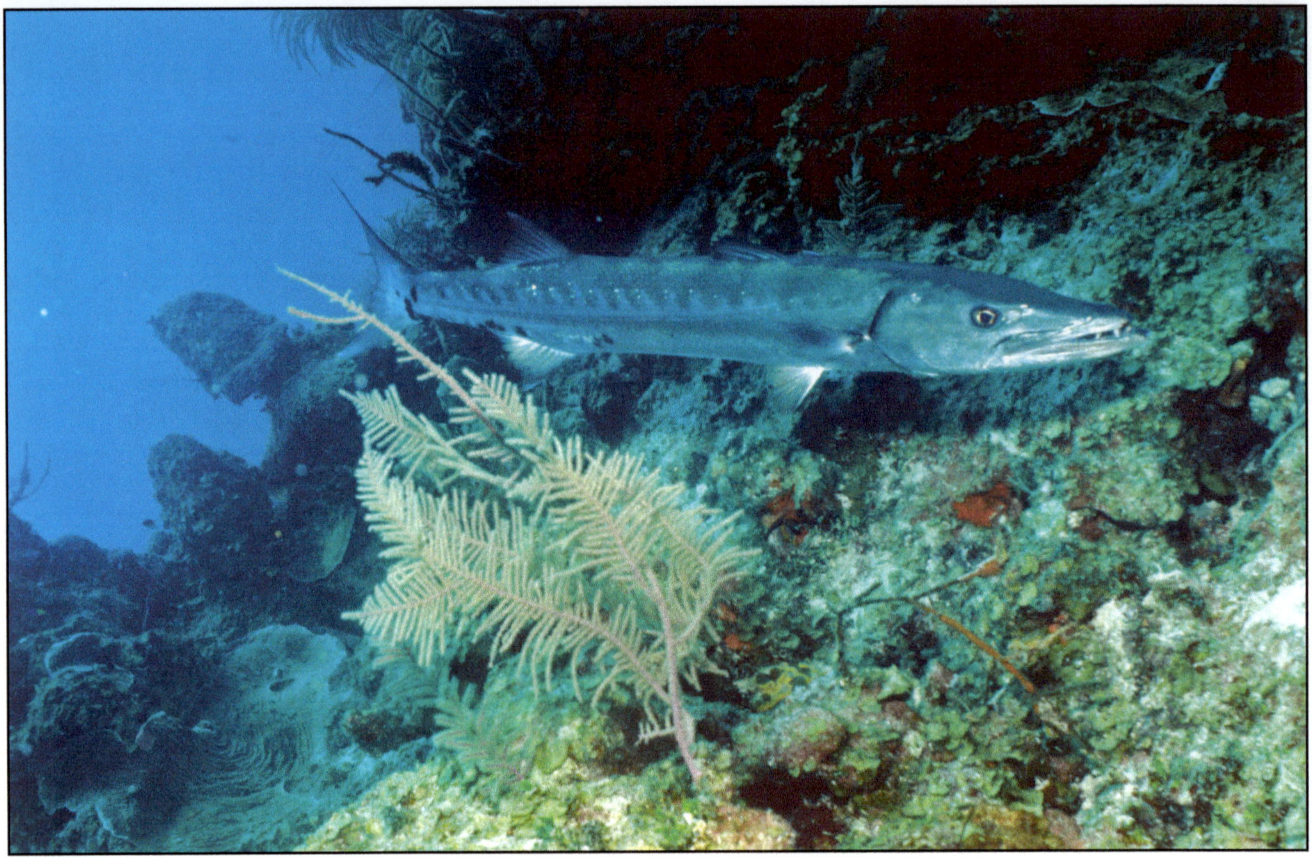

Great Barracuda (*Sphyaena barracuda*)

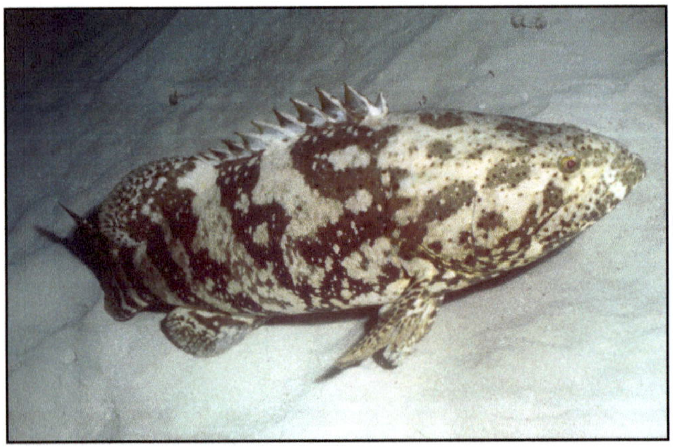

Goliath Grouper (*Epinephelus itajara*)

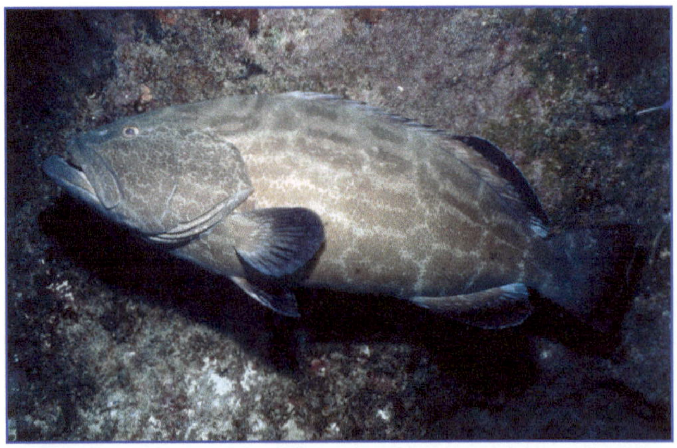

Black Grouper (*Mycteroperca bonaci*) light colour variation

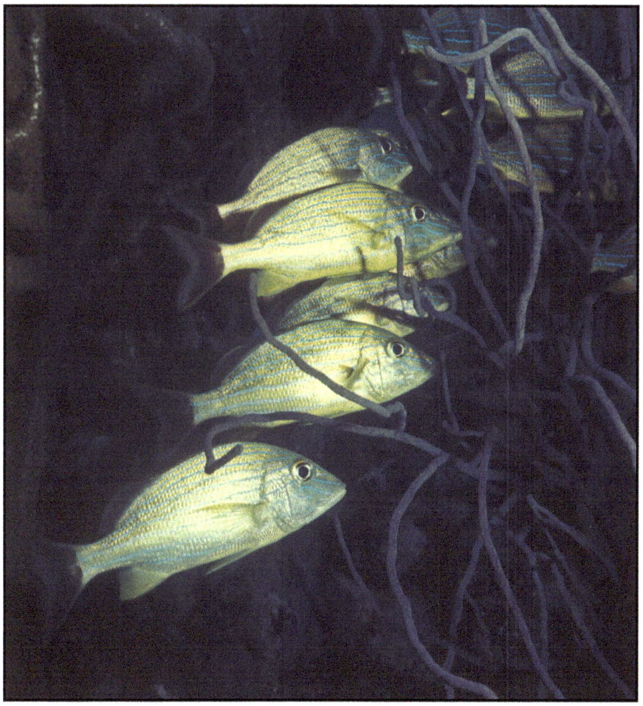

Bluestriped Grunts (*Haemulon sciurus*)

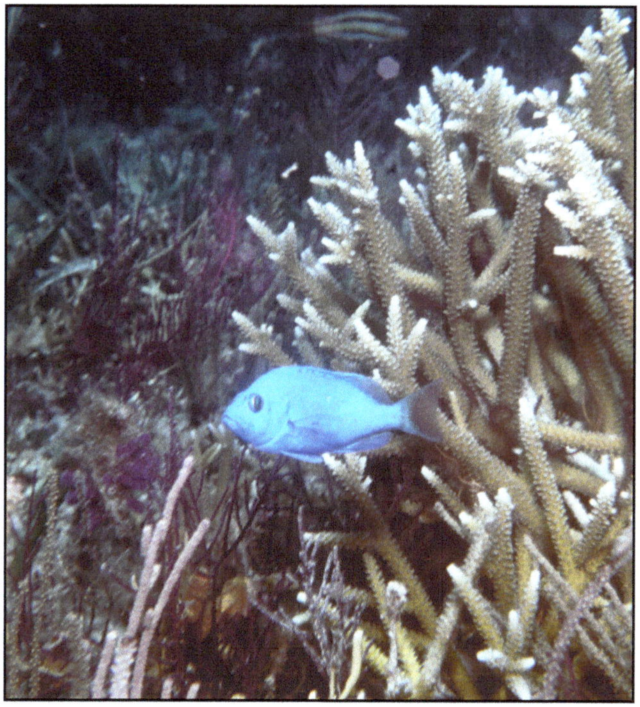

Blue Hamlet (*Hypoplectrus gemma*)

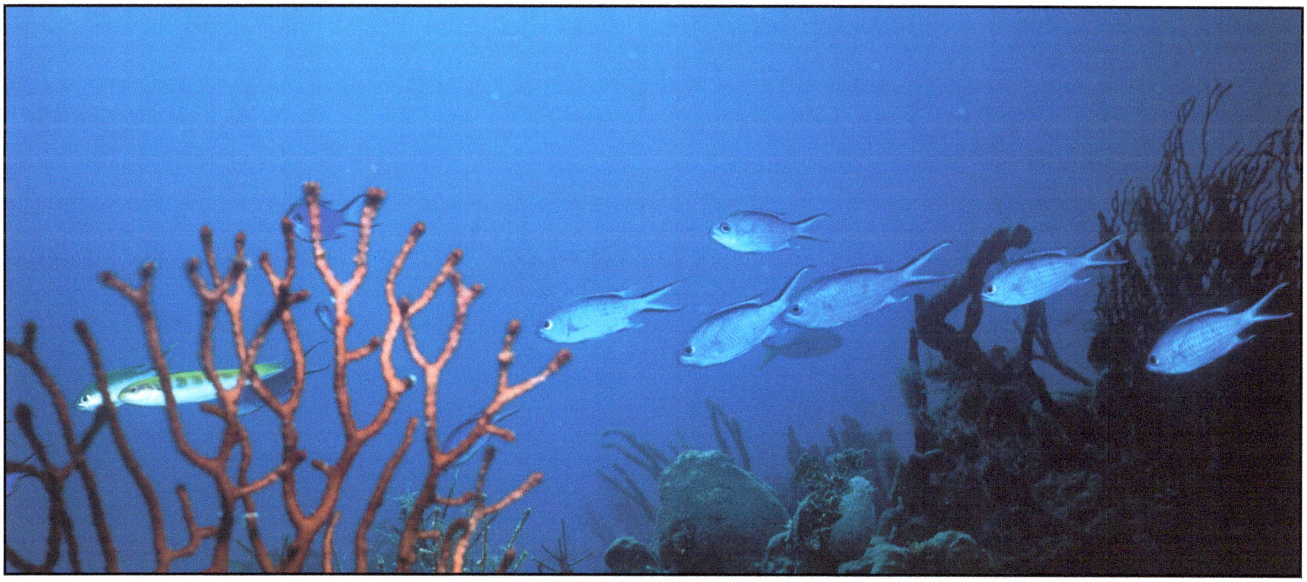

Blue Chromis (*Chromis cyanea*)

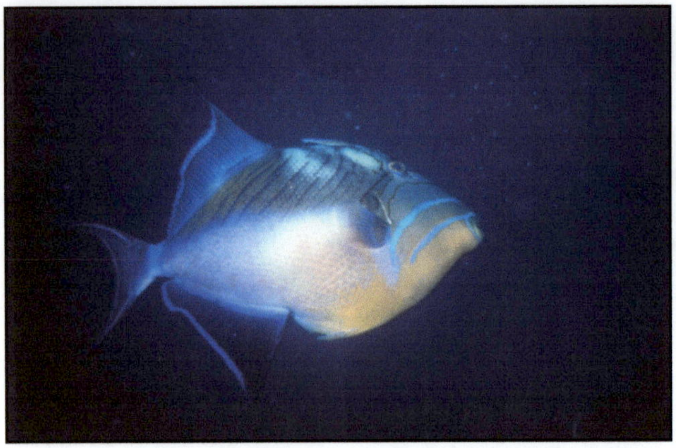

Queen Triggerfish (*Balistis vetula*)

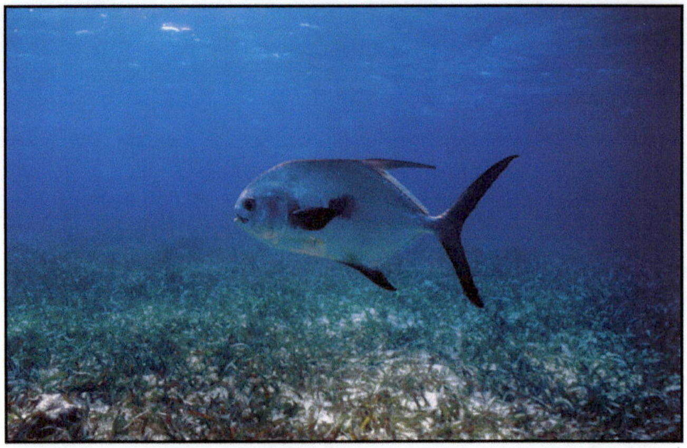

Permit (*Trachinotus falcatus*)

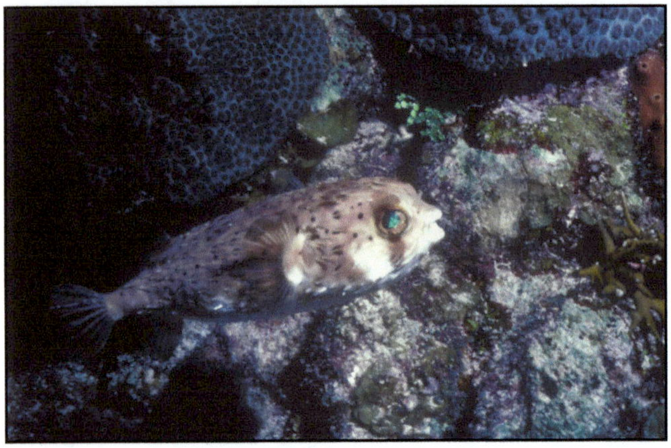

Porcupinefish (*Diodon hystrix*)

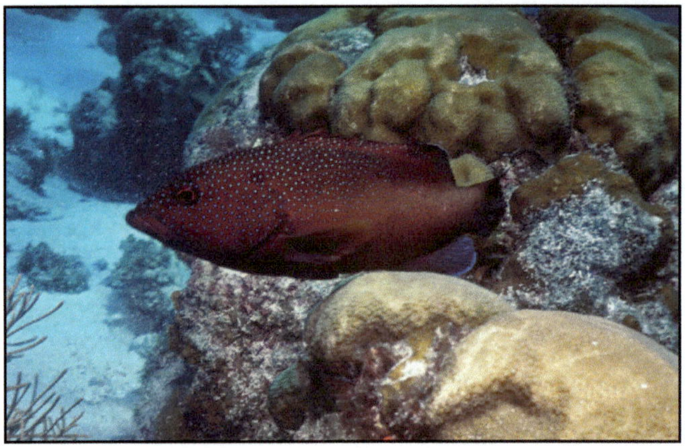

Coney (*Epinephelus fulvus*)

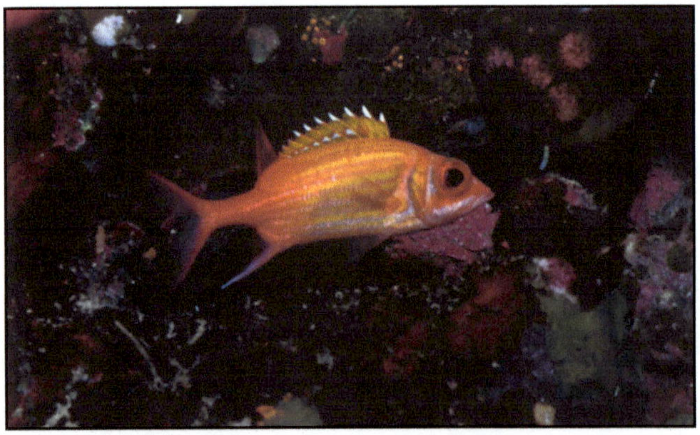

Longjaw Squirrelfish (*Holocentrus marianus*)

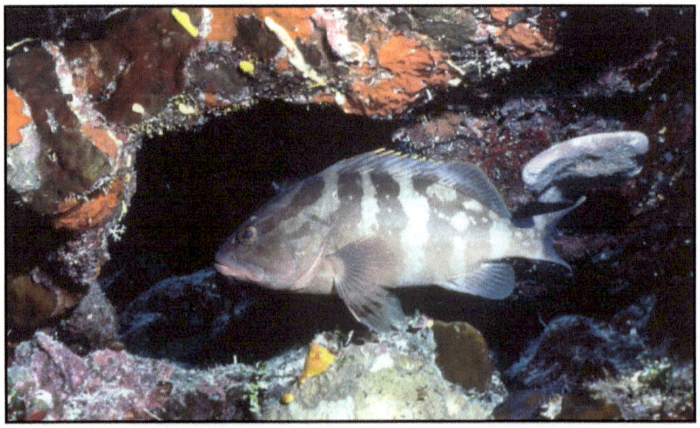

Nassau Grouper (*Epinephelus striatus*)

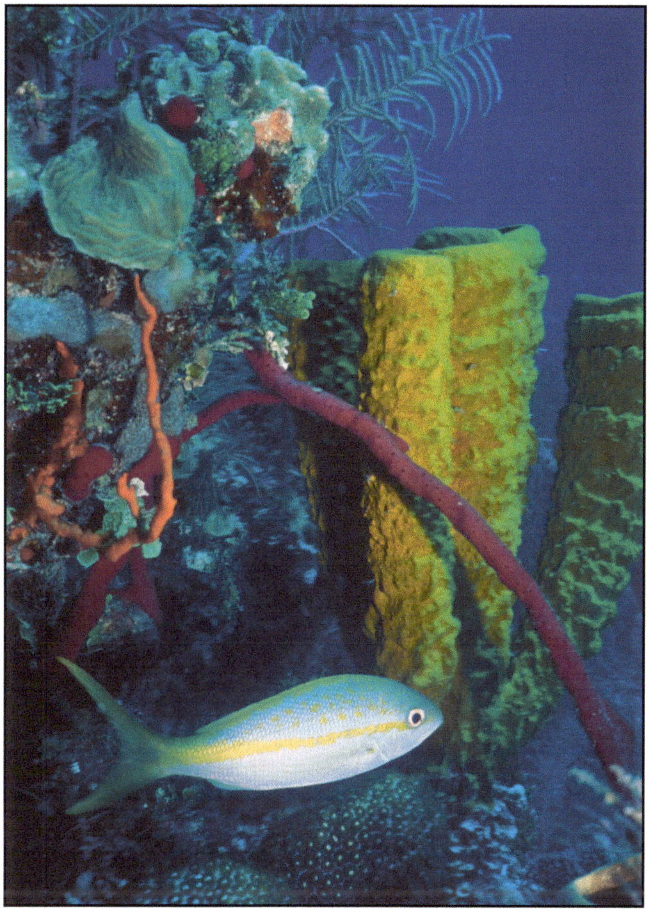

Yellowtail Snapper (*Ocyurus chrysurus*)

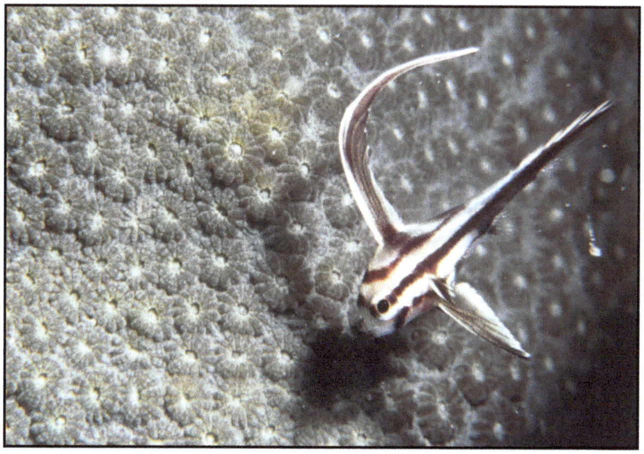

Highhat (*Pareques acuminatus*)

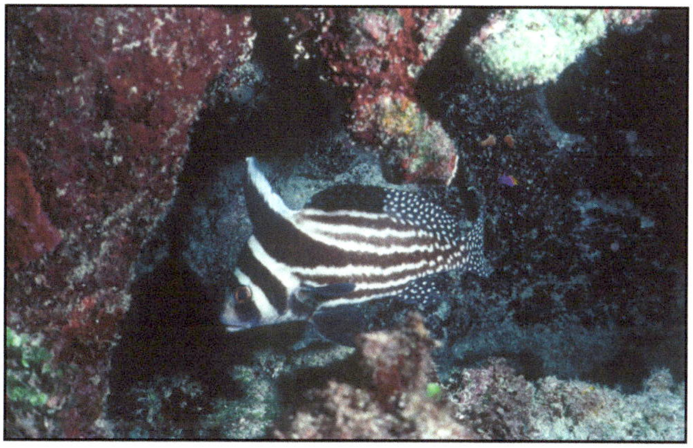

Spotted Drum (*Equetus punctatus*)

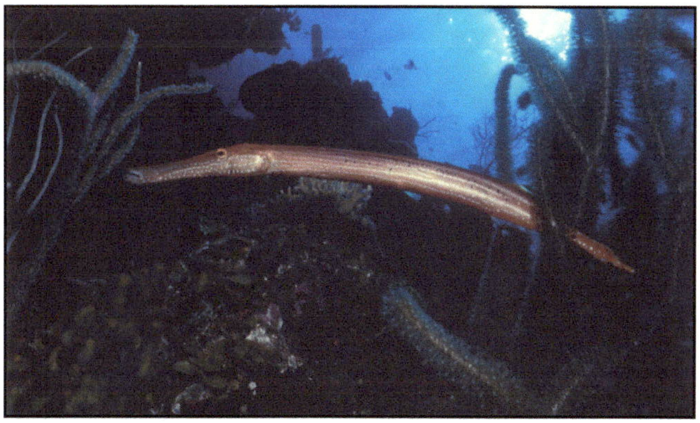

Trumpetfish (*Aulostomus maculatus*)

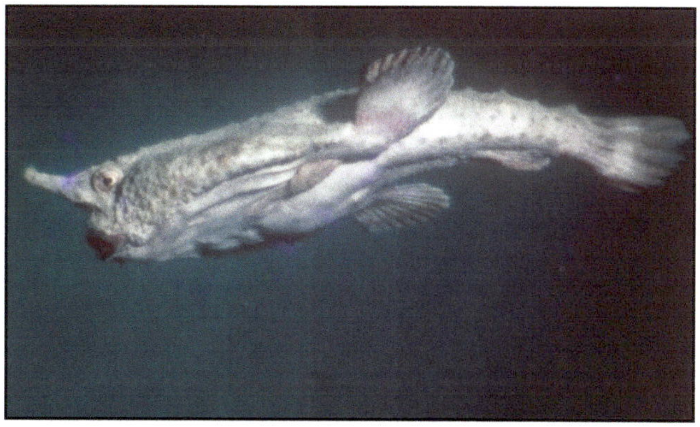

Shortnose Batfish (*Ogcocephalus nasutus*)

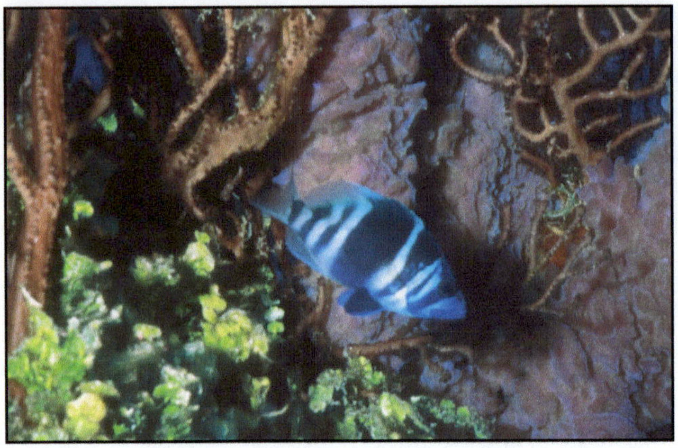

Indigo Hamlet (*Hypoplectrus indigo*)

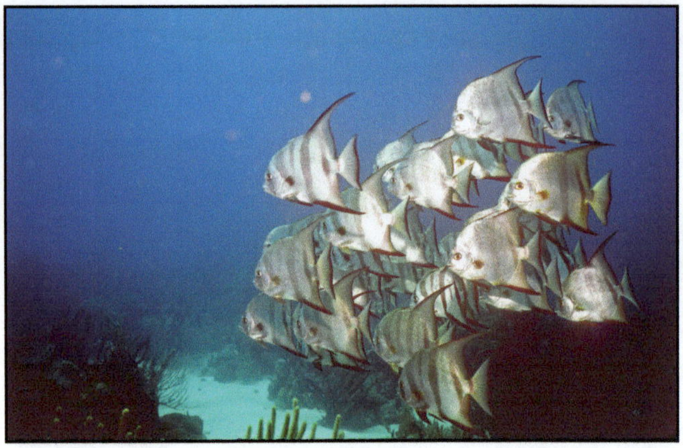

Atlantic Spadefish (*Chaetodipterus faber*)

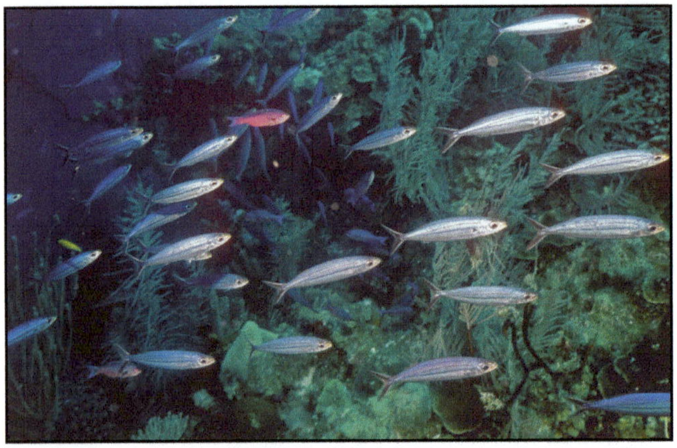

Boga (*Inermia vittata*)

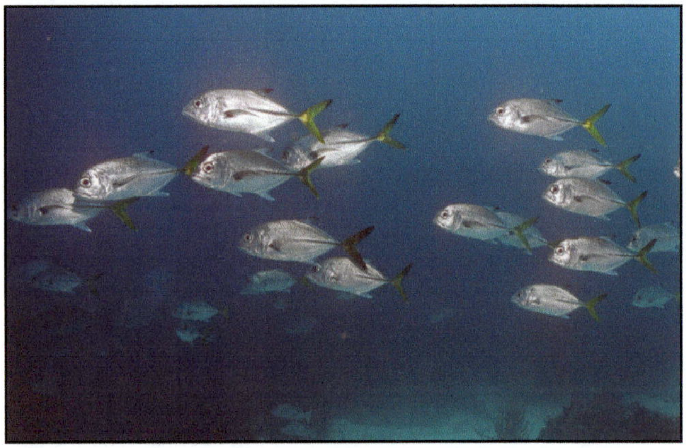

Horseeye Jack (*Caranx latus*)

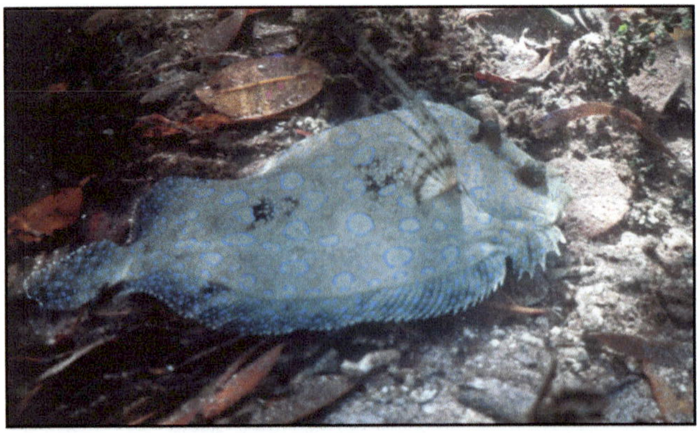

Peacock Flounder (*Bothus lunatus*)

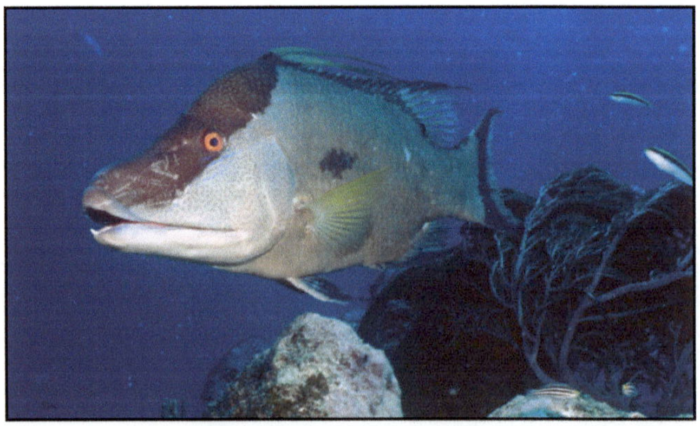

Hogfish (*Lachnolaimus maximus*)

Hammerhead

I herded my four scuba divers up to the stern of the *Isla Mia*, where willing hands were waiting to help them remove their tanks and weight-belts.

"Jim!" Henny Tromm called, "Could you go check up on Bobby? He is fishing upwind by the wall. Tell him it is time for lunch."

The *Isla Mia* was a seventy-two-foot dive boat owned and captained by Henny Tromm. It slept sixteen guests and offered week-long trips out to Lighthouse Reef, which included the Blue Hole. During this period, it was the only dive boat that offered the superb diving of Lighthouse Reef, along with a comfortable bed and good food.

Henny and I had been friends since the early 1970s when I first met him in Cozumel. For approximately eight years I worked, as needed, as the dive master of the boat.

I checked my tank and saw it still had sufficient air, so I handed Henny my camera and dropped back into the blue waters. After a fifty-yard swim, I saw a silvery stream of bubbles. Bobby was the deck-hand with various responsibilities of which fishing was not one of them. But he did enjoy it, and fresh seafood was always appreciated by the guests.

As I approached Bobby's position, I sank down to his level of about sixty feet. His bubbles were emitting from a small cave, also emitting from the same cave were two fin-clad legs which every now and again gave a kick or two. I gave the leg a slap and the fin a pull. In a few seconds, Bobby very slowly backed out.

I saw he was holding a two-foot-long hogfish to his chest. He thrust it forward towards me. I gave him the international sign, one finger pointing to the surface, then pointing to my mouth. He gave his own international sign one finger pointing to the fish, then pointing to me. Then continued by pointing to the cave, then himself, then the fish again. It was very clear that I should take the fish back and that he has a second fish in the cave, which he will take back. After a few more signals, it came clear that the second fish was still alive and it would take him a few minutes.

I took his fish and ascended to approximately twenty-five feet, where I planned to wait for him. In the far distance I noticed a dark irregular shape that seemed to move from right to left, then right again continuously. The water was exceptionally clear, but this object was hazy and beyond the point of clarity. I couldn't figure out what it was, but it was heading straight for me!

Bobby's leg was still kicking, and now and again he would pull back into the cave. That was a bad sign; obviously the fish was still alive and struggling. This time when I lifted my head, I was looking head on at a great hammerhead shark. I glanced down at Bobby. Yes, his legs were still there, but that was not what my glance focused on. What shocked me was a two foot long dead fish which I held protectively to my chest.

First, I dropped the fish and then I voided all the air in my buoyancy control pack. Both the fish and I dropped, and thankfully, the fish dropped faster. The shark was now about fifty or sixty feet away, still cruising in a straight line. He was absolutely magnificent and I was absolutely petrified!

The shark neither veered up nor down nor left nor right, but simply swam on and passed about twenty feet above my head. Then the shark moved completely between me and the midday sun. This massive fish was swimming within a golden aura with my silvery exhaust streaming by it on both sides. Damn, why did I leave my camera behind?

As the shark faded into the blue, I looked down and saw Bobby, now out of the hole, with another large hogfish. I descended and picked up the fish that I'd dropped, and we swam back to the boat. Once on the

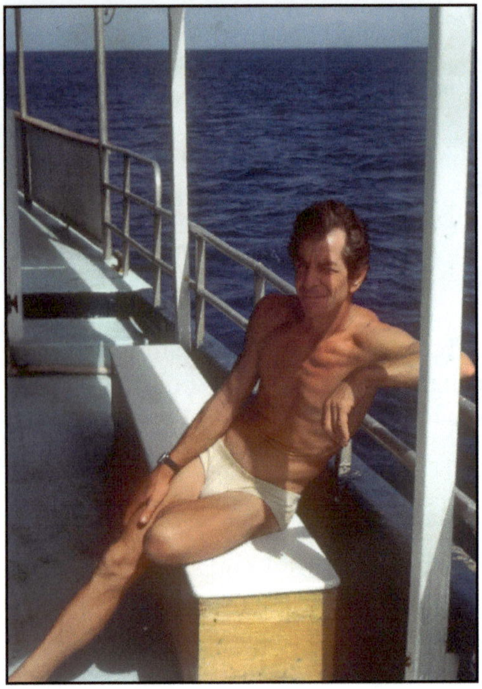

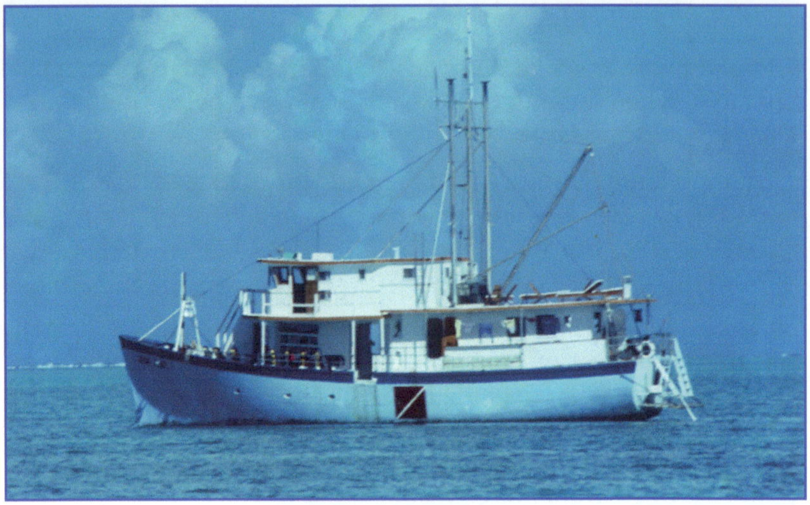

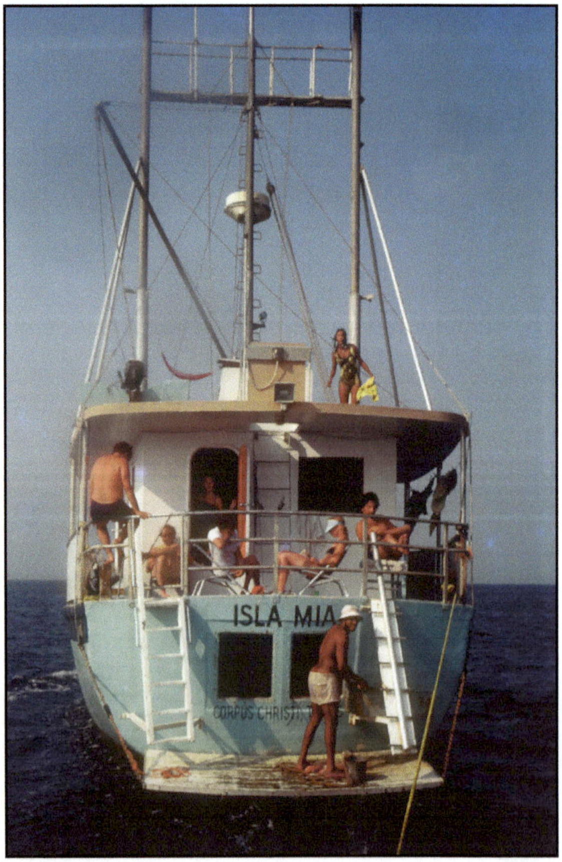

dive platform, I asked him if he had seen the shark. "Shark! What shark?" he asked.

A great hammerhead can reach a length of over twenty feet; at only fifteen or sixteen feet, mine was probably a juvenile. Several years later I read that great hammerheads feed almost exclusively on sting rays. One great hammerhead, when butchered, was found to have 96 sting ray spines in its stomach, but no scuba tanks.

As I look back over my life in what is now my seventieth year, I can recall two occasions when I believed that my life was about to end. This was the first of these occasions. The other was not to happen for another few years, but coincidentally, it happened about half a mile away from this spot.

(*top left*) Henny Tromm, Captain of the *Isla Mia*
(*above and right*) *Isla Mia*

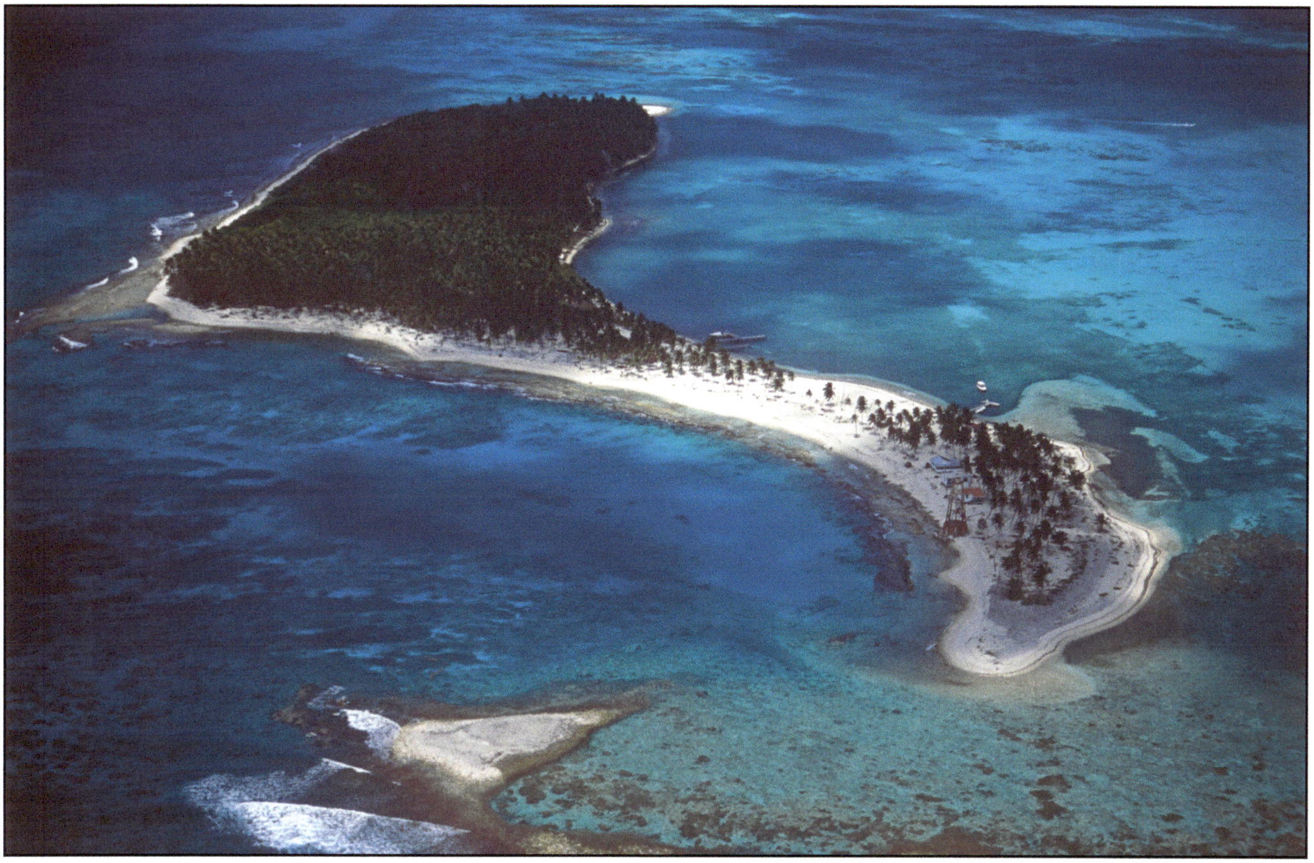

Aerial view of Half Moon Caye

Half Moon Caye

Located approximately 60 miles from the mainland is Lighthouse Reef, one of Belize's three atolls. Lighthouse Reef is 25 miles long and shelters five islands of which Half Moon Caye is the prettiest and by far the most visited. Half Moon Caye was Belize's first protected area and was gazetted in 1982 as a Natural Monument.

It is a relatively small island, located at the south-eastern point of the atoll. The area of interest on the caye is a large rookery of the white phase of the Red-footed Booby, and the only one in the greater area. The boobies share the rookery with the Magnificent Frigatebirds. It is an uneasy alliance with the frigates regularly ambushing the boobies as they return with their catch. Half-Moon Caye is a pretty island of mostly sand and coconut palms on the eastern side.

The Western section of the island is covered with the orange blooming zericote (*Cordia Sebestena*). It is here that both the boobies and the frigates nest. The caye is also a popular rest area for migrating North American migrants, both coming and going.

Half Moon Caye National Monument is not only a Sanctuary for birds, but it is also important for reptiles .

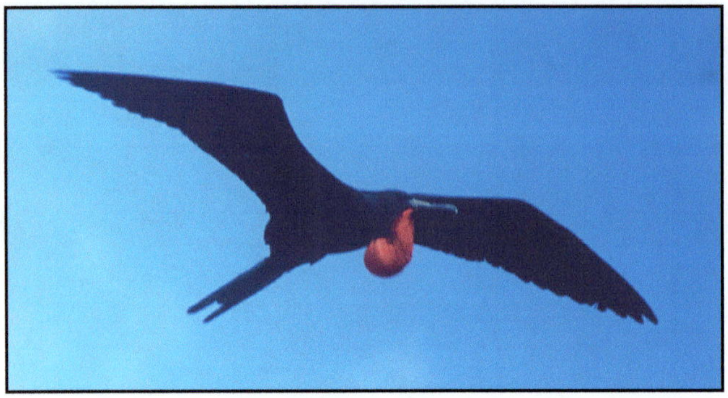

Magnificent Frigatebird (*Fregata magnificens*)

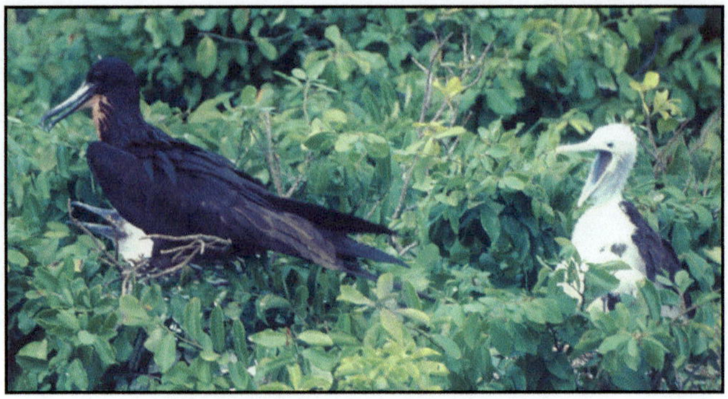

Magnificent Frigatebird and chick

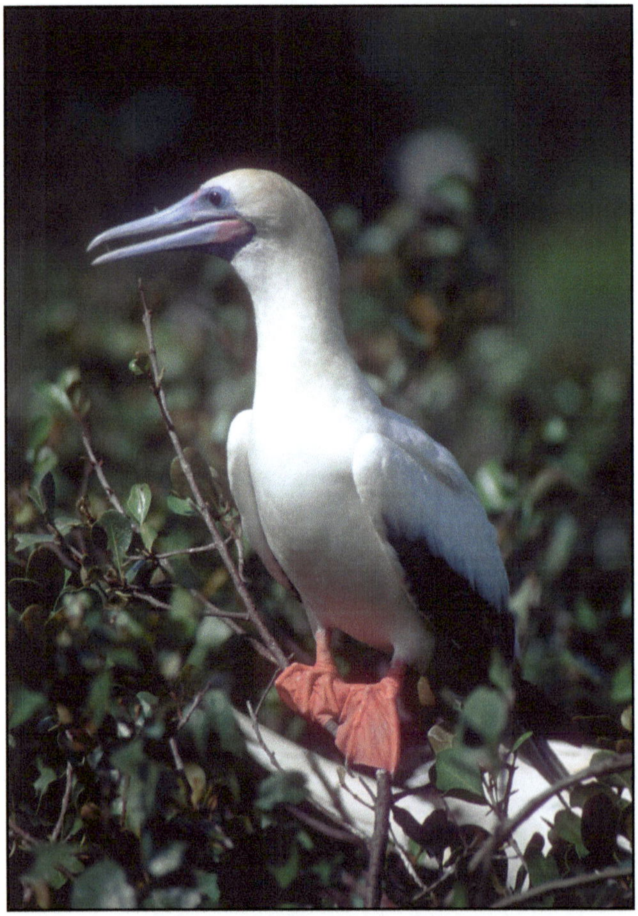

Red-footed Booby (*Sula sula*)

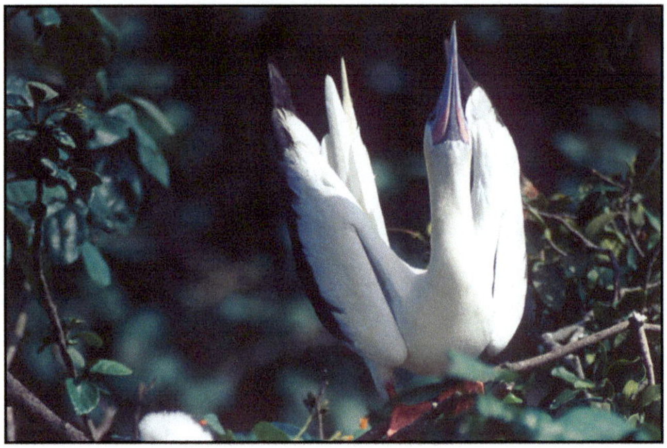

Red-footed Booby

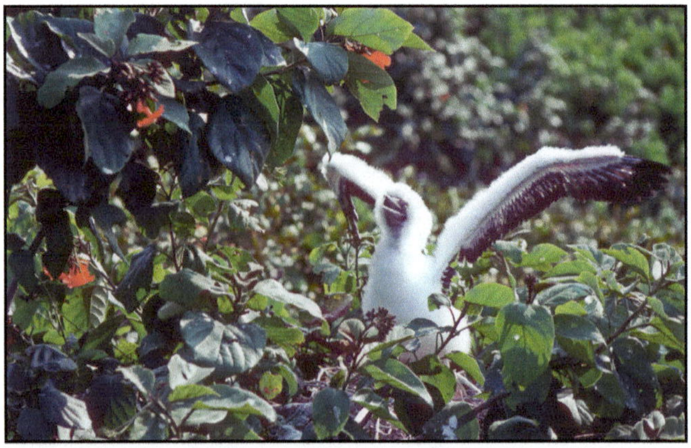

Red-footed Booby chick

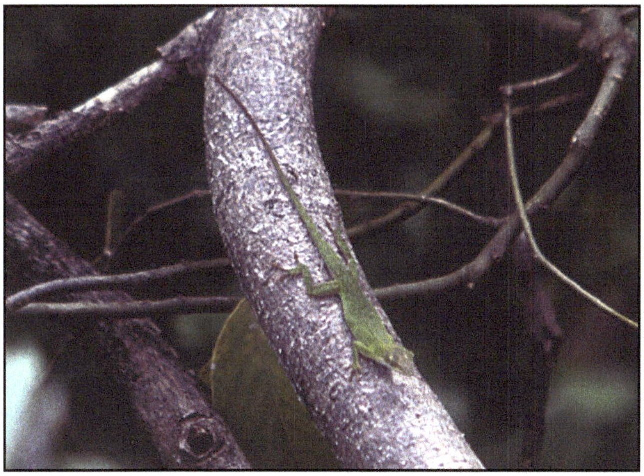

Allison's Anole (*Anolis allisoni*)

All three of Belize's sea turtles, the green, loggerhead and hawksbill turtles are listed as internationally endangered, and all three have been reported to nest on the sandy southern beach of Half-Moon Caye.

The caye has at least six different lizards. The largest of these reptiles is a reddish coloured green iguana (*Iguana iguana rhinolopha*). The wish willy or black iguana (*Ctenosaura similis*) is also quite common on the island. Two anole species are endemic (*Anolis Carolinensis*) and (*A. Allisoni*) to both Half-Moon Caye, and neighbouring Long Caye. Also the brown anole (*A. Sagrei*) found there is common throughout Belize.

Another lizard, also found on both Half Moon Caye and Long Caye is a large gecko, which I've been unable to identify. However, it appeared to be a speckled variety of the large common gecko on Caye Caulker, locally called the "Weatherman", for its loud barking cry, prior to changes in weather!

The other major attraction to tourists at Half-Moon Caye is diving the wall. Off the southern side of Half-Moon Caye a sandy bottom slopes down to coral pinnacles. These pinnacles are divided by crevasses which allow divers an easy way through to superb walls and coral gardens.

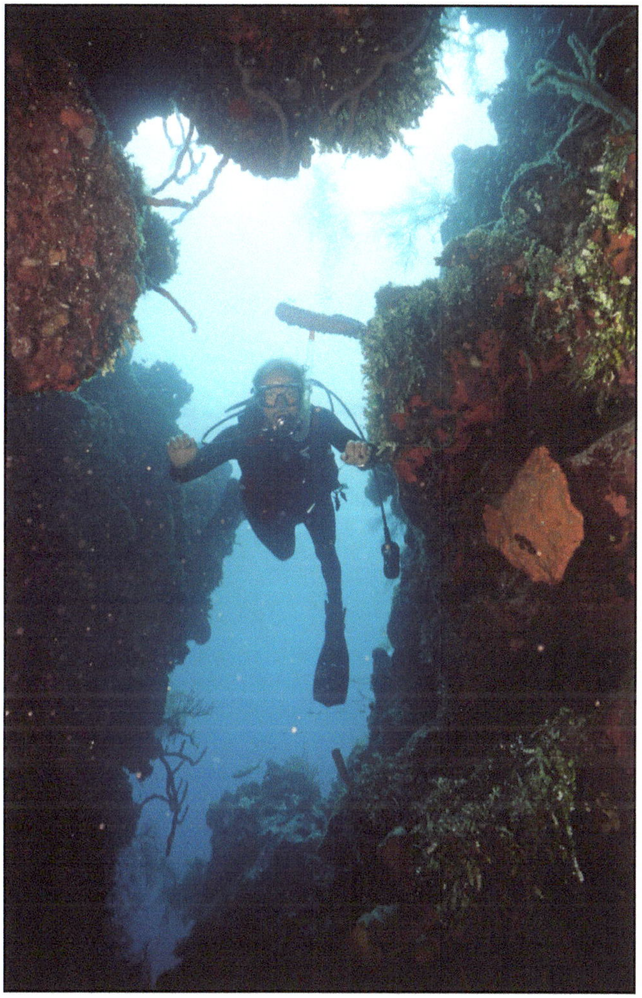

Through the crevasse at the Half Moon Caye wall.

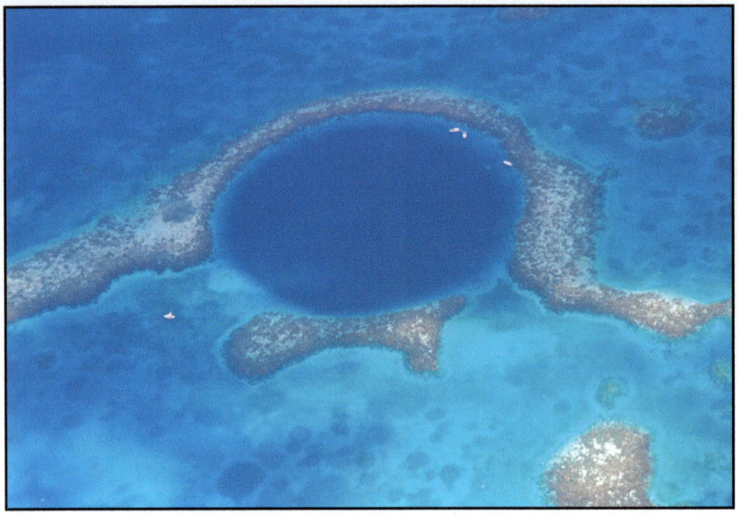

Aerial view of the Blue Hole

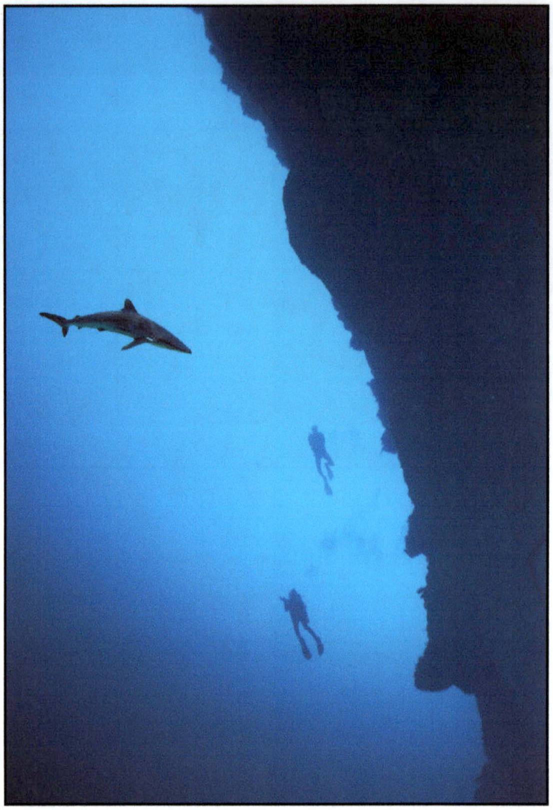

At a depth of 90 feet in the Blue Hole, divers ascend as curious bull shark watches.

THE BLUE HOLE

Approximately eight miles due north of Half Moon Caye lies Lighthouse Reef's second protected area, the Blue Hole Natural Monument, which was gazetted in 1996. It is every diver's dream to dive the Blue Hole.

From the air, one can see that the Blue Hole is aptly named. Whereas the inner waters of the atoll are pale blue-green, heavily dappled with the browns of coral against the pale yellow of sand, the Blue Hole is an almost perfect circle of a startling indigo blue, over one thousand feet in diameter. Fringing the hole is a circular reef with only two openings which face south.

As one submerges, a few curious reef sharks may appear. Some irresponsible dive boats thought feeding the sharks would add to the adventure. Even wary reef sharks can be dangerous with food in the waters. On occasions, the odd bull shark has made an appearance. Bull sharks are looked upon as being dangerous, especially near the surface.

At 30 feet the divers reach the drop-off and notice that the wall starts to slope inwards. At approximately 130 feet, the upper parts of stalactites appear. The stalactites

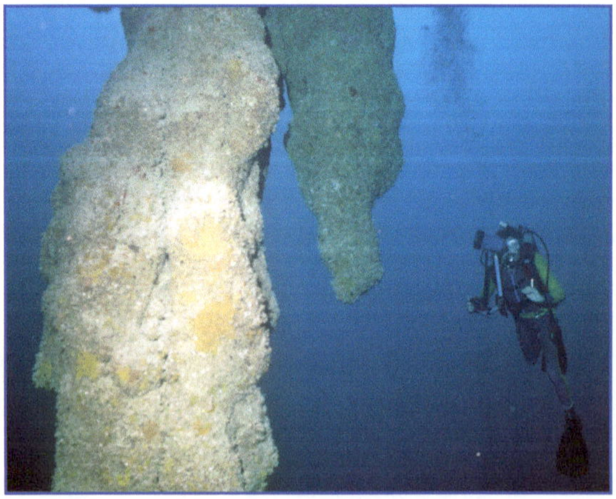

Stalactites at the Blue Hole

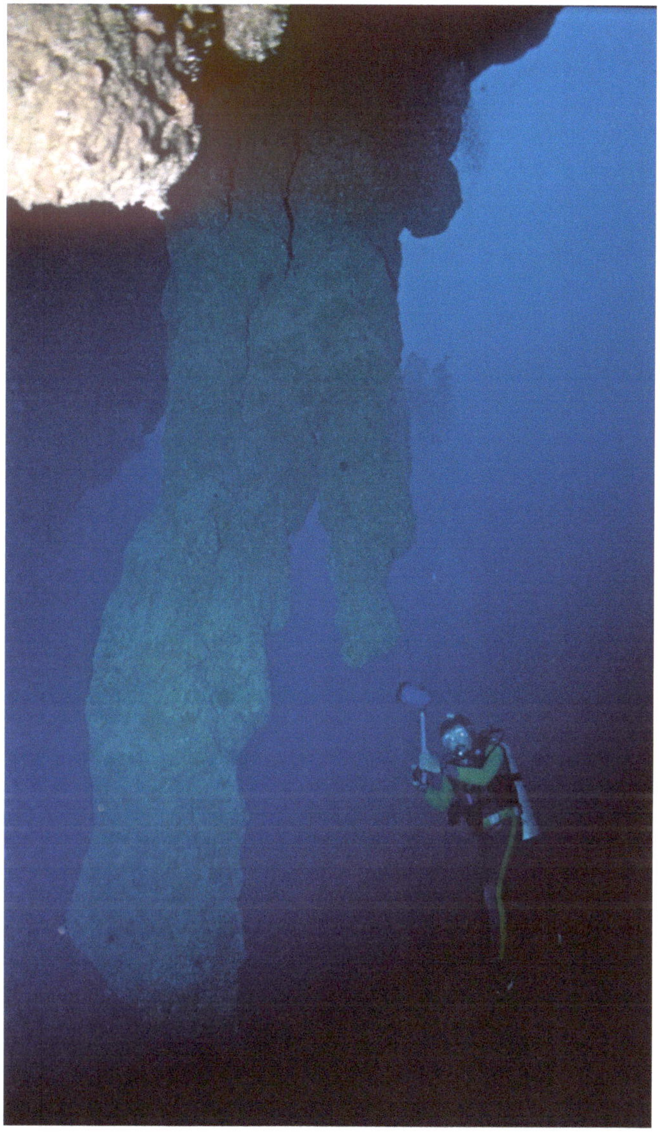

Stalactites appear at 135 feet of depth at the Blue Hole.

are massive and the wall continues to slopes away behind them, making it possible to take photos looking through the stalactites at your buddies on the other side. Nine minutes is all you are allowed at this depth – then a slow ascent to the decompression stop.

Belize's Blue Hole was first visited by Jacques-Yves Cousteau in early 1969. Co-incidentally, the same year I first visited British Honduras, as it was known then. Cousteau was the co-inventor of the "air on demand" valve, which when attached to a tank of compressed air allowed a diver to spend time underwater without a cumbersome helmet and air-hose. This apparatus is now known as SCUBA (self-contained underwater breathing apparatus). In 1950, Cousteau purchased a World War II mine-sweeper from the British Navy. The *Calypso* made its maiden expedition in 1951.

The expedition in which he visited British Honduras's Blue Hole started in 1967, was completed in 1971, and logged 140,000 miles crossing five oceans and seas. Cousteau spent several months at the Blue Hole and, using two mini-subs, reached the bottom at approximately 432 feet. The bottom of the hole was littered with rubble and limestone debris, indicating that at one time the hole was a cave with a ceiling that collapsed eons ago.

They noted that stalactites were at various depths, the deepest were quite close to the bottom. As stalactites cannot form under water, this was proof that during a previous ice age, the sea level was at least four hundred feet lower than present day.

Before the *Calypso* left, they pulled up two broken stalactites from a ledge at about two hundred feet. One was twenty feet long and weighed over one ton. The larger of the two stalactites was given to the University of Miami Marine Laboratory, and the other went into storage for the future British Honduras Museum. It was stored at the Belize Customs building.

Long Caye

In addition to diving the Blue Hole and Half-Moon Caye Wall, there are several other dive sites, which offer superb diving at Lighthouse Reef. On the west, or lee side of Lighthouse Reef, there are two other cayes, Long Caye and Hat Caye which are really world class diving sites.

As the name suggests, Long Caye is a long narrow island, with a big section of wetlands to the southeast. The west of the island, especially towards the north is pretty littoral forest. In the late 1970s, I was so taken with Long Caye and its diving potential that in the early 1980's I purchased land on Long Caye to use as overnight campsites for dive groups. These lots looked west to the dive sites, only 200 yards away. In the early 1980s, tourism was not yet the big thing it is now. But for me, this was the perfect situation for when I was not working on the live-aboard dive boat, the *Isla Mia*.

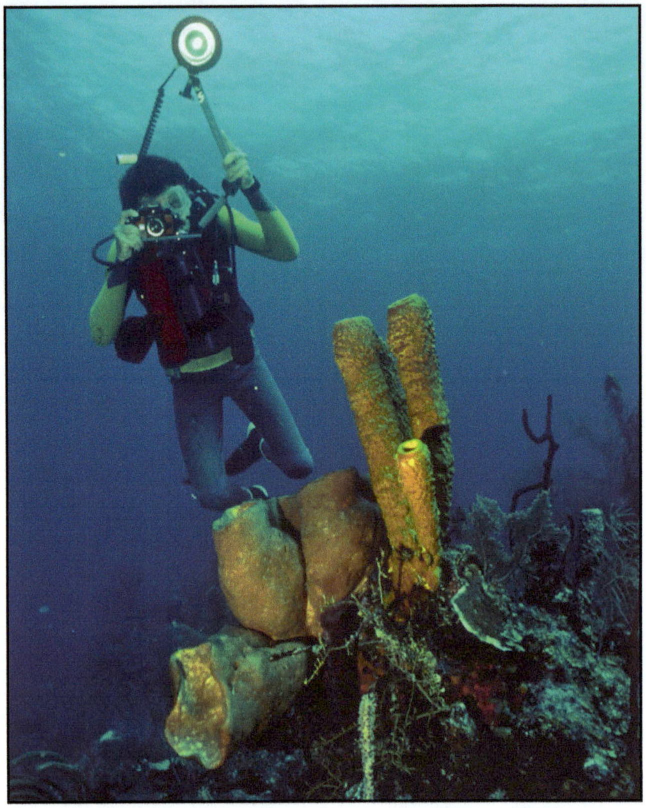

Diving the wall at Long Caye, Lighthouse Reef

The wall at Long Caye runs more than the length of both islands. So sheer is the Long Caye Wall that I could anchor my boat over the wall in thirty feet, and hang the boat over an abyss in crystal clear water.

The vertical face of the drop-off is fissured with crevasses and caves, and covered with magnificent tube sponges and black coral trees. On many occasions divers marvelled at giant manta rays cruising along parallel with the wall.

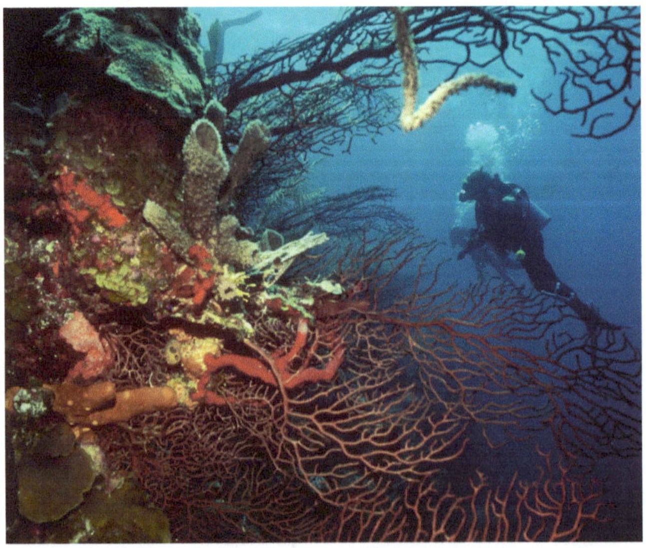

But, unfortunately, even heaven can be hell! On one particular day, Dorothy and I were hosting two extremely good friends, who were visiting from abroad. Helmut and Maria had not previously been to Belize, and although Maria was a non-diver, Helmut was a dive instructor and couldn't wait to dive Long Caye.

We had sailed out from Caye Caulker aboard my sailboat, the *Lion*, and had planned to do our first dive on arrival, prior to going ashore to cook up some food.

We sailed to the wall from the west, rather than the direction I was accustomed to – from Long Caye. As a

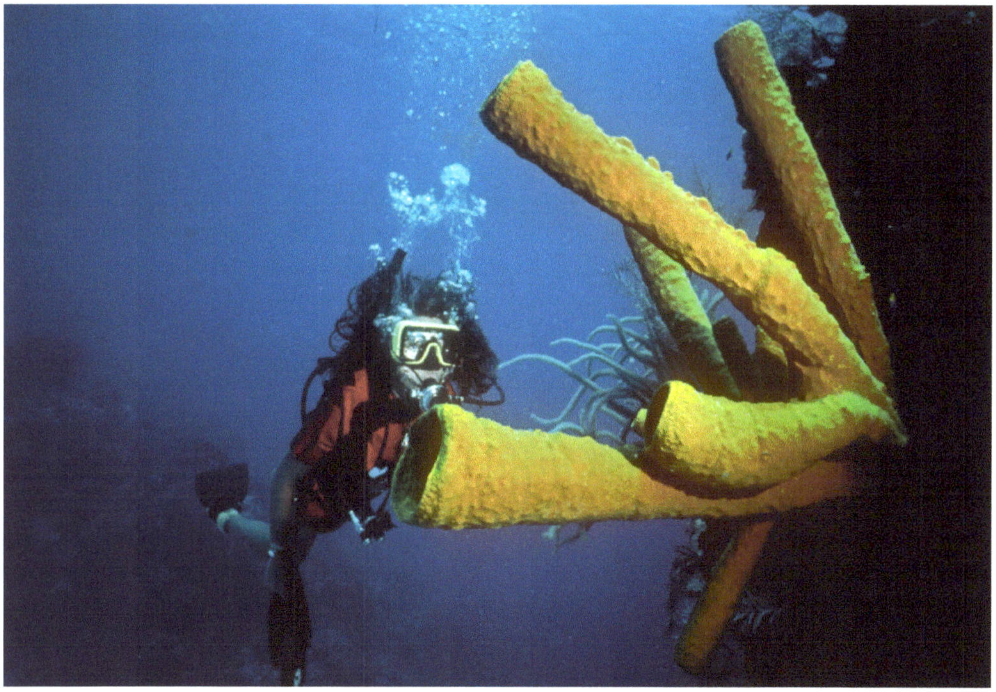

Diving the wall at Long Caye, Lighthouse Reef

result, I missed the very special site I was planning to show Helmut and Dorothy. On looking over the side I was aware that I was in the wrong spot. Having no wish to haul up the anchor, I decided that I'd just don mask, fins, and snorkel, and swim along the drop-off for I knew I would recognise it by a massive coral growing at the edge. No big deal, it couldn't be more than a hundred yards or so. I jumped in the water and started following the drop-off to the south. I found my dive site in about ten minutes. Feeling pleased with myself, I reversed my course and headed towards the boat.

I swam no more than ten seconds when on glancing into the deep, I became aware of two large ascending shadows about forty or fifty feet below. I accelerated my pace a little for I had an immediate bad feeling. The shadows kept rising. As they cleared the wall, I recognised two bull sharks!

In my twenties, I did lots of spear fishing. Some in Baja, but mostly in the Mexican reef areas of Yucatan. Sharks were always a problem when one chose to swim a quarter mile with a bag of bloody fish. But then, I had a spear gun and a shark-Billy at hand. Sharks are normally quite cowardly and a bang on their nose was usually enough to set them off on an alternate direction. Here I had no spear gun and no billy, and my present companions were at least a couple feet longer than the sharks of my youth. One I estimated to be close to ten feet, and the second, a little better than eight and a half feet.

Sharks, and other predators of the sea, rarely, if ever threaten divers at depth. However, floundering on the surface is another thing altogether, for this is where sea creatures find more oxygen, when they are injured or physically unable. Another factor was the species. Bull sharks are my least favourite fish, for they are always the ring leaders in a mixed group of predators.

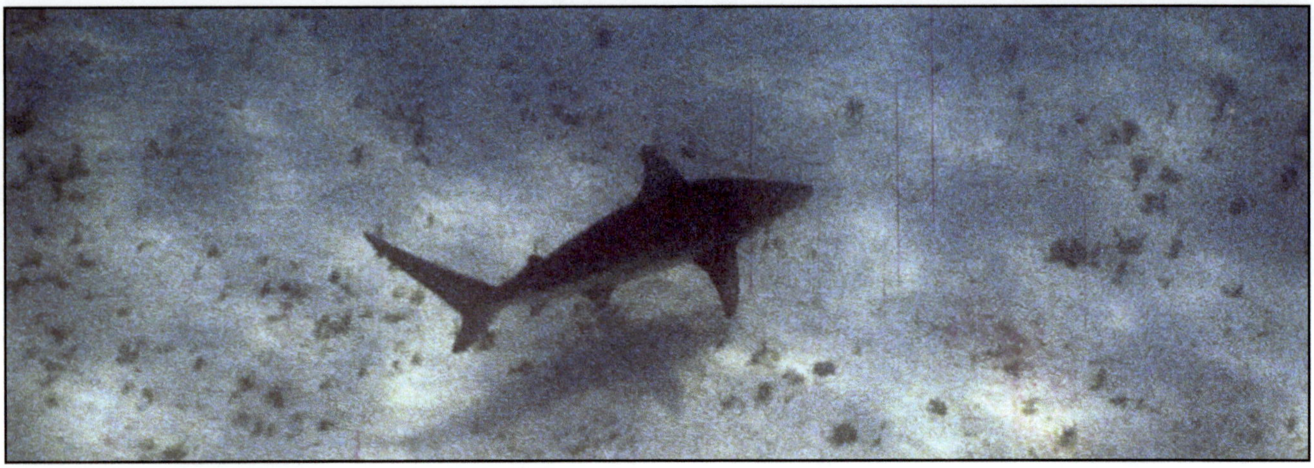

Bull Shark (*Carcharhinus leucas*)

They are actually the same species as the Lake Nicaragua shark, which has been blamed for several fatal attacks in Nicaragua. This shark is among the eight most dangerous sharks in the world. All this was racing through my mind, as I faced off these two monsters! I thought and hoped that they were probably better fed than their fresh water cousins.

With this in mind, I turned and advanced on them. I noted they were sticking together—another good sign. I told myself they were afraid to leave each other. At my advance, they retreated and dived. I turned and swam about twenty feet in the direction of the boat. They returned to the surface and me!

I again turned to face them, and I was amazed that they were still together. I was swimming on my back towards the boat, which was still a fair distance away.

When they moved into my fifteen foot circle, I flipped forwards and swam as fast as I could towards them. Again, they dove to safety.

I was swimming on my back once more. It was a little longer this time before they returned, but return they did. They had all the time in the world. I realized that I had to change my strategy. The only advantage I had was their fear of the unknown.

As I retreated on my back before them, I was hyperventilating. When I thought the time is right, I did a flip onto my belly and with my big dive fins, I accelerated towards the larger shark, as fast as I could. They both turned and dove and I followed them down. As I passed the top of the wall at about thirty feet, they were forty feet deeper, and not looking back. I returned to the surface, but kept watching the depths.

Once on the surface, I swam towards the *Lion* on my back, but spun around every two minutes. I reached the boat and as I was readying to pull myself up, I changed my mind, ducked under the keel and checked the other side. Good, all clear!

I pulled myself up and addressd my friends, "Yes; the dive site is just a fifteen minutes swim away. But I'll start the motor; I've another place in mind!"

Maria gave the game away, with only six words! "Jim, why are you so white!" So I told them of my adventure.

We only had one dive that day! Dorothy kept Maria company on the boat. I must admit that when I told my story of the sharks to my friends, I told a very abridged version. I really didn't want to spoil their trip but after all, who would believe me? Do You?

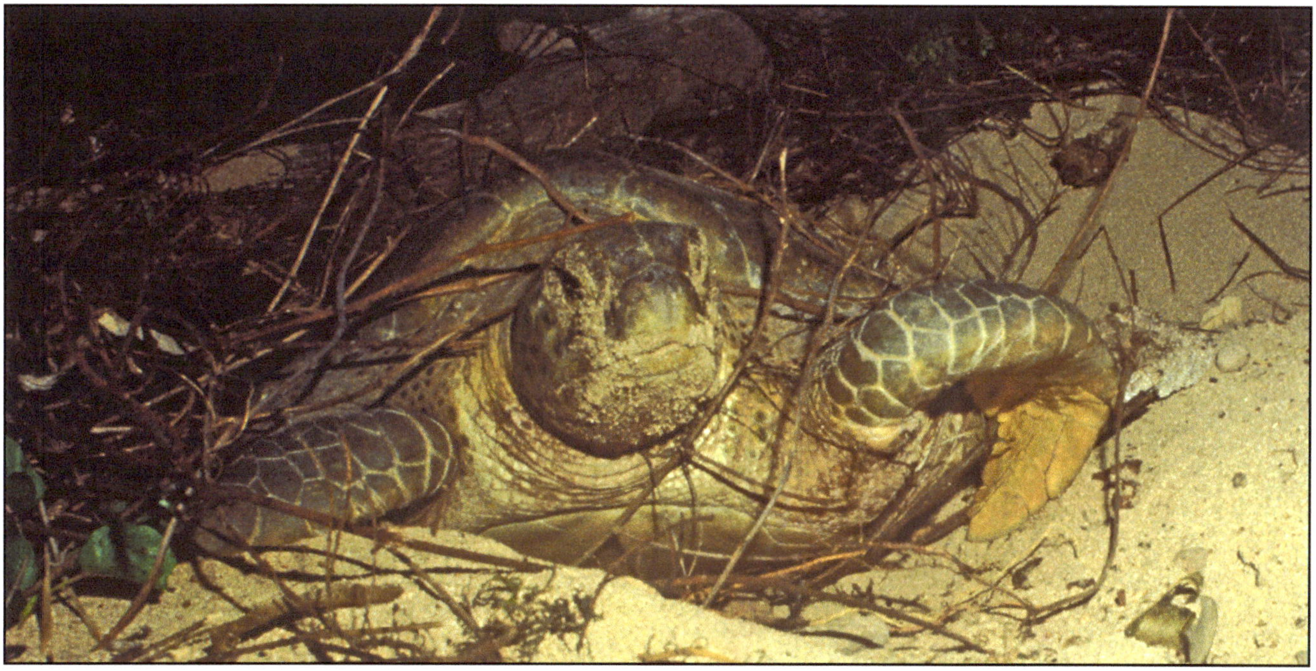

Green Sea Turtle (*Chelonia mydas*)

Sea Turtle Nesting

During 1987, I was a member of the Belize Audubon Society Half-Moon Caye/Reef Preservation Committee which was under the Chairmanship of Janet Gibson. At one of our meetings we had a visitor from northern Ambergris Caye, Greg Smith. He came to express his concerns regarding sea turtles that nest on the beach on the northern section of the peninsula from Basil Jones to Rocky Point, a distance of over ten miles. Greg was quite concerned regarding predation by both animal and human, upon the newly laid eggs and the turtles themselves.

Green turtles are vegetarians and live their lives in coastal areas, eating algae and grasses. Next in size, only to the leatherback, the green turtle is the largest hardshell sea turtle with a record weight of 869 pounds. The average weight is 330 pounds for a male, the female being slightly less.

Both the male and the female may travel hundreds, even thousands of miles. Every three or four years they return to their own birth area to breed. During that same year they may produce up to five clutches, with an average number of 110 eggs each. Gravid turtles come ashore, usually on a full moon night, to excavate their nests and lay their eggs.

In spite of its protected status, the green turtle is killed for its skin, meat, and shell. Unfortunately, nesting turtles are very predictable and human predators tend to walk the beaches after dark in search of laying turtles and, on finding them, flip them over where they will stay until dawn, when poachers return with a boat.

Animal poachers are not a great danger to laying turtles, but mammals such as racoons, coati, and skunks dig down to the newly laid eggs, eat what they can and destroy many others.

Greg approached the committee for help, mostly for volunteers to help him patrol the beaches during the nights when it seemed likely that conditions favoured turtle nesting. The committee agreed that they would seek out volunteers from among the staff and membership.

I decided to help the turtles, so I loaded up my sailboat with enough supplies to see me through three or four weeks. Greg's home is on Basil Jones Point, several miles north of San Pedro Town. I arrived shortly after the Audubon volunteers. At first Greg was a little perturbed as all his overnight accommodations were already overbooked. However, when I explained that I had brought my accommodation and food with me in the sailboat, he was all for having another body to walk the beaches.

Greg had been monitoring the turtles for several years, and had a good idea just when and where they would nest. It was my first, but not my last turtle adventure.

The volunteers stayed for one-week periods, monitoring the beaches for turtle activity along a fifteen mile stretch, both north and south of Greg's home. This area is one of the prettiest stretches of beach in Belize. In more than one spot the Barrier Reef comes in to kiss the shore, giving easy access to some spectacular snorkelling and providing the wave-swept beach needed for the turtles to nest. The beauty was marred only by a vast collection of plastic refuse—bottles, cartons, crates and styrofoam from all points north.

Greg expected a green turtle to come and lay in the near future. This turtle had already laid two batches of eggs this season, one on June 2nd and the second on June 15th. The next visit, Greg estimated would be around June 27th. My plan was to set up camp two days prior to the 27th lest the turtle show up early, so I set up a small pup tent about fifty feet directly north of the two previous nests.

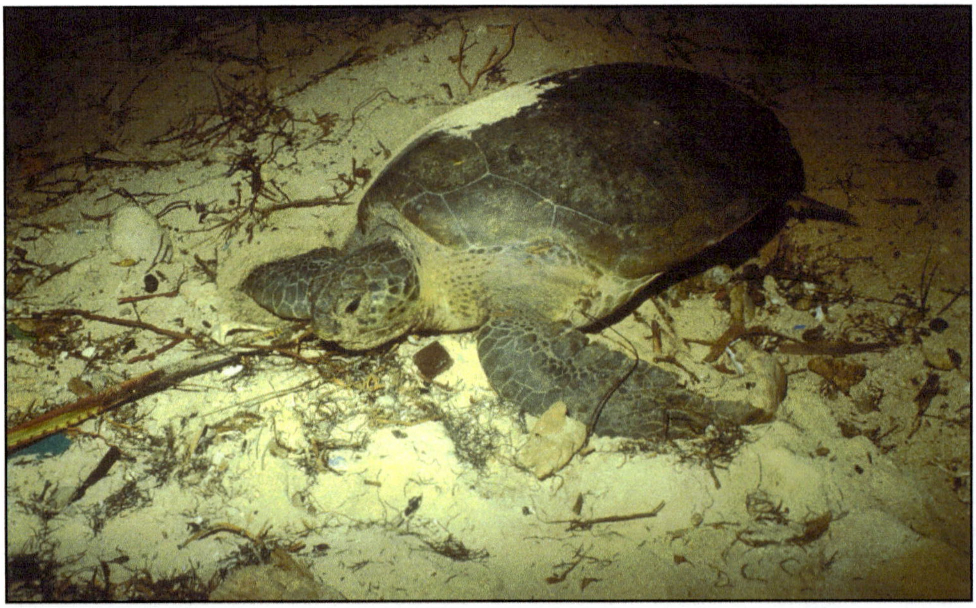

Green Turtle Nesting

I turned in early that evening thinking to sleep until 10 pm when I'd rise and walk the beach for thirty minutes. Staying awake was not going to be the problem I expected, for the wind picked up very strongly from the southeast, and the moon was close to full.

Then my tent collapsed on top of me. Wire tent pegs are useless in fine sand during a twenty knot wind. I walked the beach and collected all the plastic bottles I could carry. I filled them with wet sand and replaced my tent pegs with sand anchors. The rest of the night passed uneventfully which is to say, the turtle didn't turn up!

The next night was full moon and I had high hopes that the turtle would make an appearance this time. As I returned to my tent, my eye was drawn to a rock in the shallows. Then I rememberd that there was no rock there previously—it was a turtle!

I stood perfectly still and watched her slowly come out of the water. Fully illuminated by the moon, she climbed up on the beach and made her way south, paralleling the sea. After about twenty yards she stopped and reversed her direction. She examined her previous nests. She moved beyond me and stopped at her nest from several years ago. Yet again, she turned south for twenty yards, hesitated and reversed again and crawled back towards me. During this rather prolonged period, her passage was far from being silent, as her 300-pound body crushed various plastics and other flotsam strewn along the beach.

Her last crawl north had brought her to within eight feet of my tent, with me sitting in front of my tent in a Buddha position. I watched, as she lifted her head, now illuminated by the moon. It took very little imagination to empathize with this tired old lady, who no doubt was thoroughly disgusted with the state of today's beaches.

Before she turned, I knew there would be no egg-laying tonight. As she entered the water, I took one

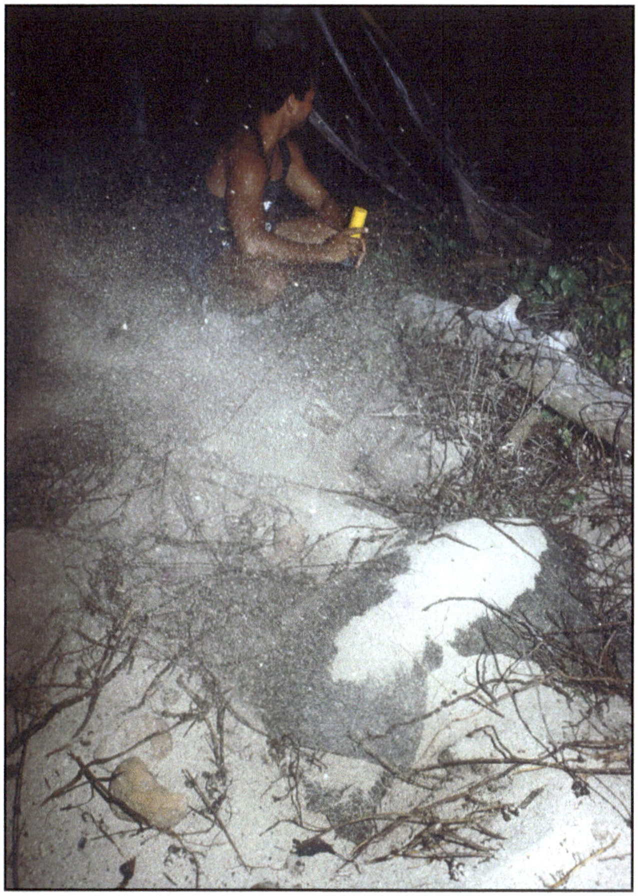

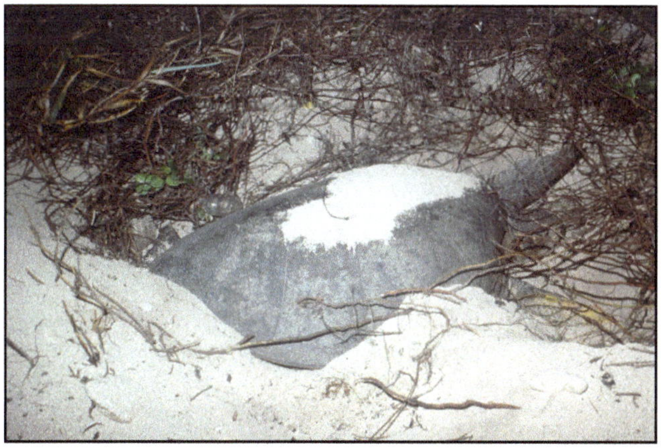

Green turtle digging her nest

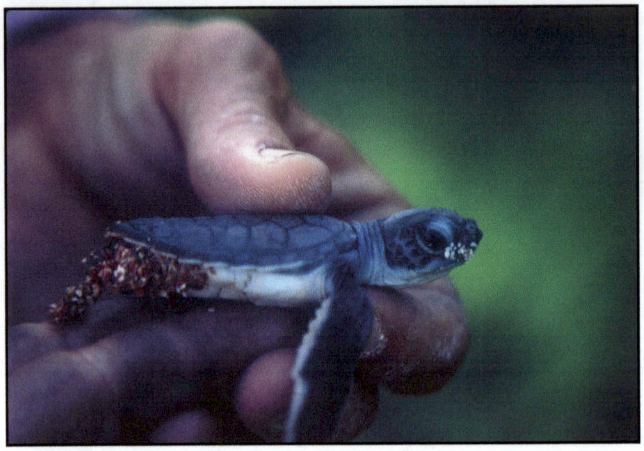

Green turtle hatching that was attacked by ghost crabs

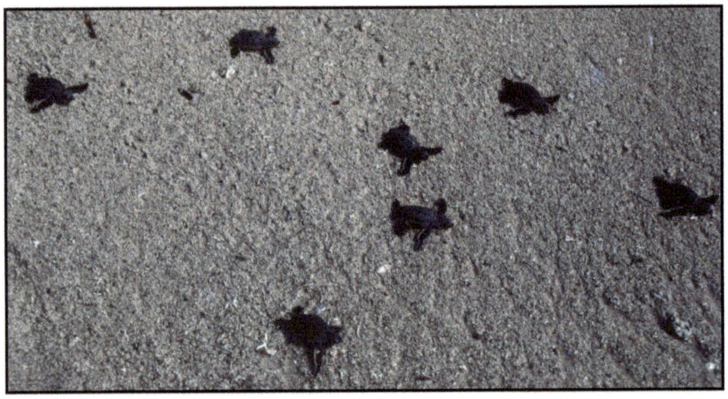

Green turtle hatchings racing to the sea

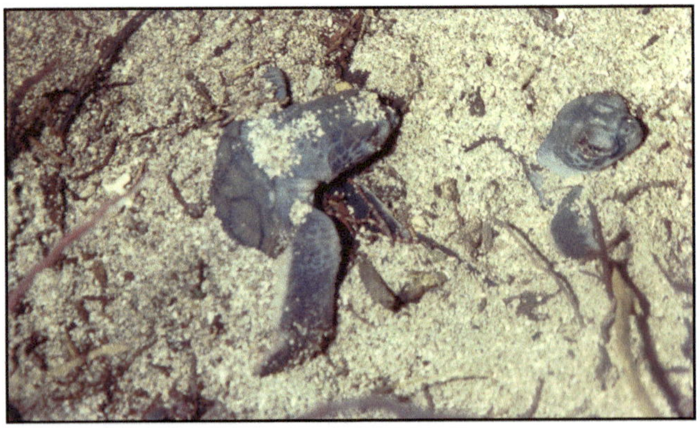

Loggerhead turtle hatchlings emerging from nest

picture of her departure, my first photo for earlier I was afraid that I'd drive her away with the flash. She was on land for exactly twenty-five minutes.

I was excited and too wide awake to sleep. It was now a little before two a.m. As I sat recollecting the previous hours. I heard from the forest behind, a repeating call, "whoo-hooo-hooo," repeated about five times, each call being shorter at the end and the beginning. It was very reminiscent of times I'd spent in northern Ontario. I recognised it as a Great-Horned Owl or something very similar.

I walked along the forest edge where the growth was far from dense, but there were several larger trees, thirty feet or higher. About fifty yards from the tent I saw it. The owl was perched on a branch with foliage to either side, maybe about twenty feet up. I moved as close as I dared and, leaning against a tree, I took a picture. The owl took off and dropped down to the base of another tree. While it was still in the grasses with my lens wide open and my Metz flash set to max, I shot another picture.

In the morning the volunteers arrived to find me frantically cleaning the beach, for I had taken the guilt of humanity upon myself. When the volunteers returned later that morning, they reported that "my turtle", had returned two hundred yards north of my camp to make her nest.

I was more than a little perturbed at this. Sure enough, there was a nest, but from my memories, this seemed more like a loggerhead nest, rather than a green turtle nest.

The 27th of July was when Greg had forecasted the old lady would return to lay her eggs, so he planned to wait up for her. I returned to my tent for a few hours sleep, for I didn't expect the turtle to arrive until midnight.

She fooled us. I awoke at 9:15 pm and when I was crawling out of my tent, I heard crashing in the bush. Then I heard Greg's excited voice crying, "Jim, she's here". Sure enough, she was already digging her nest. We both watched for about fifteen minutes, and then Greg decided to go back for the volunteers, a seven-mile round trip.

Once again I was alone with the old lady. I sat down and watched. I shot a few pictures. The eggs were already laid when Greg and the others returned. Together five of us watched as she laboriously covered up her nest. As she turned towards the sea, her eyes caked with sand, a volunteer patted the back of her shell. Then she was gone.

I returned in the middle of August as it was only right that I should see the children take their first steps. The hatching took place over a three-night period, the 14th, 15th, and 16th of August. From that nest there were a total of one hundred and two hatchlings. One died and the rest took off swimming. It was a successful effort! The adult turtle returned three more times that year, for a total of six nests.

The nest of the loggerhead that the volunteers found was not so lucky. The hatchlings started the following night when no one was around. When we arrived several hours later, they had already been found by predators. Sea birds, ghost crabs and nocturnal mammals had already eaten their fill. The surviving hatchlings were scattered in all directions. We gathered the remainder up in Greg's hat and released then into the sea.

Greg and I sailed south and continued our search for turtle nests. We interviewed a large number of fishermen and got priceless information on nesting both on the Cayes and the beaches.

We learned that different species had preferences to different nesting beaches. For example, green turtles have a definite preference to the cayes of Lighthouse Reef—Long, Northern and Half Moon Cayes. A lesser number use Ambergris Caye. Loggerheads also use Ambergris Caye and had fifty nests that year, 1987.

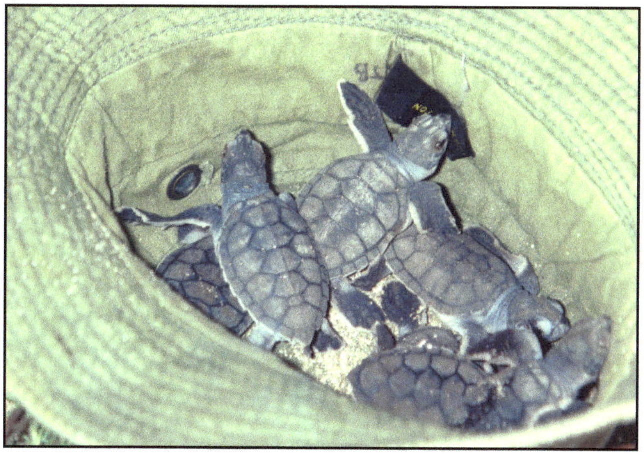

Loggerhead turtle hatchlings in Greg's hat

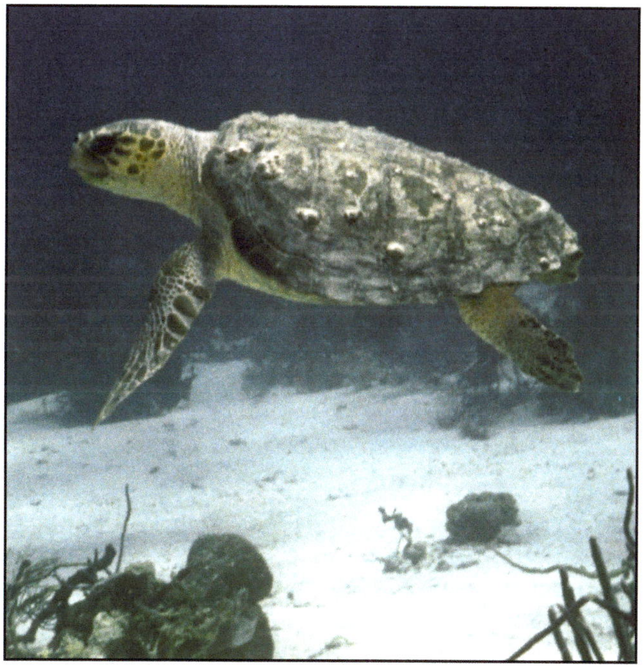

Loggerhead Turtle (*Caretta caretta*)

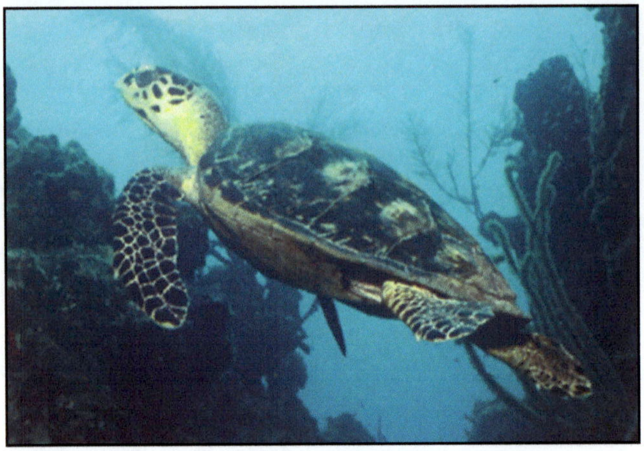

Hawksbill Turtle (*Eretmochelys imbricata*)

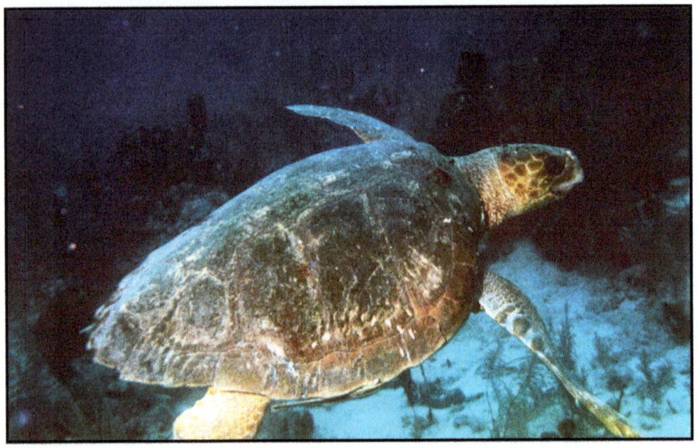

Green Turtle (*Chelonia mydas*)

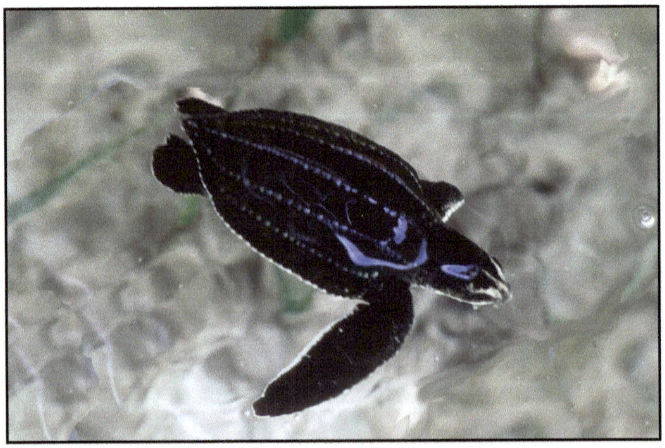

Leatherback Turtle (*Dermochelys coriacea*)

Hawksbill nests are more difficult to find but they have a marked preference to the small sand cayes of Southern Belize plus the long stretches of beach of the Placencia Peninsula. All in all, we learned a lot and hopefully altered the mindset of a few fisherfolk we met regarding sea turtles.

Belize has only three nesting turtles, but on very rare occasions, another sea turtle may be seen in our waters. Fortunately, I was one of the lucky ones!

The leatherback turtle is the largest of all the sea turtles. The shell of the leatherback is not hard, but rather like thick leather that will give under pressure. Apart from its size, the leatherback is unique in other ways. It is pelagic, spending most of its life in the open ocean. It feeds solely on jellyfish, including the deadly lions mane of northern waters, our own Portuguese man-of-war, and many others.

The most unique factor of the leatherback is its ability to withstand seas of near freezing temperatures. Like other sea turtles, the leatherback digs its nest in tropical beaches in the West Coast of Africa, the East Coast of Brazil, Caribbean Islands and New Guinea.

The closest nesting site to Belize is the east coast of Trinidad, from there the little hatchlings start swimming north, and unless they lose their bearings, they don't stop until they reach the waters off Nova Scotia in northeast Canada.

Scientists are trying to figure out how a tropical cold-blooded reptile can survive in near freezing waters, on a diet of only jellyfish which has approximately two percent of the calories that a regular fish has.

In September of 1988, a dead leatherback turtle was washed up on the beach of the village of Gwynedd, Wales. It proved to be the world's largest recorded sea turtle. It was ten feet long and weighed 2,010 pounds. It was estimated to be at least one hundred years old. The cause of death proved to be drowning caused by entanglement in commercial fishing lines. Unfortunately, fishing lines, plastic bags, and fishing nets are the leading cause of death among all sea turtles species and many other sea creatures.

On September 3rd, 1989, Greg and I took off heading south in my sailboat, the *Lion*. The plan was to visit the sand cayes and coastal beaches between Manatee Bar and Mullins River to check for the presence of turtle nesting beaches in the area.

After anchoring the boat south of Gales Point, we paddled ashore and walked south along an unspoiled beach. It wasn't long before we found turtle nesting signs. However, unlike Ambergris Caye, the dominant turtle here seemed to be the hawksbill.

The hawksbill is the smallest of the three Belizean turtles, and also the most endangered. For centuries it has been hunted for its ornate shell, which is used for jewellery.

Right away we noticed an obvious problem. Most of the nests visited seemed to be targeted by skunks or raccoons. They had dug up the nests after the turtle had laid her eggs, and devoured the upper layers leaving the shells scattered around.

It was obvious that they had devoured only a small percentage of the eggs. Greg set to work, constructing a conical cage with vertical bars, the tops being tied together and the lower ends being set into the ground about fifteen inches. Care was taken to make sure that the lower bars left sufficient space for the hatchlings to escape. We protected eight or ten nests in this manner.

When we believed that we had done all we could to save the hatchlings, we boarded the *Lion* and left. It would be close to a month before the hatchlings dug out; for that we would return to help them.

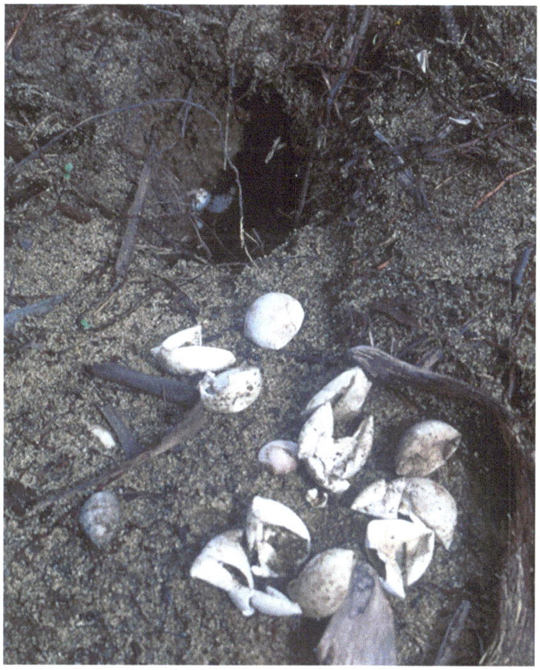

Hawksbill turtle nest after invasion by predators

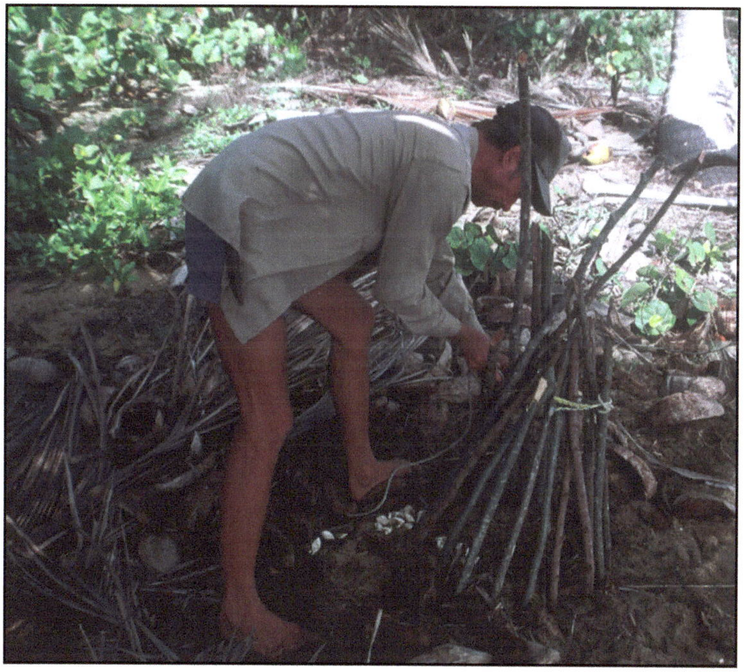

Greg Smith protecting turtle nests

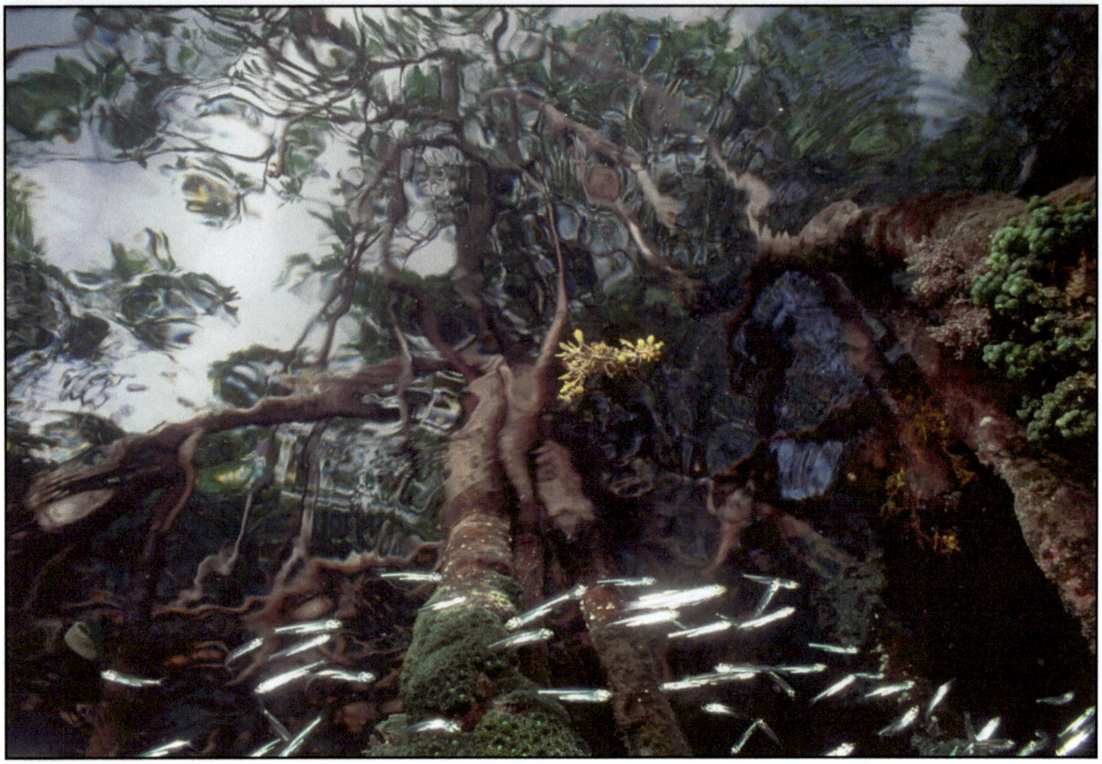

Looking up through mangrove roots, a school of sprat is evident

Mangroves

While we waited for the turtles to hatch, we did more searching and more than a little playing. We headed into the Main Channel to replenish our food supplies. Placencia seemed to be a perfect destination both for food shopping and checking the beach for possible turtle nests.

We had been sailing a couple of hours when the wind dropped and it soon became obvious that we wouldn't make it to Placencia that day. I recalled one of my earlier anchorages. I was pretty sure Greg would enjoy this location, a private and unique sanctum for marine invertebrates.

I first became aware of this little known beauty spot nine years before, while sailing at the cayes offshore of the Stann Creek District. It was already after dusk and Placencia was still two hours away, so I anchored for the night within a group of mangrove cayes near the Inner Channel.

The next morning I found I was moored in an almost enclosed lagoon. All around red mangrove trees rose out of six feet of water, forming an impenetrable screen from the rest of the world. I donned my mask and fins, lowered myself into the still dark water and was at once struck by a kaleidoscope of colour. This was my introduction to the faroes and ascidian gardens of the channel cayes. Almost ten years later, I pulled the *Lion* through a narrow opening into the same lagoon. Within minutes of lowering the sail, both Greg and I were in the water.

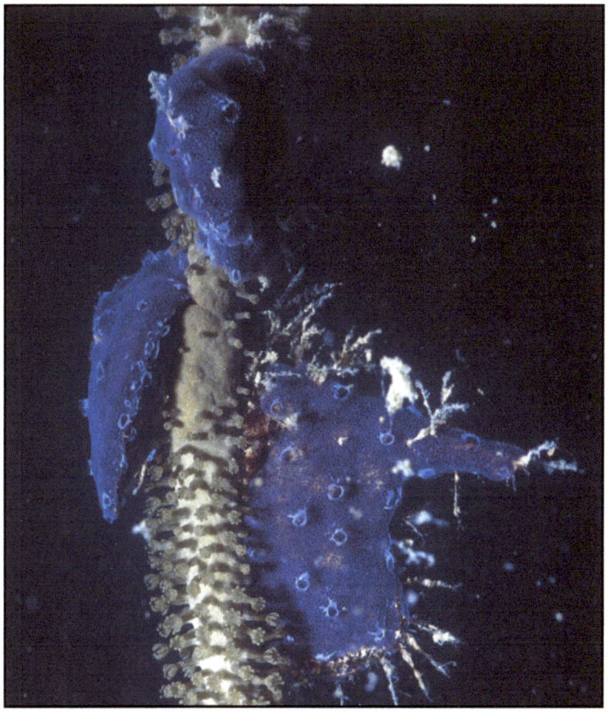

Blue Didemnid Tunicate (*Didemnum* sp.)

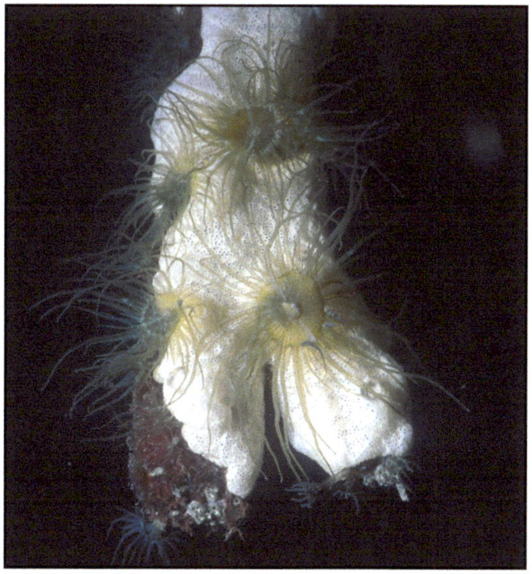

Red mangrove root colonized by white condominium tunicate (*Eudistoma* sp.) and pale anemone (*Aiptasia tagetes*)

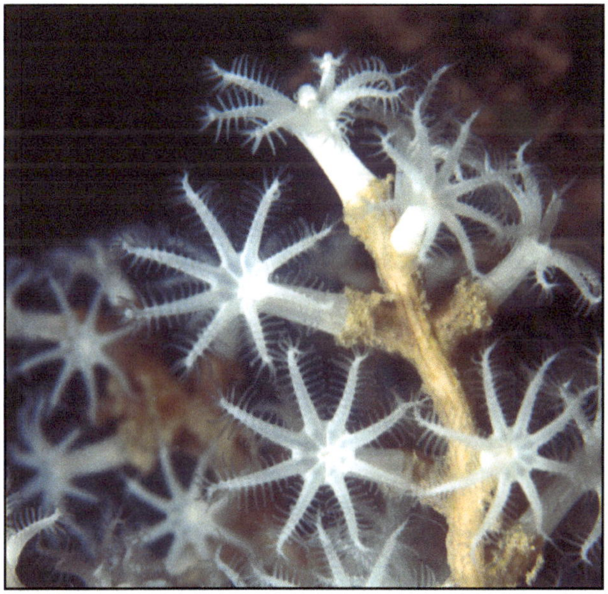

White Telesto (Carijoa riisei)

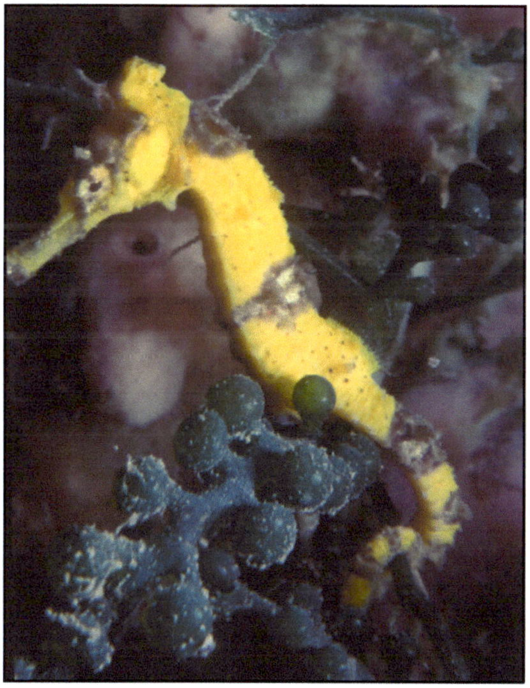

Longsnout Seahorse (*Hippocampus reidi*)

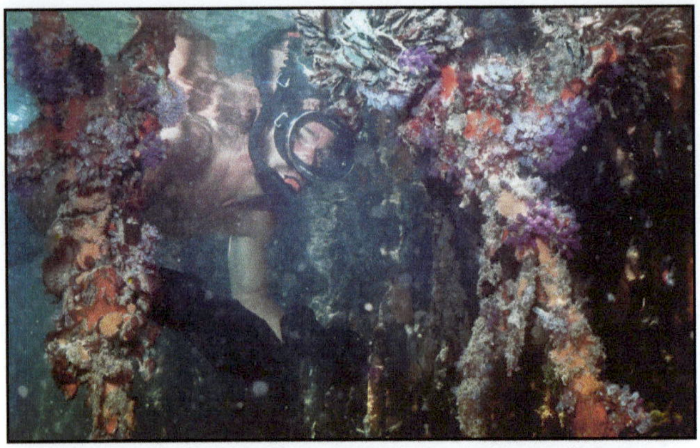

(above and below) Snorkeling in mangroves

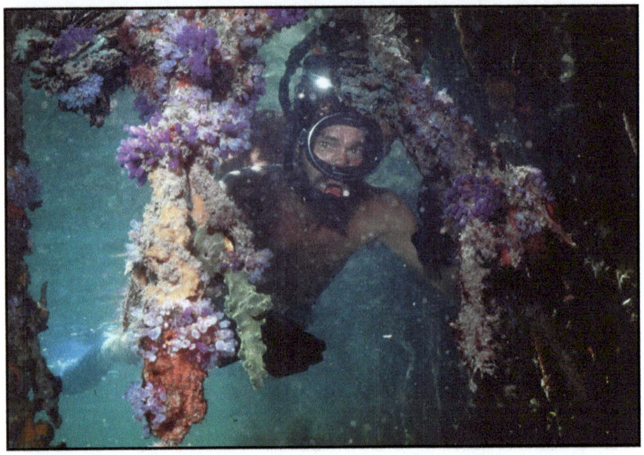

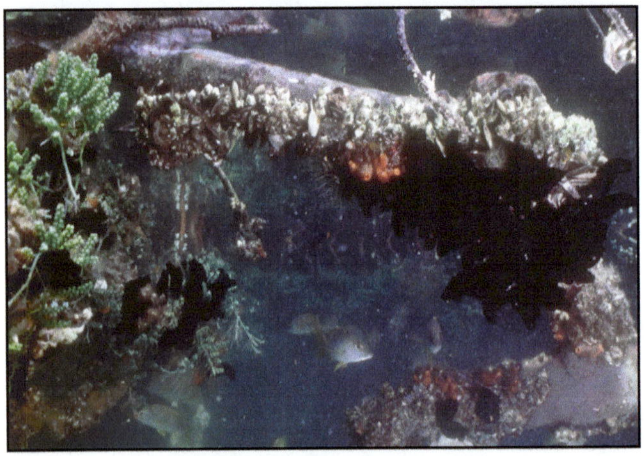

Black tunicates in mangroves

As I approached the mangrove roots, I saw sponges of every imaginable hue completely enclosing the mangrove roots which hung six feet below the surface. These sponges were festooned with tunicates of all descriptions, ranging in colour from the soft sky blue and violet of the painted tunicates, to the vivid red of the strawberry tunicates. Other tunicates were sponge like, and took the appearance of a day glow green tube.

Also colonizing the mangrove roots were the delicate pale anemone with long, tan-coloured stems and gently waving tentacles. In the shallows, the more robust sun anemones and mat zoanthids formed a dense, green carpet. I located spotted cleaning shrimps busily cleaning debris from their guardians' medusa-like bodies.

After several minutes my eyes grew accustomed to the overall mass of colour, and I began to discern the individual creatures that found shelter among the brightly coloured roots. A yellow longsnout seahorse hovered motionless alongside an identically coloured mangrove sponge, while just below a giant spiny spider crab slowly crept further back into the safety of the dense mangrove roots.

Five or six three-inch lobsters waved their antennae in unison as a scrawled cowfish swam by, stopping to nibble on a particularly enticing sponge. Deep among the vertical roots a two-inch-long barracuda hung suspended, suddenly to dart towards a group of minute fry, seizing its prey with the same ferociousness as its five-foot brethren out in the open water.

When I looked up, the colours were reflected on the under-surface of the glass-like lagoon water. The mangrove branches and aerial roots were covered with dense colonies of mangrove oysters with five-inch shells hanging just below the water surface.

On the silty brown bottom upside-down jellyfish steadily pulsated, their blue tentacles streaming towards the surface as they seemingly tried to drive their bodies into the soft mud. Now and then one rose and swam horizontally for a few yards before diving again into the silt.

I dove the eight or nine feet to the bottom to examine the turtle grass world. I saw a sea hare browsing in the grass and a flatworm gyrating a foot or so above the bottom. A roostertail conch raced (in mollusc equivalent) across the mud intent on reaching a destination only he cared about.

Out of breath, I swam up towards a silvery barrier, a large school of sprat parted and quickly disappeared as I approached the surface. Mistakenly believing that I was the cause of their hurried departure, I was totally surprised to find myself in a vast school of yellow jacks. It was not my presence that had frightened off the sprat, but rather the jacks which were in a feeding frenzy!

This Tolkien-like world is seldom seen by tourist and perhaps this is for the best, for it is infinitely more delicate than the reef corals. Even the water motion caused by a swimmer's fins is enough to dislodge large sponges which, with their cargo of tunicates and zoanthids, drift down to the bottom, only to die in the silt.

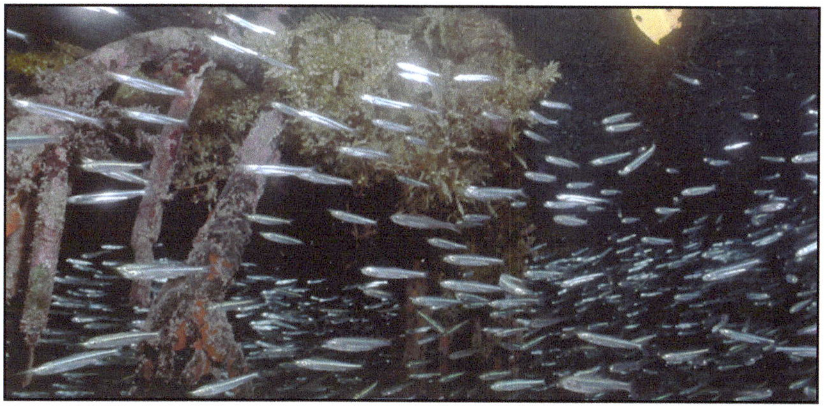

Sprat in mangroves

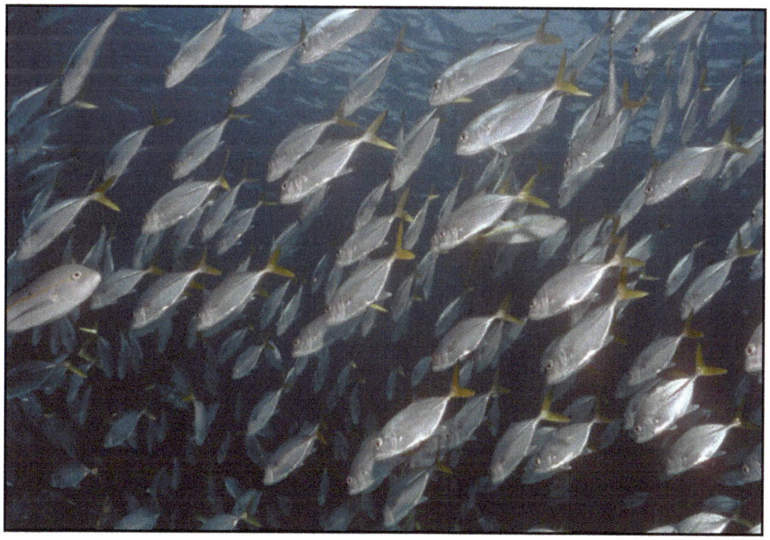

Yellow Jacks (*Caranx bartholomaei*)

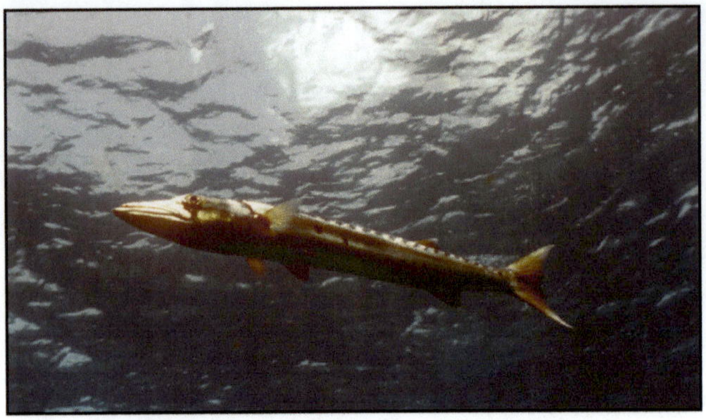

Great Barracuda (*Sphyaena barracuda*)

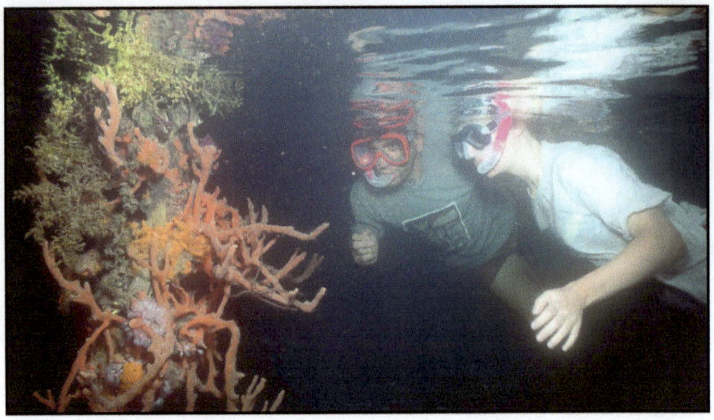

Snorkelers in the underwater Tolkien world

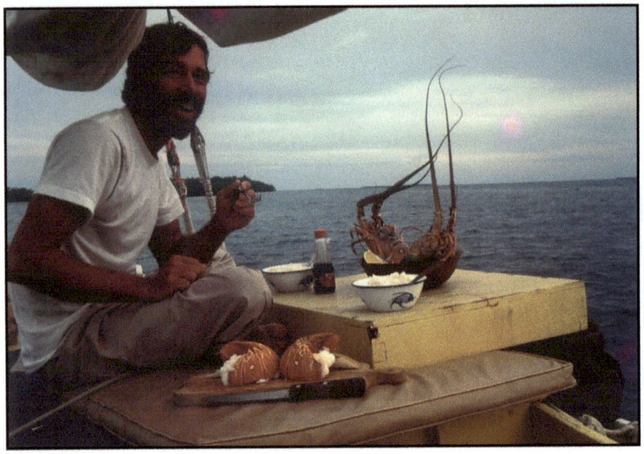

Greg roughing it onboard the *Lion*

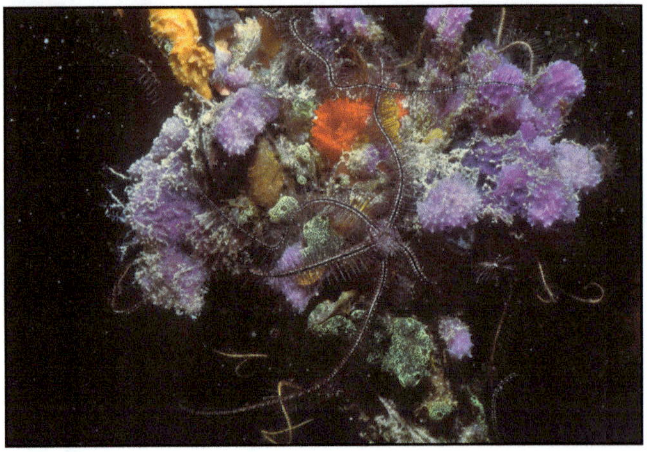

Brittle Star (Class *Ophiuroidea*) and tunicates on mangrove root

As I swam towards the boat, a six-foot barracuda glided towards me. I tried to ignore him, but remembering the ferocity of his minute brother in the mangroves, my fins beat the water just a little bit faster. As I reached the boat, he was only ten feet behind me. I clambered out.

Once aboard, Greg greeted me with a coffee. With a grin on his face, he said, "Let's stay a while!" I replied, "What about food?" To which he suggested, "Oh, we can rough it," For the next four of five days we totally explored the Pelican Cayes, plus Channel and Lagoon Cayes. Sea food was easily accessible, and turtles were all but forgotten for a while.

These islands were part of a larger system which was situated east of the Inner Channel and north of Victoria Channel. The water is very deep, in some areas reaching almost two hundred feet. Scattered in the mouth of Victoria Channel are faroes, irregularly shaped atolls bordered by fringing mangroves. Caused by geological faults, faroes are located on the continental shelf within the Barrier Reef lagoon. These unusual features are separated from each other by deep chanels. The walls of the lagoon and channel rise steeply and are studded with low, but dense corals, the most prominent species being lettuce and staghorn coral.

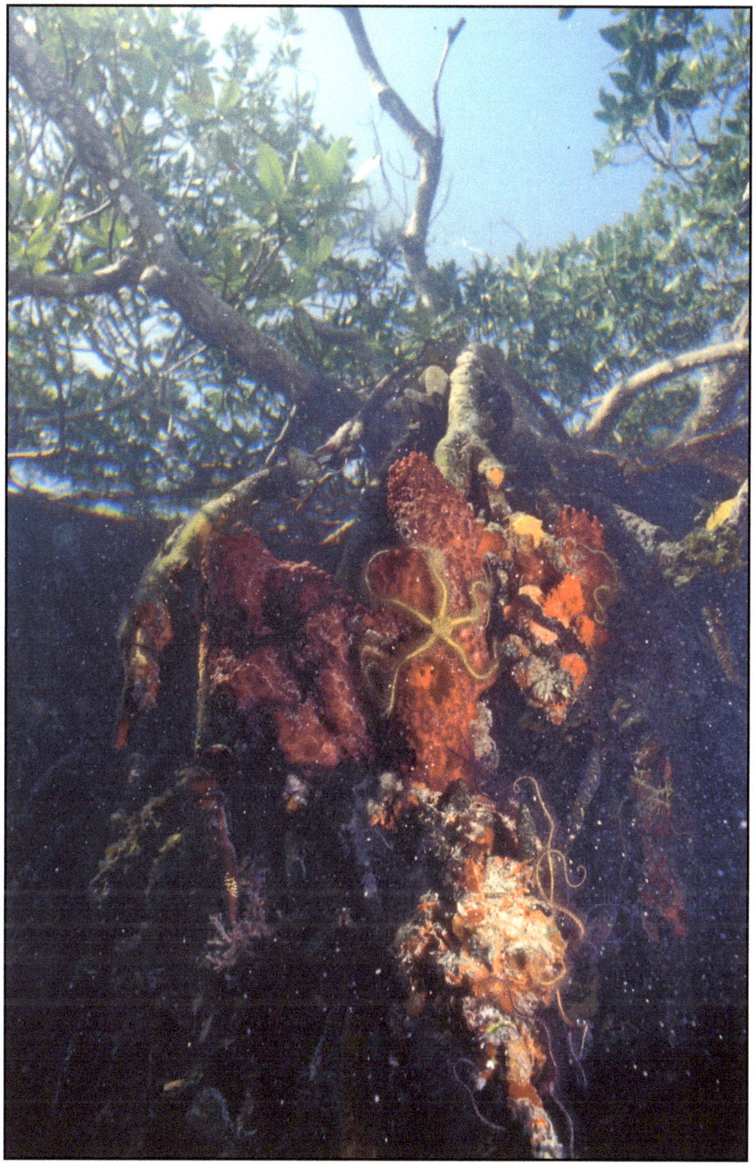

Mangrove root system

The tops of the walls form shallow ridges that can be ten to fifty feet wide and may run for two miles or more. Where the ridges are wide, red mangroves have taken root and formed awash mangrove cayes. The inner lagoons offer some the most protected and private deep water anchorages in the country. Such islands include Lagoon and Pelican Cayes. The mangrove roots hang down five or six feet in the lagoon. These roots are festooned with all types of marine life, including sponges, tunicates and zoanthids, plus the many mobile invertebrate animals who use the confusion of roots as a sanctuary.

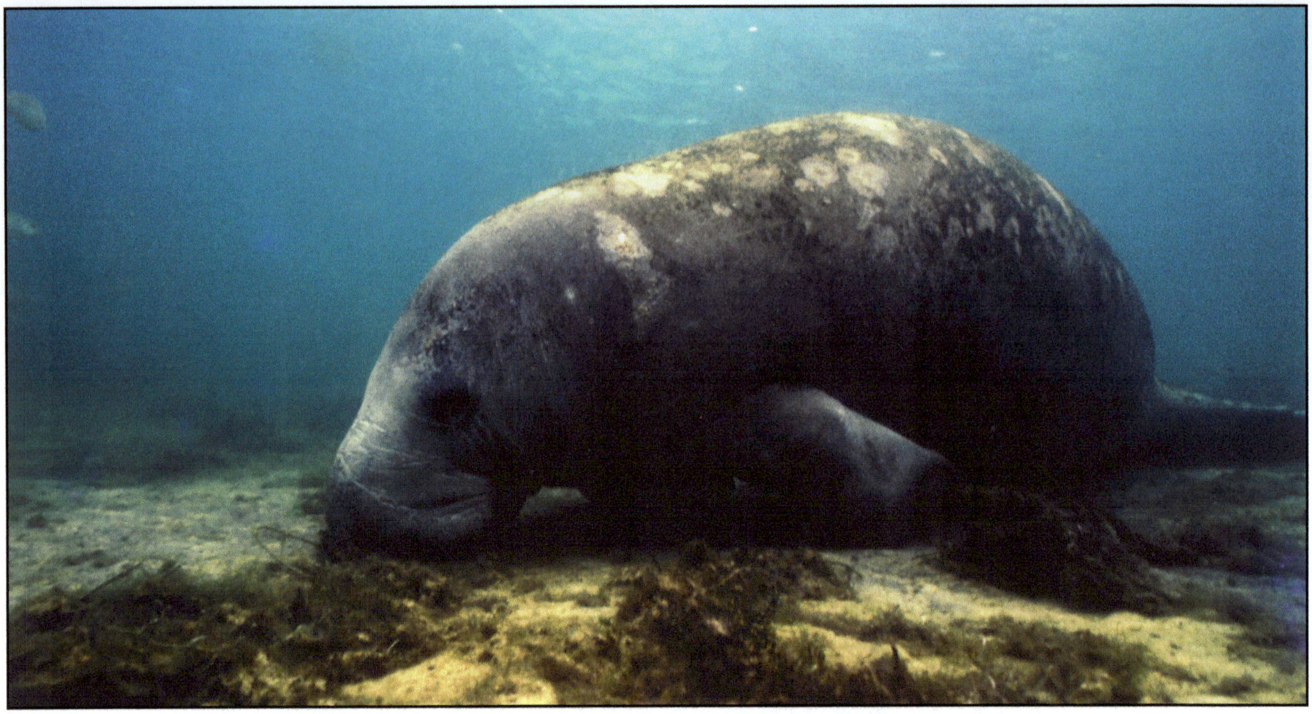

West Indian Manatee (*Trichechus manatus manatus*)

A Date with a Mermaid

The following year Greg and I left a little earlier, June 10th of 1990, to try and catch the nesting in the southern beaches. We caught the laying and made a point to camouflage the nest sites after the turtles had laid. We hoped that more of the hatchings would survive their birthdays.

In the meantime we had another quest, to locate as many manatee groups as possible. In the previous few years there had been several reports of manatees being hit and killed by the increasing numbers of powerful speed boats. The growth of the tourism industry brought more water taxis and tour boats. Reports of persons hunting manatee for their flesh, especially in the southern areas of the country, were increasing. The hunters were from Guatemala.

We had no intent on arresting anyone, but we did want to know the truth of the matter and how serious the problem was. We wanted to get an estimate of how many of these gentle giants were in Belizean waters.

My sailboat was perfect for the job. Without a motor's noise, we could ease up silently on manatee family groups and get excellent counts if the weather was at all calm. Strangely enough, manatees were most numerous near the busiest skiff routes.

Manatees are extremely intelligent creatures, but on occasions they show a great curiosity of mankind, a deadly flaw for any creature's make-up. We found one such group quite close to Belize City. Exceedingly friendly even to the point of playfulness, these were immature animals. The larger adults usually kept their distance.

Greg and I came to use this manatee area, Swallow Caye, as an anchorage, whenever we were in the region and it didn't take long for the manatees to recognise the *Lion*'s hull.

One that I believed was a female, was very friendly and visited very often, making no effort to flee when I entered the water. I called her "Maggie." Every morning after my coffee, Maggie was there. When I went in the water, she would follow close behind. I found out very soon that Maggie had both a sense of humour and loved getting her back scratched.

One day I thought I'd get a different sort of manatee picture. My thought was to sit on the bottom and take a photo of Maggie swimming above on the surface, so I put on my Velcro dive suit, mask and fins, and strapped a twelve pound lead weight belt around my waist. I chose not to wear scuba as I thought the bubbles might scare Maggie.

The water was about twelve feet deep. I figured I would have about ninety seconds before surfacing for air. I dropped and sank to the bottom. The bottom was muddy so I sat down with my legs straight out and my camera ready. No more than thirty seconds passed before I saw Maggie approach.

The script called for Maggie to swim directly over me, and then I'd take her picture! However, instead of swimming over me as I had planned, she dove towards me and just before she passed over, she twisted and settled down on my lap. I was totally flummoxed! So much so, that I forgot that I needed to breathe. When that realization hit me, I put my camera to the side, slid my arms under her and lifted her up. I've got to say that I never realised that lifting a one-thousand-pound animal was so easy!

When I recovered my breath, I dove down to recover my camera. Maggie was still sitting there looking a little perturbed! Photography over, I went for a swim, Maggie followed with a manatee grin on her face.

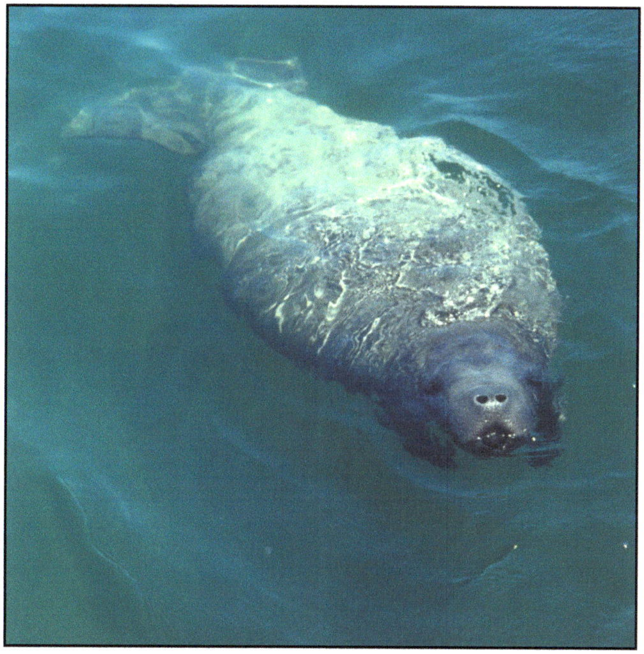

West Indian Manatee (*Trichechus manatus manatus*)

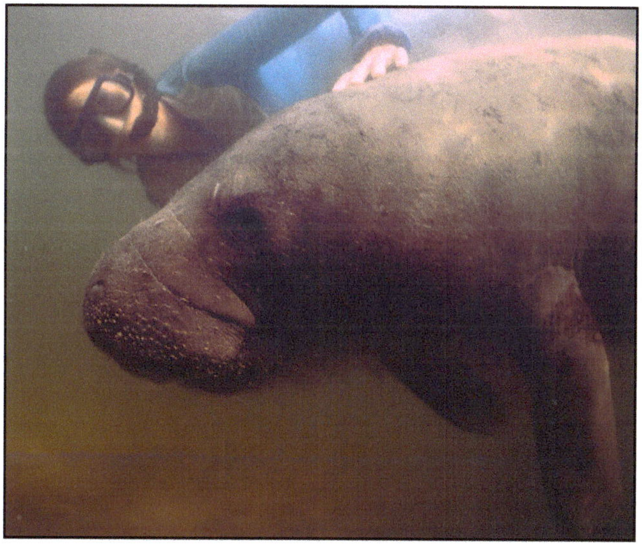

Maggie getting her back scratched by Dorothy

On yet another day, she led me to a mangrove caye and what seemed to be a solid mangrove root system. But it proved to have a tunnel running through it about two feet below the surface. She swam straight through with no hesitation; in doing so she stirred up a cloud of silt. I waited about three minutes, then followed the channel through the still cloudy water.

Above me a matrix of mangrove roots left no way through, the passage was about eight feet wide and approximately four feet high, and it turned out to be a lot shorter that I was expecting. The total distance was about 20 feet, and then it opened into a small lagoon about 30 feet in diameter. Maggie was floating in the centre. I quickly turned around to memorize the exit tunnel, for I had no wish to lose this exit, and spend the night here.

Above, the ceiling was about four feet high and the sun filtering through the mangrove leaves, lending an emerald glow to the lagoon. At another time, I may have thought it a beautiful place.

I was still clutching the edge of the tunnel when I looked around for Maggie. No problem, she was right there, except now she had a companion. This second manatee was quite a bit larger. The first thought that shot through my mind was, "she's taken me home to meet daddy!"

I turned around and made my way through the tunnel and out to the open sea. Maggie followed three or four minutes later. She seemed to have no hard feelings. Maybe it wasn't her father.

Several weeks later, I again stopped to overnight at Swallow Caye. I had arrived relatively late in the evening, so I just ate and went to my berth.

The next morning I was eating breakfast when Maggie turned up; she hung around the area until I finished eating. When I jumped in the water, she joined me for a swim. It was to be a short swim, because I had to head home, for some reason I can no longer remember.

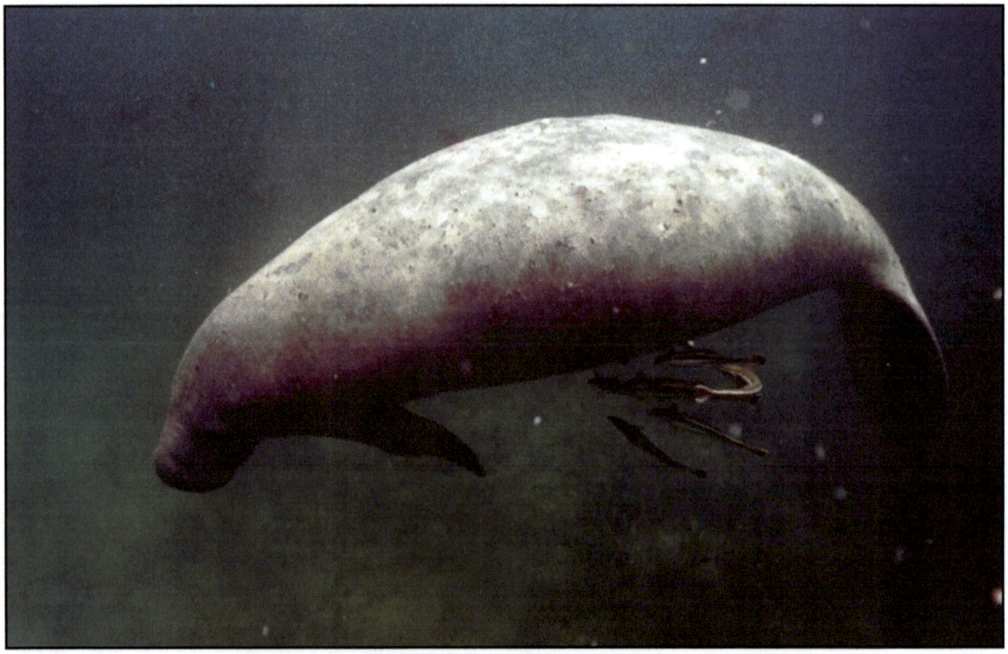

West Indian Manatee (*Trichechus manatus manatus*)

I gave Maggie as much time as I could, but eventually I swam back to the boat. She stopped in the water about twenty five feet away. As I turned around, Maggie was still watching me.

Holding the gunnels of the Lion in my left hand, I turned and waved goodbye to Maggie with my right. As I turned back to the Lion, I stopped suddenly! Was I mistaken? But no!

There was Maggie, now moving to a vertical position, her right fin was raised in the international "goodbye" gesture, just as my arm was raised seconds ago. I had time to snap a picture!

I realized that this last story about Maggie may for some be difficult to believe. It has been many years since I last visited the manatee hole, but what Greg and I didn't know when we discovered Swallow Caye back in 1990, was that we were not the only visitor to the Caye and the surrounding waters.

There was one other, and he felt in the same way about manatees as I did, Lionel (Chocolate) Heredia. He, too, kept the place secret for many years, but eventually he brought guests to visit Swallow Caye. He and his wife Annie have worked hard to provide protection for Maggie and her kin. Through the efforts of Chocolate and Annie, Swallow Caye Wildlife Sanctuary was gazetted in 2002.

The island and manatees now have full protection, along with 8,970 acres of surrounding waters. Swallow Caye is within sight of Belize City, and a day trip from Caye Caulker and San Pedro.

Chocolate still leads manatee trips from Caye Caulker and if you have any doubts about my relationship with Maggie, speak with Chocolate. For every story I can relate, Chocolate can tell you three or four more!

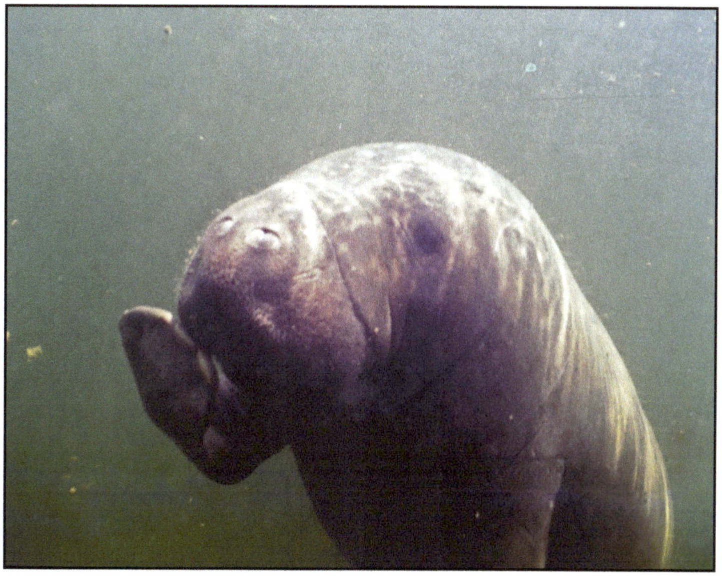

Maggie waving "goodbye"

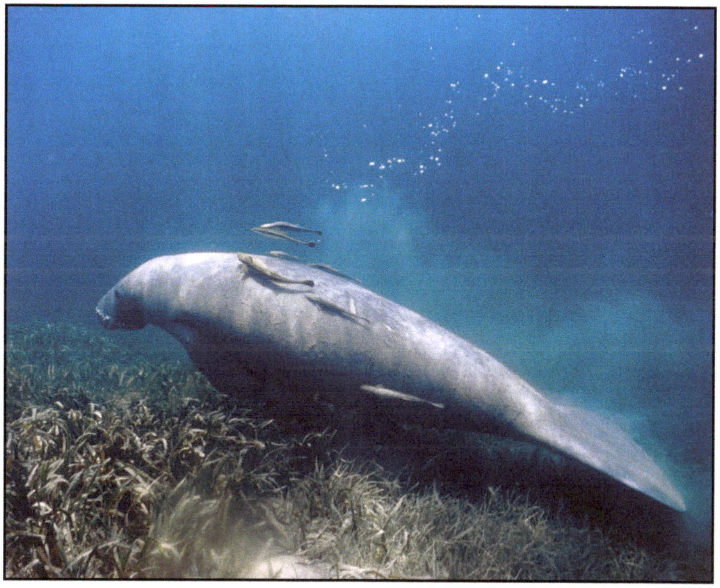

Manatee being cleaned by Remoras

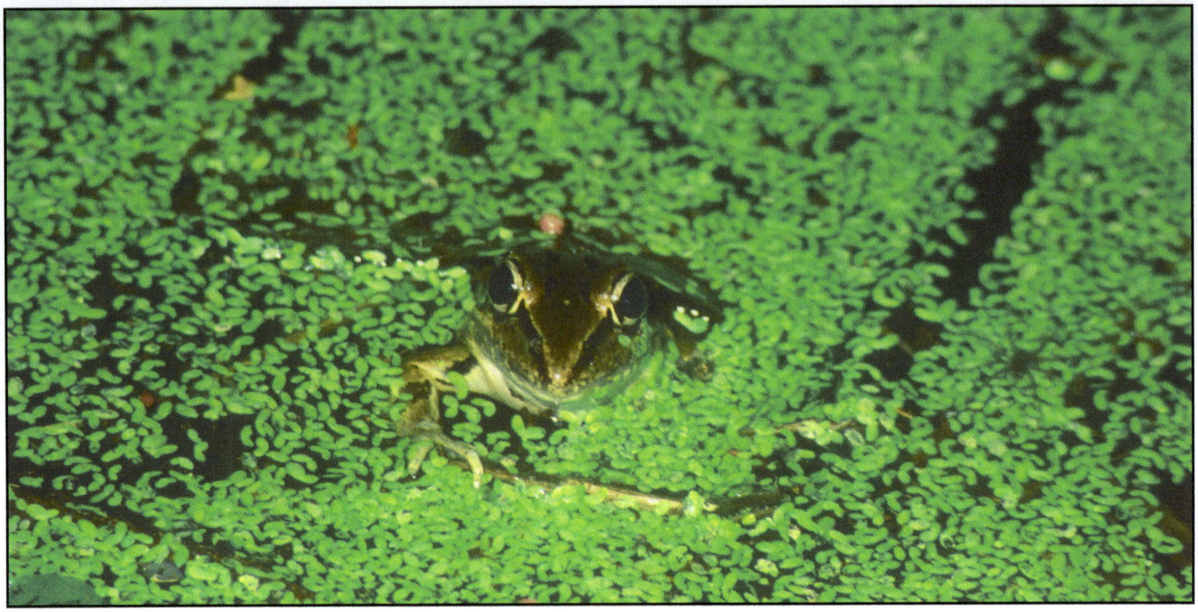

Julian's Frog (*Rana juliana*)

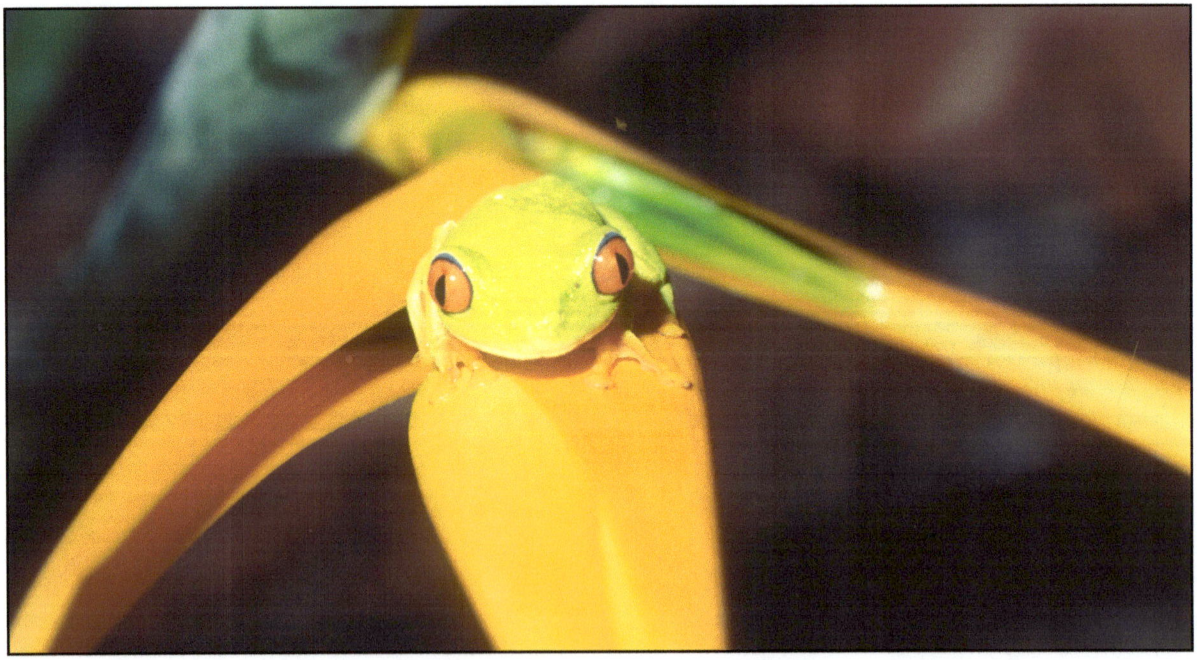

Red-eyed Treefrog (*Agalychnis callidryas*)

Chapter III
The Forests of Belize

Rains and Frogs

The tropical rains had dwindled to a heavy mist as I pushed through the foliage in the direction of that trilling clamour of sound. I abruptly broke through the clinging vines, splashing into knee deep water. The beam of my headlamp cut through the darkness and caught small hurtling bodies moving too fast to identify. Lowering the light down to the water surface, I noted that it, too, was full of constantly moving shapes. I turned around to tie a red kerchief on the branch behind me, for without a marker of some type it would be extremely difficult to find my way back through the forest when my work was completed.

I was in Cockscomb Basin Wildlife Sanctuary. It was close to midnight on a wet moonless night in June and two days earlier the seasonal rains had started, triggering a response in the arboreal treefrogs as old as life itself. Normally these seldom seen amphibians are content to remain in the canopy, far above the forest floor. But with the onslaught of the rains, they descend by the thousands to join other frogs congregating around a few chosen depressions on the forest floor that rapidly fill with water.

I moved forward, carefully placing my feet lest I tread on mating frogs that were too occupied in their love making to notice my passage. The bulk of the

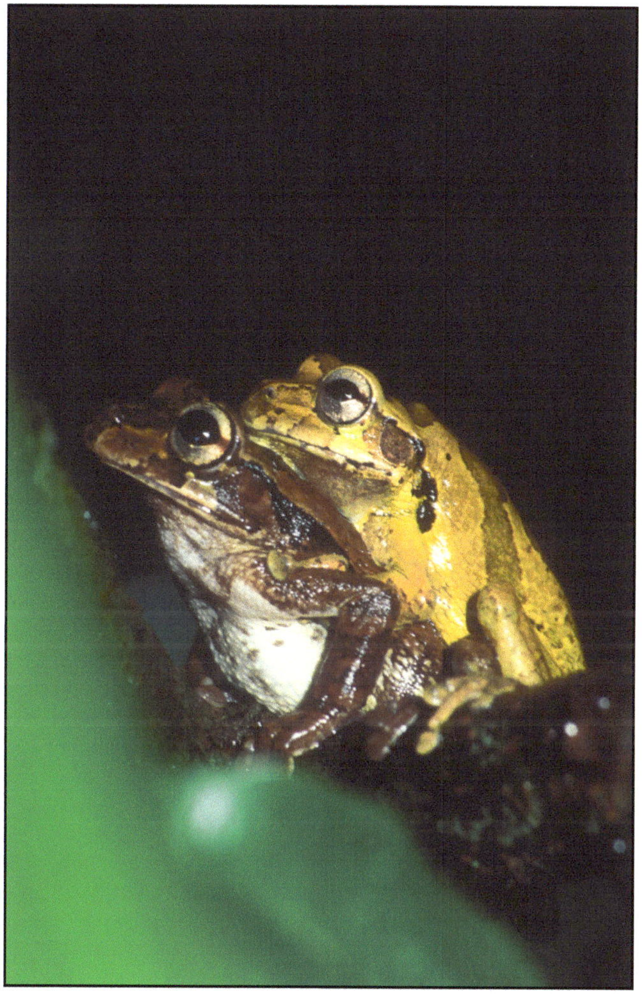

Mexican Treefrog (*Smilisca baudinii*)

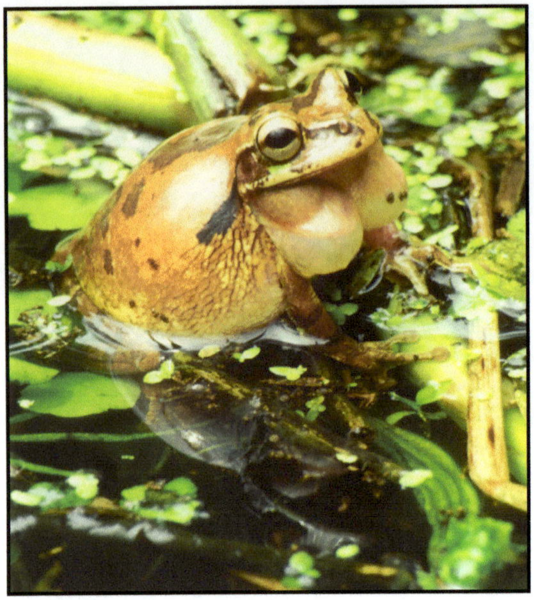

Mexican treefrog shows its twin vocal sacs

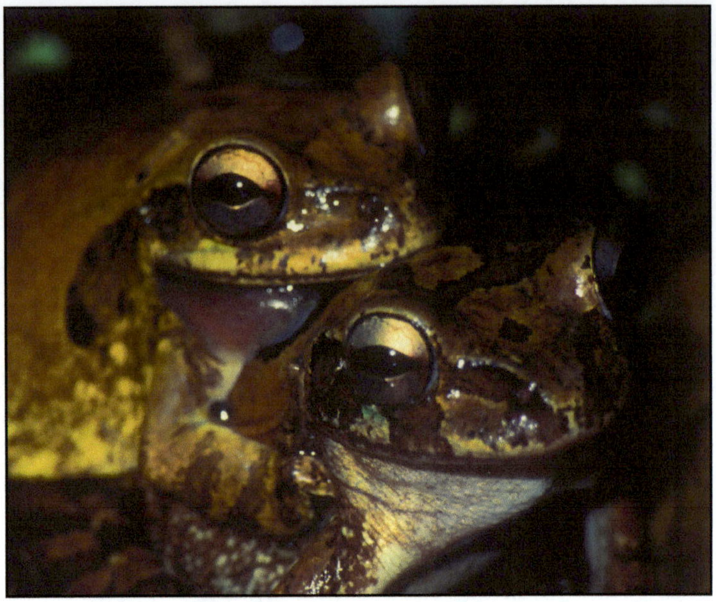

Mexican treefrogs in amplexus

frogs in the water were Mexican treefrogs, one of the commoner treefrogs of Central America. These frogs are about 70 mm long with tan and brown markings and horizontally slit golden eyes. Obvious are their unusual twin vocal sacs, with which they produce their nasal cry.

Also in the water, but not as numerous, were the 65 mm blue-spotted treefrog and the much larger pepper treefrog which has a snout-vent length of 100mm, the largest recorded treefrog in Belize. When handled, these frogs release a secretion from their glandular skin that causes skin irritation and may cause a bout of sneezing, hence the common name, "pepper" treefrog.

All the aforementioned frogs breed in a similar manner. The male grips the female while in the water. As they swim on the surface, both frogs roll forward in a 90 degree motion thrusting their rear end out of the water. Simultaneously, the female releases her eggs as the male releases his sperm. The fertilized eggs separate and spread over the surface in a film. Development to froglets takes about two months.

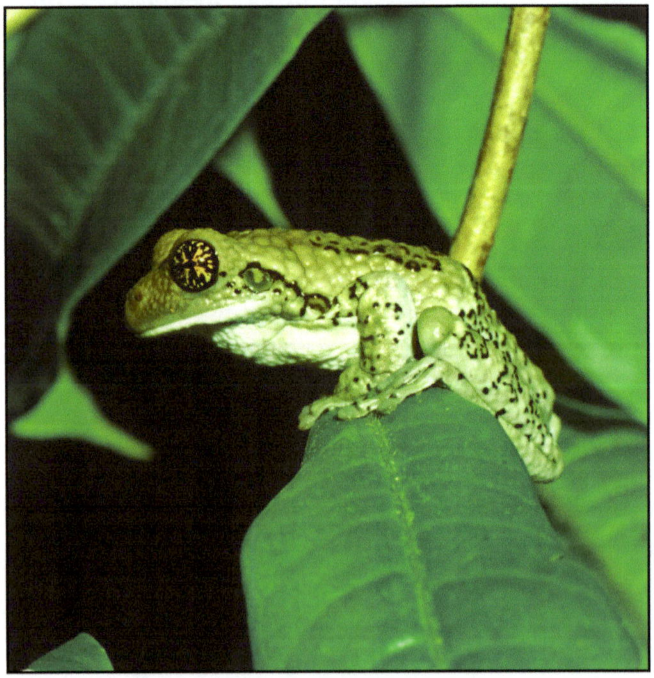

Pepper Treefrog (*Phrynobyas venulosa*)

The elegant narrow-mouth frog and the burrowing cow frog are not treefrogs, but burrow in the soft forest floor during the dryer months. According to Dr. John R. Meyer, the former chairman of the Declining Amphibian Population Task Force, the elegant narrow-mouth frog is not commonly encountered anywhere in Central America and at that time only four sightings in Belize had been reported. Like the treefrogs, elegant narrow-mouth frogs were very noticeable in this temporary pond. During the night, I observed approximately fourteen pairs of this 40mm-long frog. Its dorsal colour is a dark brown with lighter blotches and the sides are black with small white spots.

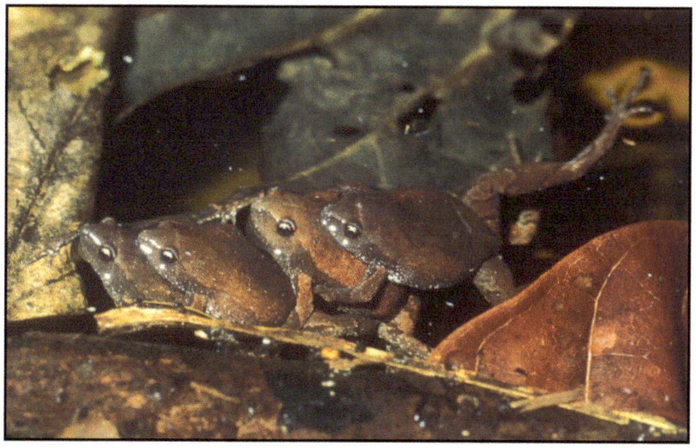

Elegant Narrow-Mouth Frogs (*Gastrophyne elegans*)

The burrowing cow frog is quite bizarre. Its formless body of about 70mm long is a bluish-black colour with a red dorsal stripe and scattered red spots. Its common name, burrowing cow frog, is quite descriptive for it spends most of its life underground in burrows that it digs itself. During the breeding season, its presence is made evident by its loud plaintiff cry that sounds rather like a sick cow. When males and females congregate during the rains this call can be heard several miles away. In Belize, many a tourist has returned from the rainforest with tales of lost cows crying out in distress. Both the elegant narrow-mouth frog and the burrowing cow frog lay groups of single eggs on the bottom of the pool.

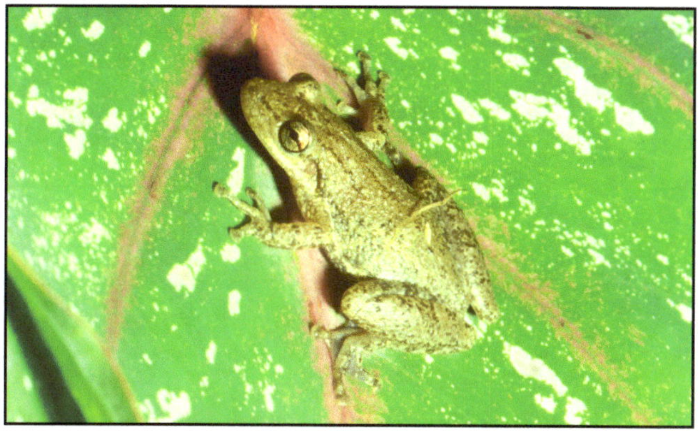

Bromeliad Treefrog (*Hyla bromiliacia*)

As I moved forward my lights caught the continuous motion in the bush. There were treefrogs of various species. These amphibians had evolved away from mating and spawning directly in standing water. Most of them were of the genus *Hyla*. They are tiny frogs, ranging from 25mm in the case of cricket treefrog to 36mm for the variegated treefrog. This genus of frogs generally lay their eggs on vegetation just above water level. In this way they avoid spawn loss by aquatic predators such as dragonfly nymph. For with the rains, the insects also breed. The variegated treefrog is quite colourful, bright yellow with brown markings, including a brown "hourglass" shape on the upper back and head.

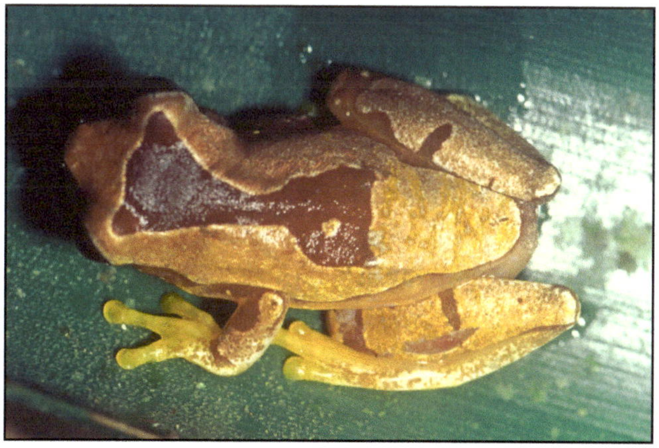

Variegated Treefrog (*Hyla ebraccata*)

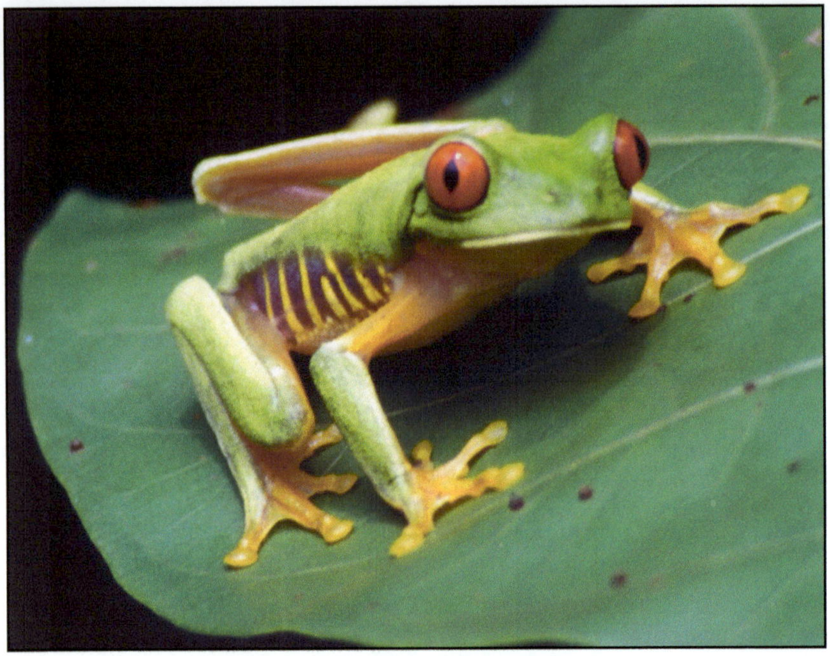

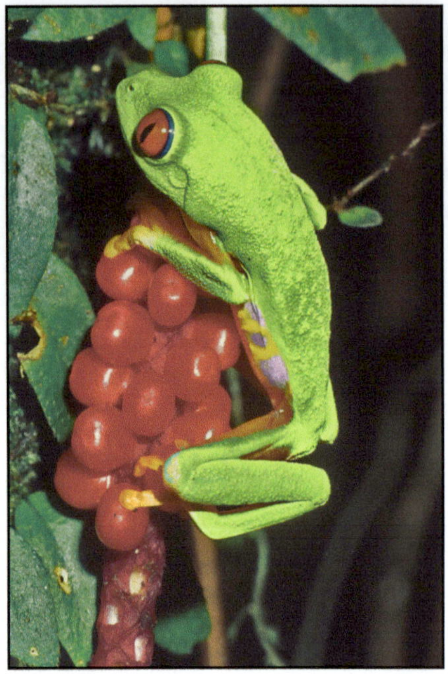

Red-eyed Treefrogs (*Agalychnis Callidryas*)

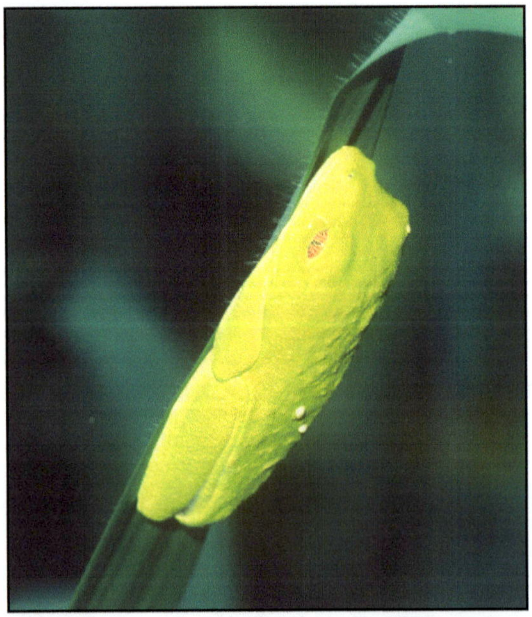

Quiescent Red-eyed Treefrog resting on the under side of leaf

Active in the upper bushes was the best known treefrog of all. The red-eyed treefrog is very likely the most colourful and most photographed treefrog in the Americas. When seen during the daylight hours, its most striking features are its eyes which are startling red in colour with vertically elongated pupils. The average size is 65mm long. With its emerald green back, red eyes, blue and yellow striped flanks it can not be mistaken for any other frog. Like all treefrogs it is nocturnal, but during the mating season they stay close to the breeding pool in daylight hours. That does not mean that they are easy to find. They enter a quiescent or resting stage with limbs pulled tight to their bodies and eyes closed. Their vibrant colours fade to a very pale lime green, so they virtually disappear on the undersides of large leaves. During the night the green was vibrant and their flank colours flashed blue. Only the eyes were disappointing, for now they were fully dilated, being black with only a narrow red rim.

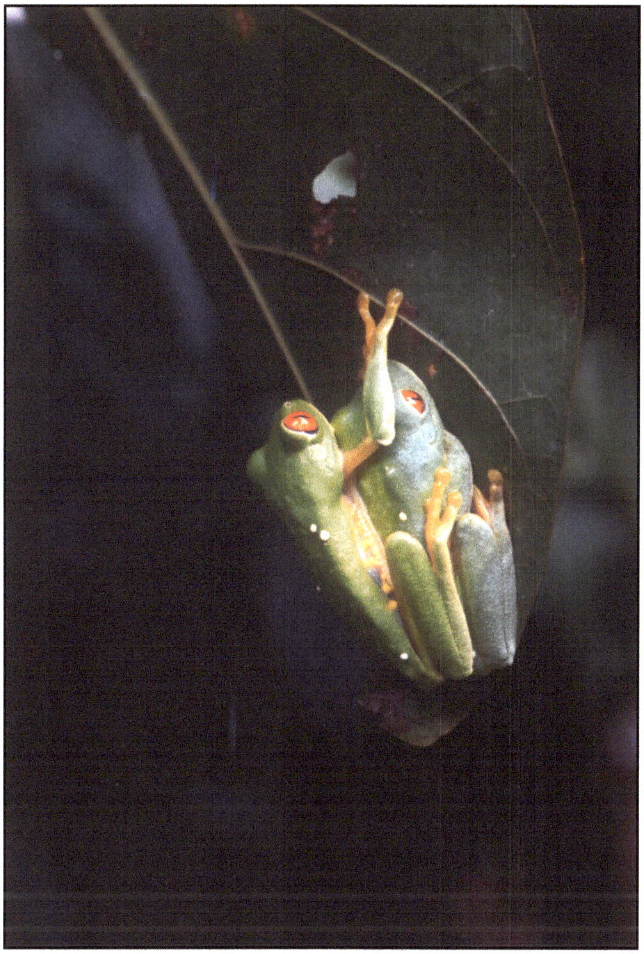

Red-eyed Treefrogs mating

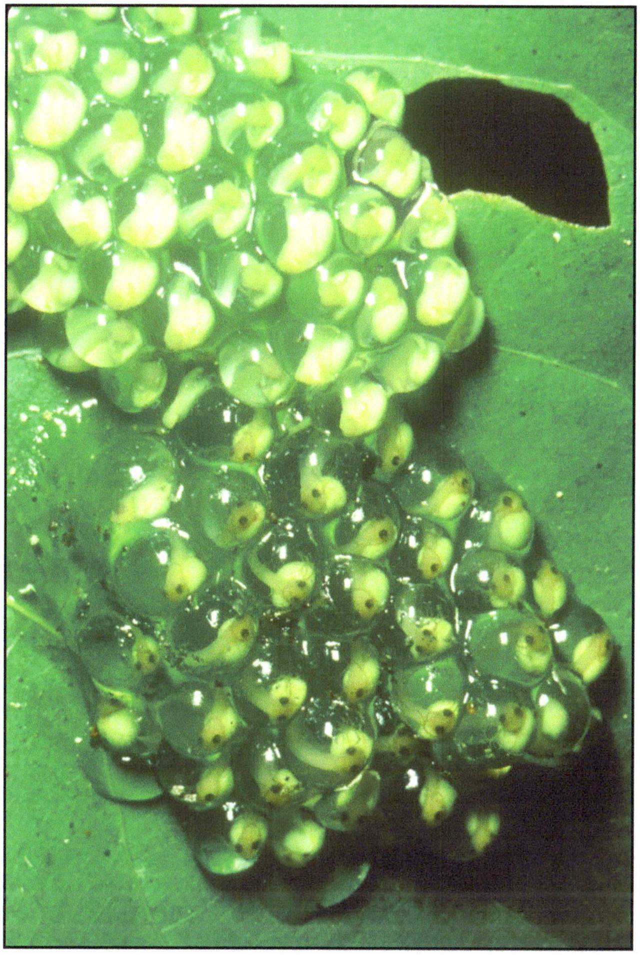

Red-eyed Treefrog spawn, lower ones more mature

Shortly after the onset of the rains, the male red-eyed treefrog sings his territorial song from an area above the temporary pool. At dusk he descends and changes to his mating call. The female answers by approaching close in front of him. The male then mounts her and tightly clamps on with his forelegs. This is known as pectoral amplexus. As one, they climb to a position over the pool. By this time she has absorbed rainwater through her skin. At the critical time she lays a small mass of eggs while releasing the stored water over them. The male then releases his sperm. The spawn, a pale jelly enclosing about forty light green eggs, adheres to the leaf. In about five days, the movement of the rapidly developing tadpoles causes the disintegrating membrane to slide to the end of the leaf tip where one by one they fall into the water below. If all goes well, the pale blue tadpoles will lead a vegetarian life for ten weeks. Then they will leave the pond as tiny froglets and feeding on small insects, slowly making their way up into the forest canopy.

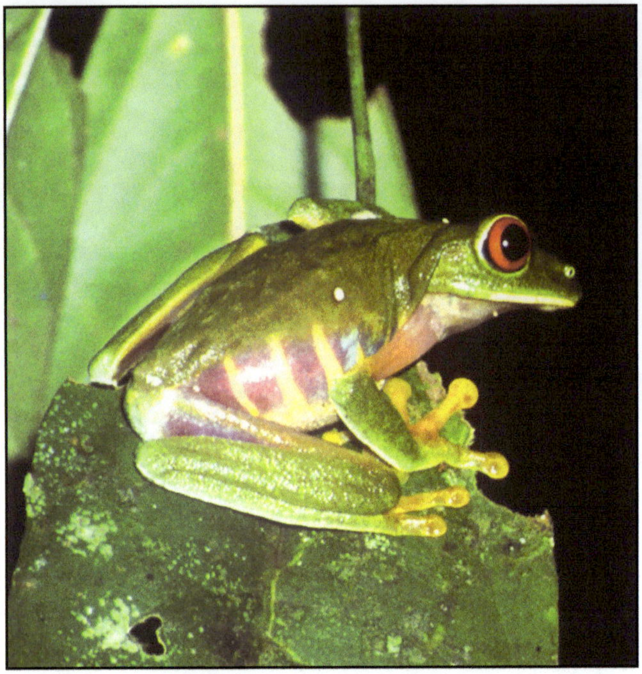

Breeding Red-eyed Treefrog,
note belly engorged with water

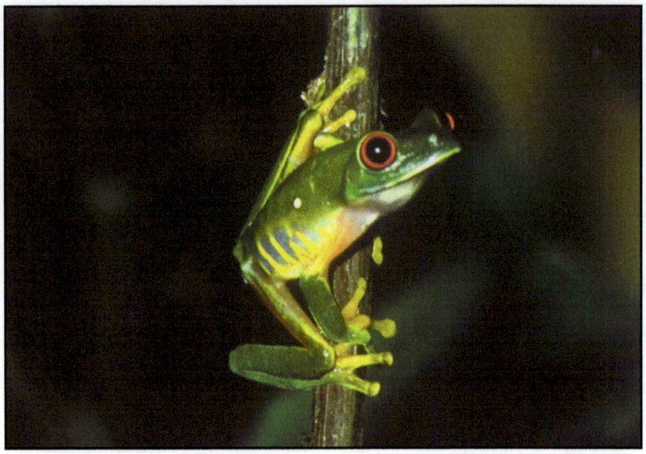

Red-eyed Treefrog at night, note large pupils

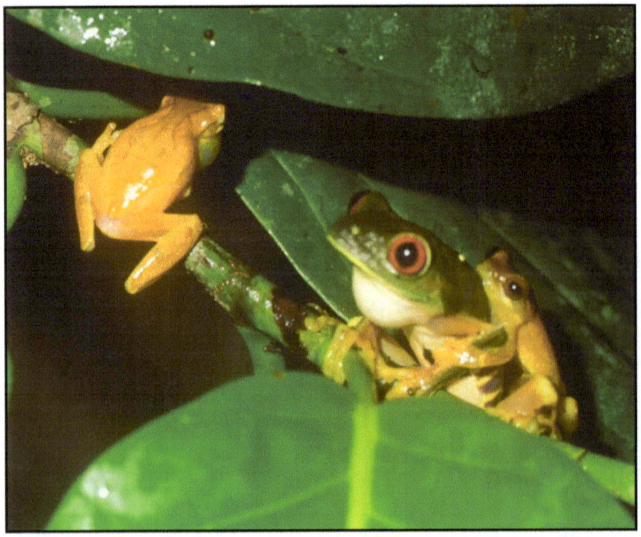

Yellow Cricket Treefrogs (*Hyla microcephla*)
and Red-eyed Treefrog

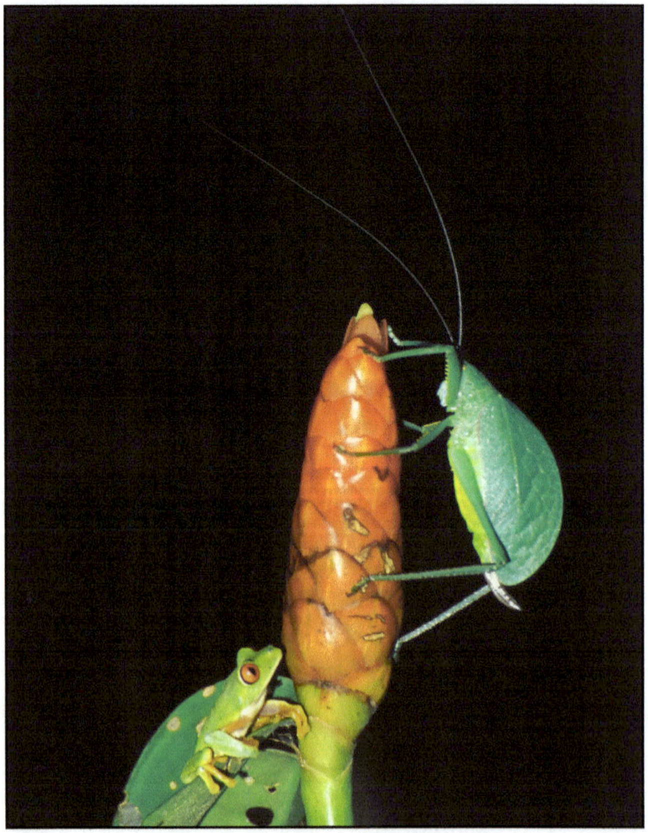

Red-eyed Treefrog and Katydid

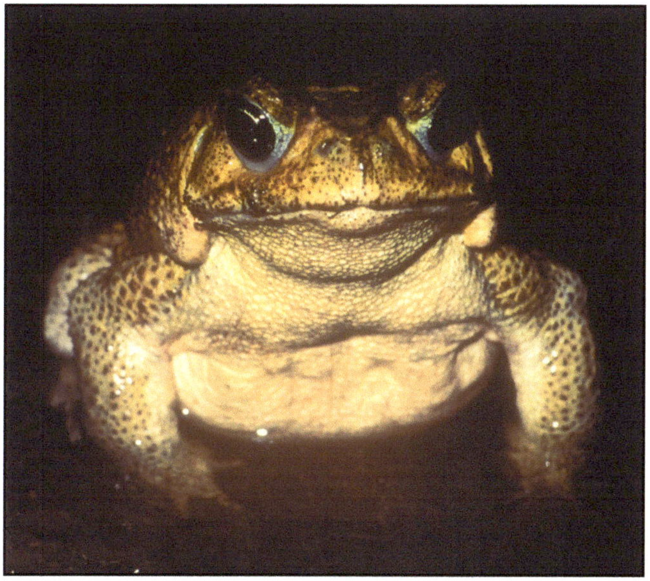

Marine Toad (*Bufo marinus*)

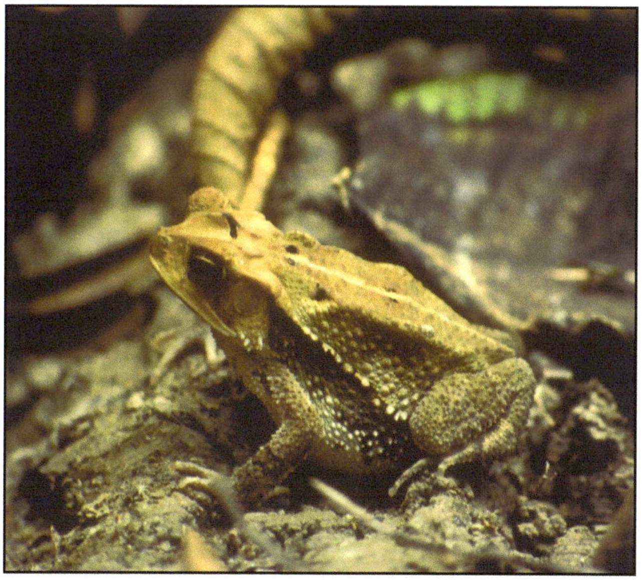

Gulf Coast Toad (*Bufo valliceps*)

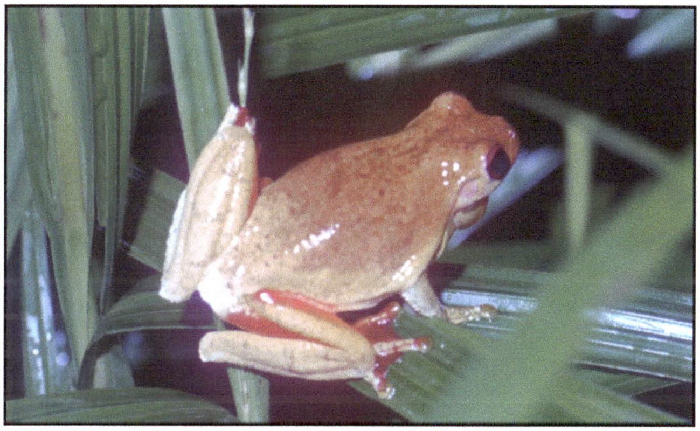

Red-footed Treefrog (*Hyla loquax*)

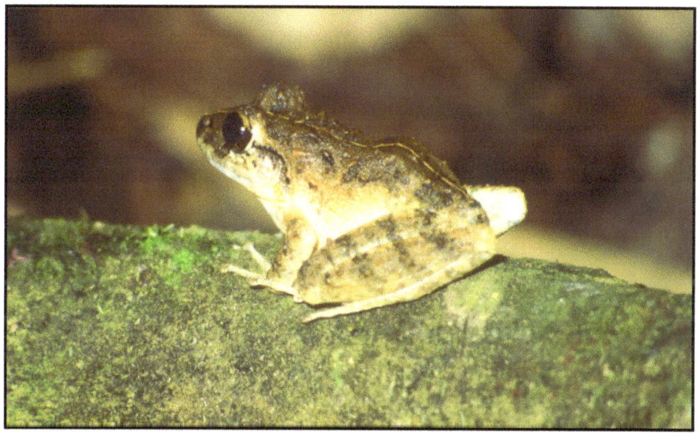

Central American Rainfrog (*Eleutherodactylus rugulosus*)

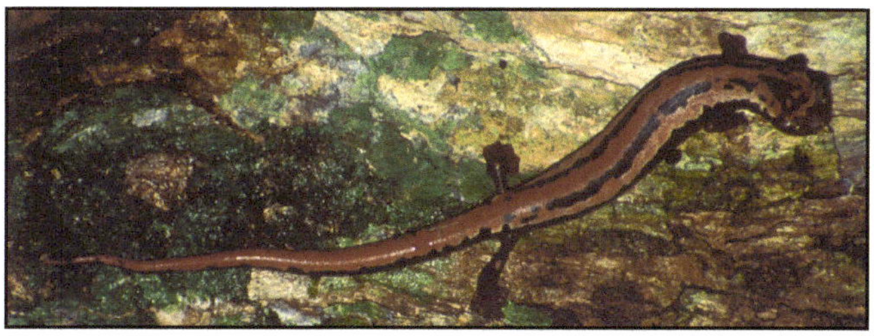

Mexican Mushroom Tongue Salamander
(*Bolitoglossa mexicanai*)

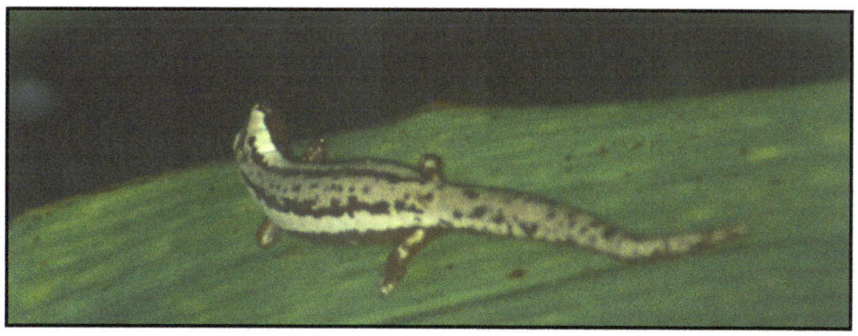

Common Dwarf Salamander
(*Bolitoglossa rufescens*)

Treefrogs have many predators including snakes, jungle cats, and even bats. The fringe-lipped bats can actually locate frogs by zeroing in on the male's call. When treefrogs are threatened, they can leap as far as thirty times their body length. As the red-eyed treefrog takes off the predator is presented with a flash of its colourful flanks. But when the frog lands on a leaf, its legs are drawn in and only the leaf green colour shows, puzzling the predator. This leaping defence is also advantageous because it breaks the scent trail, which is so important to frog-eating mammals and snakes. Earlier that day I had witnessed the diurnal green tree snake partaking in a feast of a Mexican treefrog. These treefrogs tend to be active in daylight when mating.

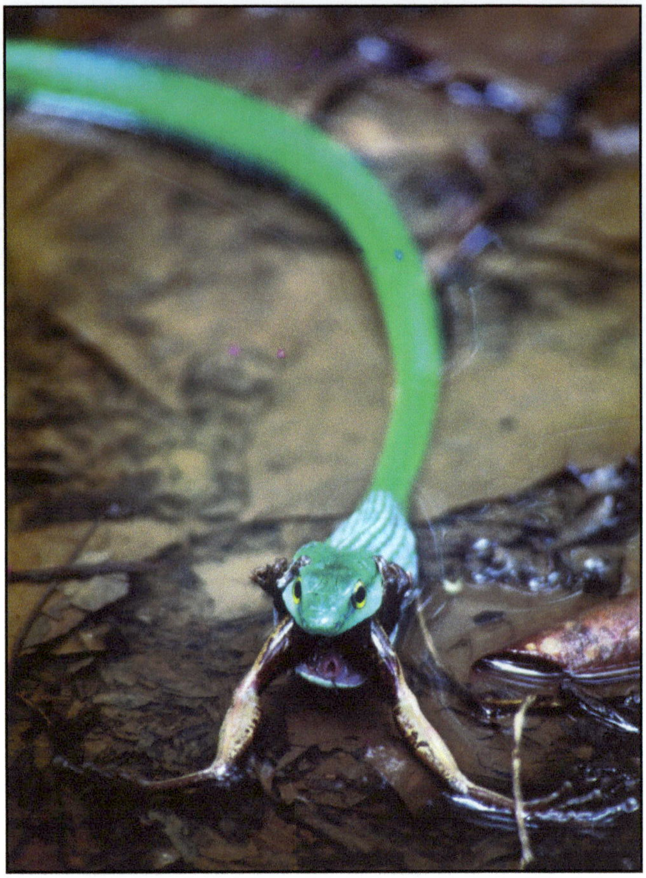

Green Tree Snake (*Leptophis ahaetulla*) swallowing a Mexican Treefrog

Another time I was relaxing with a book on the porch of La Milpa, one of two large recreation areas in the Rio Bravo Conservation and Management Area, owned and operated by Programe for Belize in northern Belize. My attention was drawn to a large pale treefrog that jumped up onto the deck about four feet from me. I had no sooner focused on the treefrog, when it took off with another leap. This time it landed on the leg of one of two large tripods, each bearing a telescope, for the use of bird watching guests.

I watched as the frog quickly scaled the tubular leg until it reached the telescope. There it sat, looking around for its next port of call. As the frog went over its options, another actor arrived on the scene in the form of a four-foot-long speckled racer, a snake that delights in dining on amphibians.

The snake made its way directly to the tripod, then started climbing, but it had a problem. It couldn't exert sufficient pressure to grip the tubular aluminium leg without sliding down. If at first you don't succeed, then try another method. Next time it climbed up the leg only until it reached the angled brace, then it stretched across the brace and started on the upper section and upwards to the telescope itself.

Meanwhile, the frog was watching worriedly. Fortunately the racer had reached the other side of the scope platform and was looking in the opposite direction. The frog didn't wait for the snake to correct his mistake and made another leap. This time to the dining room wall, about 30 inches away, then quickly scaled the vertical wall until he was twelve feet high, grinning down on the hungry snake.

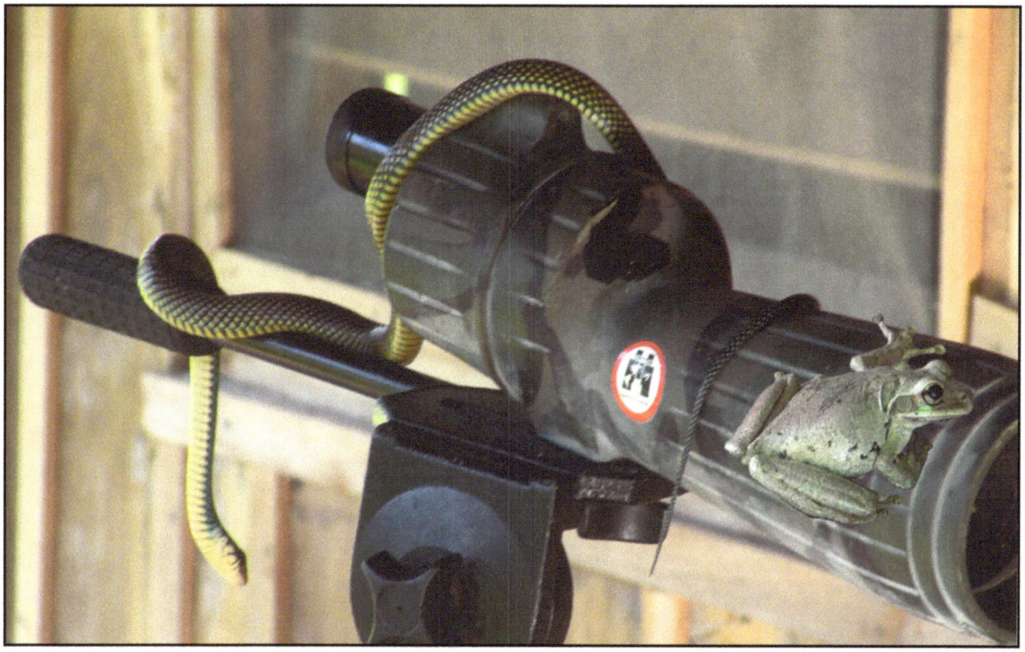

A Mexican treefrog escapes a speckled racer.

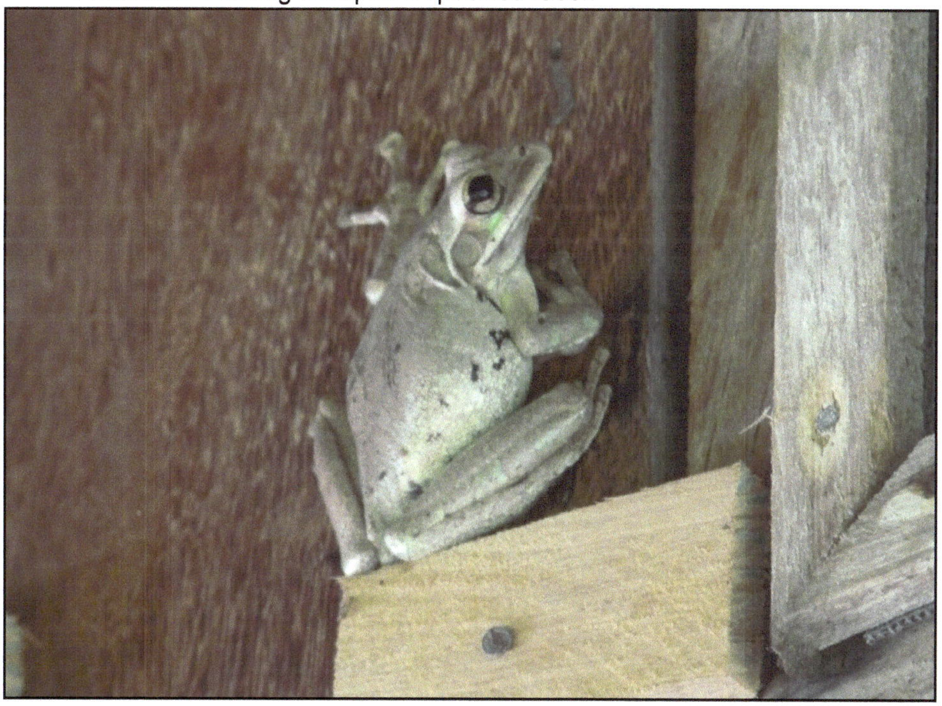

Snakes

Having had my fill of photographing treefrogs for the moment, I directed my spotlight to the upper foliage. I was interested in filming predators who were taking advantage of this amphibian orgy. I was looking for the nocturnal cat-eyed snake which, although normally a frog eater, tends to ignore them in favour of frog spawn during breeding time. I followed a cat-eyed snake as it slowly and deliberately examined each leaf in its path. When it came across frog spawn, it studied the spawn from every angle before daintily swallowing a small portion at a time.

Standing in knee deep water, I found it difficult to focus on my subject. Each time I raised my 300mm lens, a raindrop immediately smeared it. As I cleaned the lens my breath created an instant haze that I didn't realize until I tried to refocus. That night I witnessed three cat-eyed snakes before the rains became torrential, forcing me to retreat to a drier environment, promising to return.

Early the next morning, I found a speckled racer in the final stages of swallowing a Mexican treefrog. It took another half an hour. Then I caught a glimpse of movement above my head. It was the extremely slender, twig-like snake, Neotropical vine snake, also known as a tie-tie snake (*Oxybelis aeneus*). The tie-tie snake is also a frog eater. The interior of their mouth is purple to black in colour.

Belize has approximately sixty species of snakes which range from the 15-inch *Adelphicos quadrivigatus* which feeds on earthworms, to the 12-foot boa constrictors that will feed on anything it can overcome. There are eight venomous snakes in Belize that are dangerous to man.

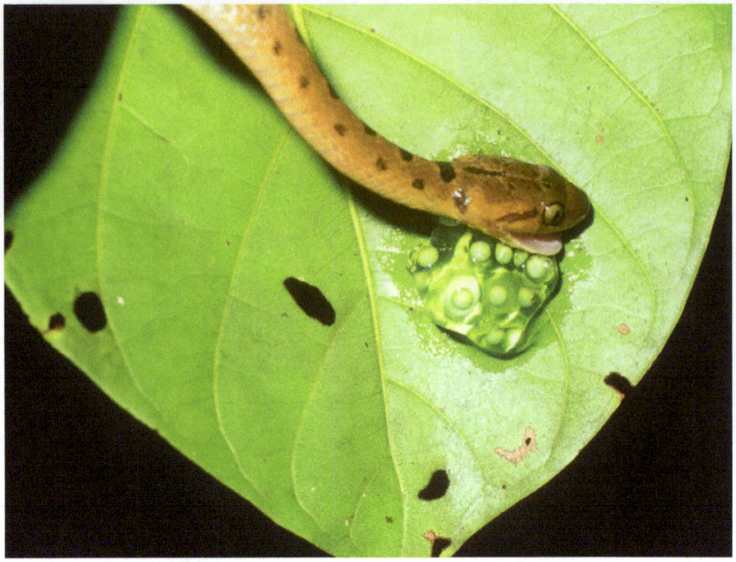

Cat-eyed Snake (*Leptodeira septentrionalis*) eating red-eyed treefrog spawn

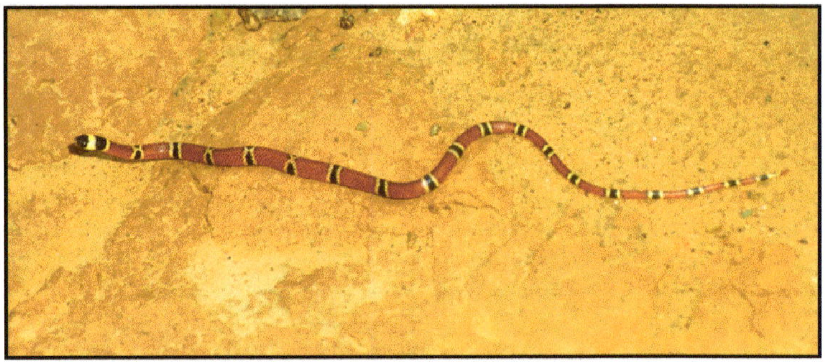

False Coral Snake (*Pliocercus elapoides*)

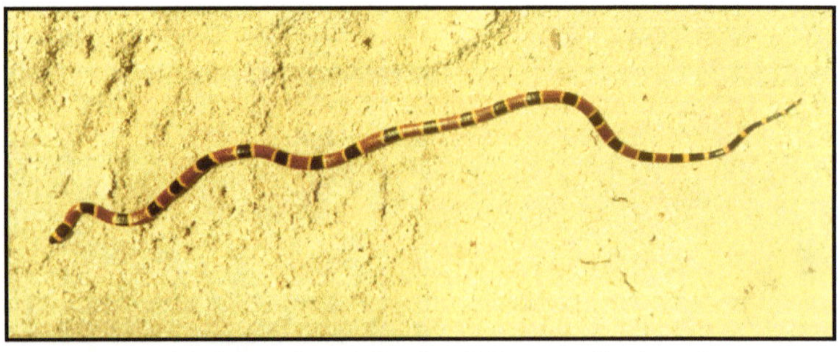

Variable Coral Snake (*Micurus diastema*)

The two coral snake species have an exceptionally strong neurotoxin that will cause a breakdown of the nervous system. But coral snakes would rather hide than fight. They are beautiful snakes that spend much of their time hiding under dead bark, deadfall and even in the interior of leaf-cutting ant nests. On night walks they may be seen on the trails.

The other six dangerous snakes are all pit vipers. The venom of pit viper attacks the blood system and related tissues. Both coral snakes and pit vipers can be deadly and medical attention should be sought as soon as possible. Of the six vipers in Belize, by far, the most dangerous is the fer-de-lance, or the yellow-jaw tommygoff. Their danger is not necessarily because of their potent venom, but rather the aggressiveness of the snake itself.

I have had several experiences with venomous snakes, and more by luck than common sense, have escaped injury. A short walk from the Cockscomb Basin Wildlife Sanctuary Park Headquarters is South Stann Creek which I enjoyed swimming down. On the left is a small tributary that I followed back to the Currasow Trail. On this particular occasion, I found a dead red brocket deer in the creek. The animal had been bleeding from its mouth, nostrils and around the eyes, a sure sign of a pit viper bite. I returned to the Park Headquarders and told the wardens about the dead deer. They were interested in making a display.

Six month later I returned during the June rains and, like before, I swam down the river. On the way up the creek I stopped to check on the deer's bones. As I was feeling around for the bones, a movement on the bank caught my eye. It was a large fer-de-lance, and I could see that it wasn't in a good mood. Its eyes were white and skin was in tatters. Every year or two all snakes shed their skin and emerge as a new handsome snake. But during

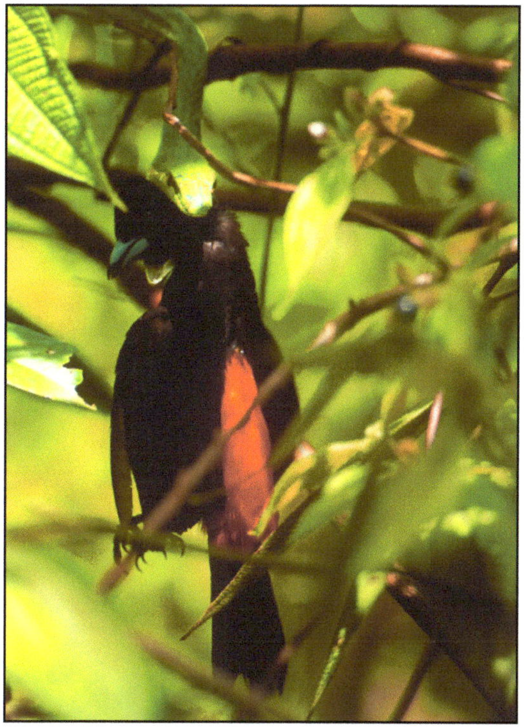

Green Tree Snake (*Leptophis ahaetulla*) eating a Passerini's Tanager

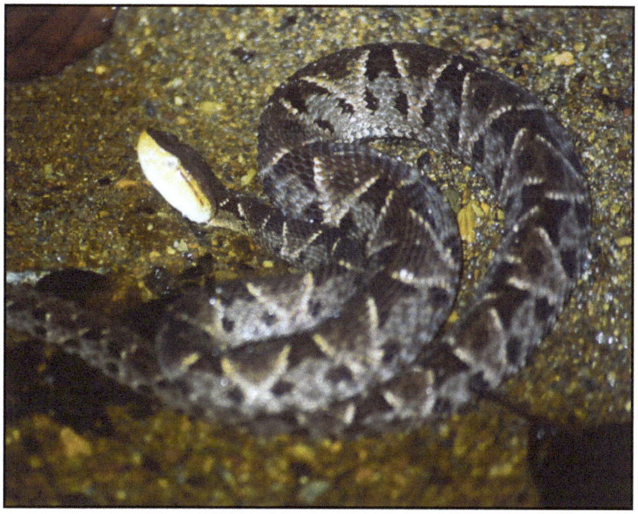

Fer-de-Lance by the edge of the river (*Bothrops asper*)

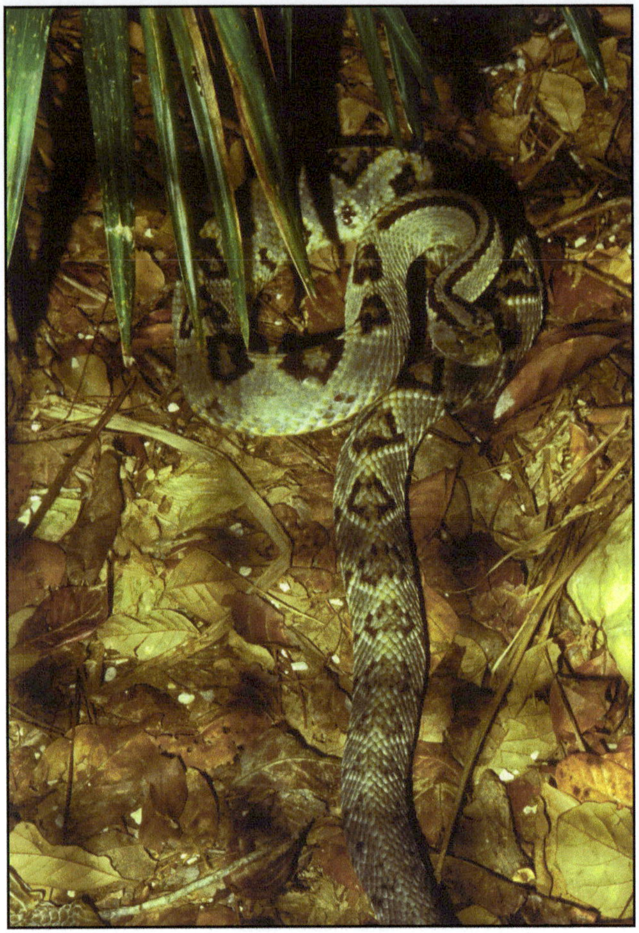

Yucatan Rattlesnake (*Crotalus durissus*)

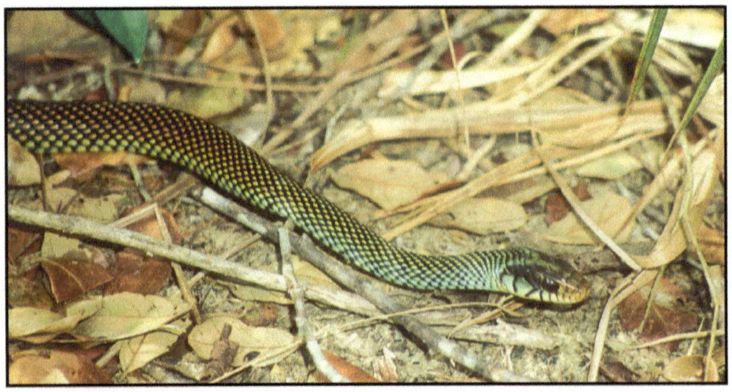

Speckled Racer (*Drymobius margaritiferus*)

this period, they are blind with dead skin over their eyes and invariably have a very bad disposition.

I was standing in about thirty inches of water, and immediately began to shuffle backwards. The snake took to the water and, sensing where I was with its darting tongue, headed towards me. My feet were bare and the creek-bed had large round boulders. The snake was gaining. I changed my tactics. I stood up and started splashing water in its face. It turned upstream, and I increased the deluge I was throwing in its face. It gave up and left the creek where it had entered and kept going until it reached a large tree and disappeared into the root system.

My mind went back to the previous January. The brocket deer had been killed by a fer-de-lance. No doubt it had a cosy little home in the root system of the tree. I don't believe I've swam up that creek since!

On another occasion I had another snake experience in Cockscomb. This time I was doing a night walk with a tourist on the Gibnut trail. On night walks, I always carried two headlamps, for there is nowhere as dark as a rainforest on a moonless night. A spare might never be used, but it is nice to know it's in your pocket.

We had walked about ten minutes, when my light picked up a large snake in the middle of the path ahead. My first impression was that it was a boa constrictor for it looked to be about eight feet long and heavy bodied. It was still about ten or twelve feet away, so I shone my bright light on it. It didn't like this and immediately moved fast towards us. When it was eight feet away, I realized that it wasn't a boa and with a push, told the tourist to run. He was slow to start, but he did eventually move, I dropped my headlamp to the ground and followed my companion, pulling out my emergency light as I ran.

Early the next morning, when I picked up my headlamp, the snake was nowhere to be found. I don't know if abandoning my light was the right move, but since then, I've heard of at least three occasions when a tommygoff went for a headlamp.

My last encounter with a fer-de-lance was very brief. I was trying to locate Red-capped Manikins in the canopy and walking back and forth with my eyes to the heavens. I never did find the manikins, but I did notice a twitching snake, belly up in the middle of the path! I stopped and turned it over with a stick. It was a fer-de-lance, only about a foot and a half long. I was so concentrated in finding my manikin, that I had stepped on the fer-de-lance's head and, without realizing, I had broken its neck!

To this day, I still wince when watching a group of birders wandering along a trail, their eyes to the skies, and their feet ploughing through dead leaves. The fer-de-lance is by far, the most aggressive and dangerous snake in Belize, and even a juvenile can be deadly!

The boa constrictor is the second largest snake in the Americas, but north of Panama, a boa rarely exceeds twelve feet in length with a weight of sixty or seventy pounds. Boa constrictors are relatively common in Belize, but they are no danger at all to man. It seems to me that boa constrictors are more common in the Cayes, than on the mainland.

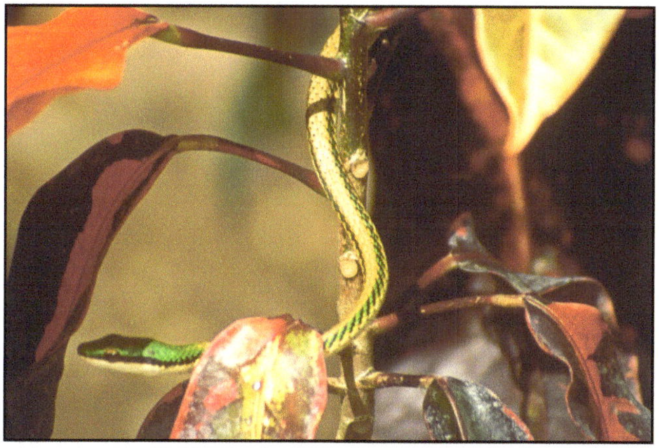

Green-headed Tree Snake (*Leptophis mexicanus*)

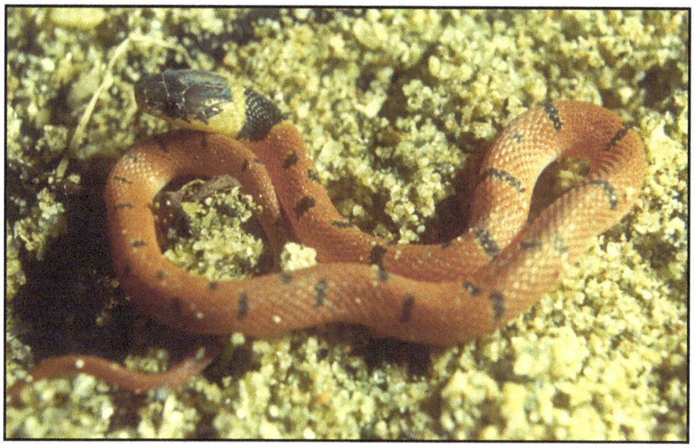

Red Coffee Snake (*Nina sebae*)

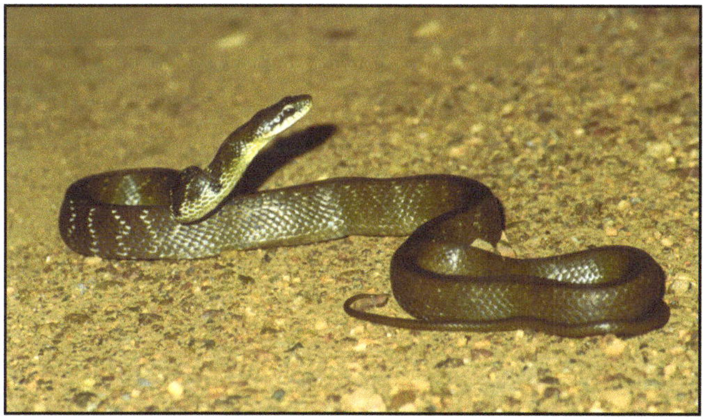

Tropical Whipsnake (*Masticophis mentovarius*)

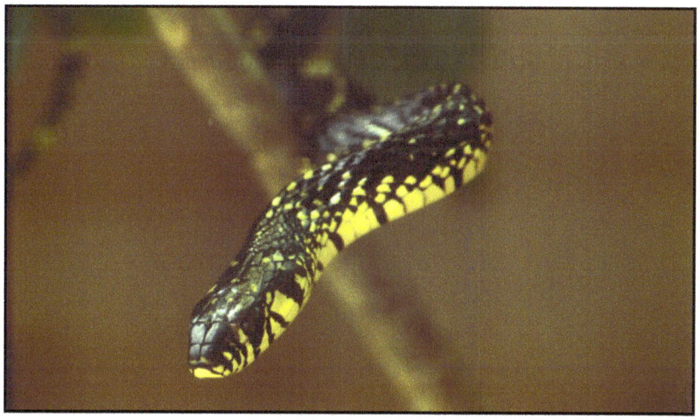

Tropical Ratsnake (*Spilotes pullatus*)

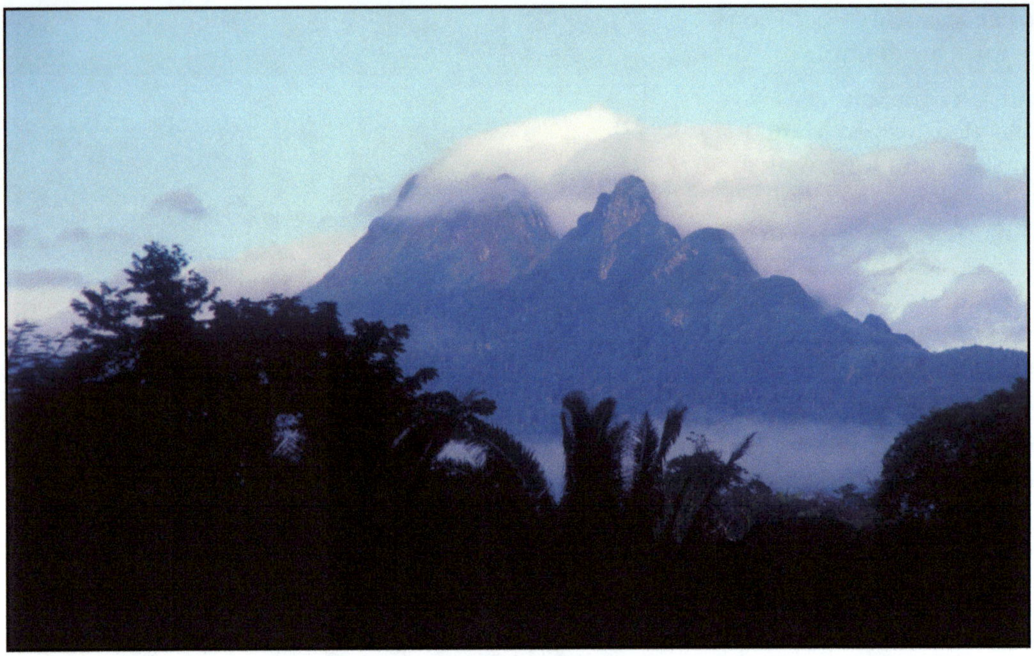

Victoria Peak east side (*above*), west side (*below*)

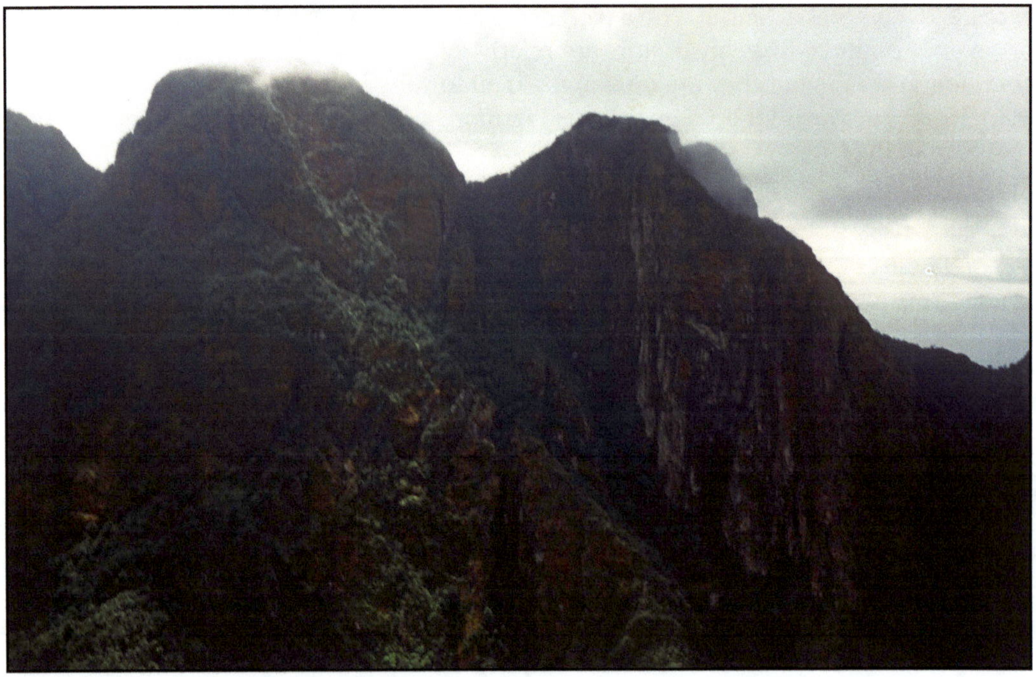

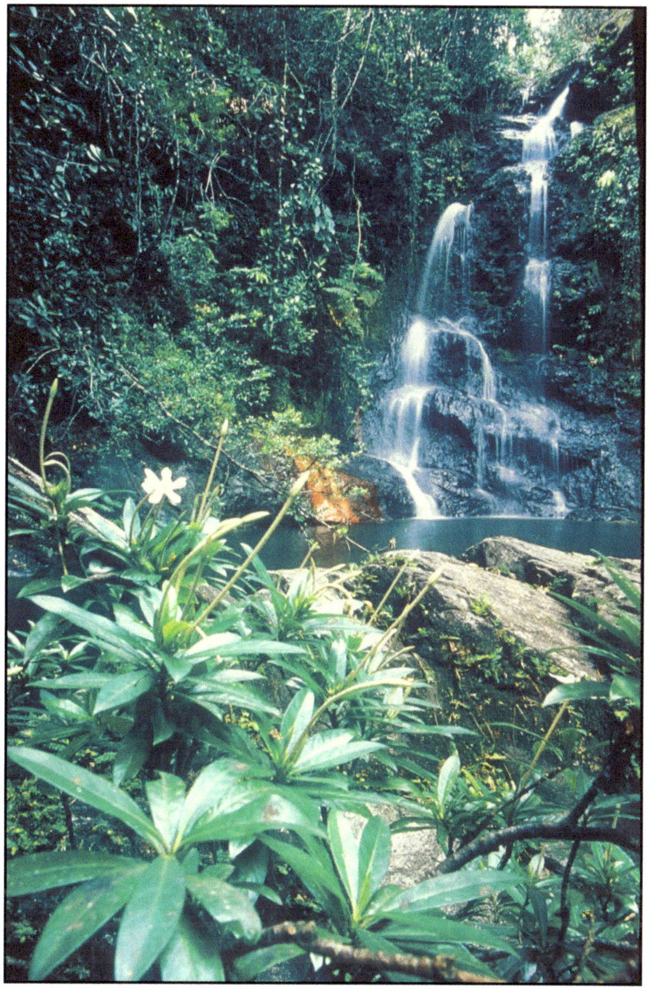

Upper Tiger Fern Falls at Cockscomb

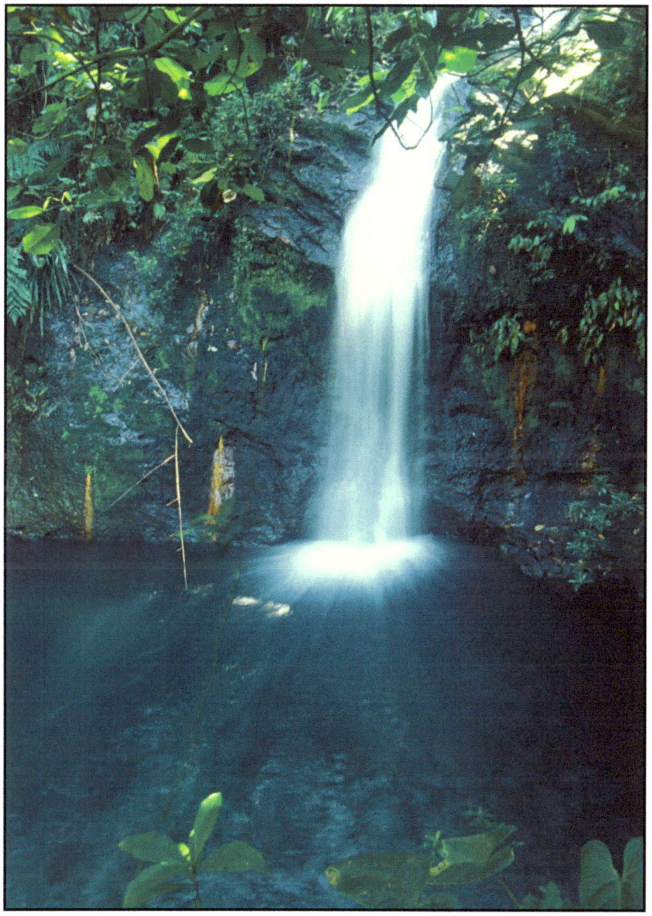

Lower Tiger Fern Falls at Cockscomb

Cockscomb

Cockscomb Basin Wildlife Sanctuary was first established as a sanctuary in 1986. It was a long hard struggle which began in March of 1983 by a young zoologist by the name of Alan Rabinowitz who was connected to the New York Zoological Society. Alan's specialty became jaguars, and for the following two years, he trapped and studied Cockscomb's big cats, using radio collars.

When the Government of Belize first asked the Belize Audubon Society to manage the Sanctuary, it was only 3,648 acres. This was nowhere near sufficient to protect the number of jaguars that Rabinowitz had established existed in the area, so in 1990 the area was expanded to 102,000 acres. Seven years later it was once more increased by another 25,000 acres. The Cockscomb is now a well established park with thousands of annual visitors passing through its gates, many of whom hope to see a jaguar.

But there are plenty of tamer pursuits, such as the colourful forest birds, butterflies, and, of course, treefrogs. With the exception of squirrels, brocket deer, peccary and howler monkeys, mammals are hard to see in a rainforest.

For the past twenty-four years, Cockscomb has almost been my second home. As a wildlife photographer, it matters little what species I come across. A bug, a bird, or a boa, I find them all interesting! I have been luckier than most regarding wildlife sightings, possibly because I take my time and don't walk continuously. If I come across a rest area, I use it. If I remain motionless and quiet, it is amazing how much wildlife will come to me.

For example, about twelve years ago, I was walking along Cockscomb's Wari Trail. I stopped at a pool, a former ox-bow. It is quite picturesque with kaway trees along the opposite bank. I sat down on the grassy bank with camera on hand and waited.

I didn't have long to wait. Within minutes a pygmy kingfisher alighted on a kaway root. After a few minutes it dove into the water; then emerged with a small fish in its bill.

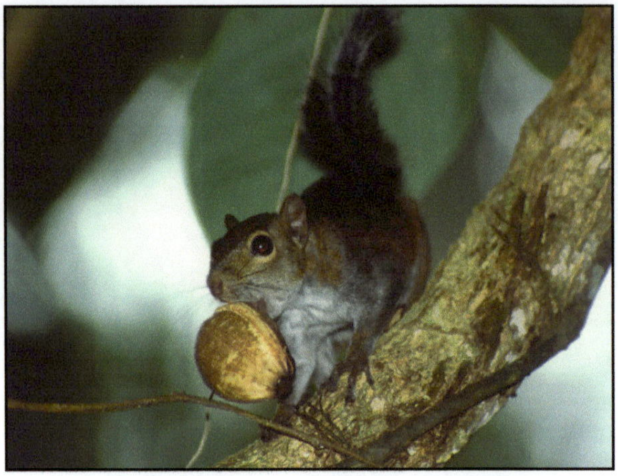

Deppe's Squirrel (*Sciurus deppei*)

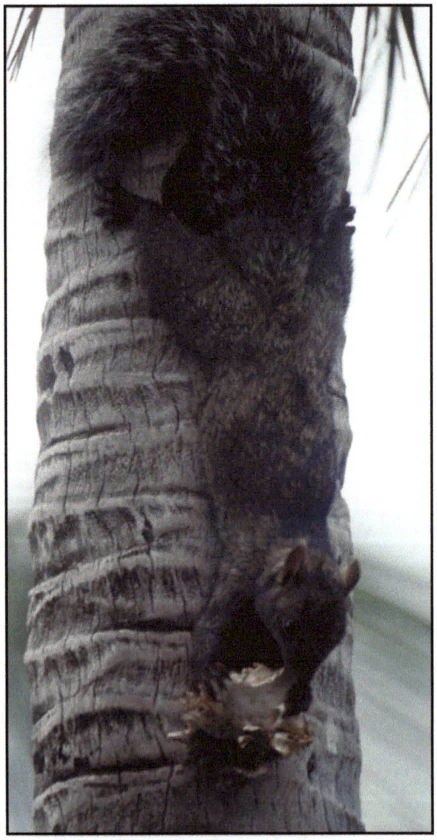

Yucatan Squirrel (*Sciurus yucatanensis*)

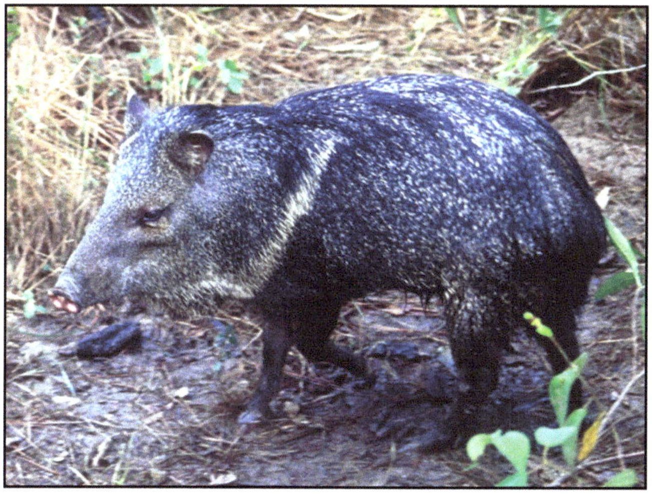

Collared Peccary (*Tayassu tajacu*)

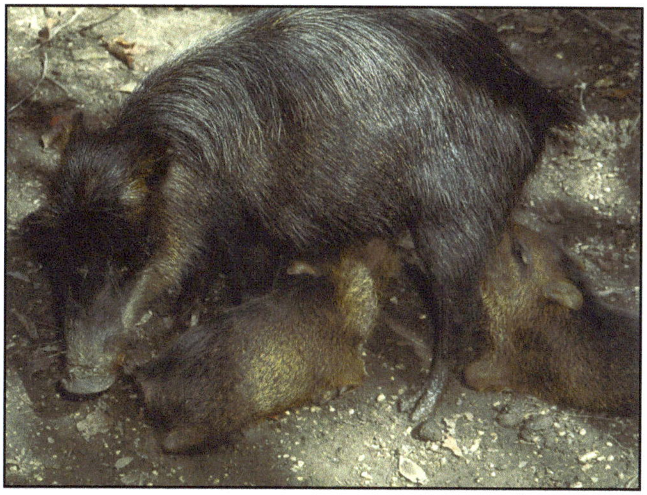

White-lipped Peccary (*Tayassu pecari*)
— The Belize Zoo

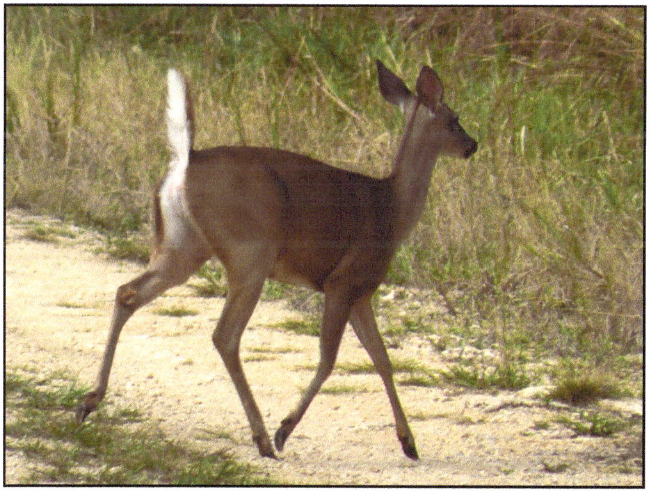

White-tailed Deer (Odocoileus virginianus)

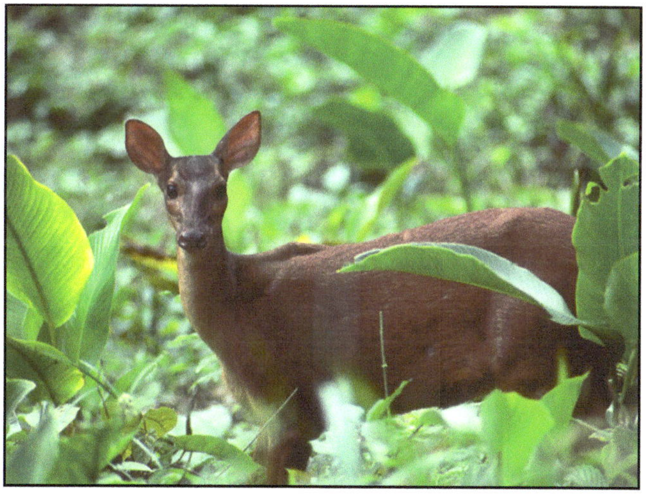

Red Brocket Deer (Mazama americana)

After taking photos, I turned off my flash. Unfortunately, that was my mistake. Seconds later, I noticed movement about eight feet to the left of the kingfisher. A smallish mammal was emerging from a fissure in the kaway roots. It slowly moved out, and glided over the buttress roots just above the water surface. I checked my camera; the flash had still not recycled!

There was no choice but to wait and watch. It was a rich reddish-brown, with white facial markings. The tail was about eight inches long and of the same coppery colour as the body, except for a black tip. Its belly was a lighter tan and the legs were very short. I recognised it as a weasel. The battery ready-light lit up just as its tail disappeared in between the roots. Later I checked my reference book to find it was a long-tailed weasel, a new species for Belize!

The more time spent in the forest, the higher the chance of seeing wildlife. Black howler monkeys are usually found close to rivers. They broadcast their presence by their extremely loud, even frightening cry. Peccary are usually heard before being seen. There are two species of peccary. The smaller of the two is the collared peccary, which usually weighs no more than 45 pounds and has a head-body length of about 35 inches. They are always found in groups. My own experience in Cockscomb is that they have either ignored me or run. The other species is the white-lipped peccary, which is much bigger than the collared-peccary and can weigh closer to 80 pounds. It has a reputation for fierceness, though I have never experienced it.

We have two deer in Belize, the red-brocket and the white-tailed deer. The red-brocket deer is a delicate little animal, quite suited for passage through dense forest. It is reddish-brown in colour and can be seen quite commonly in Cockscomb's campground. The white-tailed deer is more commonly seen further north and near open areas.

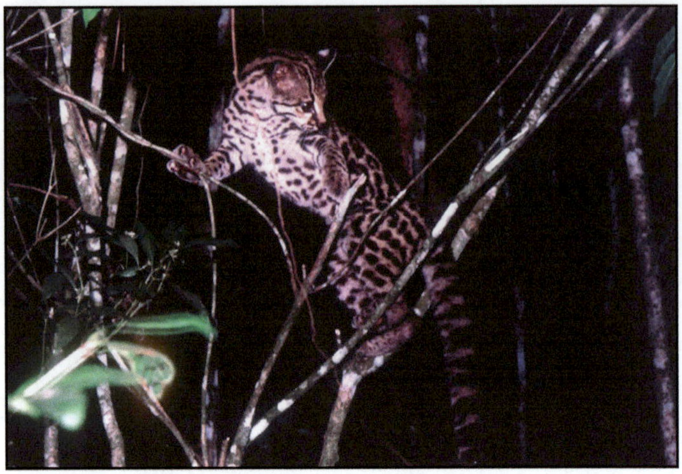

Margay (*Felis weidii*)

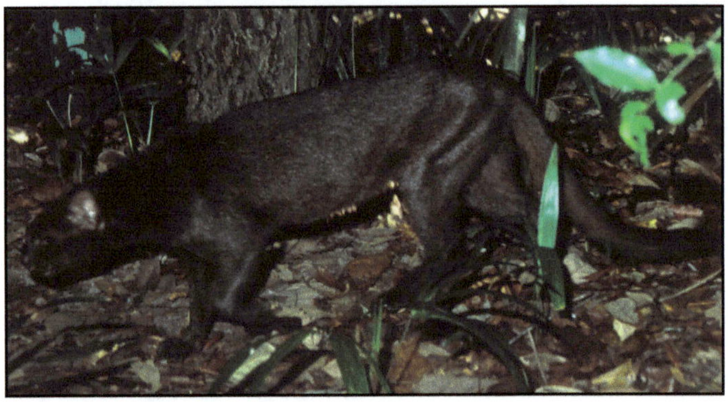

Jaguarundi (*Felis yagouaroundi*)

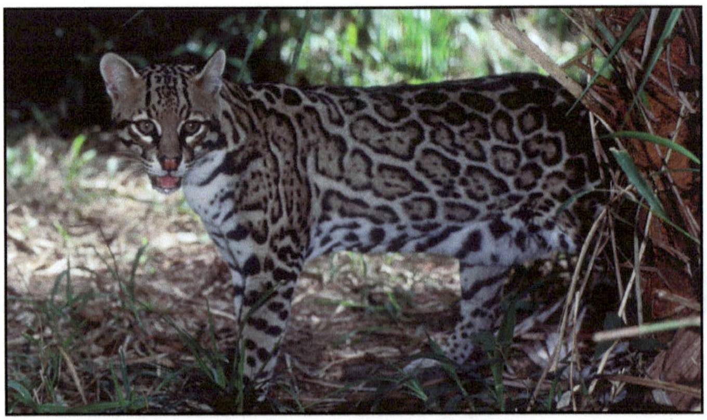

Ocelot (*Felis pardalis*) —The Belize Zoo

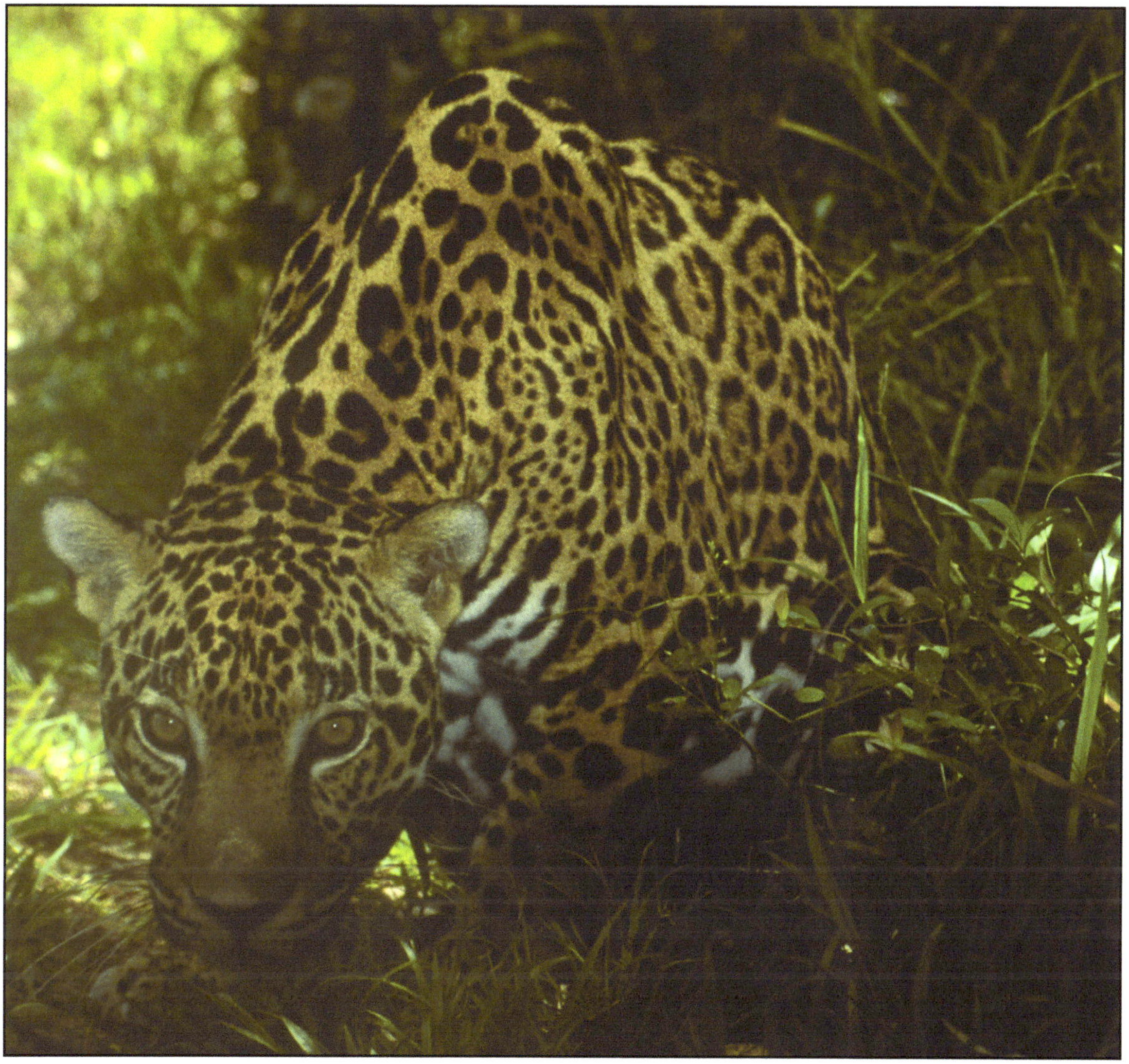

Juvenile Jaguar (*Panthera onca*)

Forest Cats

Belize's carnivores are seldom seen and most especially the smaller ones, such as the margay, the jaguarundi and ocelot. Only once have I photographed a margay in the forest. Several times I have seen jaguarondis close to and crossing roads. I have only seen an ocelot one time in the wild. On the other hand, I have seen jaguars seven or eight times in Belize forests and pumas twice.

One night about 10 pm I was leading a group of about three persons on a night walk along Victoria Peak Path. We had barely been walking five minutes

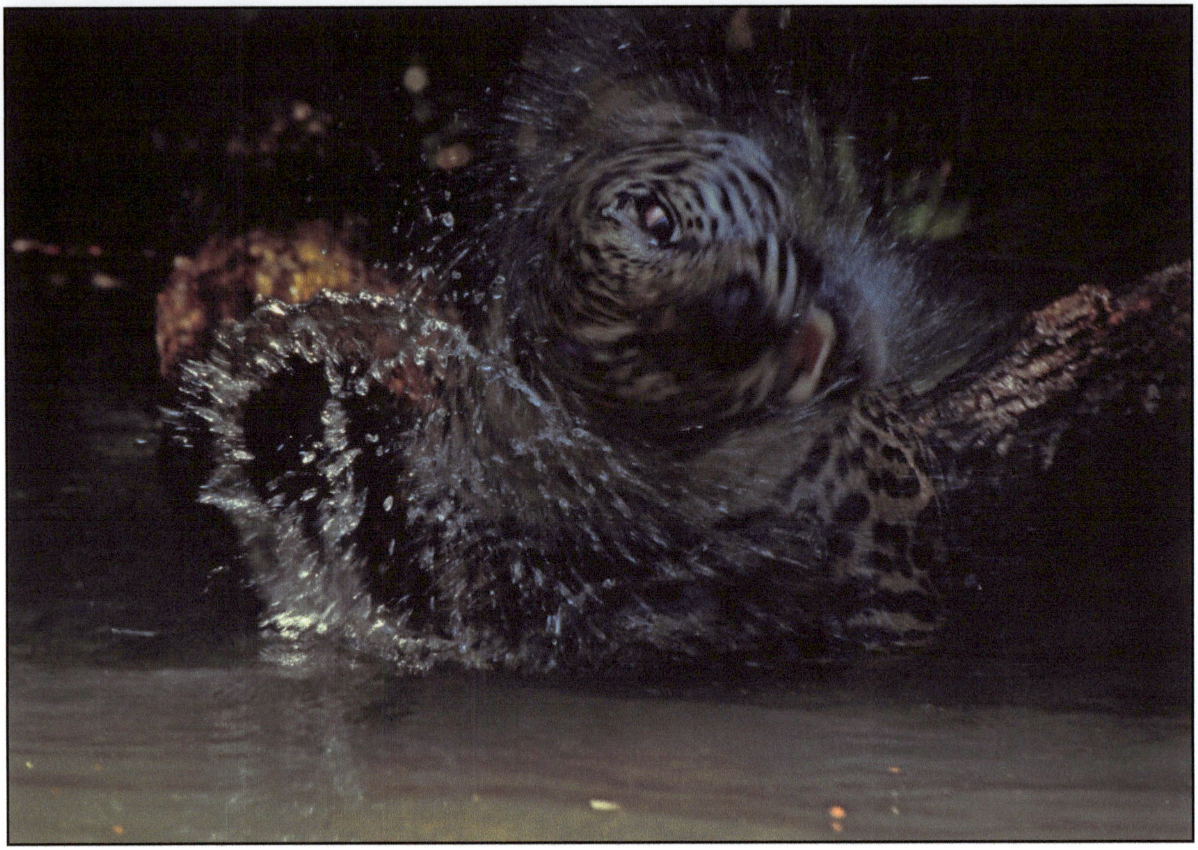

Jaguar bathing (*Panthera onca*) —The Belize Zoo

when we approached the bridge on the path. I was in lead position when my headlamp picked up eye shine on the bridge. I halted my group and slowly moved ahead on my own, sans light. I stopped at what I believed was a safe distance away and put my headlamp back on---then off again!

It was a jaguar! He was a little off centre on the bridge, hunkered over, in the act of relieving himself. I explained to the group that we had to wait a few moments. I guess I could have used the stool sample as part of my spiel, and showed the group what the jaguar had ate last, but it was still steaming in the cool night air, so I suggested that we wait until our return trip. But when we returned forty-five minutes later, the remains were all but gone, and several large beetles were cleaning up the last remaining crumbs.

Another time in Cockscomb Basin I was alone on the Victoria Peak Path at the last junction of the Warrie loop. As I was approaching the left turn into the Warrie loop, I noticed some movement of the lower section of a large bush that made the southwest corner of the trail junction. I stopped and waited and saw another set of twitching leaves. I walked a little closer and there was heavier movement in the bush, which was accompanied by a low growl.

I backed away and moved into the Warrie Trail and curved around the bush which was now about twenty-five feet away. From here I could see a young jaguar, probably not much more than fifty pounds. It was alone as far as I could see, but I could not be sure that its mother wasn't snoozing a few yards away. As it quietly watched me, I took a photograph , then I turned and walked slowly down the Warrie Trail.

I had several other jaguar sightings in Cockscomb and two in La Milpa, one on the trail to the ruins, and the second was sleeping on the access road to Gallon Jug. But most controversial sightings were of black jaguars, the melanistic variety. At 8:25 am on April 25th, 2002, I was driving out of Cockscomb with my friend Rad, his sister Margie, and his aunt and uncle. We were about two miles beyond the park gate when we saw a large animal climb up onto the road on the left side, a little less than a hundred yards ahead of us.

It was obviously a large cat, but it was pretty dark and not easy to identify. As we slowly coasted toward it, we saw that it was a young black jaguar. It walked to the centre of the access road and stopped almost exactly in the middle. We let the vehicle roll towards it until we were close enough to examine it in detail. It was broadside to us with its head turned watching us with no sign of fear. Its tail was lifted above horizontal, so from the lack of evidence to the contrary we concluded it was a female. She decided we were close enough and continued to walk to the embankment where she disappeared into the bush.

We all agreed on the species and the sex. We all saw the black rosette pattern that was visible on its charcoal grey fur, indicating that she was a black jaguar. In Maya Center, we stopped to report the incident to the park director, Ernesto Saqui. Unfortunately, Ernesto was not home, but we did speak to his wife Aurora, who told us that a tourist had already reported the same thing, but Ernesto had difficulty believing the tourist's statement. In the following months, several persons claimed to have sighted the black jaguar, both tourists and locals. Ernesto's brother, Lindo Saqui, drove his truck off the road to avoid hitting one.

But all reports were doubted by Bart Harmesen, the Wildlife Conservation Society jaguar researcher in Cockscomb. For years the black jaguar debate continued. I had numerable discussions with Bart, but he was adamant that no black jaguars lived north of the Darien Gap! Bart worked with jaguars using camera traps in Cockscomb and he had never caught an image of a black jaguar. But still, the odd tourist reported seeing a large black cat-like animal.

In April of 2008, I was walking the Tiger Fern Trail when I met a middle-aged tourist couple from Canada. They were visibly excited and told me that they had just seen a black jaguar, only minutes before, crossing the trail. Two days later, two very excited teenage girls ran up to me and asked if I was the man who had seen a black jaguar. They had been returning with their parents on the Tiger Fern Trail. As they approached the main access road, they had the thrill of a lifetime!

The girls were ahead of their parents by several yards, and separated from each other by a couple of yards. As the leading girl became level with a tree, a black jaguar jumped down and ran ahead of her.

She screamed out to her sister, "Look, look, look!" Her sister replied, " No, you look, look." The first girl turned around in time to see, two black jaguar cubs running towards her, and onward, after their mother. The girls' worried parents came along and saw all three cats.

That same afternoon, I checked out their story. I found the tree with several scratch marks on the trunk and disturbed dirt on the trail. I waited up for Bart to return that night and told him what had transpired that day—not only a black jaguar sighting, but a black jaguar with two black cubs!

I could tell by his demeanour, that he couldn't bring himself to believe this. Anyway I suggested that he put out a camera trap in the area, and told him that

I would be walking that trail early morning for a few hours in the hope of seeing and photographing the mother and cubs.

Early the next morning I walked back and forth on the first mile and a half of the Tiger Fern Trail a couple of times. The third time I had more or less given up and turned off my camera flash. Then I heard rustling of dried leaves to my left. I looked but nothing was visible. Then I caught a glimpse of something moving and it seemed to be moving in a direction that would bring it across my path, about thirty feet or more ahead of me.

I moved ahead about ten feet and stood behind a large tree. It was the black jaguar, but without the cubs. When she was fully in view, I pressed the shutter release; there was a click, but no flash. I quickly I turned the flash on ... three, four seconds and the ready light came on. However in that four seconds, the jaguar had moved across the path. I pressed the shutter release as she moved into the jungle on the other side. This time there was a flash, but I knew that she was beyond the flash limit—no second chance, she was gone!

When Dorothy processed my films, the one with the jaguar was pushed as much as we dared. But nothing visible except the eye, a fang, and the tongue—white and pink, in a frame of black, edged by the little greenery of the bush. We won a little more in the slide scanner and a little more in the print. I won't say it was the worst photograph I ever took, but it certainly was the worst one I ever kept!

One week later I returned to Cockscomb and gave prints to the staff and the two jaguar researchers. Back on the Tiger Ferns trail, I noted that Bart had placed a camera trap at the creek, so maybe he did believe there was a black jaguar. Later Bart admitted, "it does look convincing."

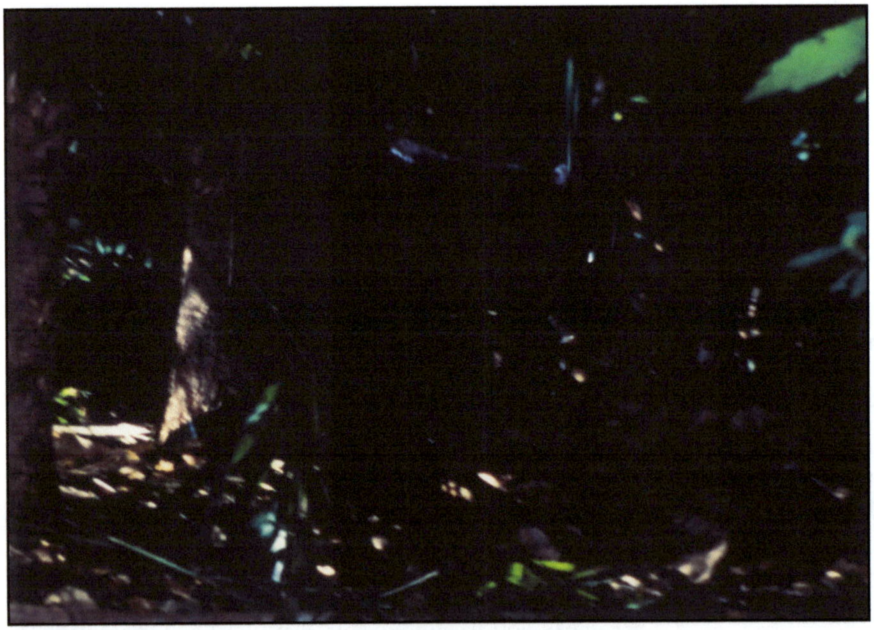

Black Jaguar (original photo)

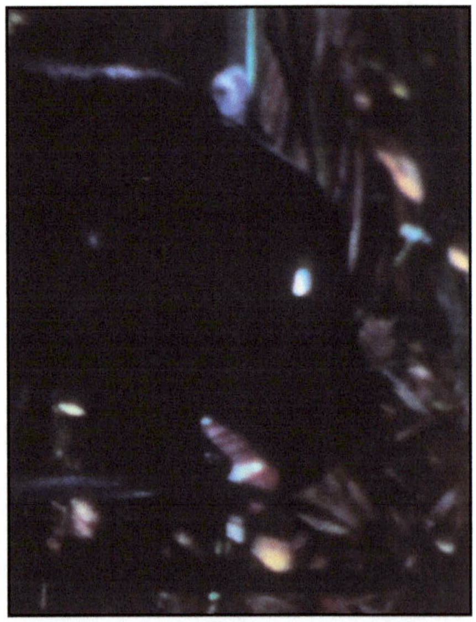

Black Jaguar photo lightened and cropped

Next to the black jaguar, my most remembered cat sightings were of pumas. I was guiding a small birding group when we came across a surprised puma just off the trail. It gave us a snarl and then moved into the forest, but not before I snapped its picture.

On another occasion I was walking on the Victoria Peak Path with my friend Rad. We had gone a little less than a mile and had just walked around a curve in the trail, when the pair of us came to a sudden stop. Coming toward us, and no more than fifty feet away, was a very large male puma.

The puma, too, came to an abrupt stop. The three of us, Rad and I on one side and the puma on the other side, stood motionless waiting for the other to move out of the way. For a couple of minutes, but seemingly much longer, we stood eye to eye, and then the puma gave in and turned around. He started walking along the trail in the direction he had come and, after a short wait, we followed. He sauntered in a casual way without a backward glance. Just as I was thinking that all three of us were planning to climb the Outlier, he turned, then checked us out for several seconds, as if to say, "Are you guys still with me?" then he slid into the forest on the left side. As the three of us walked the trail, I noted that we passed two camera traps on the right side of the trail. The next time I saw Bart, I asked if he had got pictures of a big male puma on his cameras. He said "yes, I did! How did you know?" he asked me. "Well, because I was walking right behind him."

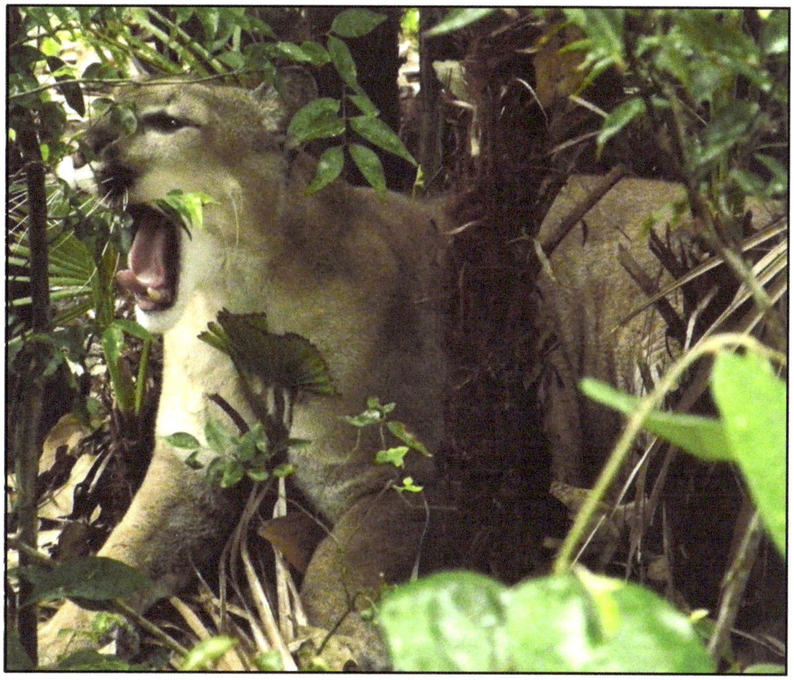

Puma (*Felis concolor*) surprised by my birding group

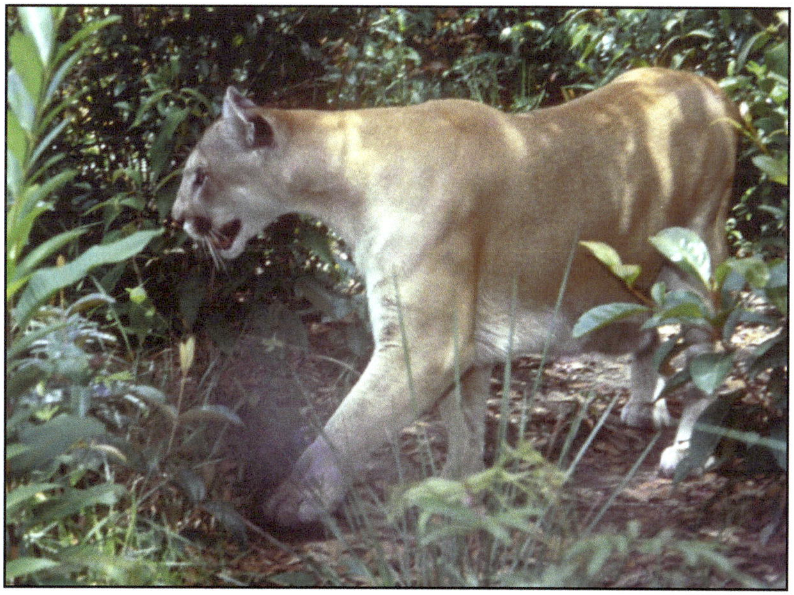

Puma (*Felis concolor*) on Victoria Peak Path

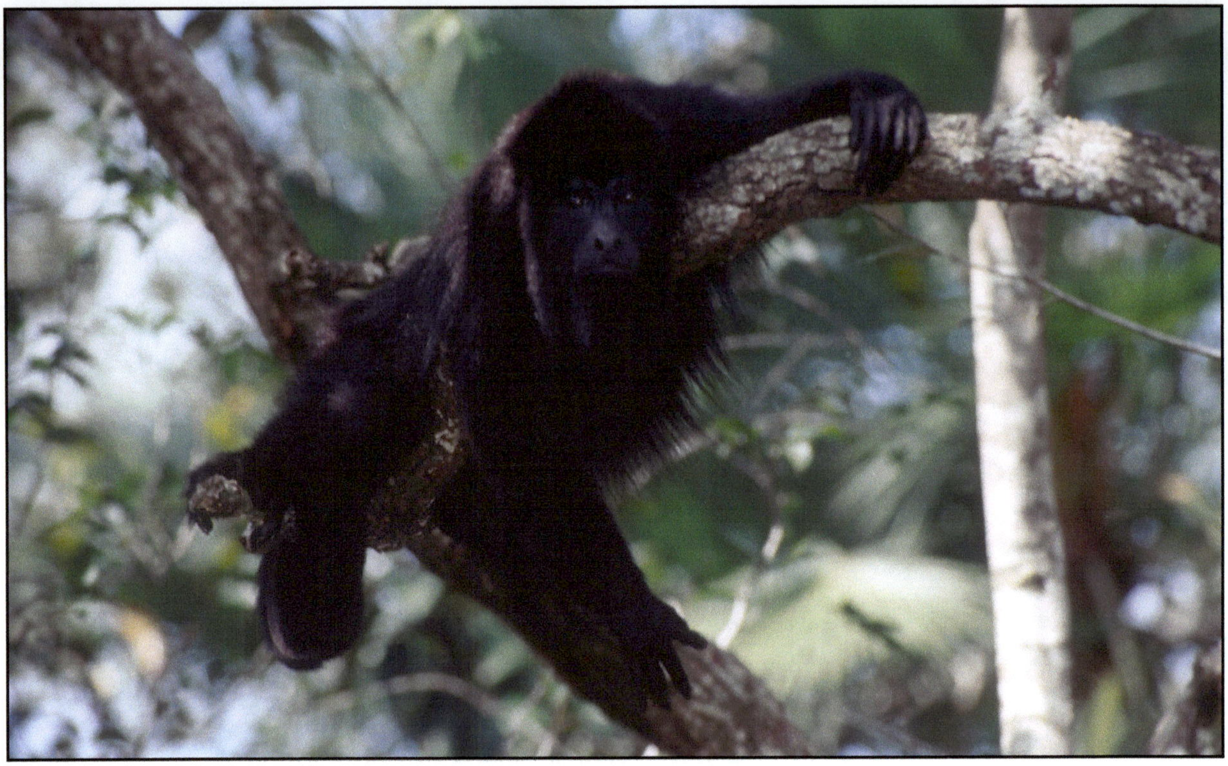

Black Howler Monkey (*Alouatta pigra*)

Howlers

In the early days of the Cockscomb Basin Wildlife Sanctuary, there was one important item missing, there were neither spider nor howler monkeys in the forest. There was a single lonely male howler in the trees close to Maya Centre Village who howled incessantly, no doubt trying to attract others of his race.

Black howler monkeys had lived in the Cockscomb Basin, but in the 1950s a yellow fever epidemic killed many of them and Hurricane Hattie in 1961 destroyed their habitat so none survived. In February of 1992, researchers from Wildlife Conservation Society (WCS) and Community Conservation Consultants began a re-introduction of the Mexican black howler monkey into Cockscomb Basin Wildlife Sanctuary. I was pulled in as a photographer to document the project.

Prior to moving the monkeys, much planning and studies were made, concerning the best time in the year for the actual translocation. The availability of the monkeys diet of ripe fruit was important, as was available riverine locations as howler monkeys have a definite preference for flowing water below them. Spacing was also important, so different troops would have at least a mile of territory from the next howler troop.

The monkeys were captured in the Community Baboon Sanctuary, an area well known in Belize, for having a multitude of black howlers. Capture was by the means of an anesthetizing dart fired from an air rifle. The monkeys then lost consciousness and fell into a net.

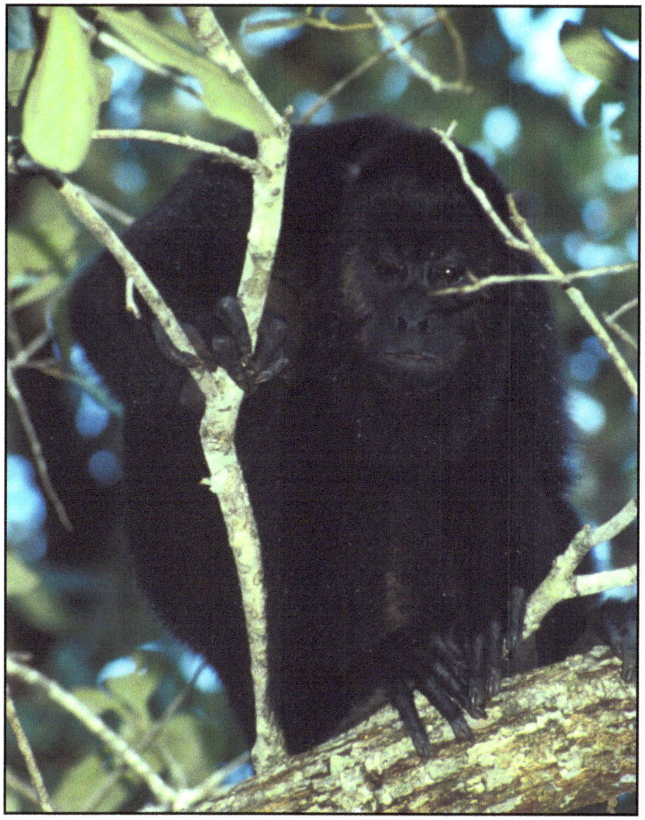

Black Howler Monkey

They were examined by a veterinarian who checked out their vital signs and removed parasites. These parasites were mostly botfly larvae, the scourge of the rainforest. While still unconscious, they were fitted with radio transmitters. Then the complete monkey troop was taken to Cockscomb either by van or helicopter.

Once at Cockscomb, each individual troop was placed together in a large acclimatization cage built within the forest, close to a waterway and the monkey's favourite fruit trees. They were kept in the acclimatization cages for approximately three days and fed daily with wild fruit. At night, the cage was guarded by the park wardens, lest some large predators had a craving for a midnight snack.

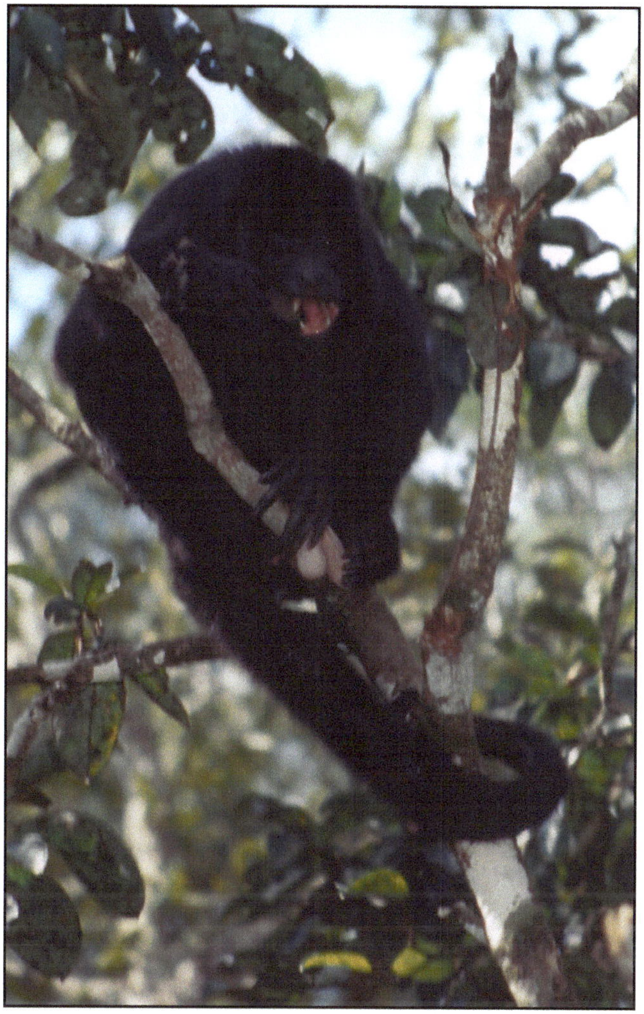

Black Howler Monkey

After three days the troop was released into the wild. The howlers were monitored by radio for another two years. During the three-year project, sixty-five howler monkeys were released into Cockscomb. These consisted of fourteen original troops.

Most troops settled down quite close to their release areas and gave tourists a thrill either on spotting them, or a scare on hearing a large male directly overhead.

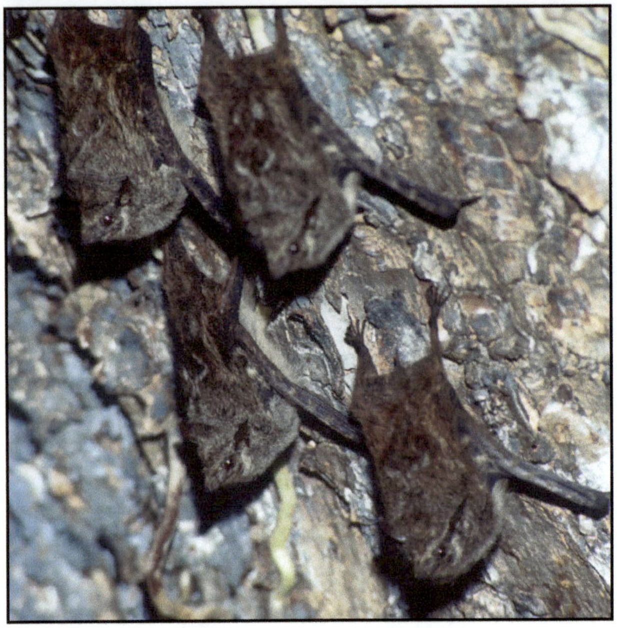

Little Brown Bats (*Myotis migricans*)

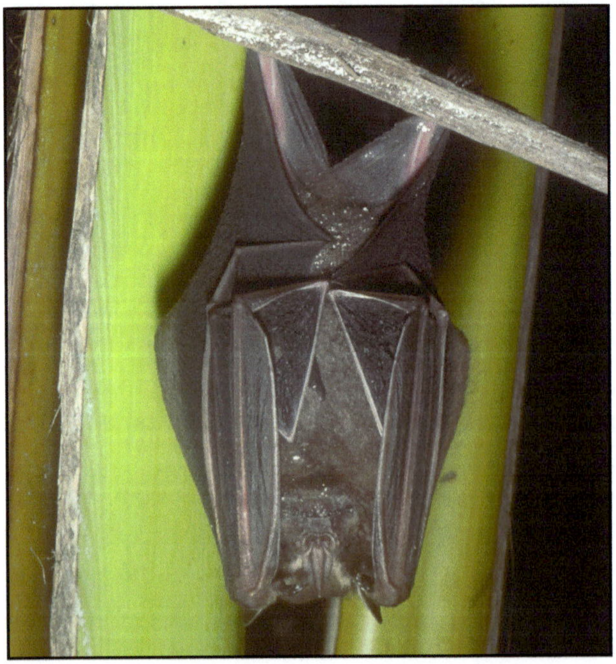

Short-tailed Fruit Bat (*Carollinae* sp.)

Night Visitors

Since I spend much of my time overnighting in rainforest areas, visits from wild creatures are expected. Most forest mammals are nocturnal and have an excellent sense of smell for meat, fruits, even bread. Only canned foods are safe.

Over twenty years ago I was sleeping in an old building with lots of previous use. It was around two am. when I was awakened by Dorothy, who was stroking my bare shoulder. Then I realized that Dorothy was home in Caye Caulker.

I couldn't see who was stroking my bare shoulder. Slowly I stretched out my arm to get my flashlight and turned it on. I was looking into the face of a snake, no more than fifteen inches away! I stayed motionless; the beam from the flashlight was focused on its head. As I watched, its pupils began to constrict.

I recalled a piece of wisdom from Arturo in Amazonas. If a snake's pupil is round, it is non-venomous, but if the pupils are vertically split, it is venomous. As I watched, its pupils closed until it was merely a narrow vertical slot. It was venomous! I grabbed my pillow, rolled backwards and simultaneously pushed the pillow onto the snake's upper body.

My roll carried me right off the bed, but I regained my balance almost immediately. Once upright, I started looking for the snake. It wasn't under the pillow, or under the mattress. I checked under the bed and saw there was a hole about four inches by ten through the outer wall, at floor level. The snake was gone!

The next morning I checked out my reptile book and easily identified it as a cat-eyed snake, which is a nocturnal creature, an eater of frogs and lizards, and only mildly venomous.

Another time I was sleeping in quite an up-scale establishment; the windows were well screened and the doors were lockable. I awoke with a tickling sensation on my legs and crotch. As I sat up, the tickling changed

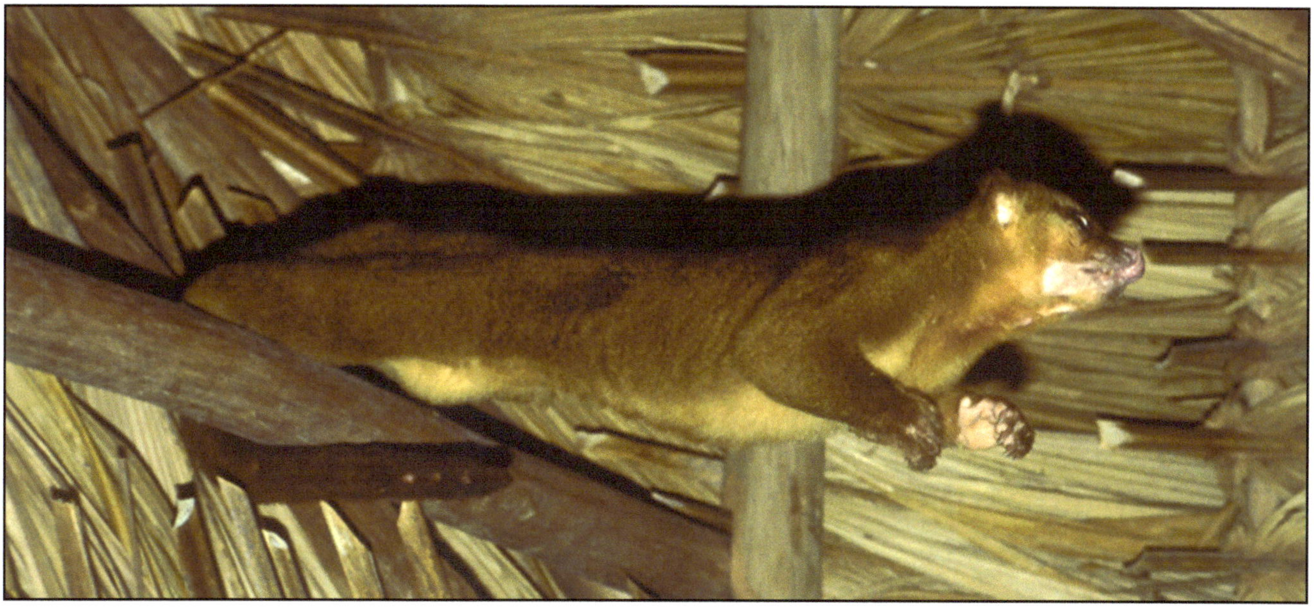

Kinkajou (*Potos flavus*)

to a stinging sensation. By the time I turned the lights on, the stinging was approaching agonizing. The bed, the floor, and my lower body were covered with ants! I grabbed my shirt from the bedpost and started beating off the ants. The more I beat, the more they climbed my legs. Then I saw the reason why! The floor was a two-feet-wide column of marching ants that stretched under the door.

I knew that there was only one thing I could do! I ran into the shower and turned on the hot water as full and as hot as it would go. Gradually the ants washed off me. The hot water on my stinging legs felt great and the ants slowly disappeared down the drain. But gradually the pain of the ant stings was being replaced by the pain of scalding water! I turned the hot off, and put the cold on.

My bed was still dark with ants; the bed to one side was about thirty percent covered, but the third bed was still pristine. I ran across the room, killing a few luckless ants and jumped onto the third bed.

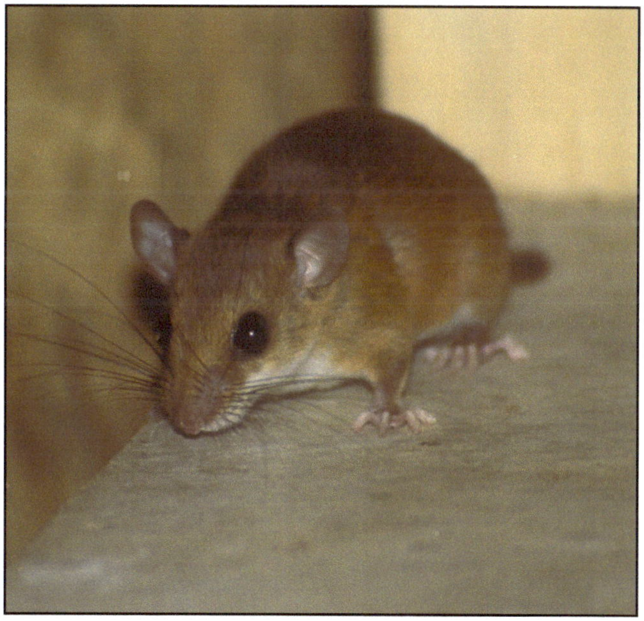

Vesper Rat (*Nyctomys sumichrasti*)

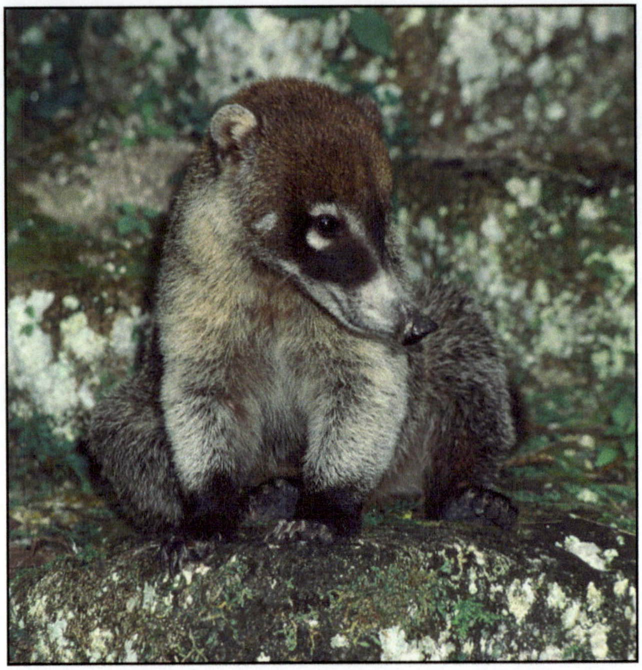

(*above and below*) White-nosed Coati (*Nasua narica*)

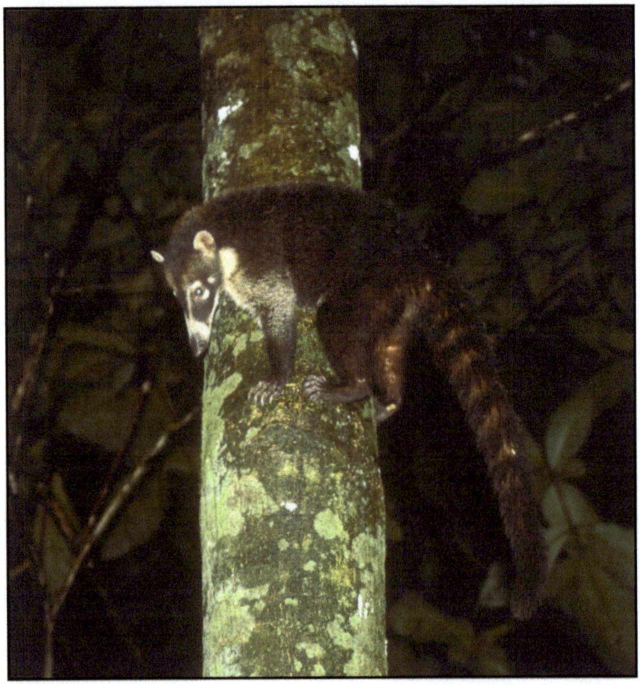

Gradually my room was cleaned of all bugs and spiders, more thoroughly than any room service could have managed. The ants stayed until 10 am the following morning!

I told the manager of my night adventure. He was sympathetic and informed me that it was not the first time. It had happened before to an acquaintance of mine and a well known birder. No doubt he will recognise himself when he reads this book.

On another occasion the cabin where I was staying was quite spacious, so I brought all my supplies in from my truck. I placed the food stuff on a shelf, and after making myself a meal and consuming it, I went to bed.

The next morning I noticed that I had a visitor during the night. A banana had been partially eaten, and a loaf of bread had been nibbled at one end. Obviously, I had a rodent visitor during the night.

I had no wish to carry my supplies back out to the truck, so I figured that I'd make a deal with mousy. I put all my food supplies in my backpack and hung it from a roof crossbeam. Each evenings, I left a little offering to the "Rodent God," usually half a slice of bread and a piece of fruit.

This worked very well for the next four days; my offering was accepted; and my food was untouched. I knew very well that he or she could have easily chewed through my line holding the backpack.

But on the fourth day, I had done an exceptionally long hike, and was pretty exhausted when I returned, and promptly fell asleep. I neglected to eat supper and, what was worse, I forgot to leave my offering.

It was approximately midnight when something woke me up, and I saw a small rat sitting on the palm of my hand. She definitely wasn't the Norwegian rat that mankind is so used to eradicating around the world. She was quite petite and a beautiful orange-red colour. As I moved to get up, she jumped off my hand and ran up to the ceiling.

I remembered that there had been no offering to the "Rodent God" this evening; I also remembered that I had not eaten supper myself!

So I lowered my back pack and took out the makings of a ham and tomato sandwich for myself with a slice of papaya for desert, then I made my offering which was accepted very quickly. She ate the papaya and then carried off the bread, probably to her young outside in a tree nest. I later identified her as a Vesper rat, which nests in the tree hollows.

The following night I stayed awake to photograph the rat. She arrived about 9:30 p.m. after accessing the room from the roof and climbed down the slats of the window. I took a couple of photos as her eyes darted between me and the overripe banana that was tonight's offering to the "Rodent God"! After she was sated, she left via the roof. I took the remains of the banana out and left it on a tree branch. The next morning it was gone. Probably she did have a family to feed.

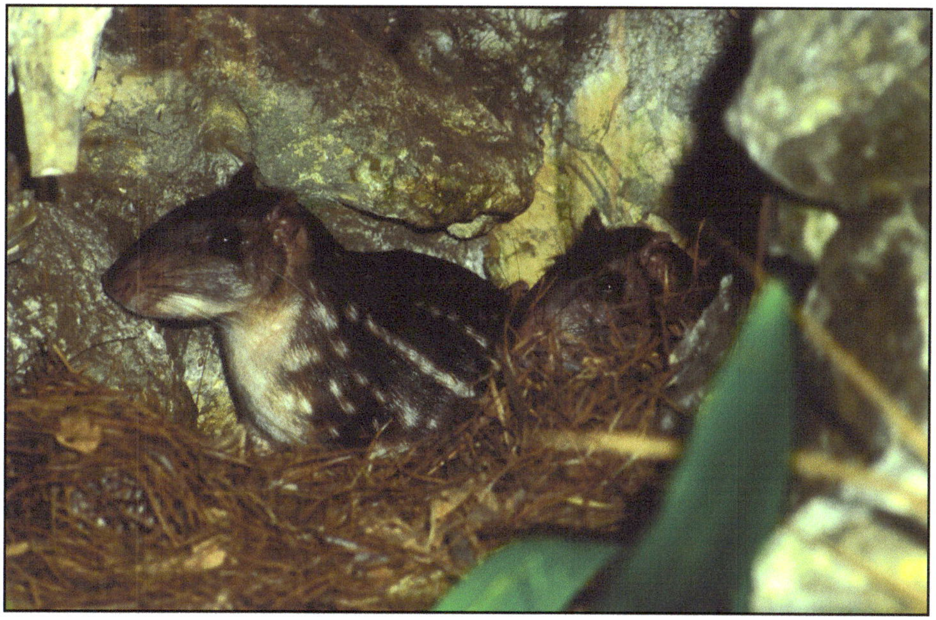

Paca (*Agouti paca*) —The Belize Zoo

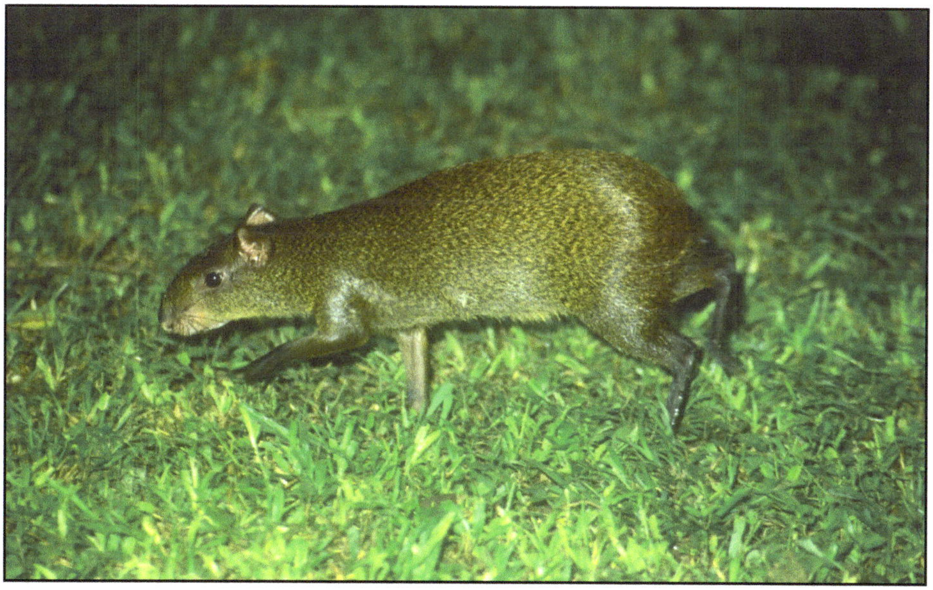

Agouti (*Dasyprocta punctata*)

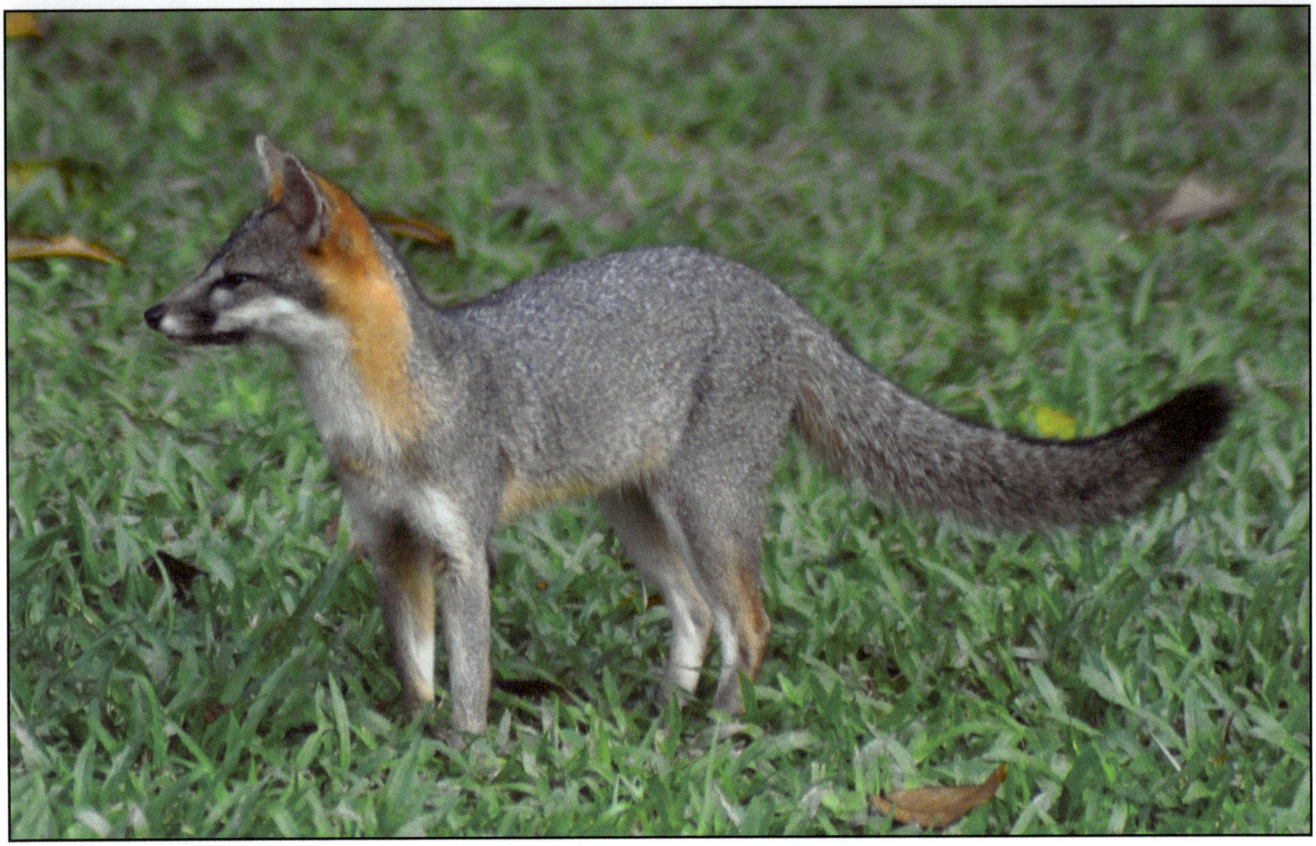

Gray Fox (*Urocyon cinereoargenteus*)

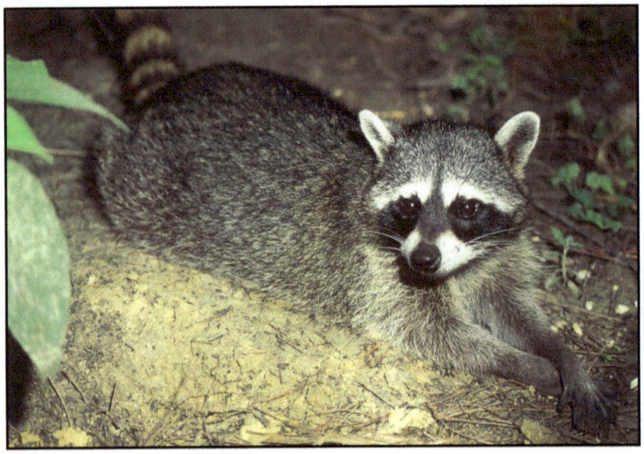

Raccoon (*Procyon lotor*)

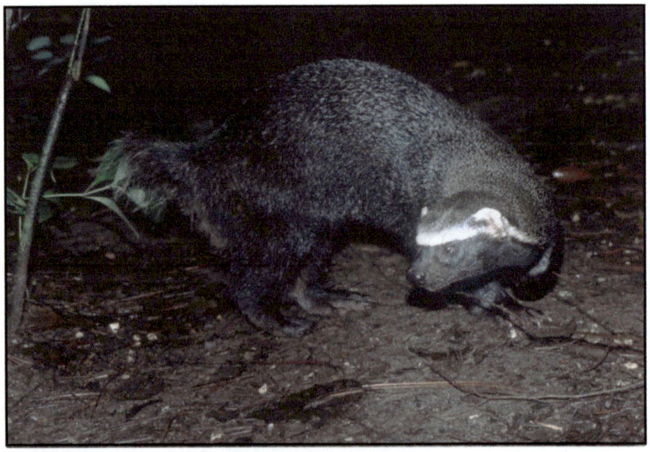

Grison (*Galictis vittata)* —The Belize Zoo

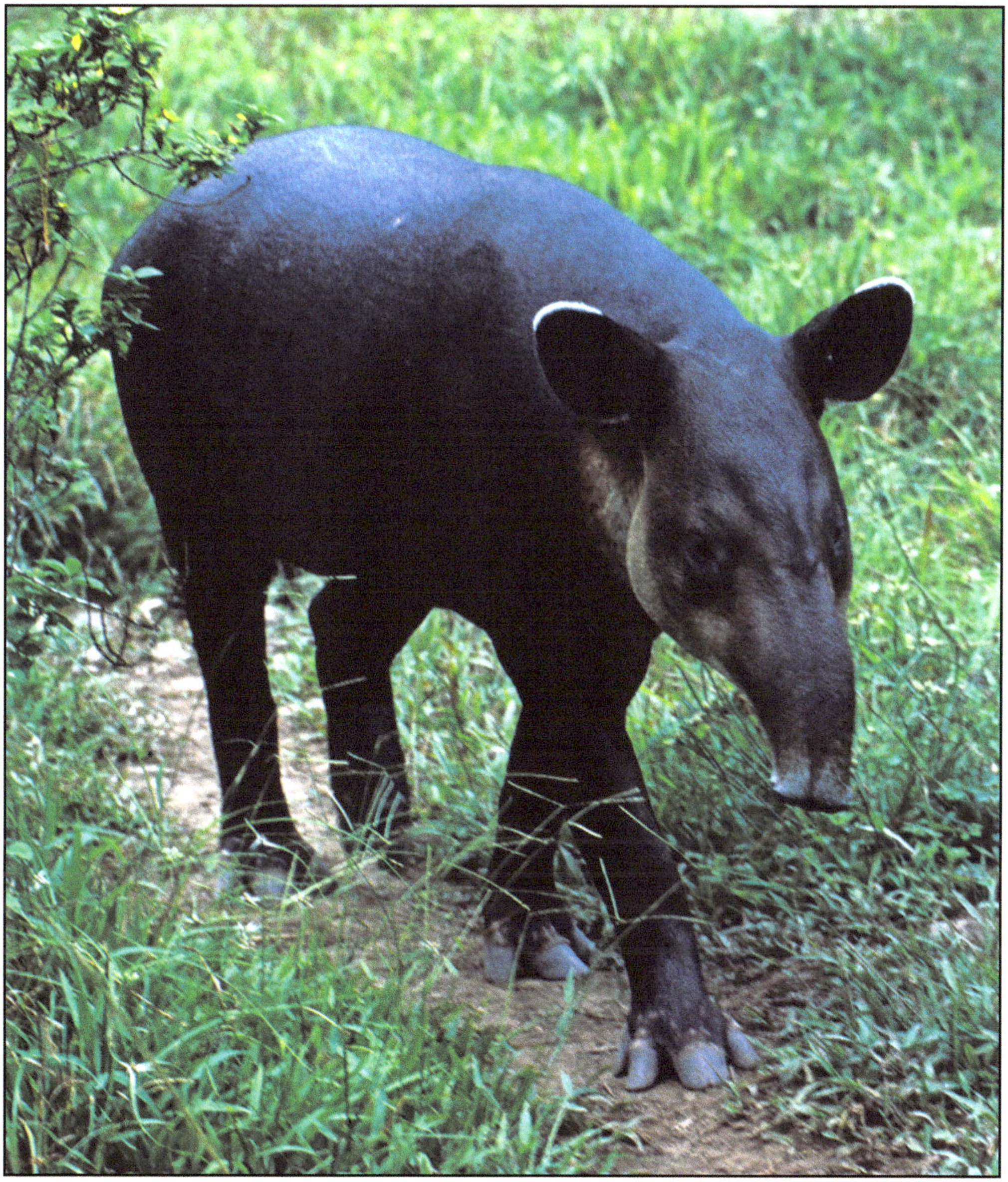

Baird's Tapir (*Tapirus bairdii*)
—The Belize Zoo

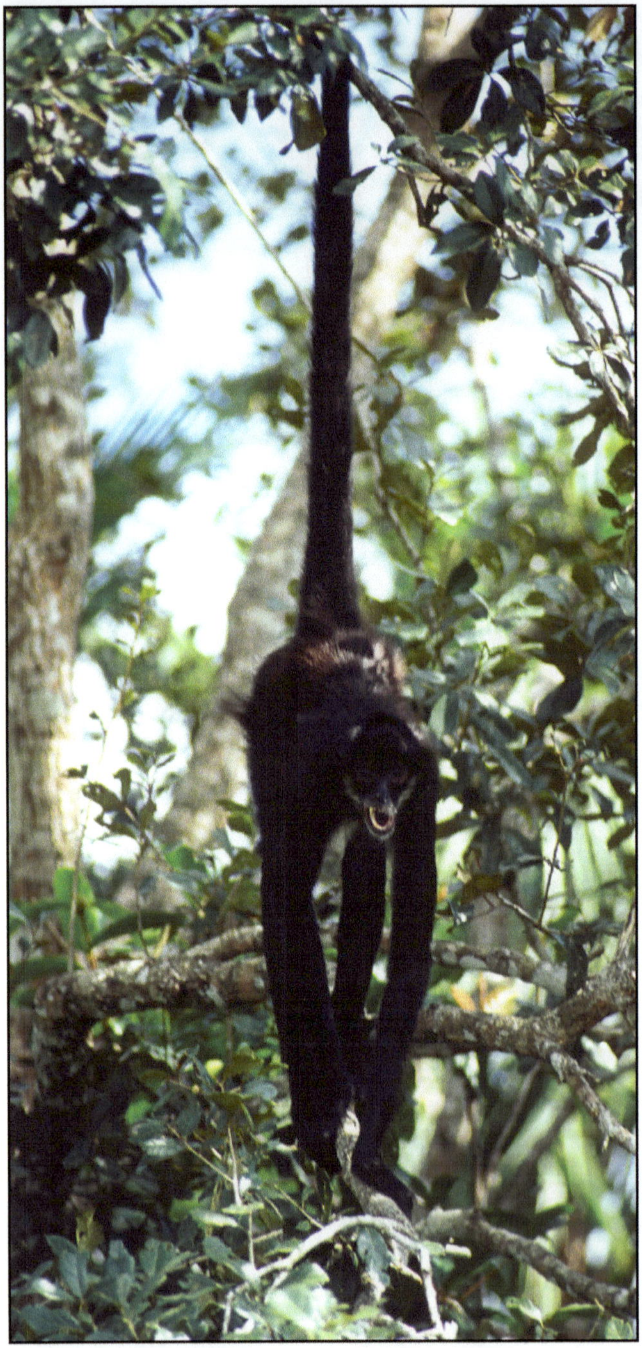

Spider Monkey (*Ateles paniscus*)

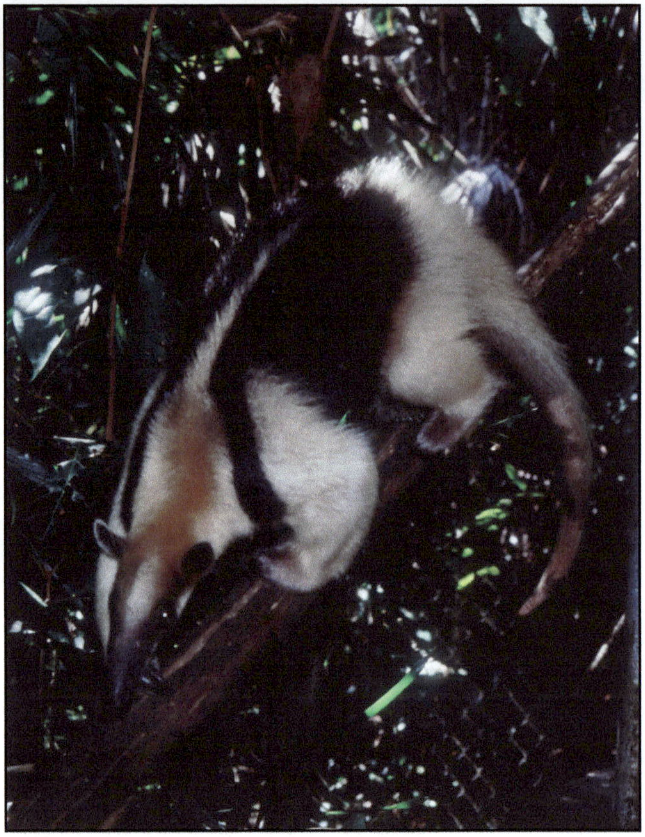

Tamandua (*Tamandua mexicana*)
—The Belize Zoo

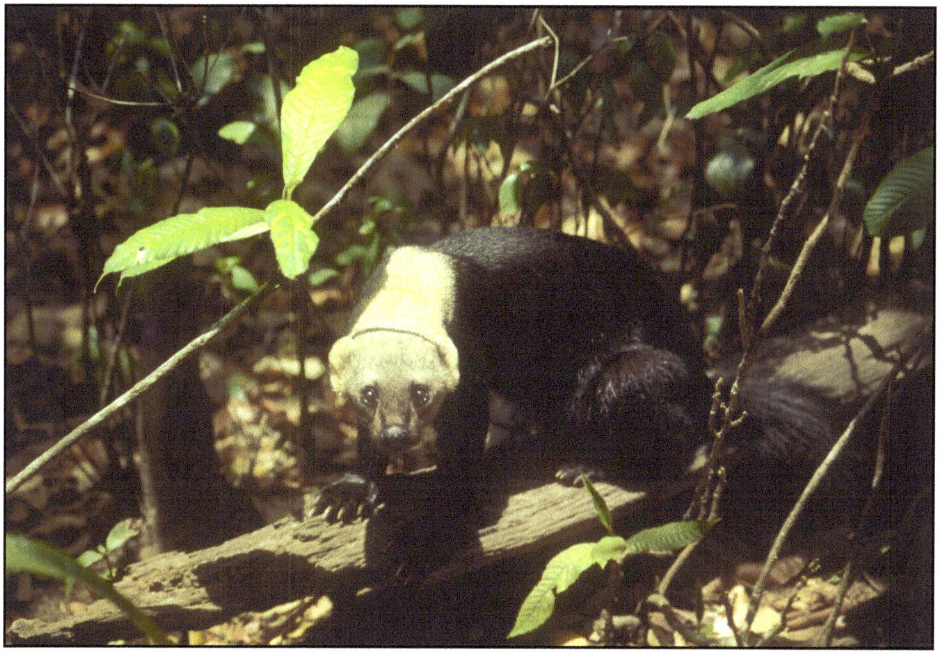

Tayra (*Eira babara*)

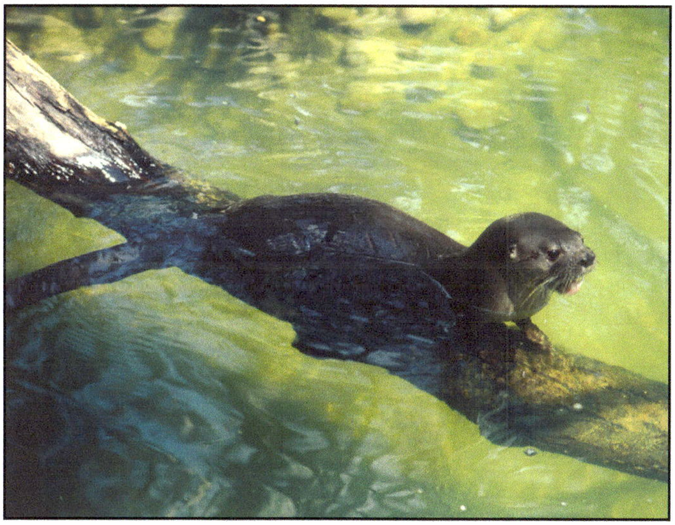

(*above and right*) Southern River Otter (*Lutra longicaudis*)

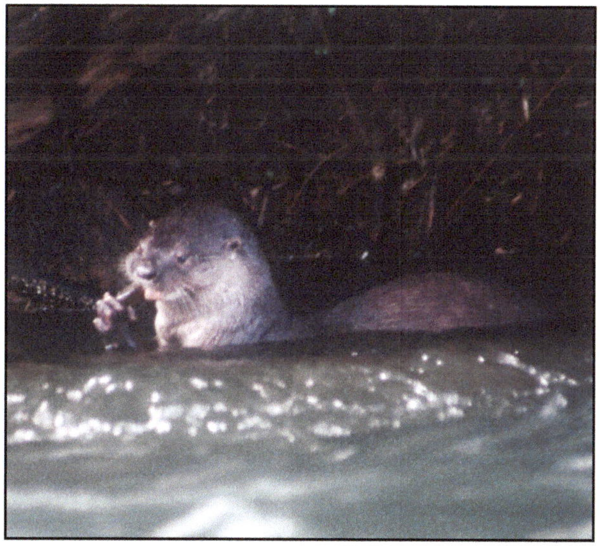

Unknown Mammals

It is highly unusual for a layman to come across a new species for an area or a country be it a bird, mammal, reptile or fish, and have that sighting accepted by the experts in that particular field; unless that layman just happened to have with him a camera and the speed of mind to shoot a photograph of that said creature.

The black jaguar was seen by myself and four friends, plus several tourists and still was not accepted for over four years, until yours truly managed to photograph it. Since then, I have seen two other large mammal species in Belize that seemingly do not exist here.

I was walking through a privately owned forest area when this animal came through the bush, less than five feet from my right leg. It stopped and, without any fear, it looked up at me for about ten seconds, then unhurriedly walked across the trail and disappeared into forest on the other side.

The trail was about seven feet wide, but the closest focussing distance from my lens was slightly over nine feet, so I got no picture. I recognized what it was instantly. Even from my childhood books, I was familiar with the bushdogs. While in Amazonas, I had seen one in captivity. It was a bushdog (*Speothus venaticus*) not to be confused with a tayra (*Eira barbara*), the Creole name of which is "bushdog."

The bushdog is a social but secretive dog-like mammal that is very seldom seen, even by the forest Indians. It is found as far north as Panama. After my black jaguar experience, I did not report my bushdog sighting except to the owner of the property, and a friend, also a photographer.

The other mammal, I did photograph. I had been photographing birds at Clarissa Falls in Cayo. I had been quite pleased the previous day in shooting photos of various species of birds fishing in the falls. My only problem was that my presence was making the birds wary and difficult to photograph.

So I decided to build a hide, and early the following morning I set up to wait in a cramped hole in the ground. It was still dark when I noticed the outline of an animal swimming low in the water, crossing the river from the other side. It was swimming directly towards me and I took it to be an otter.

I readied the camera/flash on its silhouette and waited. Only its head was visible while it swam. When it was eight feet away, I took the first picture, then three more as it came into the shallows and I was photographing it in its entirety. It was still too dark to make out the details of the creature as it clambered over the little island where it jumped back into the Mopan River above the falls.

After Dorothy processed the film I still could not identify the creature. It looked a little like an opossum, and a little like a cat. But, mostly it didn't look like either.

Those photographs have been examined by four professional zoologists, and quite a few naturalists. To this day, the creature from Clarissa Falls is still not identified. If you can identify it, I'd appreciate if you can tell me what our mystery animal is?

Unknown Mammal

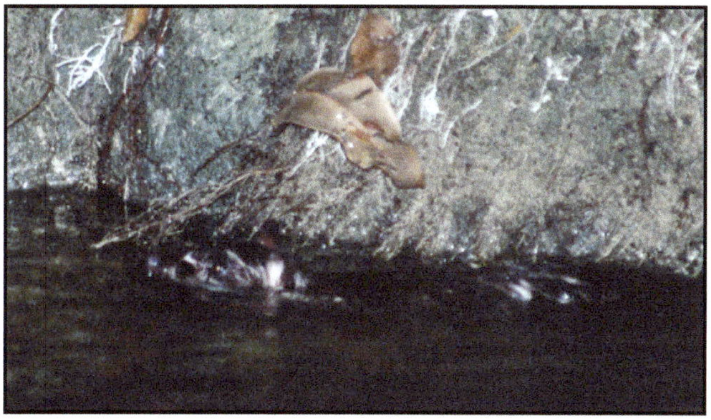

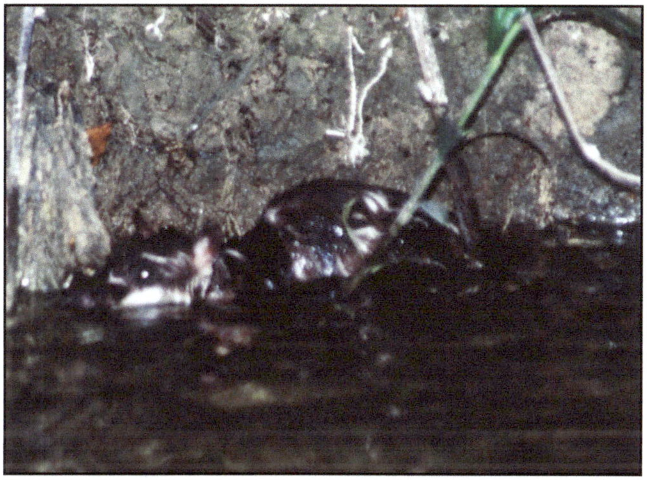

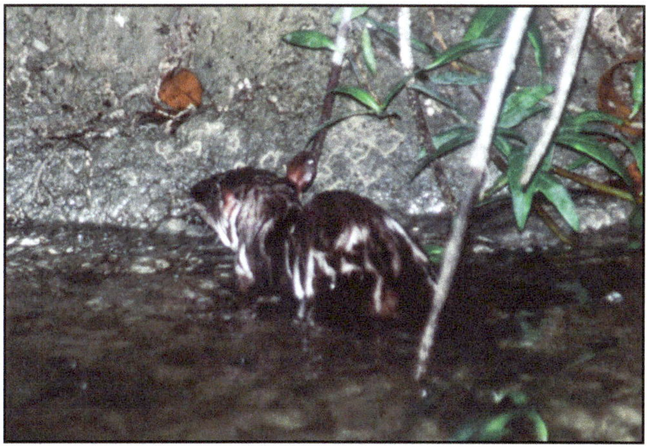

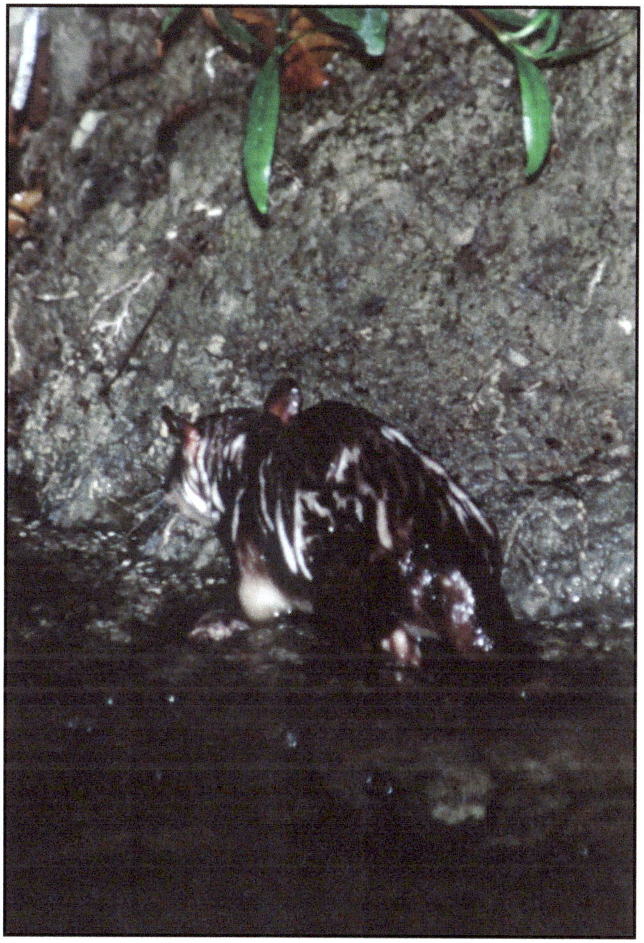

Can you identify this mystery mammal?

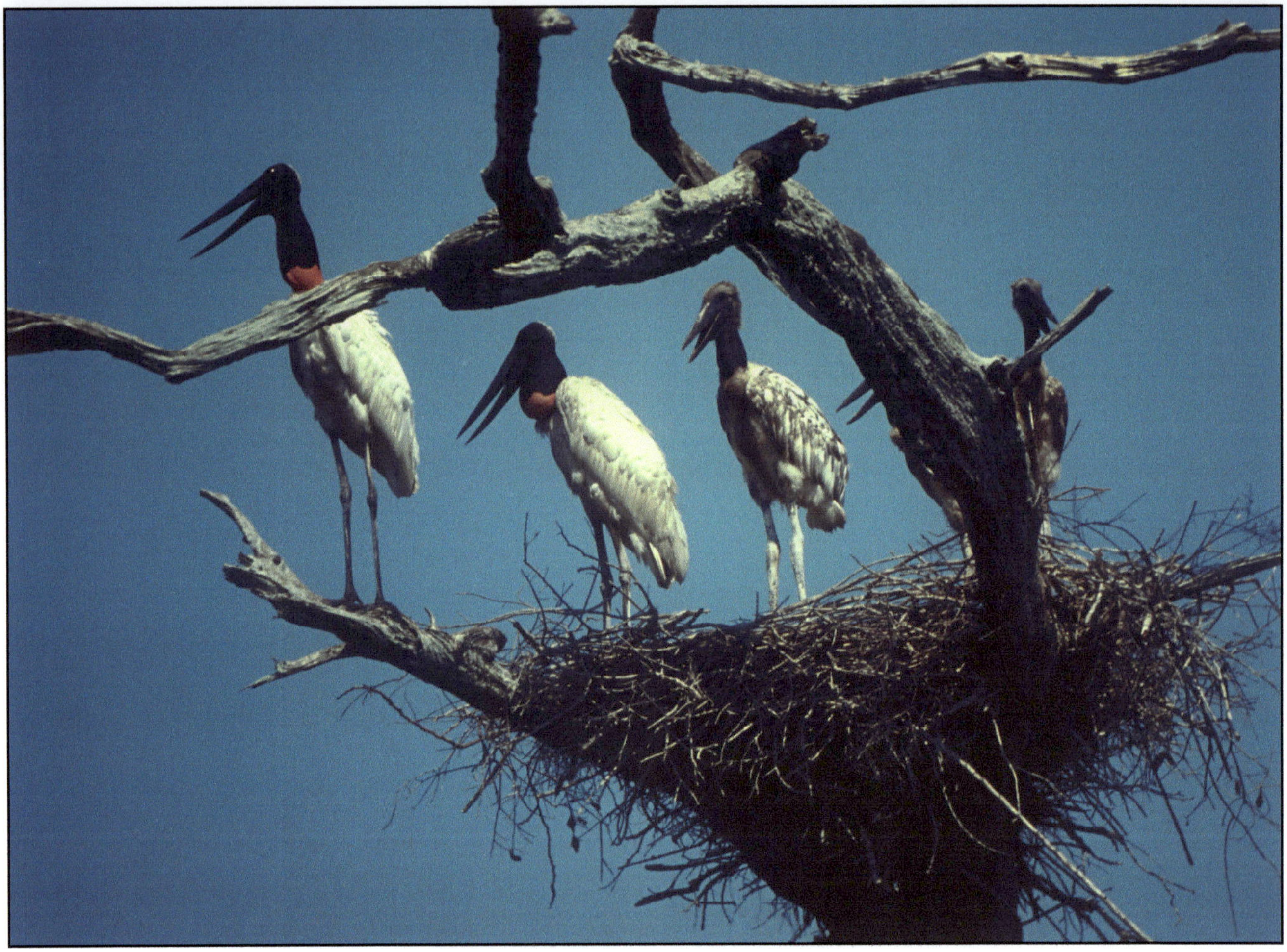

Jabiru nest with two chicks and parents

Birds

As I write, 586 species of birds have been positively identified in greater Belize! I personally have never started a "Life List", an objective almost all bird enthusiasts enjoy with a passion. I do admit, as a wild life photographer, I keep a record of all the species I photograph, be they mammals, reptiles, birds or bugs. However, this is more as on aid to business than a Life List.

I very seldom carry binoculars, but almost always carry a camera. Binoculars are too much of a distraction, as my main purpose is to get as close as possible to my chosen subject. For this same reason, I seldom go out with groups, be it in forest or field, or forty feet under the sea. When I worked as a naturalist guide in the 1990s, I had to curb my enthusiasm to get close to the subject of interest, lest it upset a paying customer.

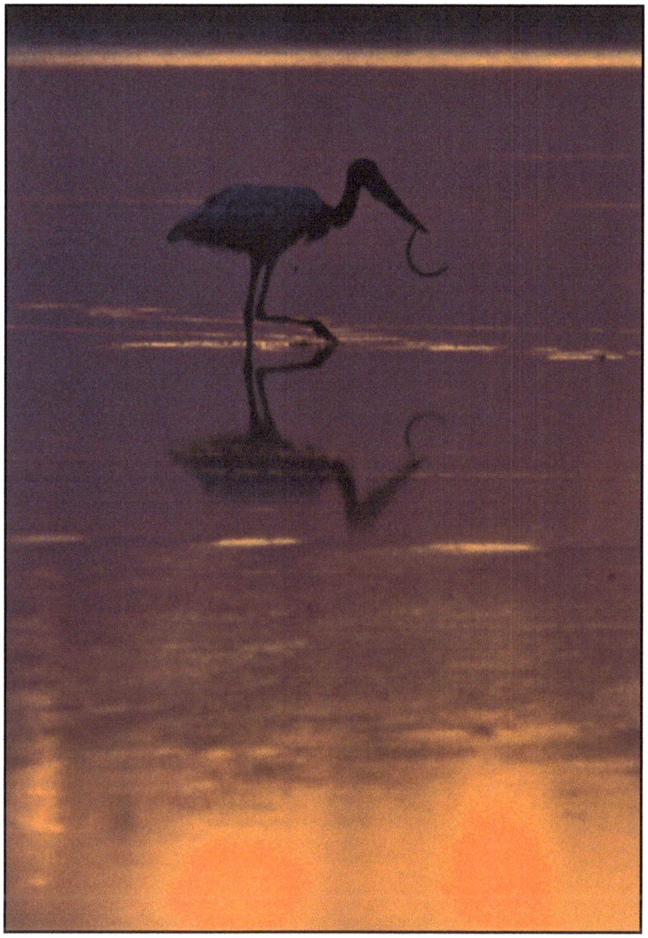

Jabiru catching a snake at sunrise

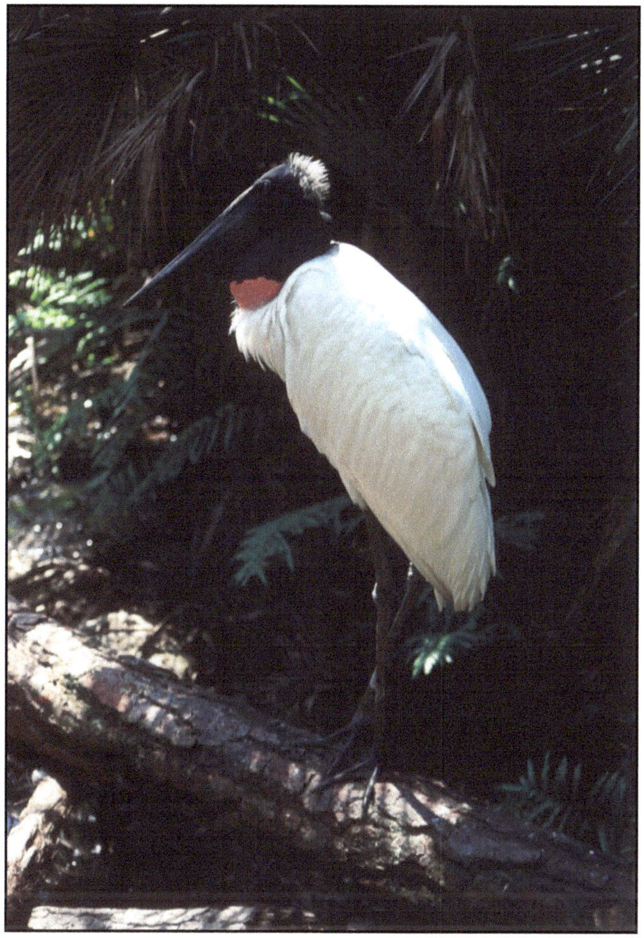

Jabiru (*Jabiru mycteria*)

There are many choices of birding destinations within Belize. My favourites however are Cockscomb Basin Wildlife Sanctuary and Crooked Tree Wildlife Sanctuary.

Crooked Tree is close to mile thirty on the Northern Highway and, as far as aquatic birds are concerned, it would be difficult to imagine a higher density of waders and swimming birds anywhere else in Central America. The best time to enjoy Crooked Tree is March and April when the lagoon waters drain to a point where it is shallow enough to allow the waders, such as, the Wood Stork, Jabiru, Glossy and White Ibis, Northern Jacana, and most species of herons and egrets, to feed quite close to shore, allowing the visitor some excellent photo opportunities.

At the same time, your concentration may be interrupted as Limkins and Gray-necked Wood-Rails pass within yards, and Purple Gallinules, coots, and moorhens swim by only yards from the shore.If this is not enough, a short walk will take you back into the village where one is certain to see Vermillion Flycatchers, trogons, and tanagers of various species.

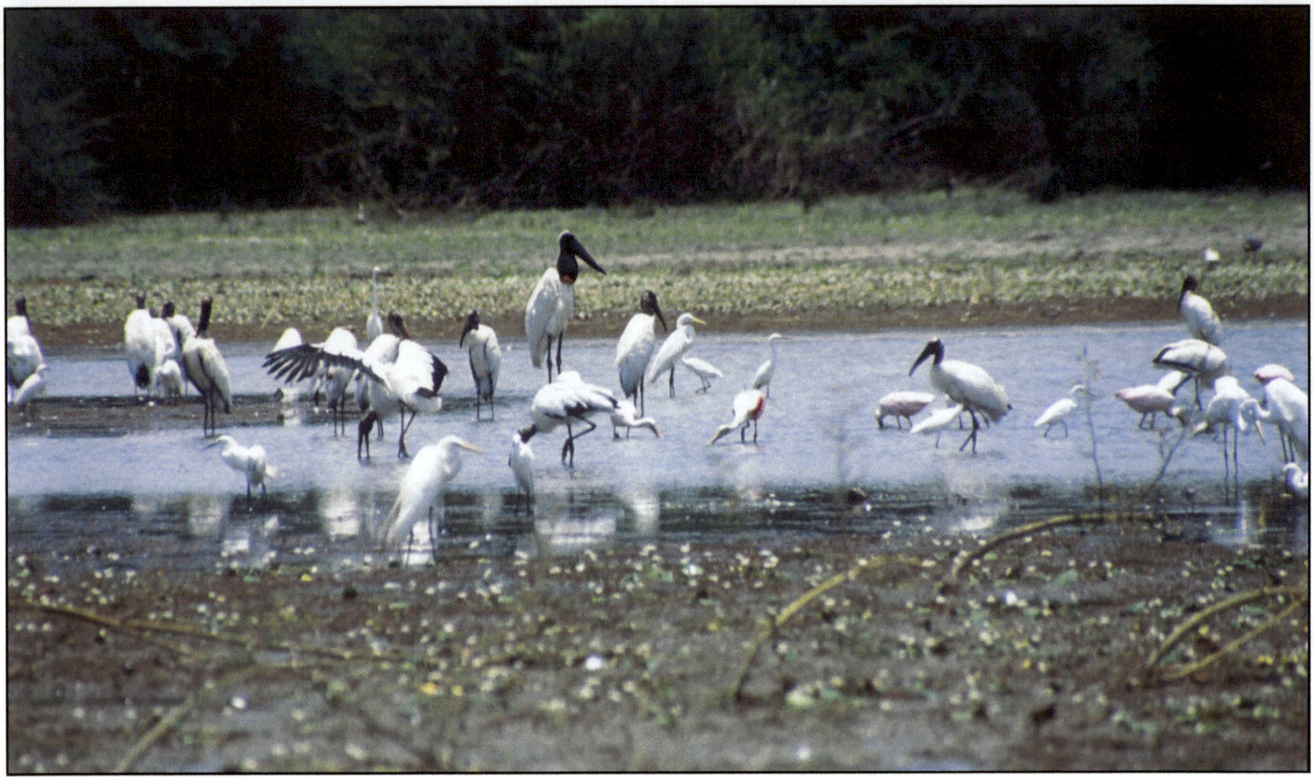

Waders at Crooked Tree

Still looking for more? Then it is possible to rent a small boat, with or without a guide. A visit to Black Creek is certain to please you, for here the Black-collared Hawk is normally seen, along with Agami Herons, Pied-billed and Least Grebes, Mangrove Cuckoos, and several species of kingfishers and woodpeckers.

Cockscomb Basin Wildlife Sanctuary is a tropical moist forest and hosts a multitude of bird species. There is an excellent trail system, and I think it is fair to say that many visitors to Cockscomb are birders.

Other forested parks managed by the Belize Audubon Society include Guanacaste National Park and the St. Herman's Blue Hole National Park, both of which offer birding opportunities.

The largest protected area in Belize is Rio Bravo Management and Conservation Area of 245,822 acres. Rio Bravo has two tourist destinations; La Milpa is forested and Hillbank is wetlands, both of which have excellent birding.

Mayflower Bocawina National Park is off the Southern Highway about sixteen miles north of Maya Centre, the Cockscomb junction. It is seven thousand acres, an extremely rugged area with some fantastic hiking to spectacular waterfalls and excellent birding.

In addition to the aforementioned, Belize has many other protected areas which are open to the public, including wildlife sanctuaries, natural monuments, and archaeological reserves, all of which offer excellent birding.

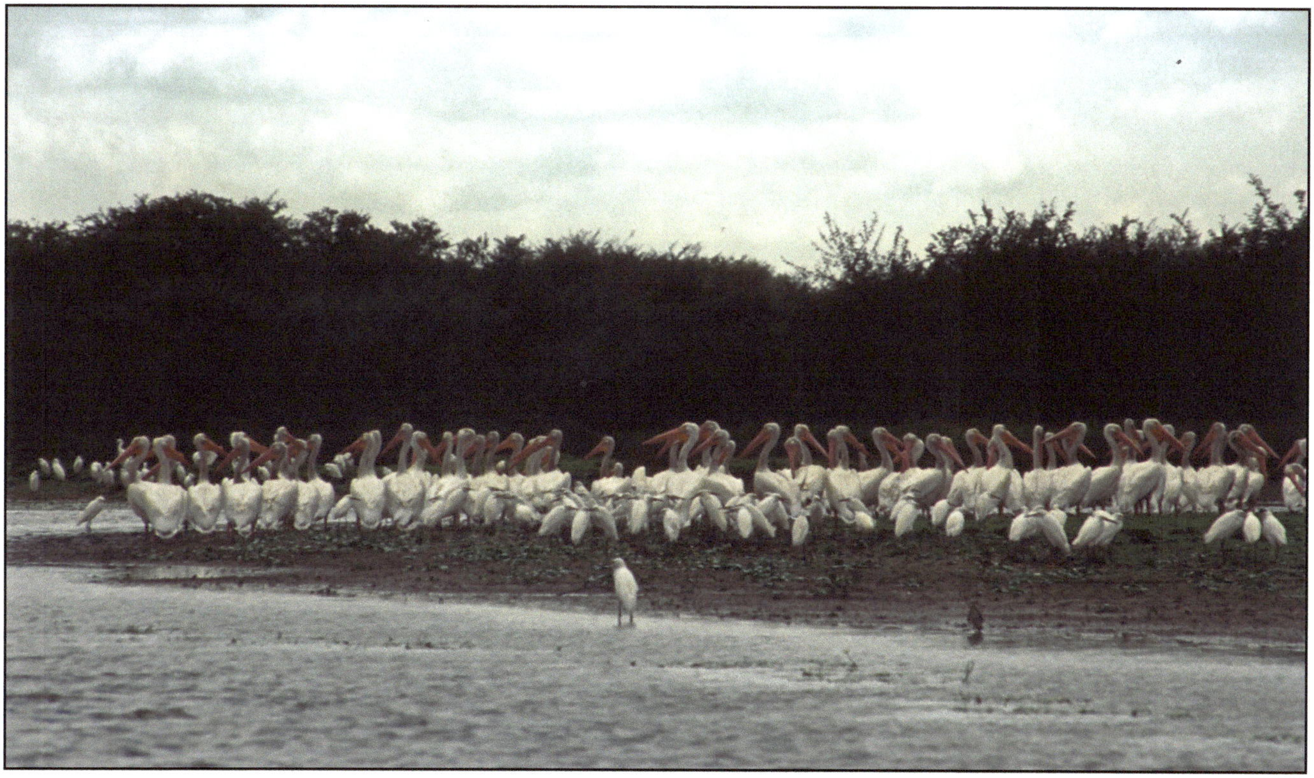

American White Pelicans (*Pelicanus erythrorhynchos*)

As a wildlife photographer, I much prefer my bird subjects to be colourful. Certainly our migrant warblers and several other migrants are not lacking in colour, especially during the Fall and Spring migrations on Caye Caulker. In Cockscomb while I am on the nightshift with frogs and snakes, in between naps I shoot pictures of birds by day. It is really the hummingbirds that are my favourite subjects. Taking photographs of birds, I try to freeze motion, and hummingbirds' reflexes are very swift and interesting. And their colours are so beautiful.

Raptors are another one of my favourite subjects, but they are more difficult to find and photograph. With few exceptions they are very aware of things being pointed at them and take off at the slightest hint of danger.

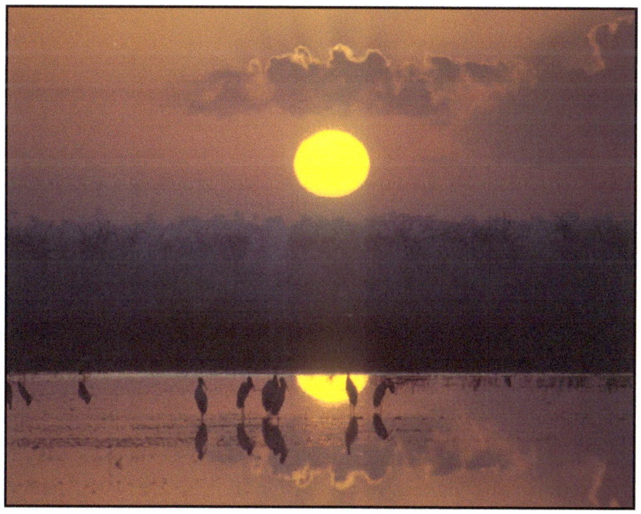

Dawn at Crooked Tree

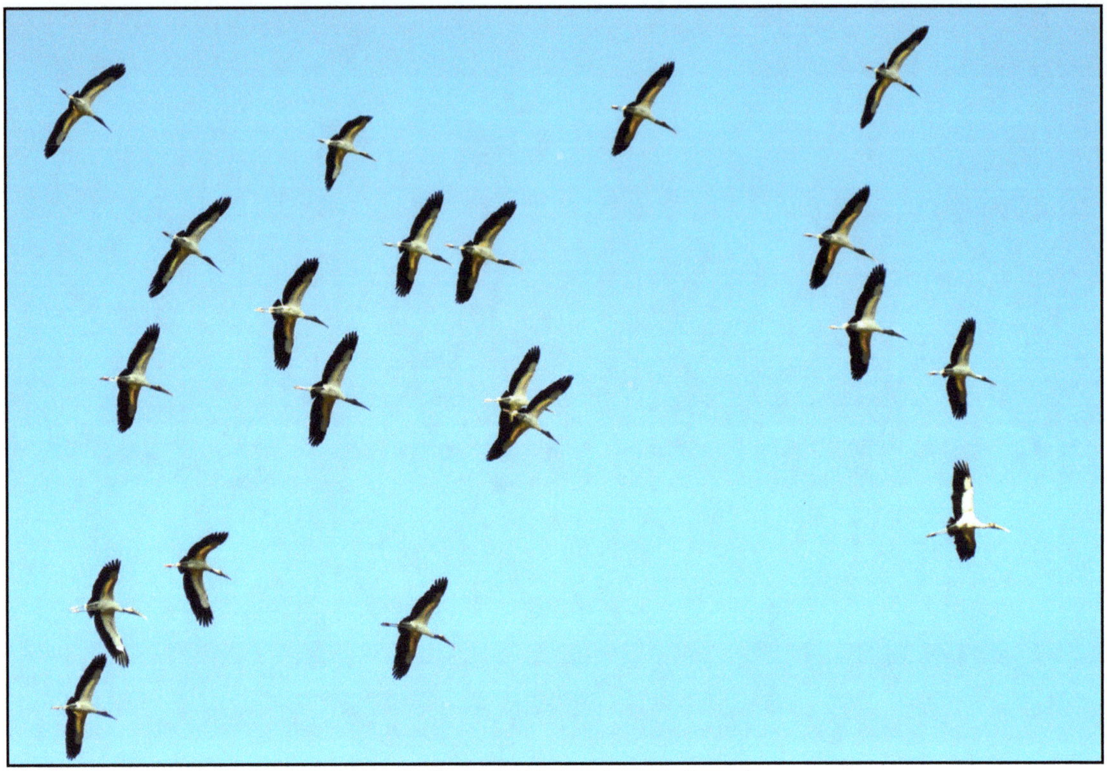

Wood Storks (*Mycteria americana*)

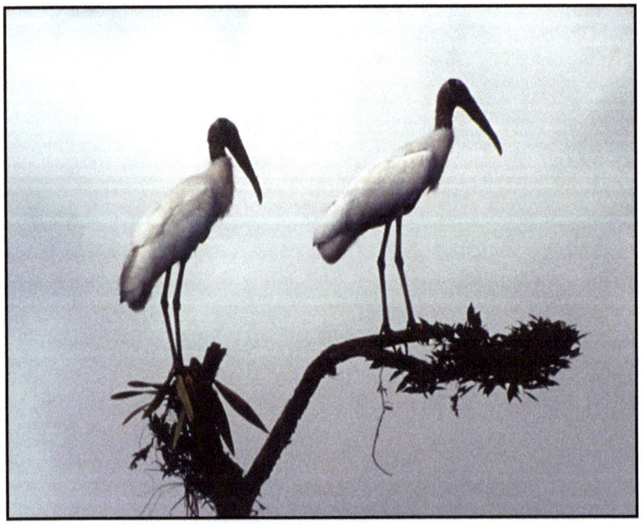

Wood Stork pair

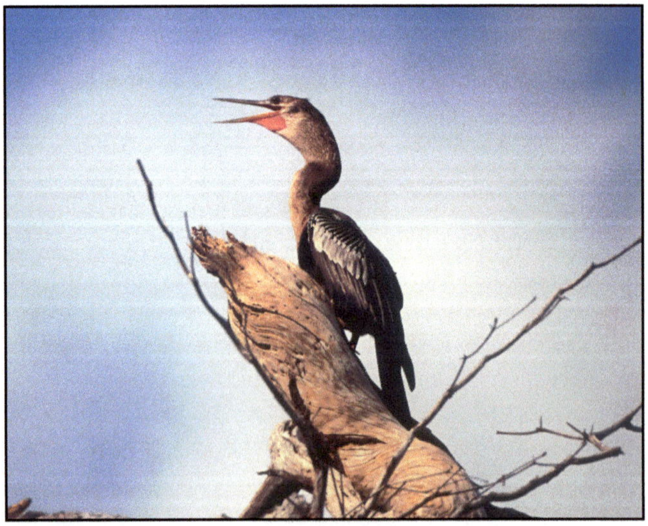

Anhinga (*Anhinga anhinga*)

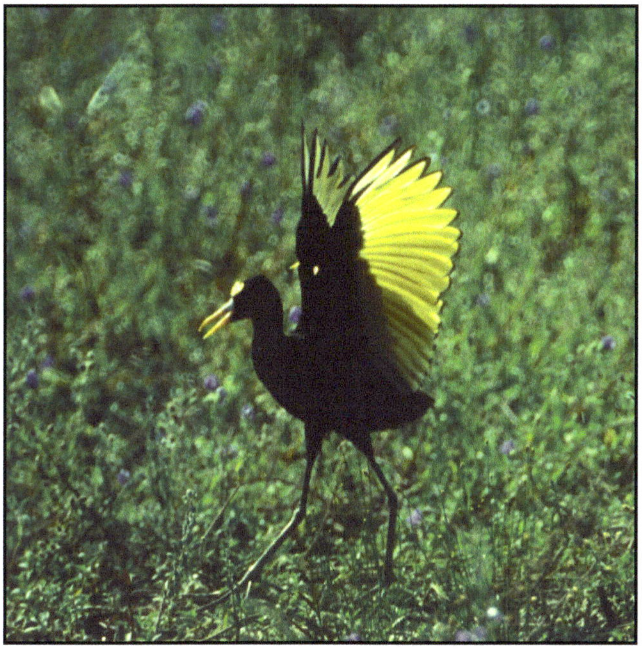

Northern Jacana (*Jacana spinosa*)

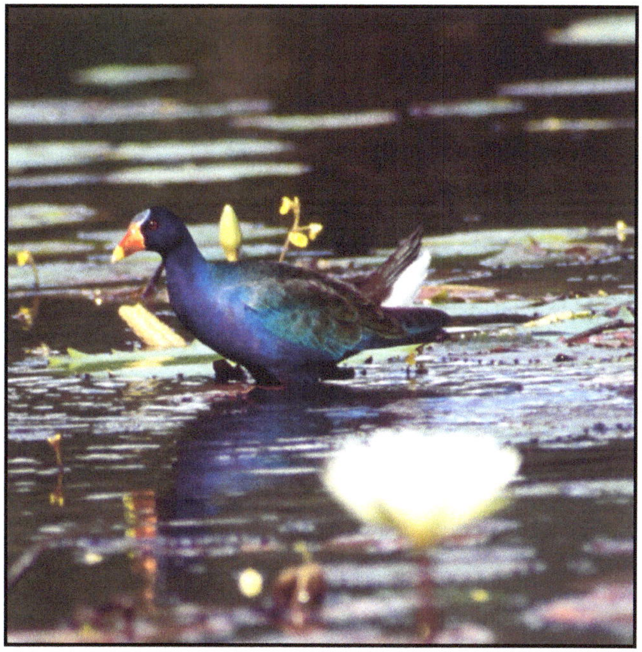

Purple Gallinule (*Porphyria martinica*)

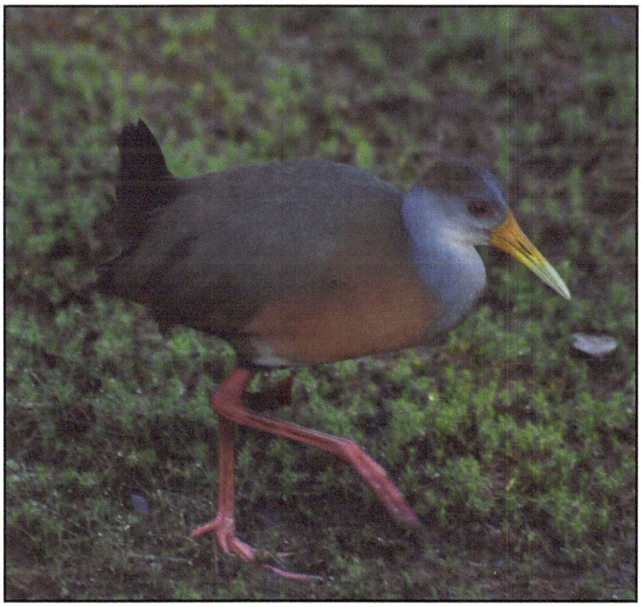

Gray-necked Wood-Rail (*Aramides cajanea*)

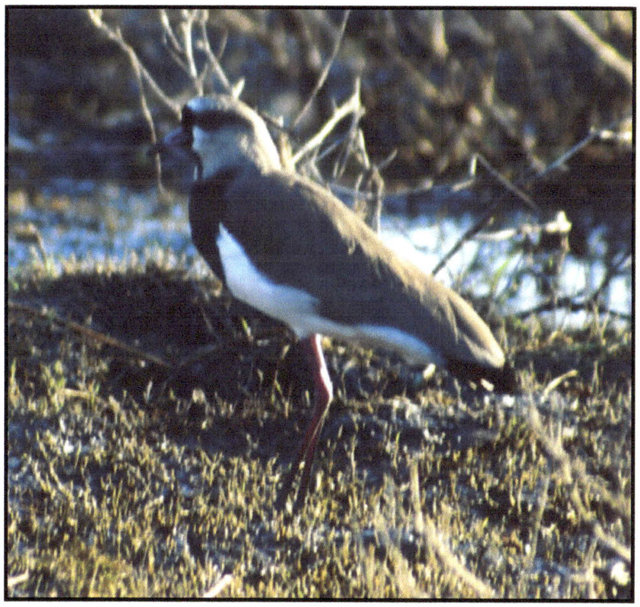

Southern Lapwing (*Vanellus chilensis*)

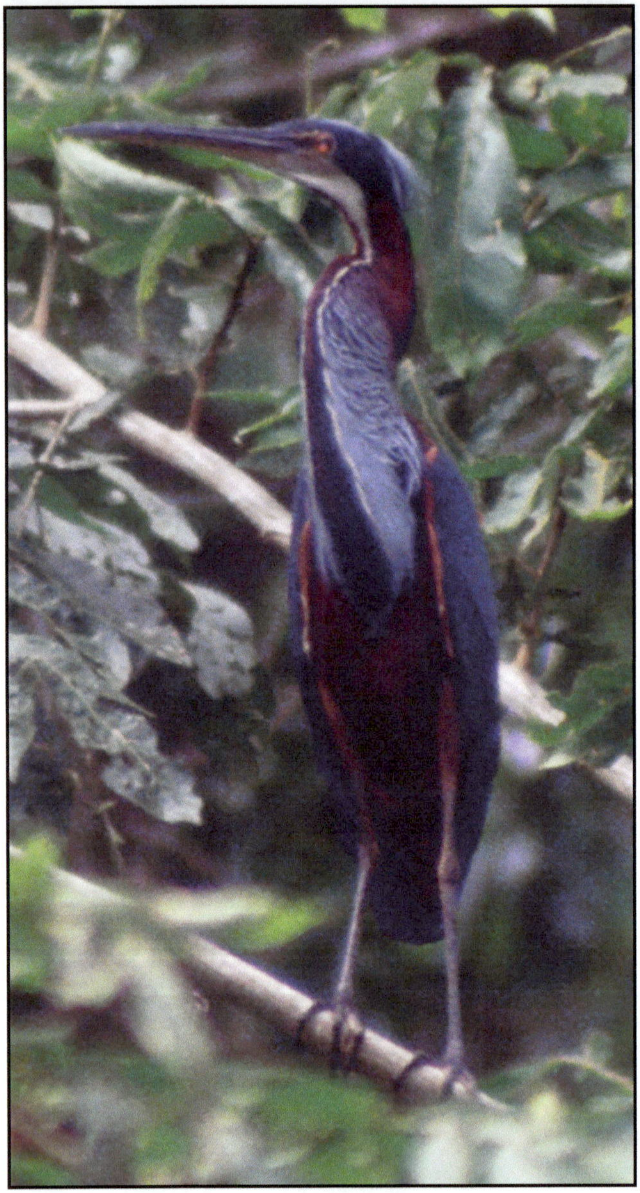

Agami Heron (*Agamia agami*)

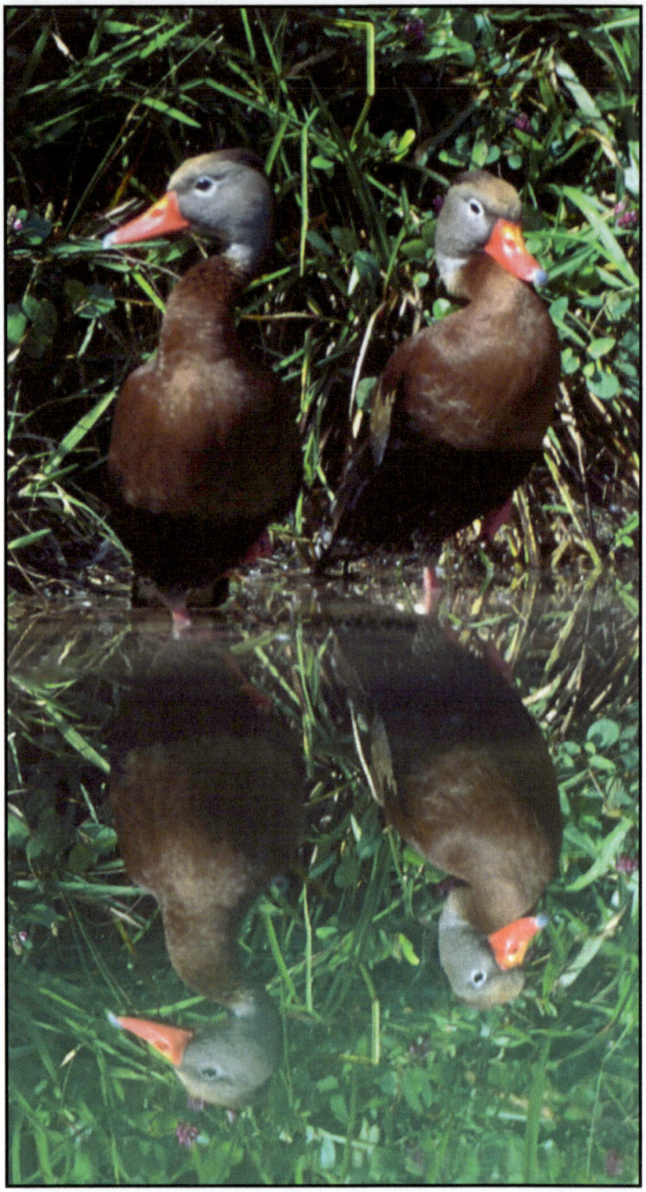

Black-bellied Whistling Ducks (*Dendrocygna autumnalis*)

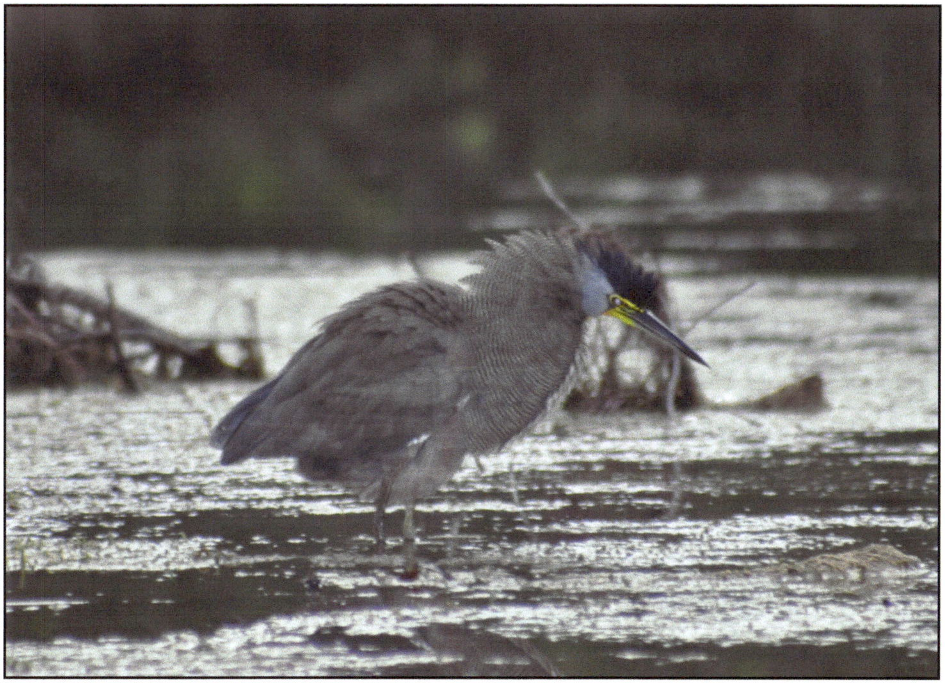

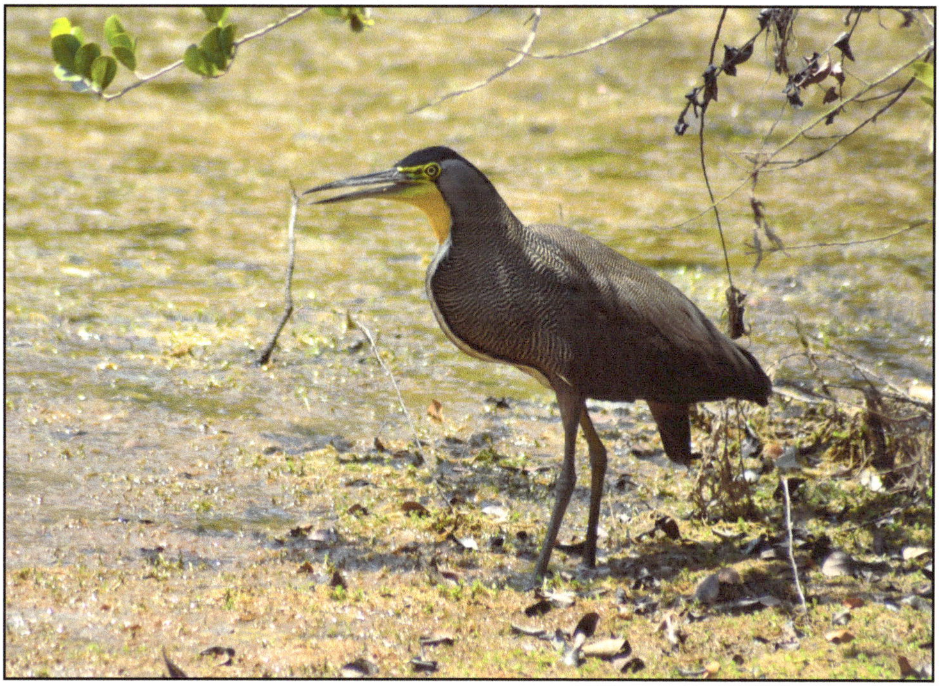

Bare-throated Tiger-Heron
(*Tigrisoma mexicanum*)

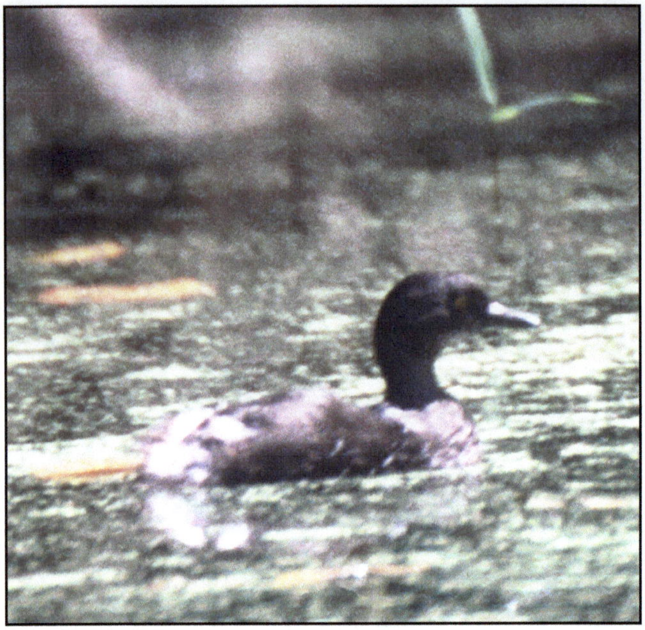

Least Grebe (*Tachybaptus dominicus*)

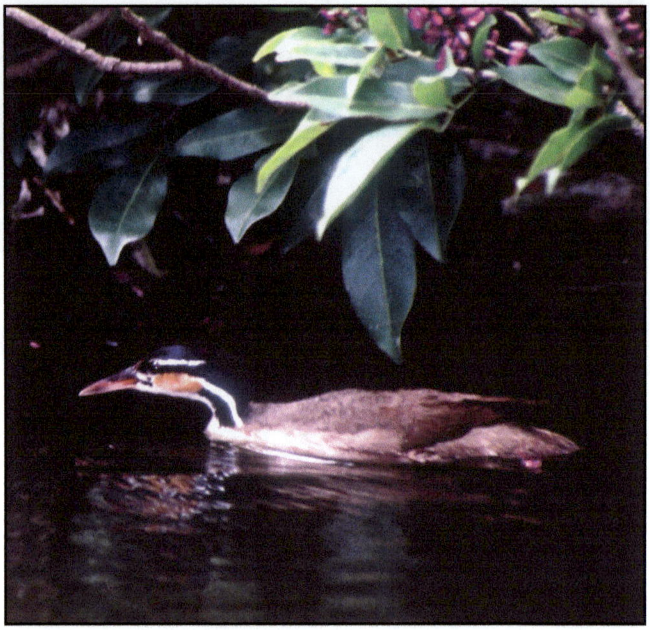

Sungrebe (*Heliornis fulica*)

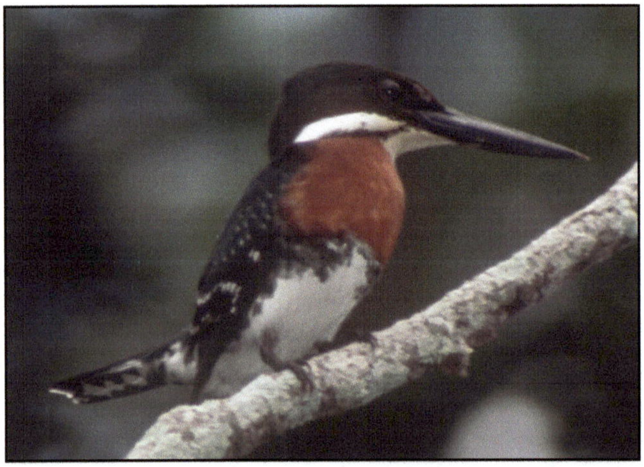

Green Kingfisher (*Chloroceryle americana*)

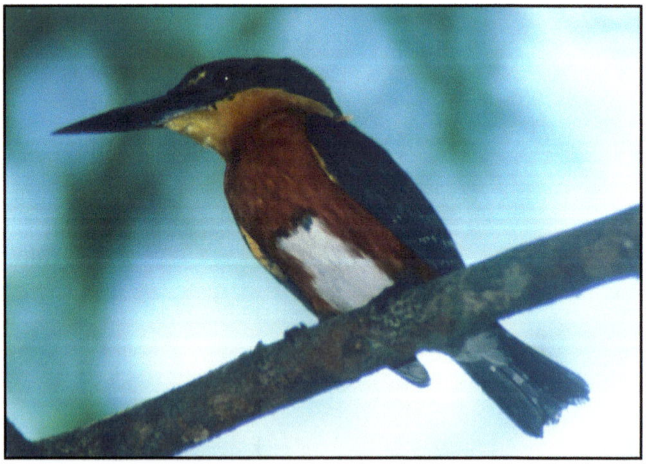

American Pygmy Kingfisher (*Chloroceryle aenea*)

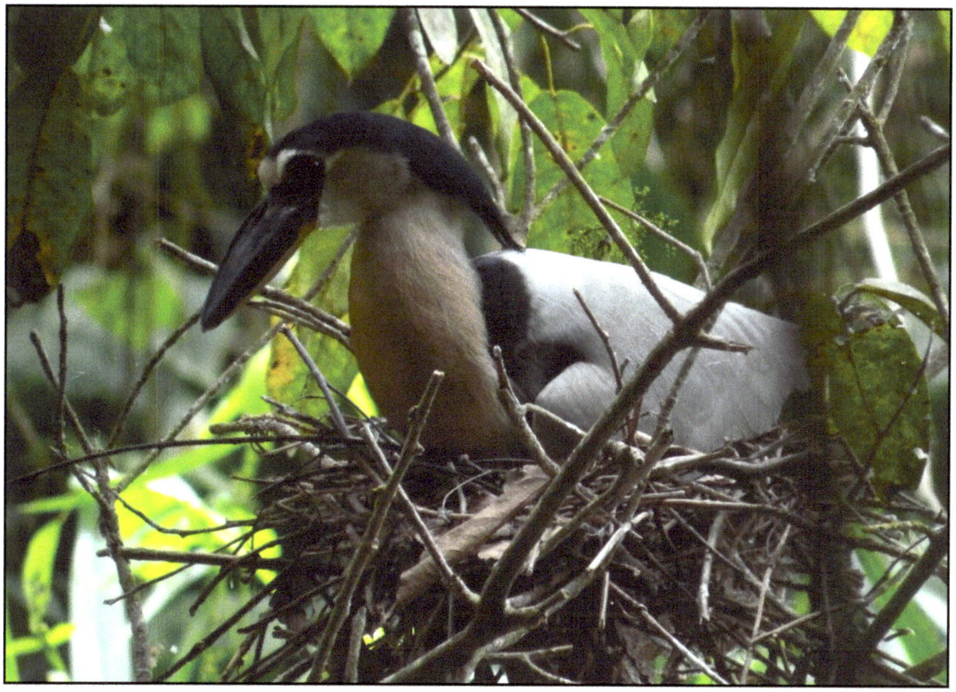

Boat-billed Heron (*Cochlearius cochlearius*)

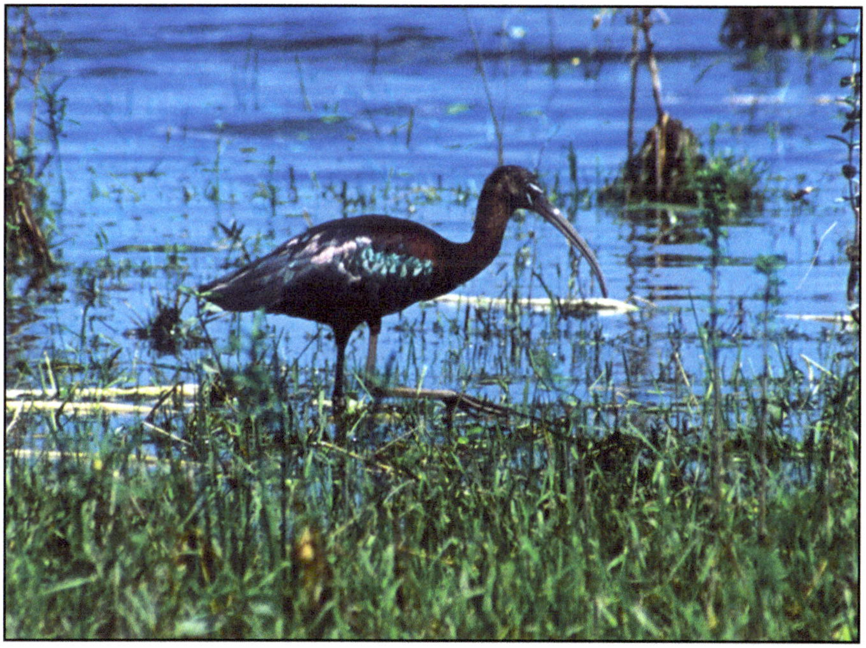

Glossy Ibis (*Plegadis falcinellus*)

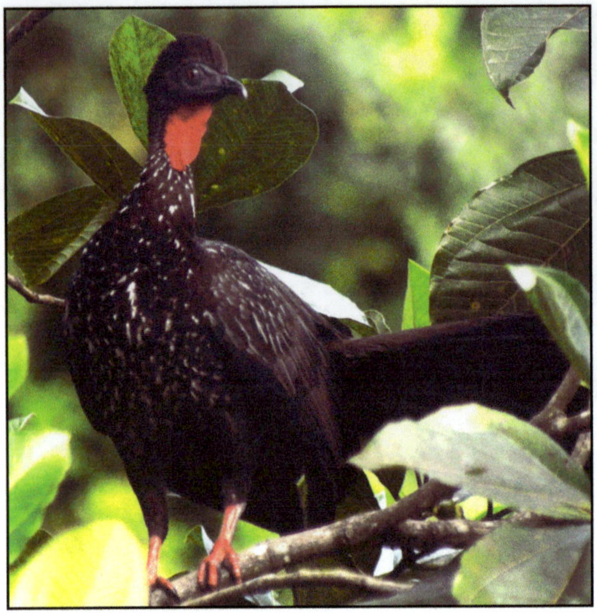

Crested Guan (*Penelope purpurascens*)

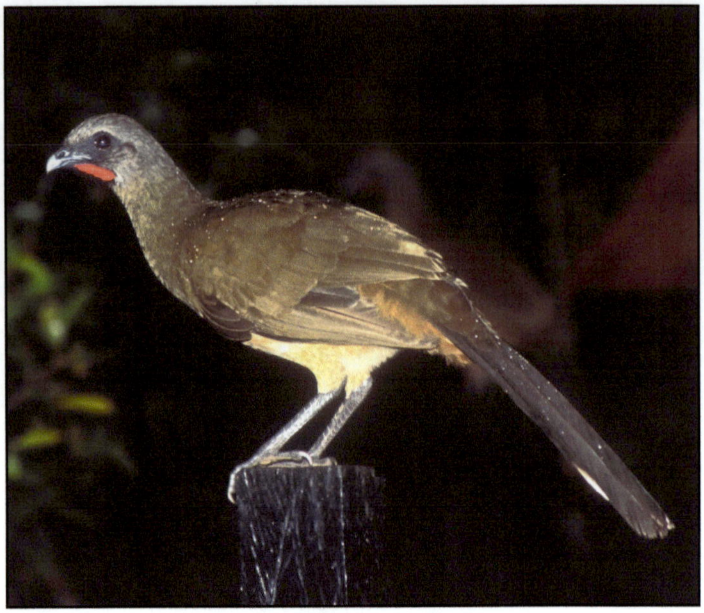

Plain Chachalaca (*Ortalis vetula*)

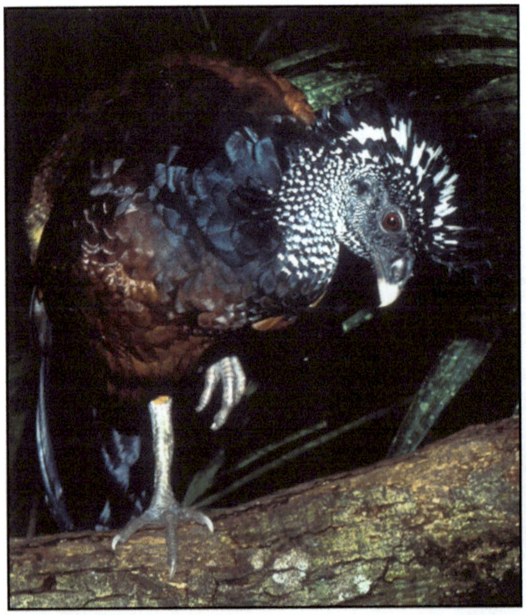

Female Great Currassow (*Crax rubra*)

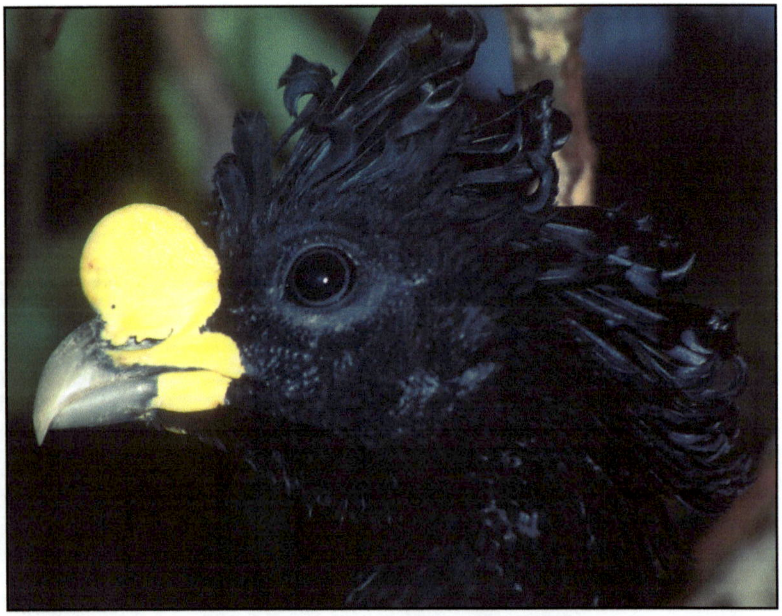

Male Great Currassow (*Crax rubra*)

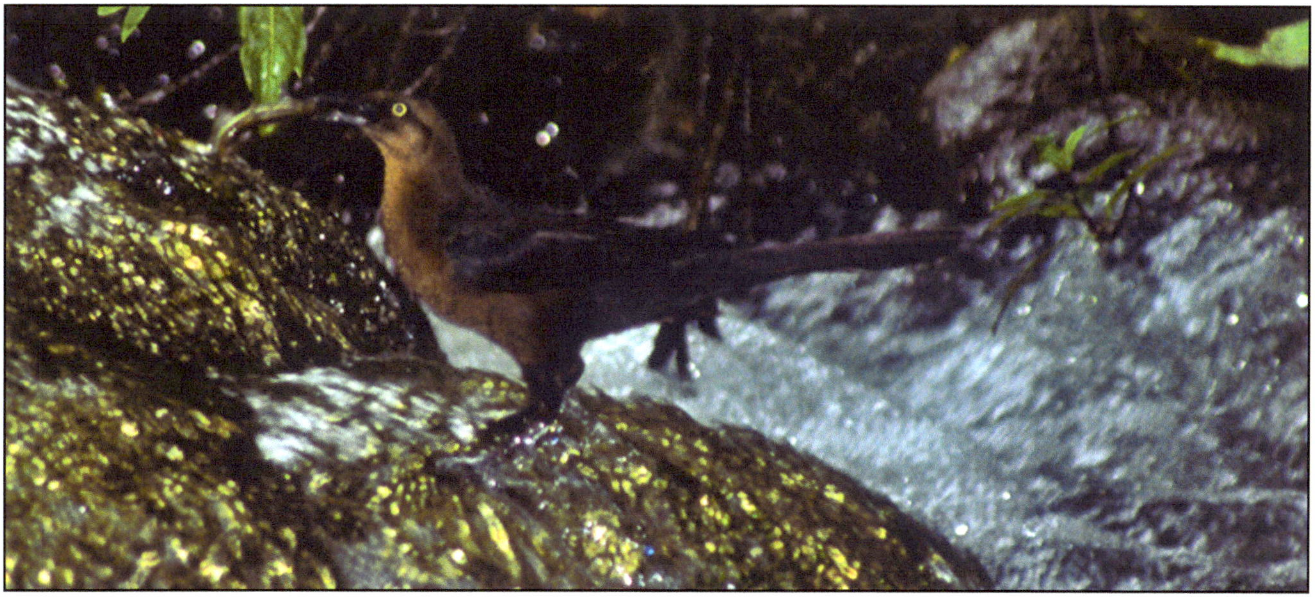

Female Great-tailed Grackle (*Quiscalus mexicanus*) catching fish

Great-tailed Grackles

Although grackles are far from being my favourite bird, I must reluctantly give them their due. They never seem to want for food. The Clarissa Falls on the Mopan River, although relatively shallow, are quite fast moving. During the summer months many small four-inch-long fish travel upstream as salmon are apt to do in the more northern rivers.

These small fish attract birds and I showed up, camera in hand, to document this occurrence. I noted two Green Herons, one Snowy Egret, a Bare-throated Tiger Heron, and four Great-tailed Grackles were already at the site. In the three hours that I watched, I took note of which birds were the most accomplished fishers. The green herons got one fish between them, the tiger heron also got one, the snowy egret got two fish, but the grackles succeeded in catching five fish between them!

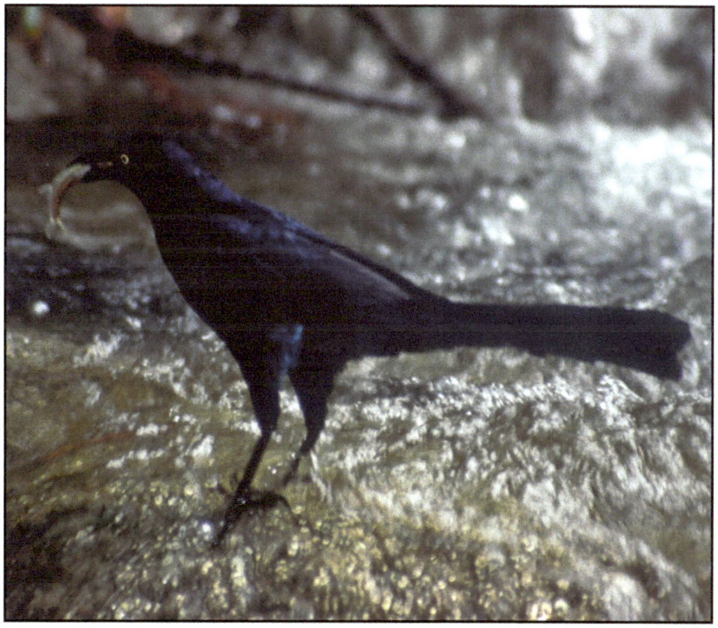

Male Great-tailed Grackle catching fish

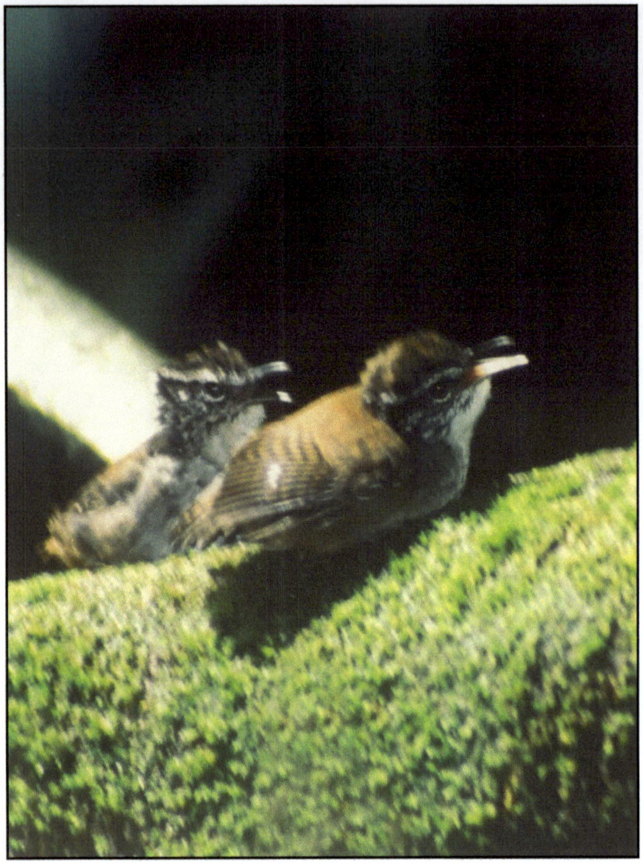

Gray-breasted Wood-Wren chicks

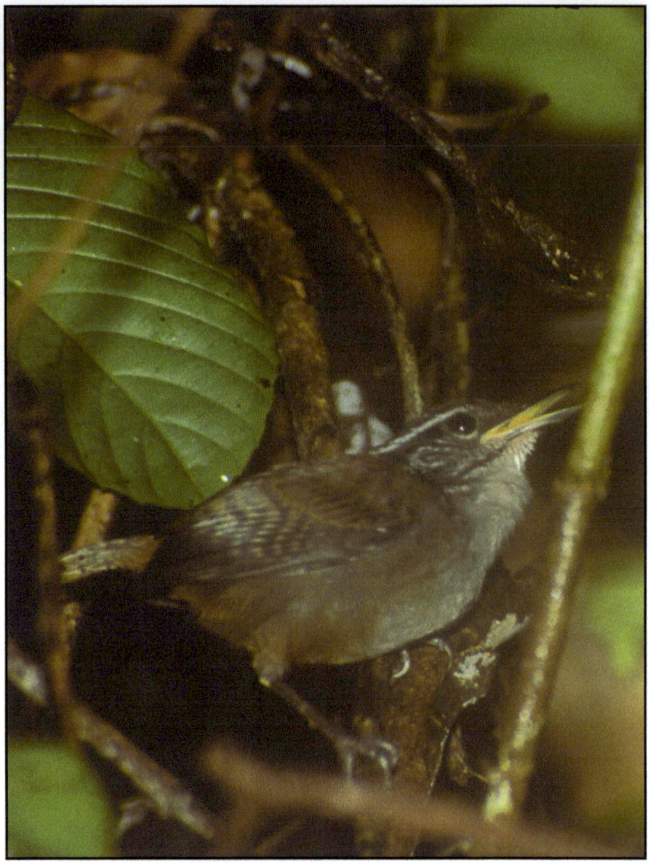

Gray-breasted Wood-Wren
(Henicorhina leucophrys)

Gray-breasted Wood-Wren

The Outlier is a small peak just east of Victoria Peak. A trail had been constructed by Raleigh International in 1995. Dorothy and I had climbed it, and although quite tough we had made it to the top and totally enjoyed the view and the Elfin Forest at the summit. I made another visit to the Outlier on the 31st of August, 2005. The weather was clear, but humid. As I climbed, it seemed to get hotter and more humid. "Cloud Forest" was an apt description!

Suddenly a small bird flew in front of me, and landed on a twig, only four feet away. I backed away; my closest focussing distance was nine and a half feet. I managed one shot, and then it flew towards me again. I recognised it as a wood-wren, probably a young one because its breast was grey, rather than white. As I walked up the slope, the bird flew to the ground in front of me, but I ignored it and walked on. I became aware of a twittering sound further up the hill. Pretty soon I came upon a large piece of deadwood by the side of the path; sitting on it were two wren chicks.

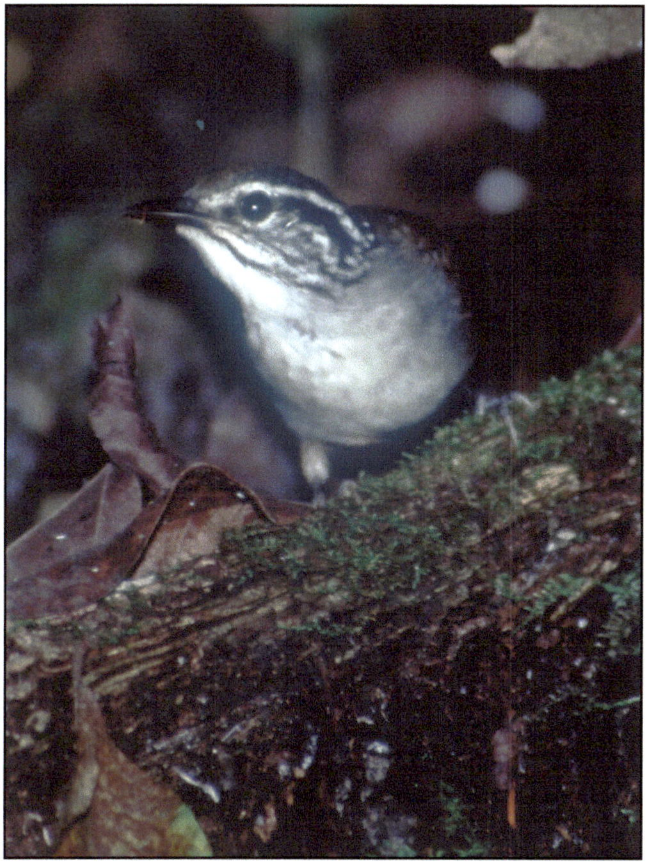

White-breasted Wood-Wren (*Henicorhina leucosticta*)

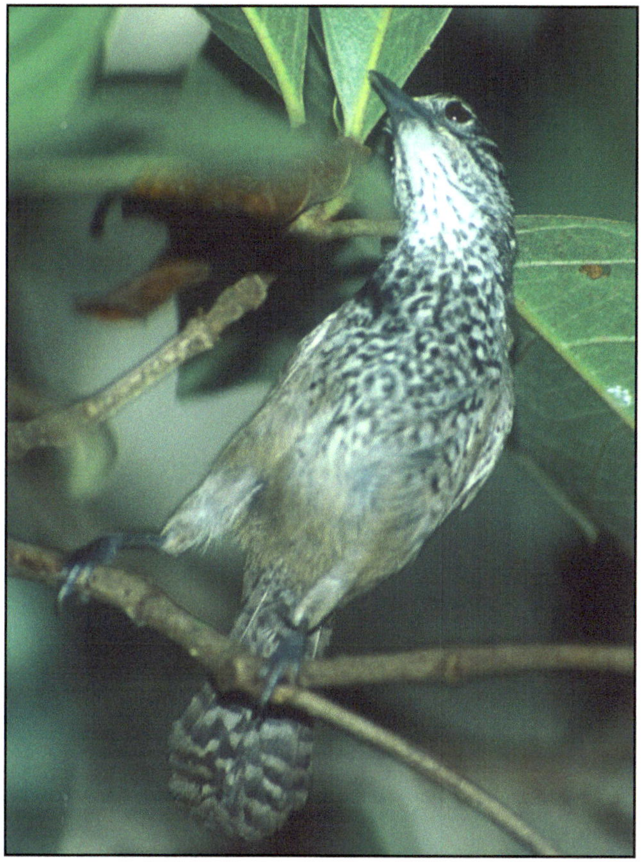

Spot-breasted Wren (*Thryothorus maculipectus*)

The puzzle solved! I took three quick pictures of the wren chicks, then walked on.

I hit the summit fifteen minutes later. On the way down I stopped to take a picture of the spectacular scene. But the camera didn't work! The camera was less than a year old. I tried my flash and it didn't work either, even with new batteries. Obviously cloud forest is death to electronic photo equipment.

I stored the film in the freezer and forgot about them for two years. When I saw the slides, I checked the bird book; it looked good for a Gray-breasted Wood-Wren that had not yet been documented in Belize and was only known in the Mexican and South American highlands. I sent the photos to Lee Jones, a friend and the authority on birds in Belize, who consulted with other birding experts. Because the two chicks were in close vicinity of the single bird, it was identified as the Gray-breasted Wood-Wren as opposed to a juvenile White-breasted Wood-Wren. To cut a long story short, the Grey-breasted Wood-Wren was declared to be a new species for Belize.

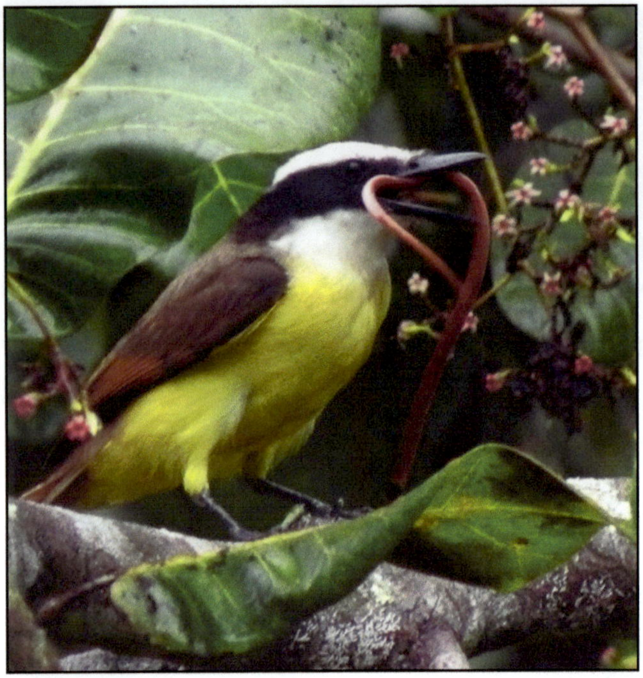

Great Kiskadee (*Pitangus sulphuratus*) eating a snake

KISKADEES

Great Kiskadees are common flycatchers in Belize. However, as far as I'm concerned, the word flycatcher is quite erroneous. On several occasions I've come across kiskadees feeding not on flies, not even large insects, but reptiles.

My first sighting of a reptile-eating kiskadee was in Cockscomb Basin Wildlife Sanctuary. I was wandering around the campground when I noticed a small commotion in the surrounding bush. On investigating I found a kiskadee holding a snake of the mussurana group, a slender rear-fanged snakes of the *Clelia* genus. I watched them wrestling for about five or six minutes, then the snake seemed to tire, at which point the bird flew off with the snake body dragging behind on both sides. It was in late June, shortly after nesting season; my supposition was that the bird was taking the snake back to hungry chicks.

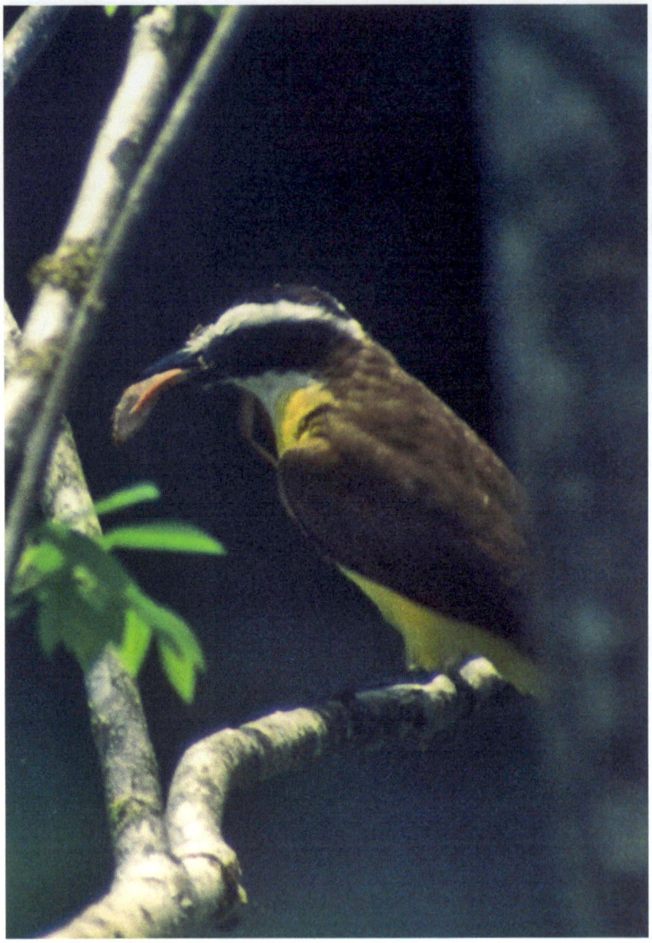

Great Kiskadee eating a lizard

On another occasion I witnessed a kiskadee overpowering a lizard and not a very small one, at least compared to the bird. The lizard appeared to be a small basilisk. Like the first incident, this one also flew off with the body.

I also witnessed a third incident which unfortunately I couldn't photograph. Again it was a lizard, possibly a large anole, which was devoured on the spot by a Great Kiskadee.

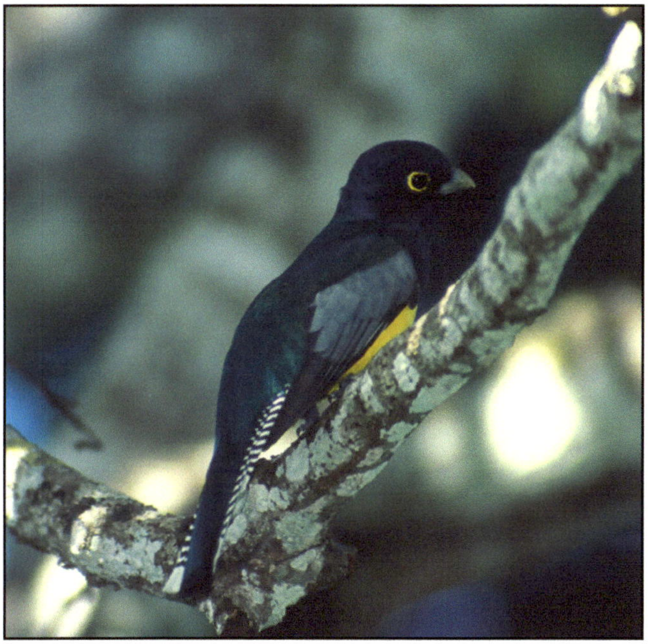

Gartered Trogon (*Trogon caligatus*)

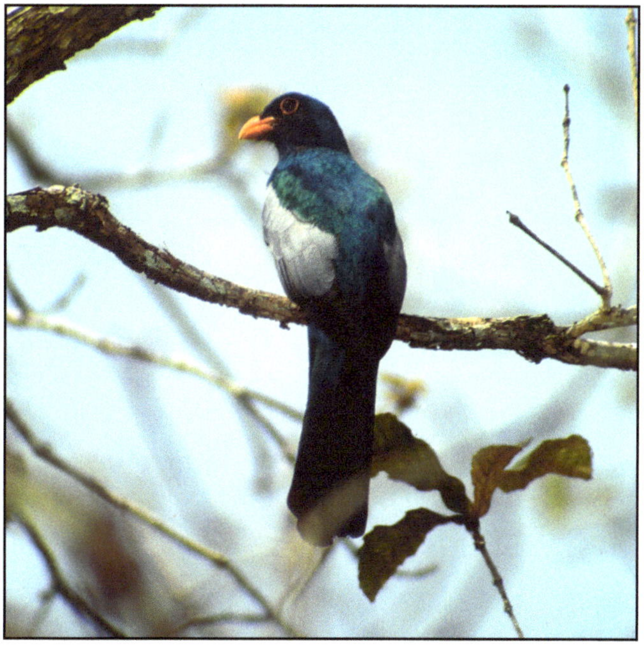

Slaty-tailed Trogon (*Trogon massena*)

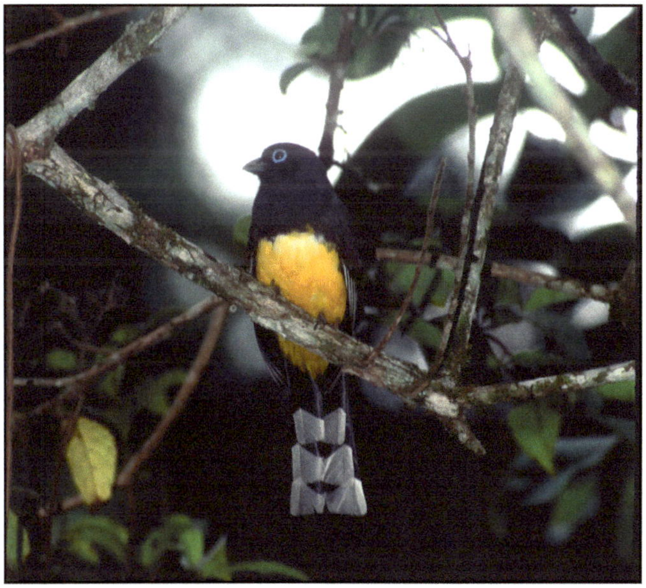

Black-headed Trogon (*Trogon melanocephalus*)

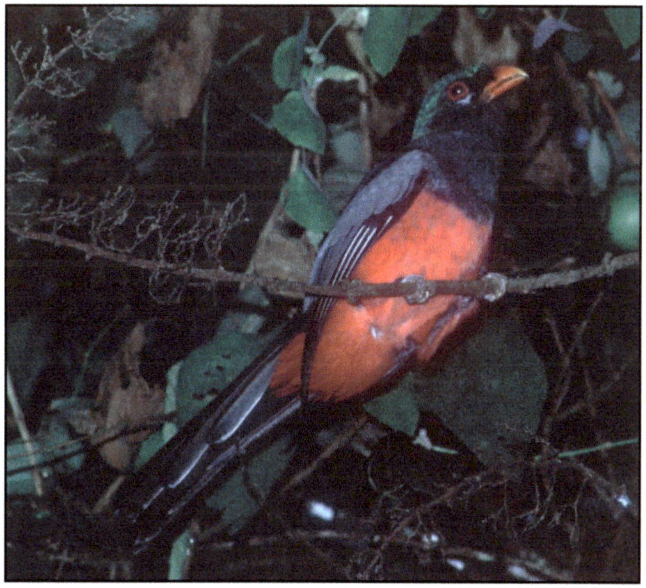

Slaty-tailed Trogon (*Trogon massena*)

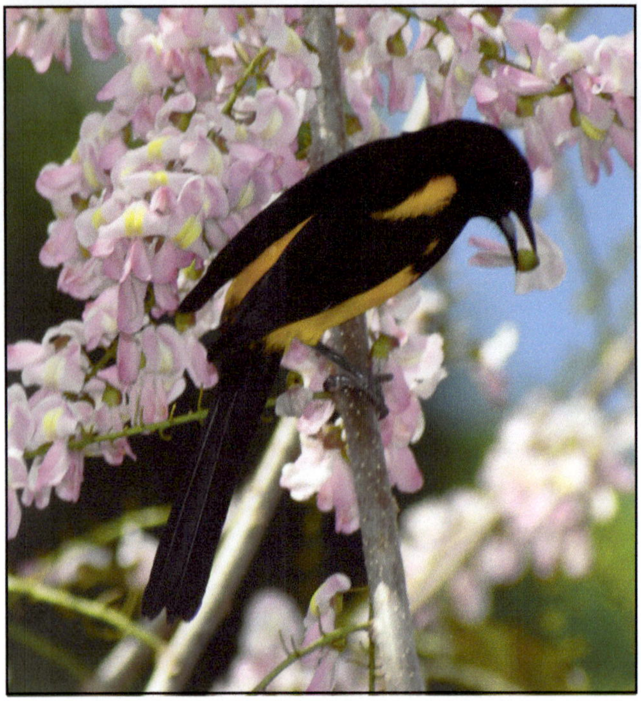

Black-cowled Oriole (*Icterus prosthemelas*)

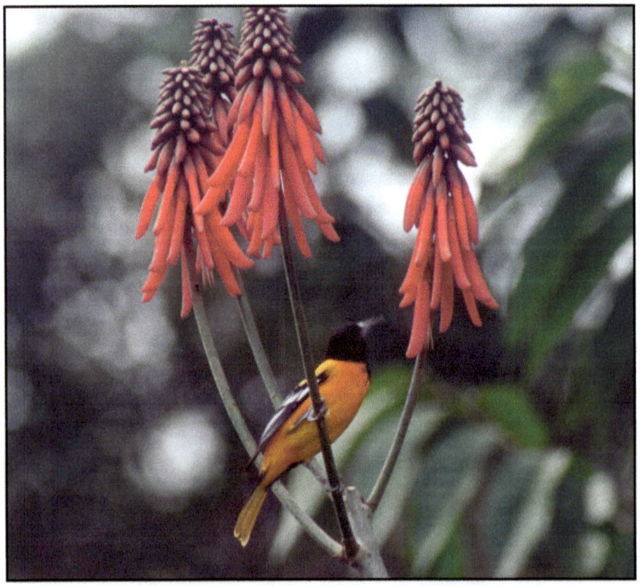

Baltimore Oriole (*Icterus galbula*)

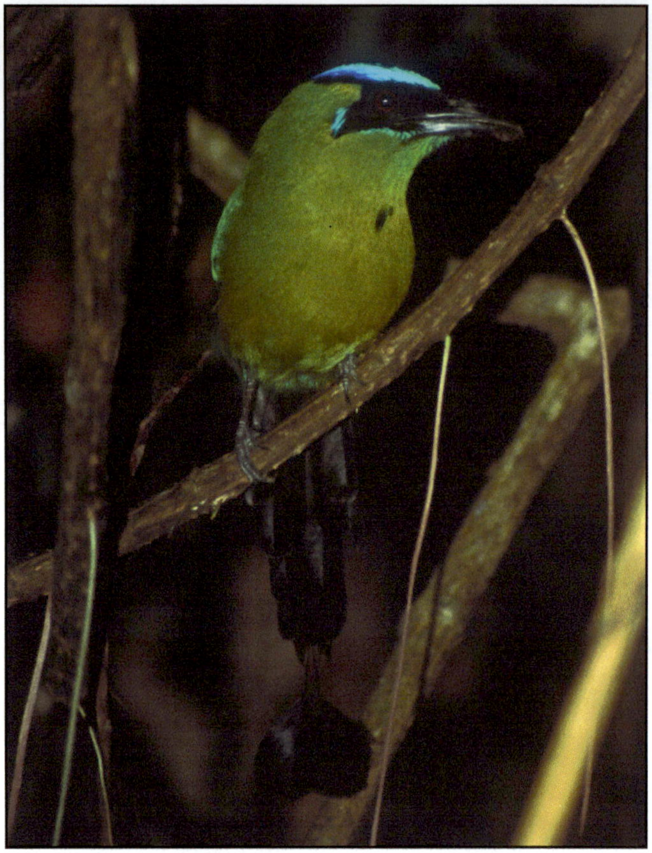

Blue-crowned Motmot (*Momotus momota*)

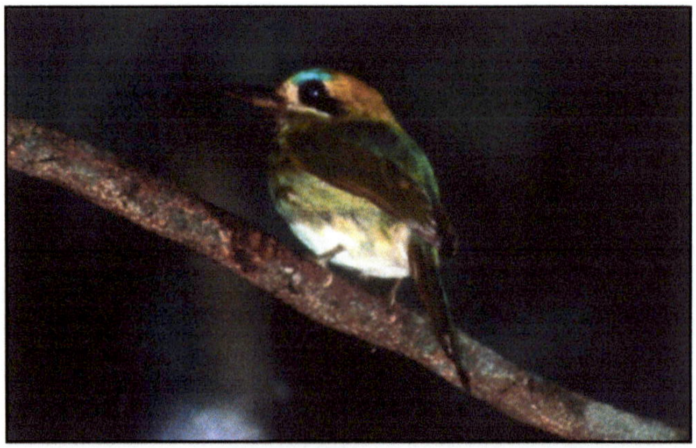

Tody Motmot (*Hylomanes momotula*)

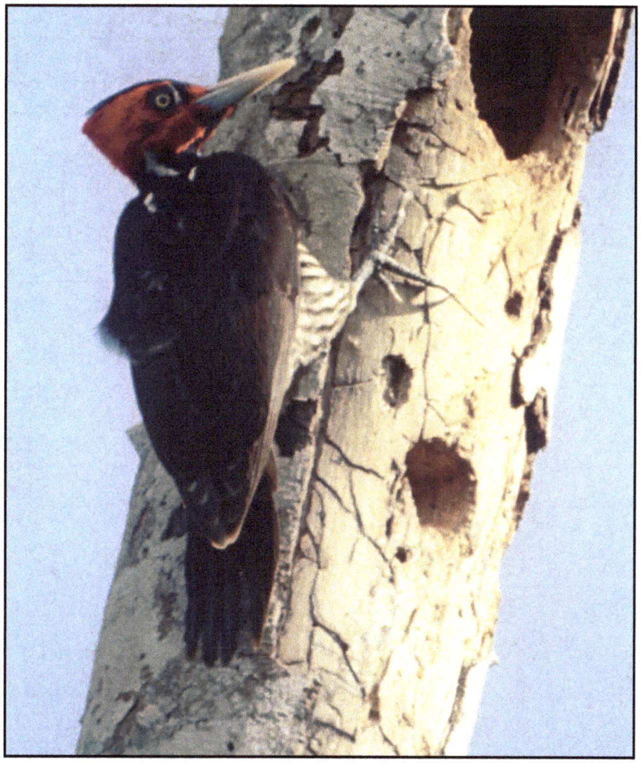

Pale-billed Woodpecker (*Campephilus guatemalensis*)

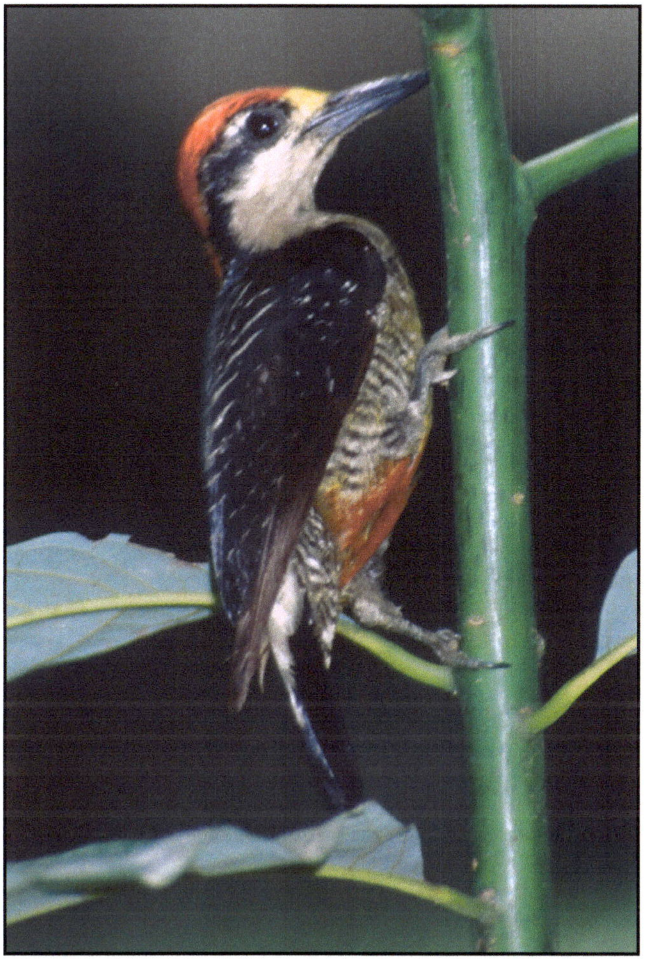

Black-cheeked Woodpecker *(Melanerpes pucherani*)

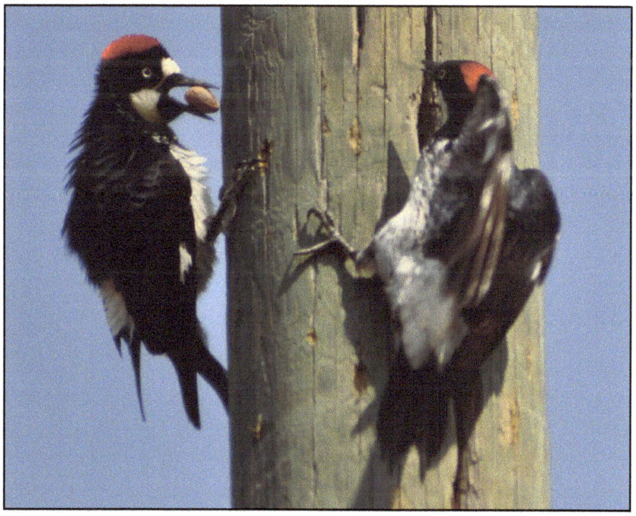

Acorn Woodpeckers (*Melanerpes formicivorus)*

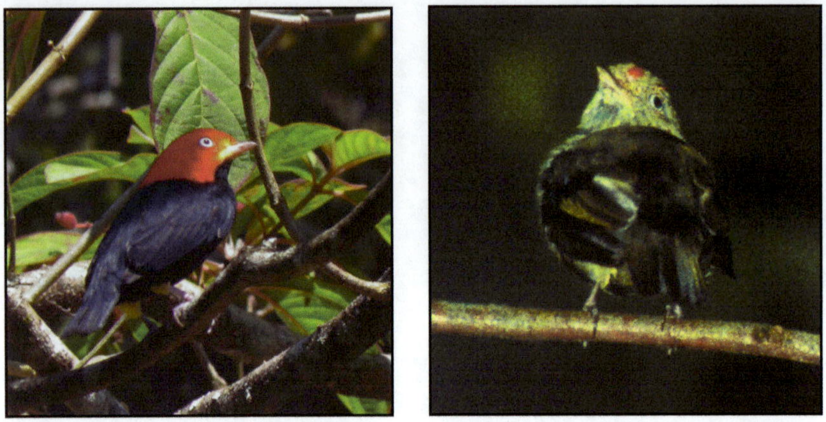

Red-capped Manakin (*Pipra mentalis*) adult male (*left*) juvenile male (*right*)

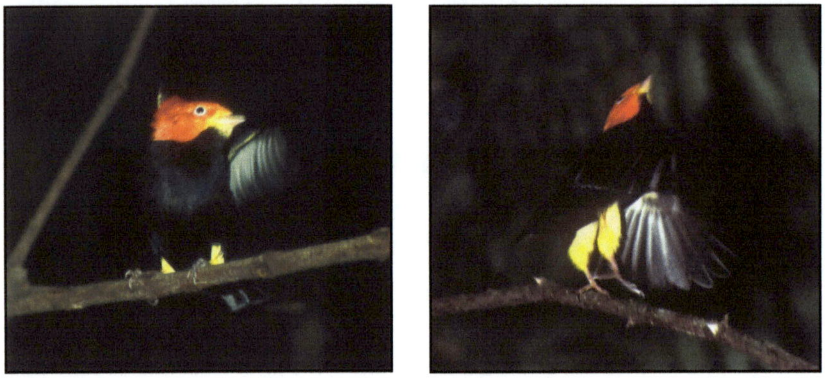

Red-capped Manakin wing slapping (*left*) and dancing (*right*)

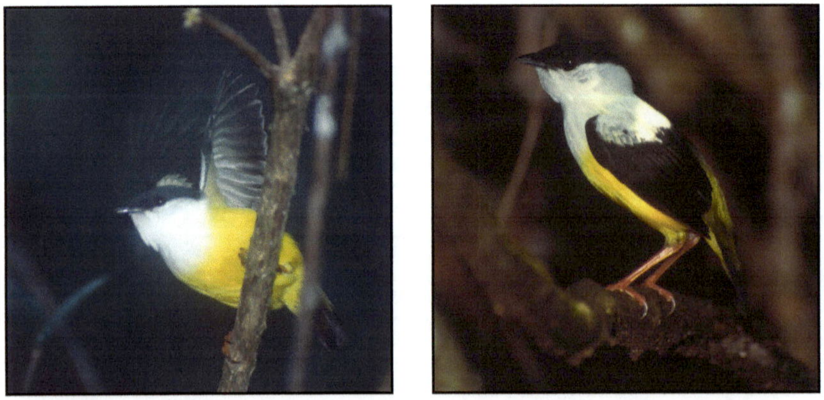

White-collared Manakin (*Manacus candei*) wing slapping (*left*)

Manakins

One of my favourite forest bird species are the manakins. In Belize there are two species of manikins, both equally interesting. My personal favourite is the Red-capped Manakin. The male of which is a small black bird with a bright red head. The other manakin species is the White-collared Manakin. The males of both of these small birds are quite gregarious, especially around mating time.

The male Red-capped Manakins get together and perform a dance in an attempt to attract the females, who, sad to say, are quite dowdy. The dance of the male White- collared Manakin is seldom seen, for they tend to dance in the privacy of a forest clearing well hidden by the forest itself.

However, the dance of the male Red-capped Manakin can be quite easily viewed via binoculars as they perform on a horizontal limb of a large forest tree, usually about thirty feet above the ground.

In addition to the dancing, both manikin species indulge in another habit, wing slapping. While wing slapping, the male manakins get together in small groups. In wing slapping the wings come back very rapidly above their backs to make a very loud slapping noise, which will leave the uninitiated puzzled and searching for the cause in the forest. Both species of manikins are relatively common in the forests of Belize.

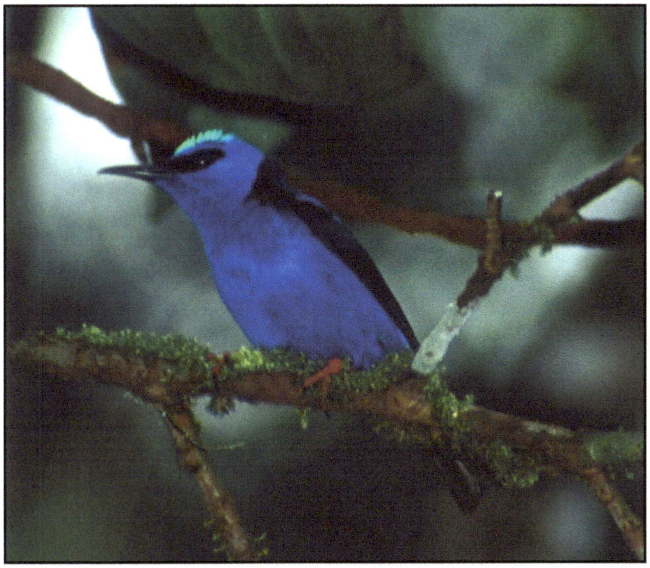

Red-legged Honeycreeper (*Cyanerpes cyaneus*)

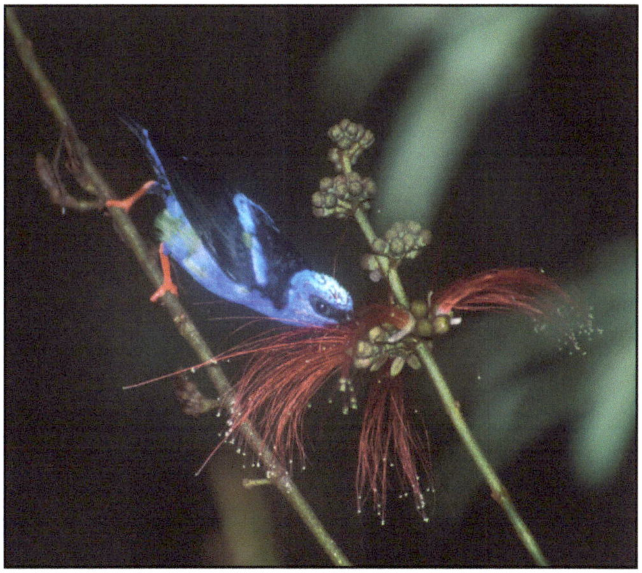

Juvenile Red-legged Honeycreeper

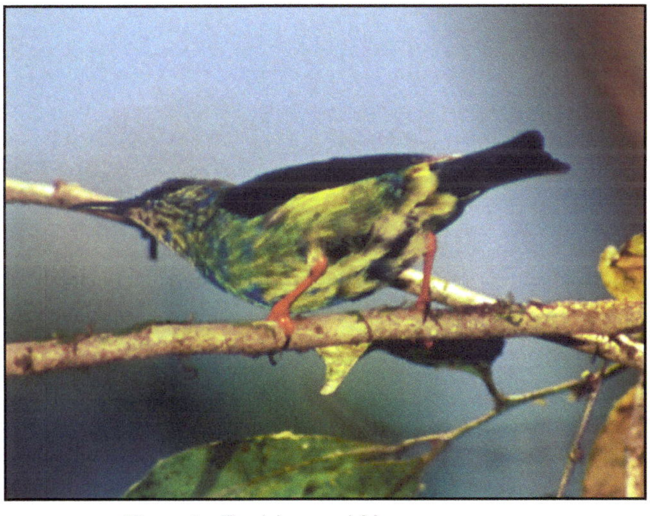

Female Red-legged Honeycreeper

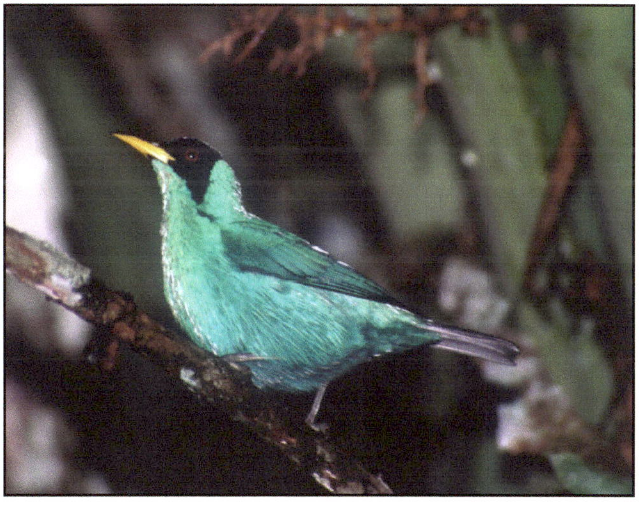

Green Honeycreeper (*Chlorophanes spiza*)

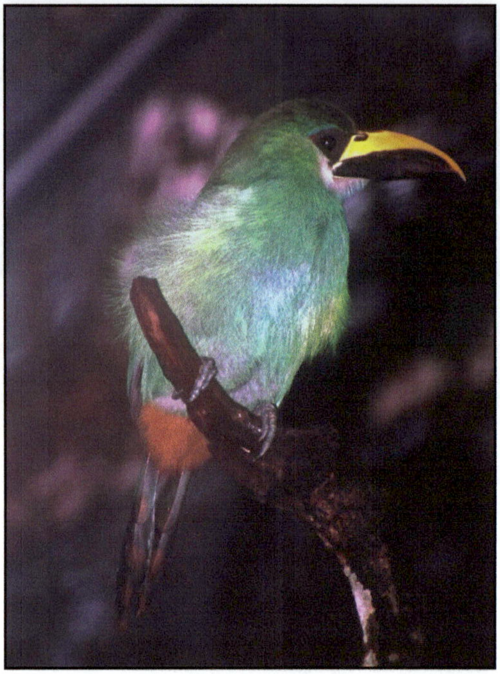

Emerald Toucanet (*Aulacorhynchus prasinus*)

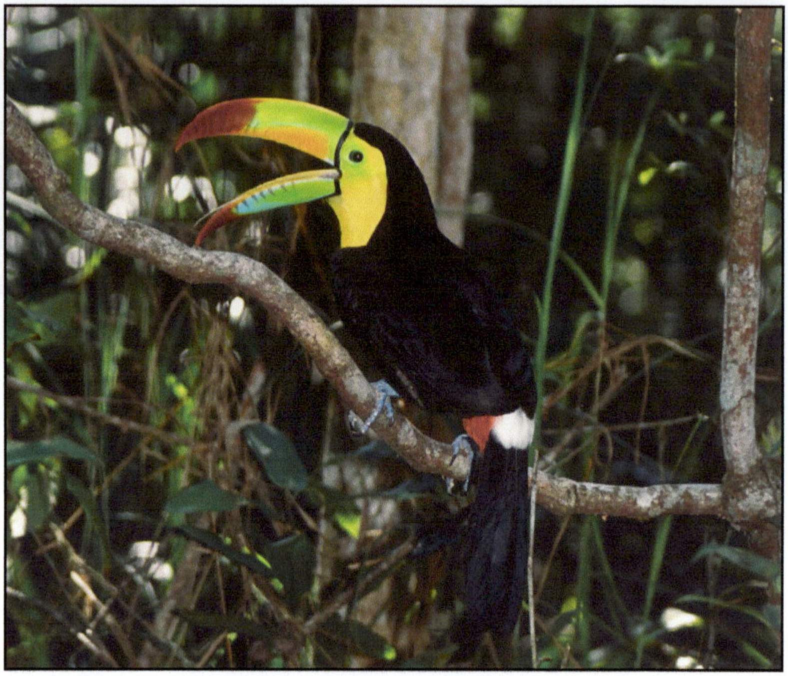

Keel-billed Toucan (*Ramphastos sulfuratus*)

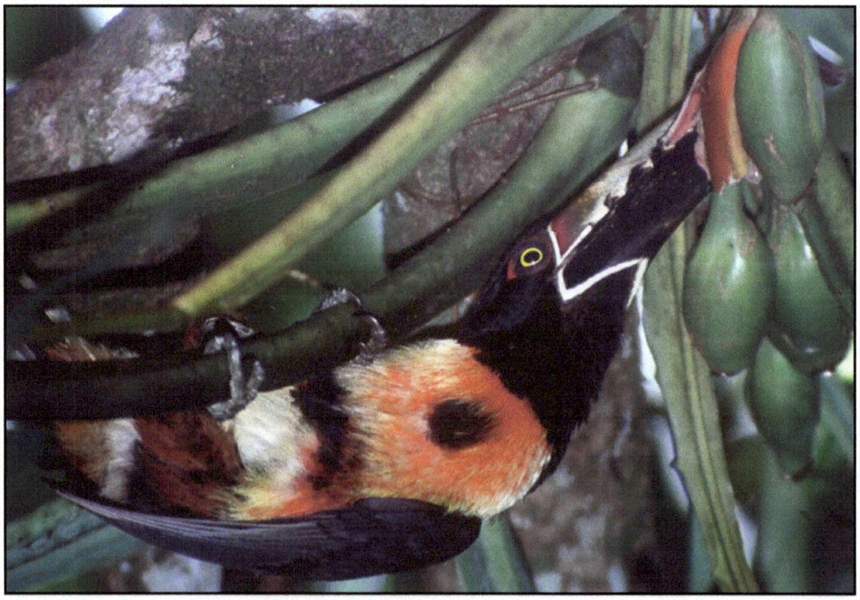

Collared Araçari (*Pteroglossus torquatus*)

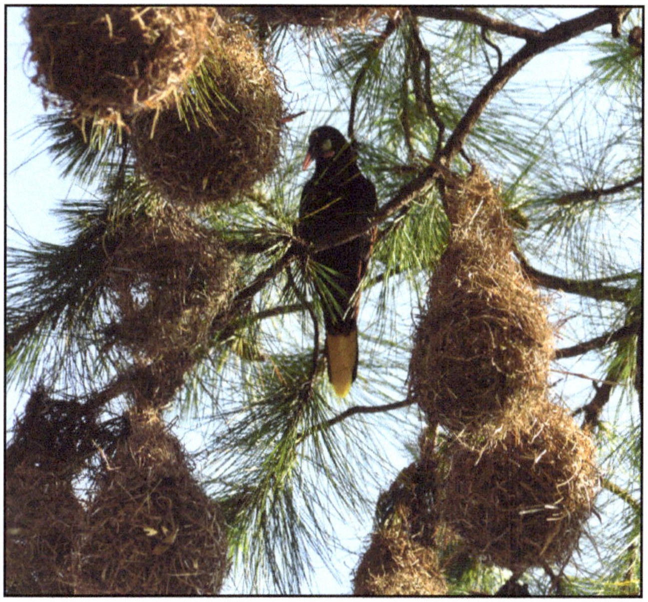
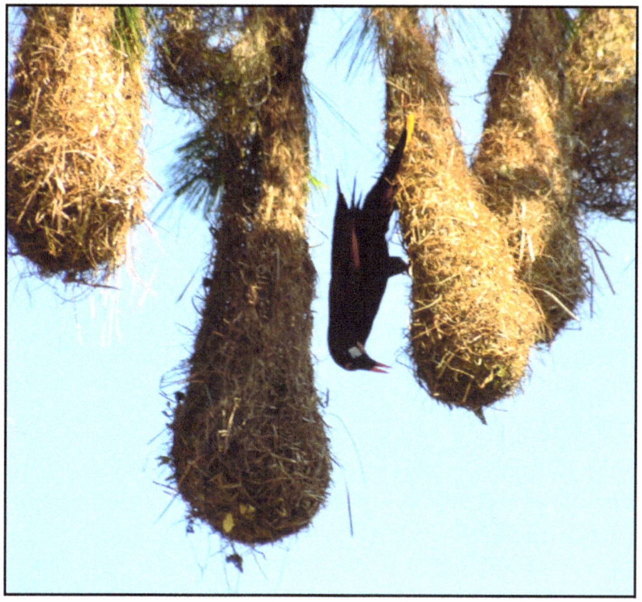

Montezuma Oropendola (*Psarocolius montezuma*)

Montezuma Oropendola

If the manikins are the dancers of Belize's bird world, then the oropendolas are the architects and builders. Like the manikins, there are two species of oropendolas. The more common of the two is the Montezuma's Oropendola, and the other is the Chestnut-headed Oropendola.

Oropendolas are colonial birds that build their nests in a single large tree. The nest could be described as a pendulum, hence the name oropendola. There may be over thirty nests in one tree and the nests are normally built by the females, which are less than half the weight of the male.

The colourful oropendola is a large icterid with a noticeable yellow tail and a predominantly pale blue head with a large black and orange-red bill.

Montezuma Oropendolas have an extremely complex relationship with another icterid, the Giant Cowbird, a brood parasite. The cowbird will attempt to lay an egg in the oropendola's nest. On some occasions oropendolas will not make a distinction between its own and the cowbird's eggs even though it is a different colour. These eggs are incubated along with the oropendula eggs. However, on other occasions the oropendola will be aggressive towards the cowbird and remove the cowbird egg from its nest.

The reason for the two different receptions has been discovered by researchers. When oropendolas are not aggressive towards the cowbird and happily adopt the cowbird chick, they have a good reason. The cowbird chick will capture and eat botfly, even grooming the oropendula chicks. If left to their own means, the botfly would lay its eggs in both the oropendola and the cowbird chicks.

Where the oropendolas were aggressive towards the cowbird and pushed out the cowbird egg, researchers found that the oropendola nest was close to either a wasp nest or a bee hive. Botflies avoid both these insects. In the absence of botflies, a cowbird has no purpose in the nest.

HUMMINGBIRDS

In the middle of June, 1986, in Cockscomb Basin Wildlife Sanctuary, just prior to 4:40 am, I was readying to leave the pond where I'd done a photo shoot of a treefrog orgy. The mosquitoes had been steadily building up with the approach of dawn. I did have some allies; the bats were also increasing in numbers and circling my head where the mosquitoes were already fat with my blood.

I noticed that one circling bat seemed different. I followed it with my headlamp until it landed on a branch about twelve feet away. It wasn't a bat, but rather a hummingbird. I unpacked my camera, flash, and spotlight which I attached to the underside of the camera. The hummingbird was still sitting on the branch. I focused and took one picture. The bird flew off. I waited, but it didn't return. Two weeks later, when my film was processed, I came across the hummingbird slide. I was surprised to see that it had red eye-shine. This is usually only seen in nocturnal animals and birds, such as cats, owls and crocodiles.

Although I was familiar with several hummingbirds, I knew none by name. And it stayed that way until the late 1990s when I purchased *Birds of Mexico and Northern Central America* by Howell and Webb. So the slide of my red-eyed "nocturnal" hummingbird remained unidentified for twelve years and then I likened it to a Violet Sabrewing, possibly a juvenile male. Was the red-eye normal? About twenty years later, I took a picture of an adult Violet Sabrewing at dusk that also showed obvious red-eye. In the many hummers I've photographed, these are the only ones I've seen with red-eye.

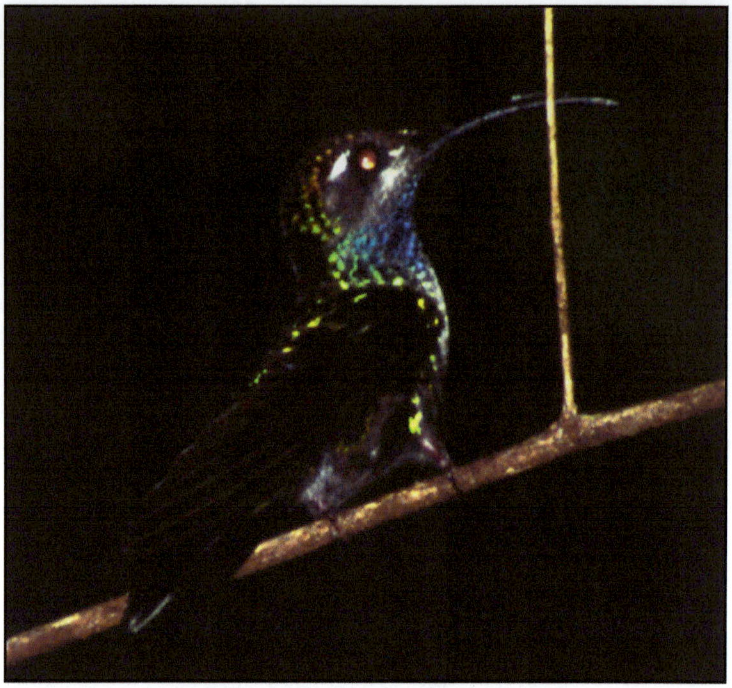

Juvenile Violet Sabrewing at night

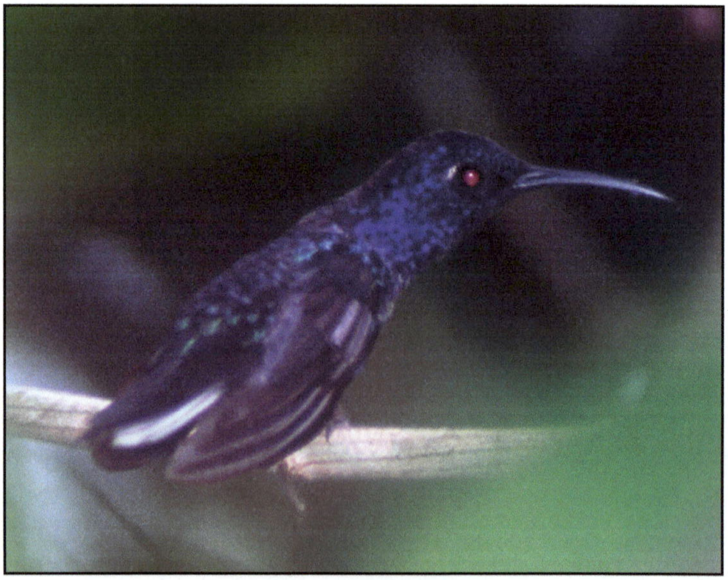

Violet Sabrewing at dusk

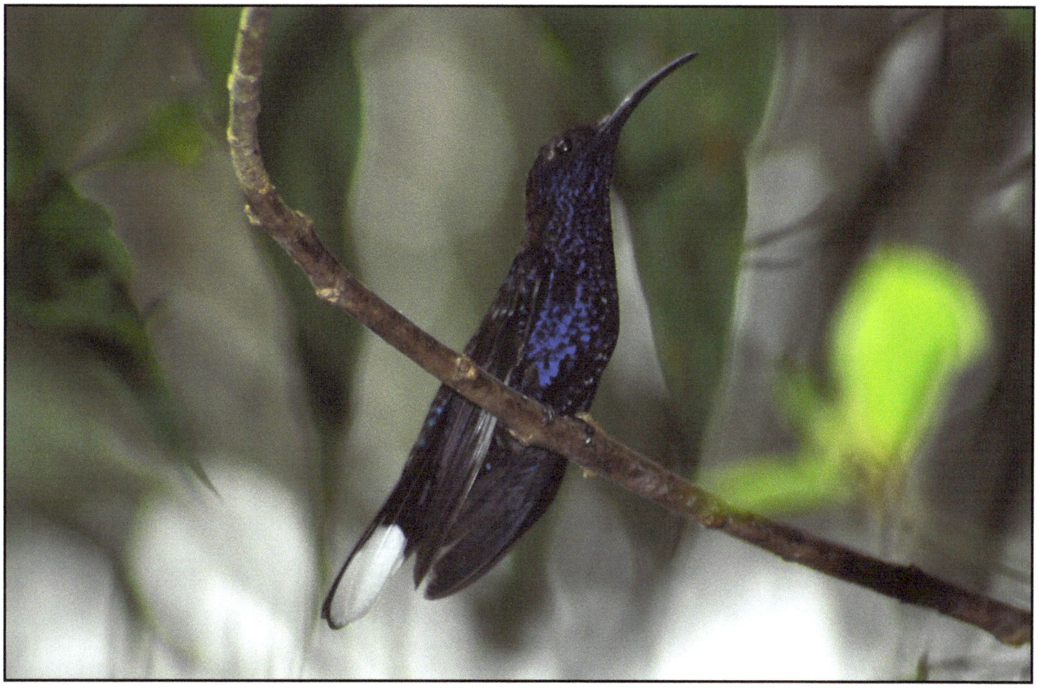

Violet Sabrewing (*Campylopterus hemileucurus*)

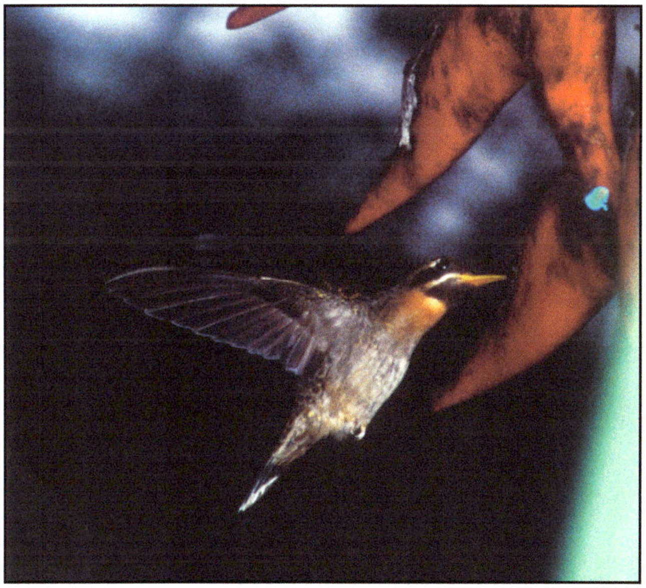

Band-tailed Barbthroat (*Threnetes ruckeri*)

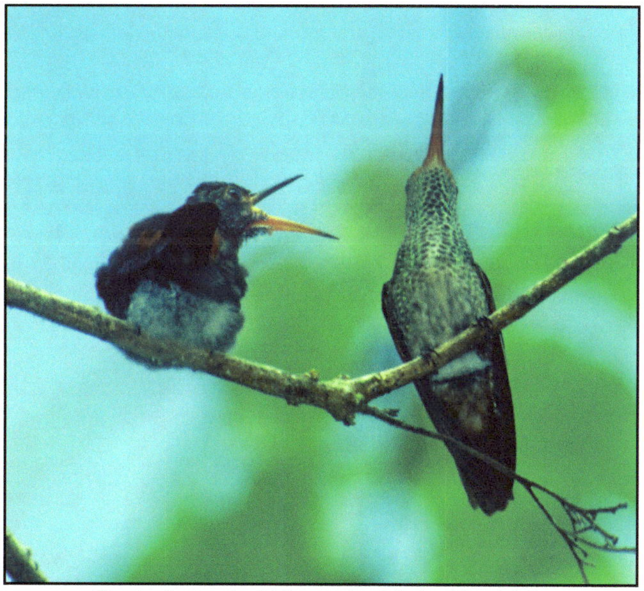

Scaly-breasted Hummingbird with chick (*Phaeochroa cuvierii*)

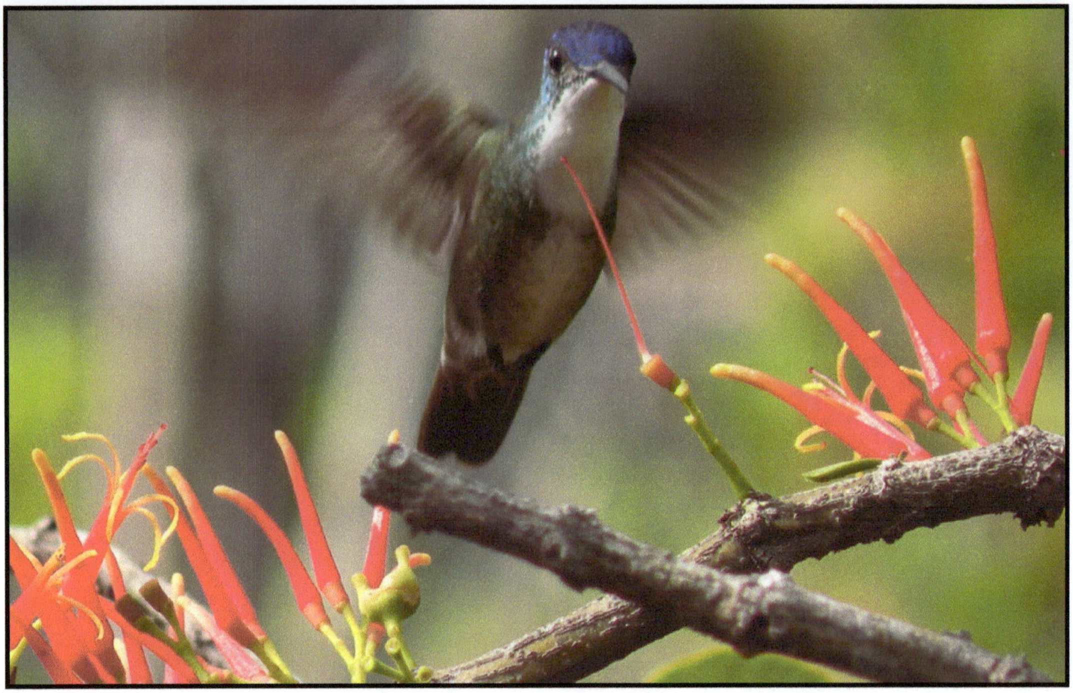

Azure-crowned Hummingbird (*Amazilia cyanocephala*)

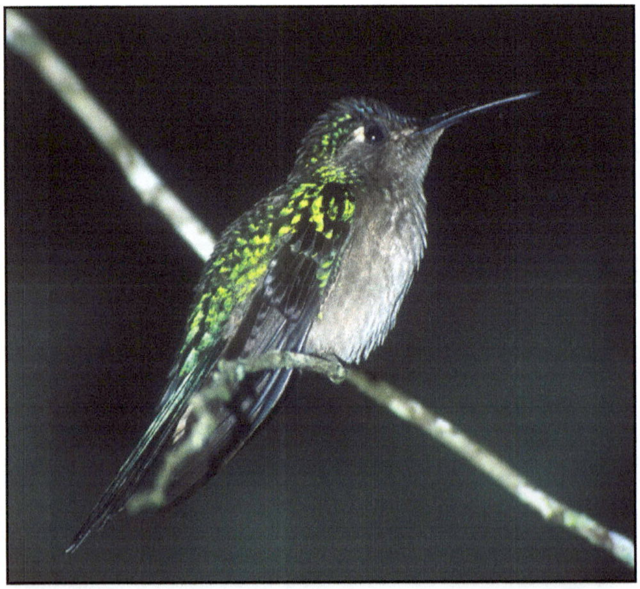

Wedge-tailed Sabrewing (*Campylopterus curvipennis*)

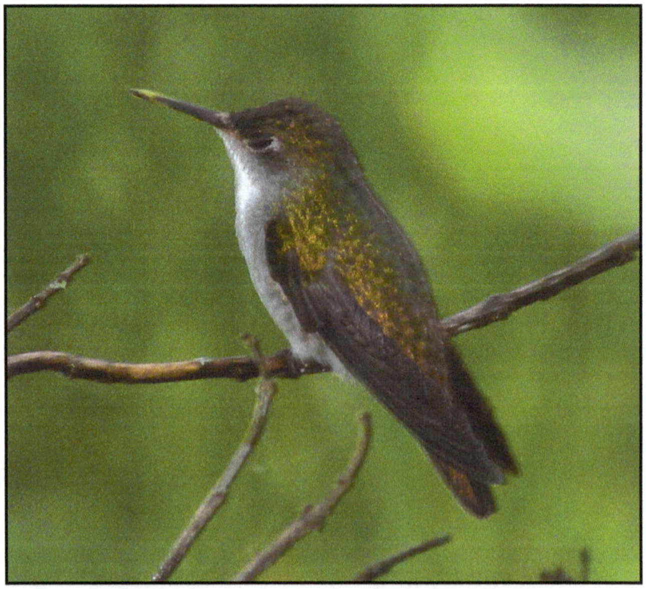

White-bellied Emerald (*Amazilia candida*)

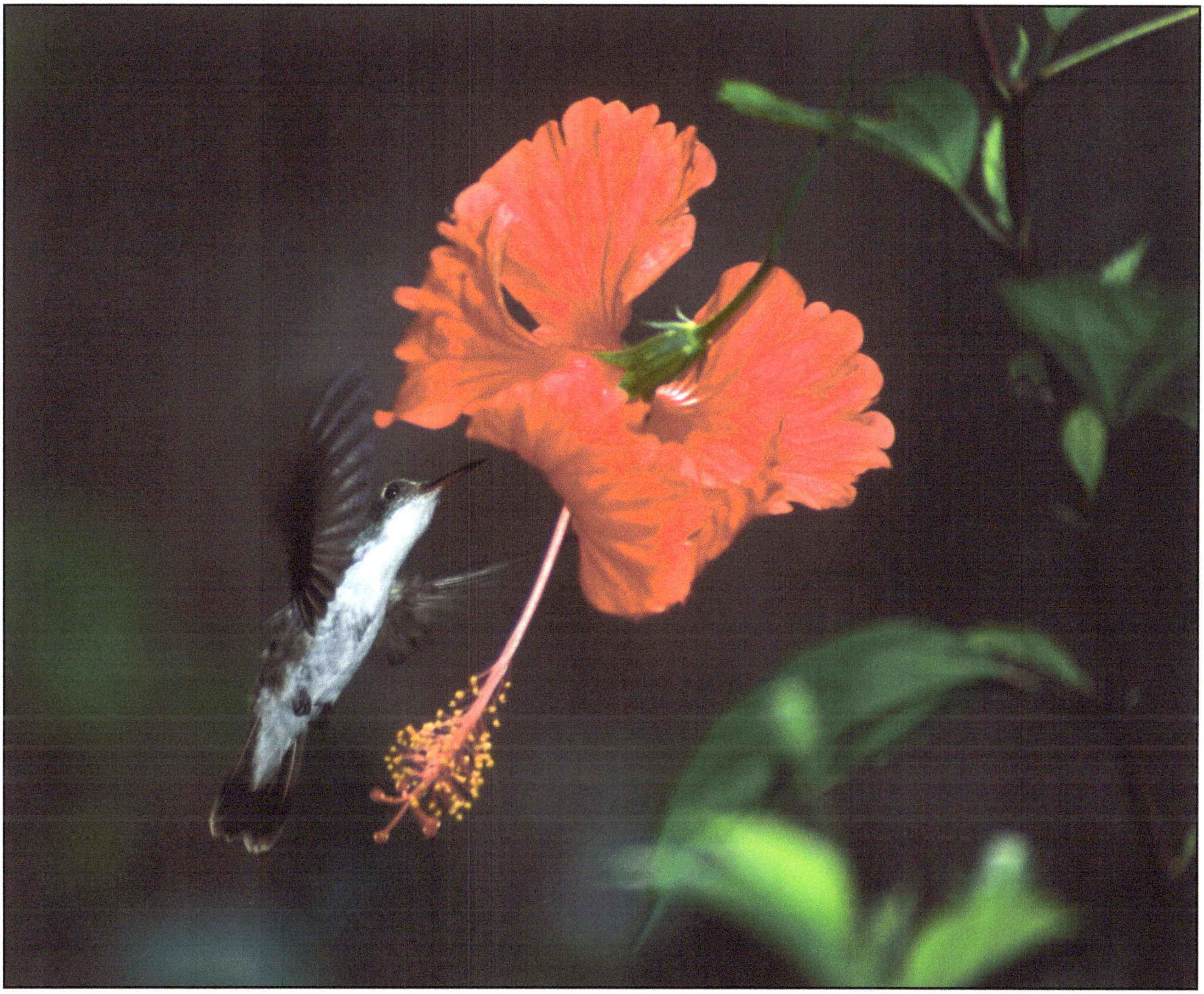

White-bellied Emerald (*Amazilia candida*)

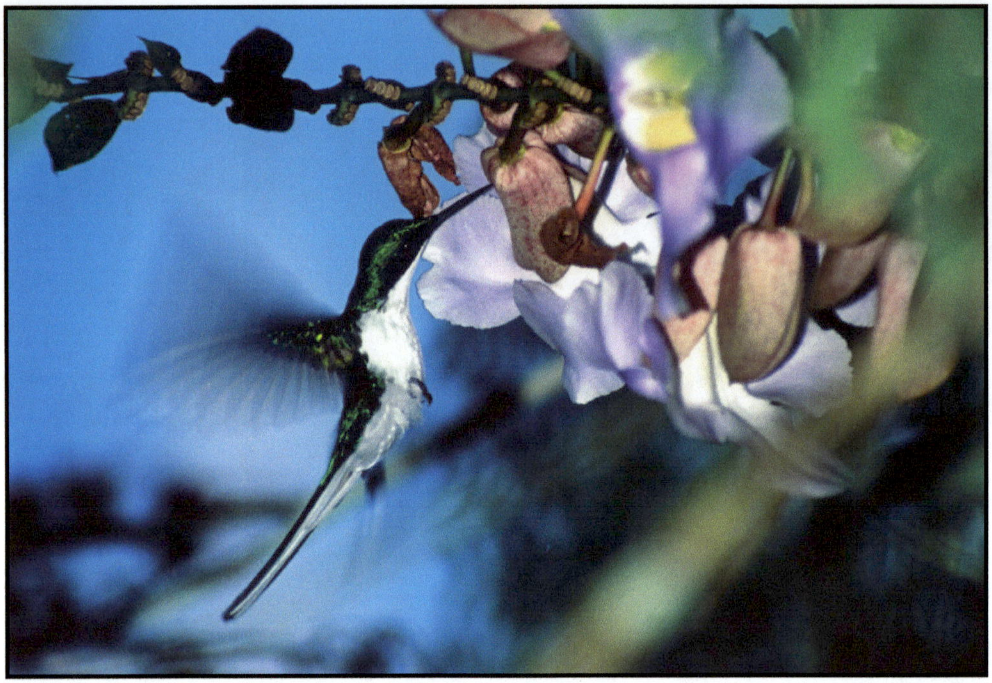

Purple-crowned Fairy (*Heliothryx barroti*)

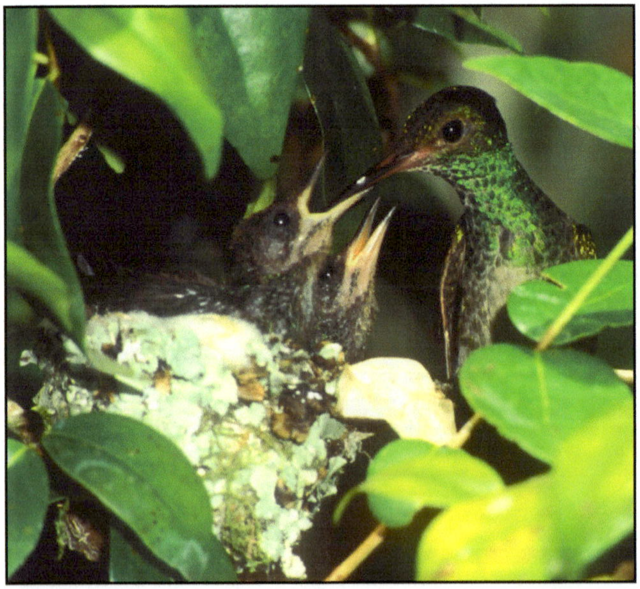

Rufous-tailed Hummingbird parent feeding two chicks

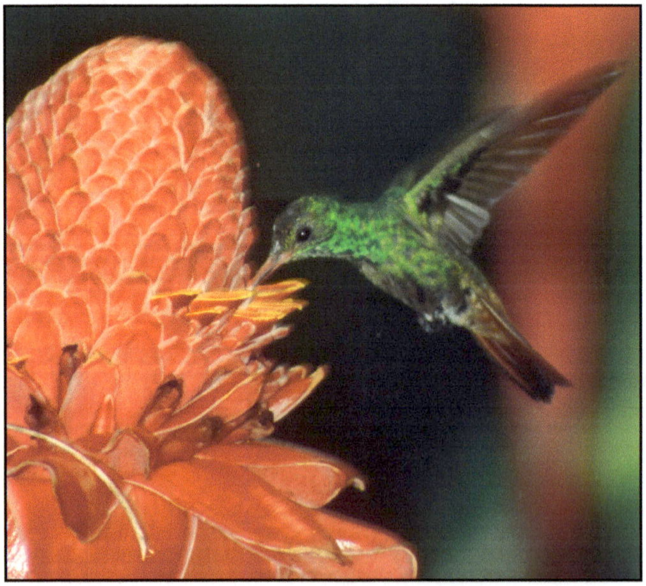

Rufous-tailed Hummingbird (*Amazilia tzacatl*)

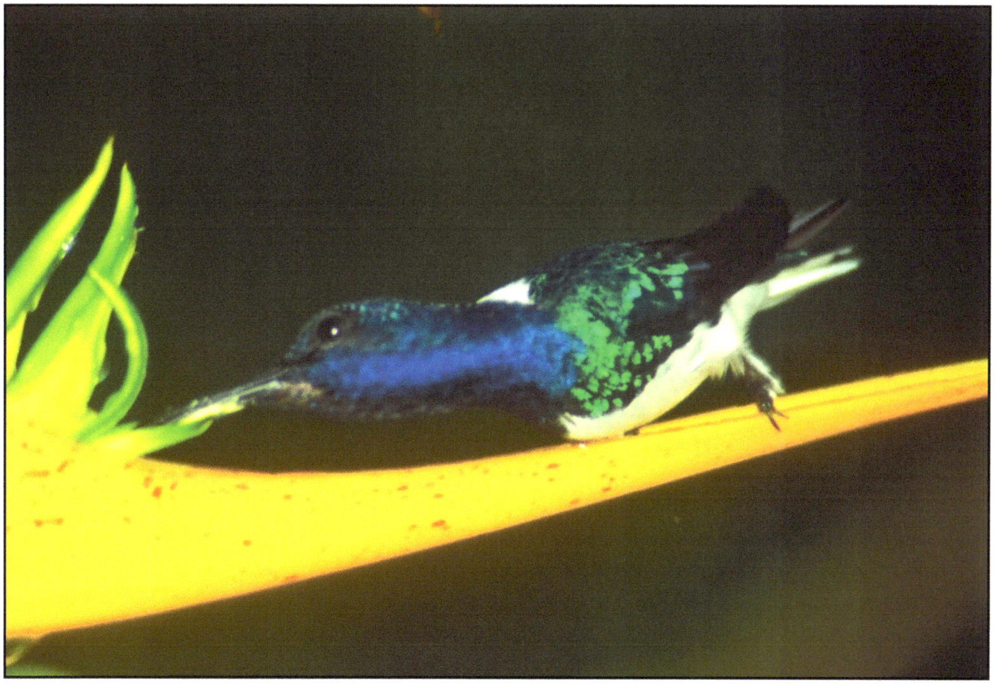

White-necked Jacobin(*Florisuga mellivora*)

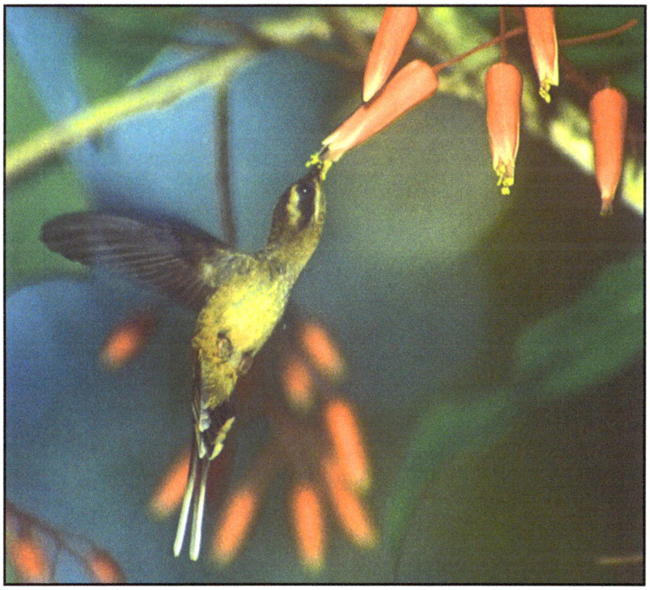

Long-billed Hermit (*Phaethornis longirostris*)

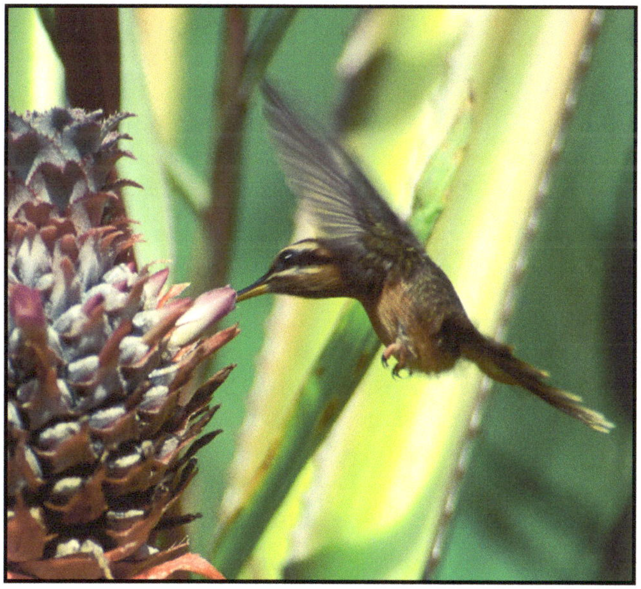

Stripe-throated Hermit (*Phaethornis striigularis*)

Golden-hooded Tanager

I recall one June day in the Cockscomb campground, when I was wandering around looking for something interesting to photograph. I almost stepped on a small chick that had obviously fallen out of a nest. This little bird was so ugly, it was beautiful. I had absolutely no idea what species it belonged to. There were no adult birds around lamenting its loss, as they are sometimes apt to do. I looked around and examined each tree for signs of a nest but to no avail.

I knew that I would feel guilty forever if a hungry fox, snake, or raptor happened to come by and satisfy its hunger pangs with this ugly unclaimed chick, so I figured I would give it an even chance. I lifted the tiny creature and placed it on a broad tree branch about eight feet off the ground.

Two hours later I returned and, to my amazement, the chick was still standing exactly where I left it, but standing on the same branch was an adult Golden-hooded Tanager. I went as close as I dared and took a picture. I had no sooner done so, than a second adult tanager landed on the branch, and fed a grub to the chick. I felt great! I had been fully expecting an empty branch, and now I had a family unit. I took several photos of the family group and left.

Early the next morning, I again visited the tree; one adult and the chick were still there. Later, that afternoon all three birds were waiting. Soon after I had to leave, but I returned about two months later, and out of curiosity went down to the campground. The tree was still there, but it was now empty. I turned to walk away when two Golden-hooded Tanagers flew over my head and alighted on the same branch. For anyone who asks, do birds have memory? I can definitely answer, "They sure do!"

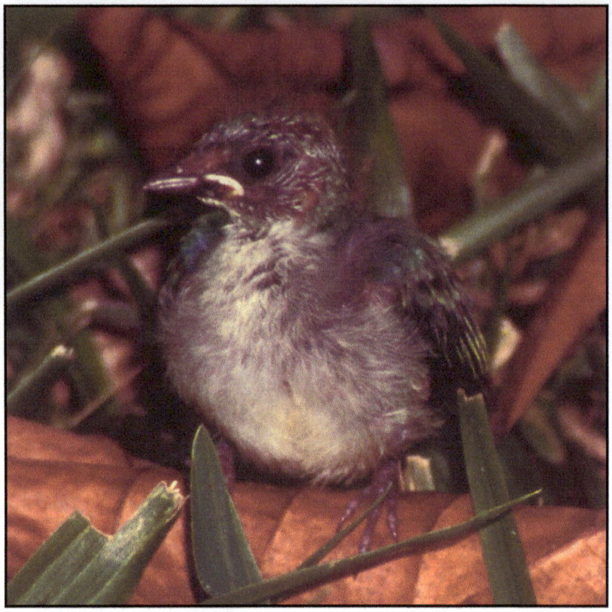

Golden-hooded Tanager chick on the ground

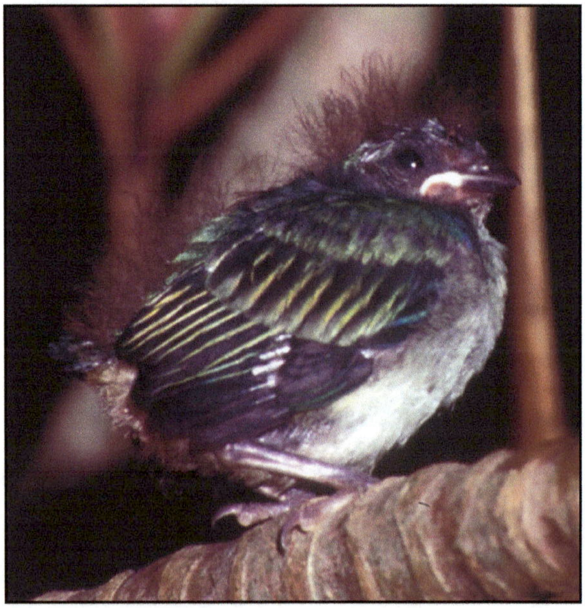

Golden-hooded Tanager chick placed on the tree branch

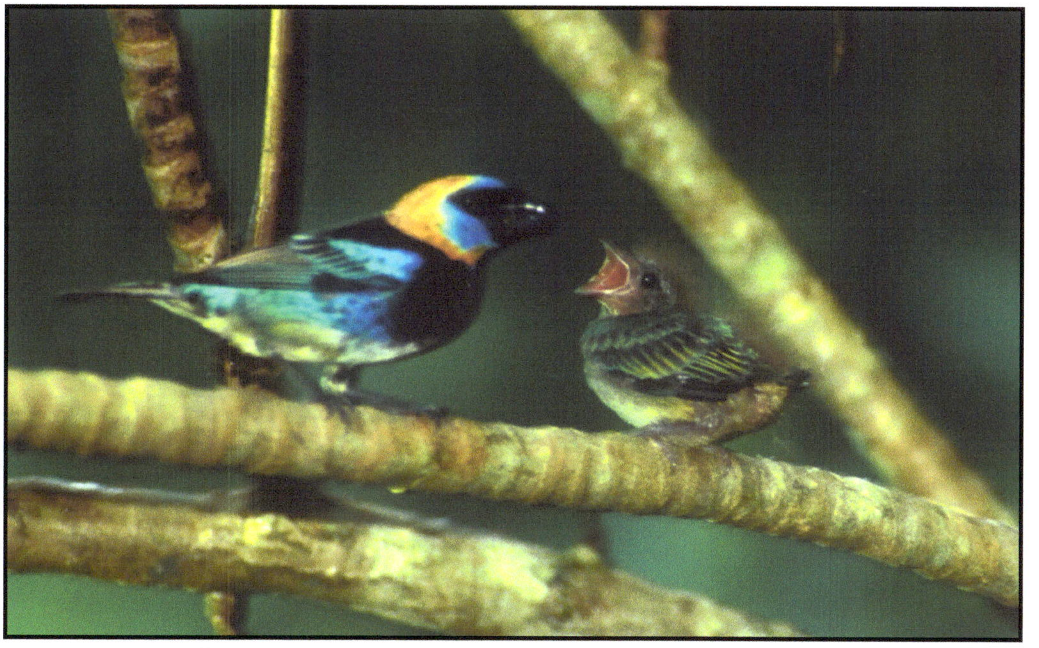

Golden-hooded Tanager (*Tangara larvata*) feeding chick

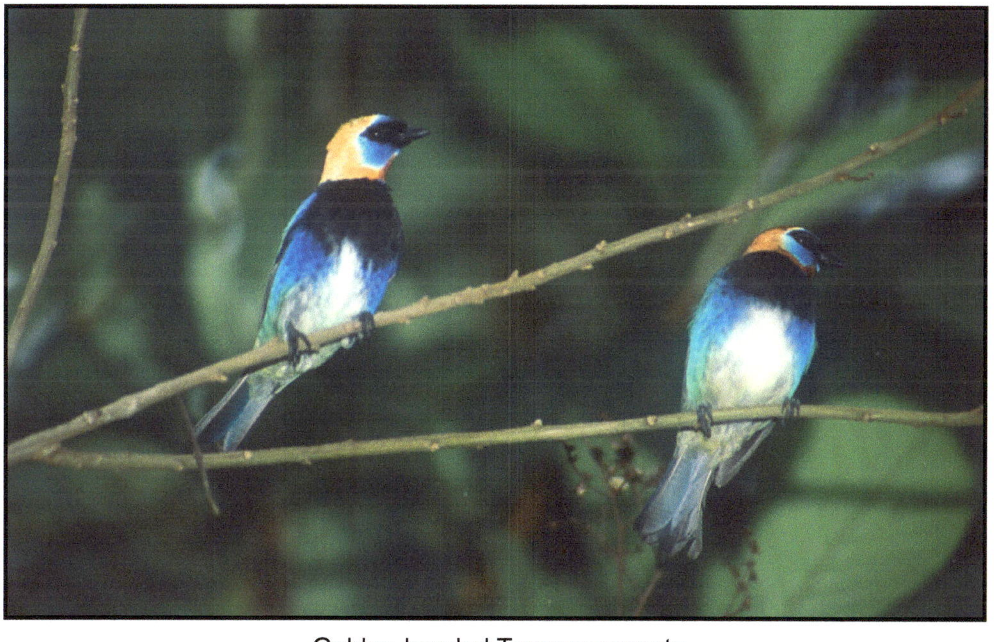

Golden-hooded Tanager parents

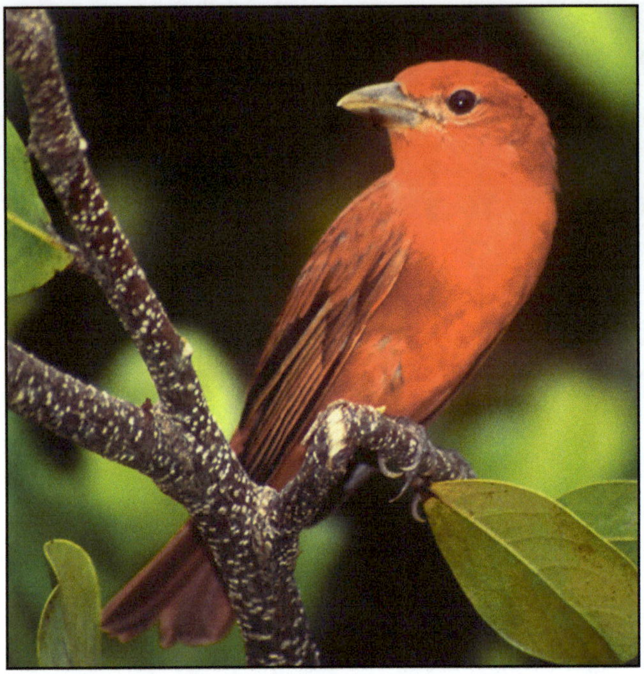

Summer Tanager (*Piranga rubra*)

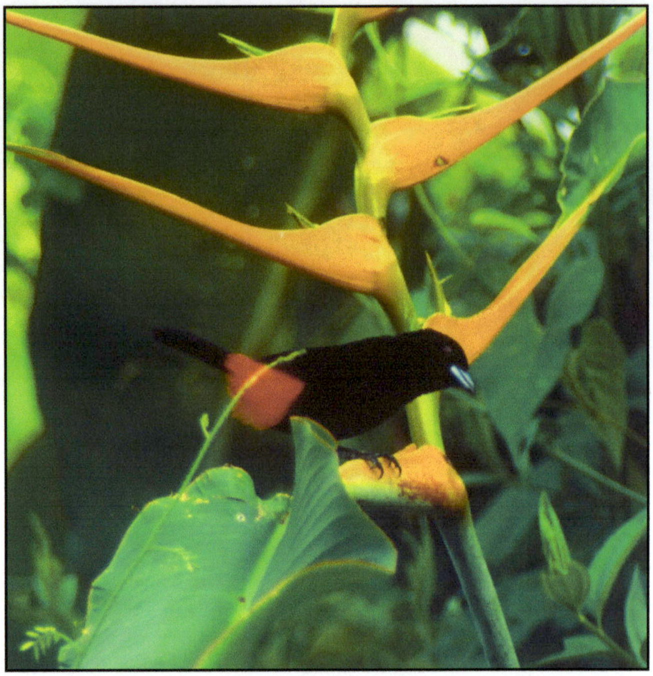

Passerini's Tanager (*Ramphocelus passerinii*)

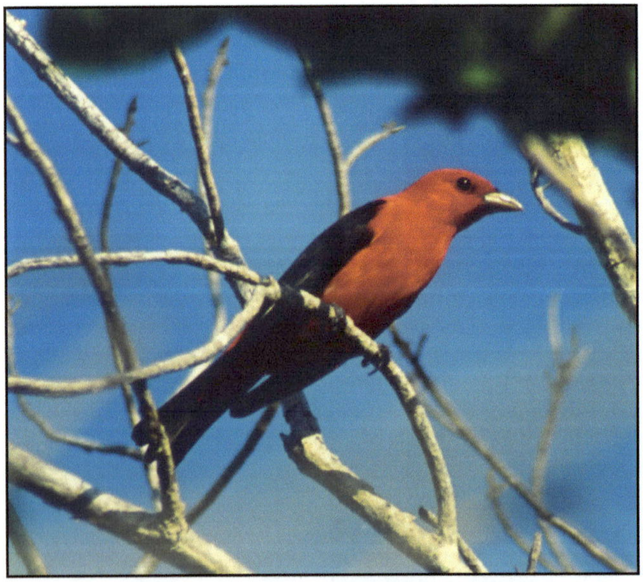

Scarlet Tanager (*Piranga olivacea*)

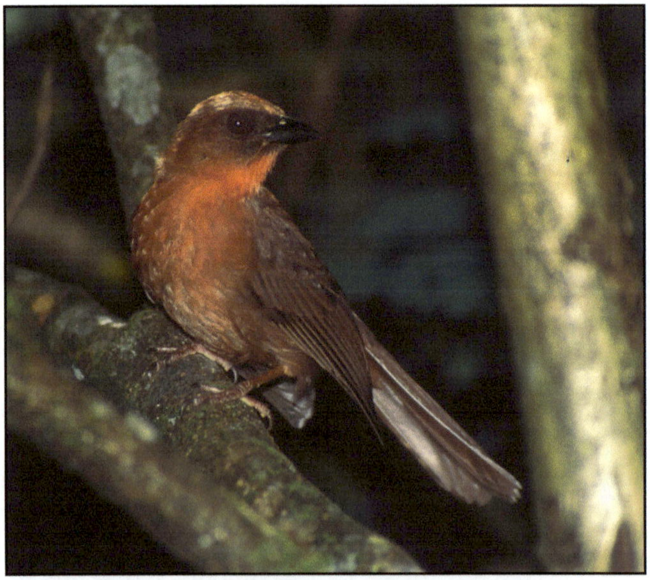

Red-throated Ant-Tanager (*Habia fuscicauda*)

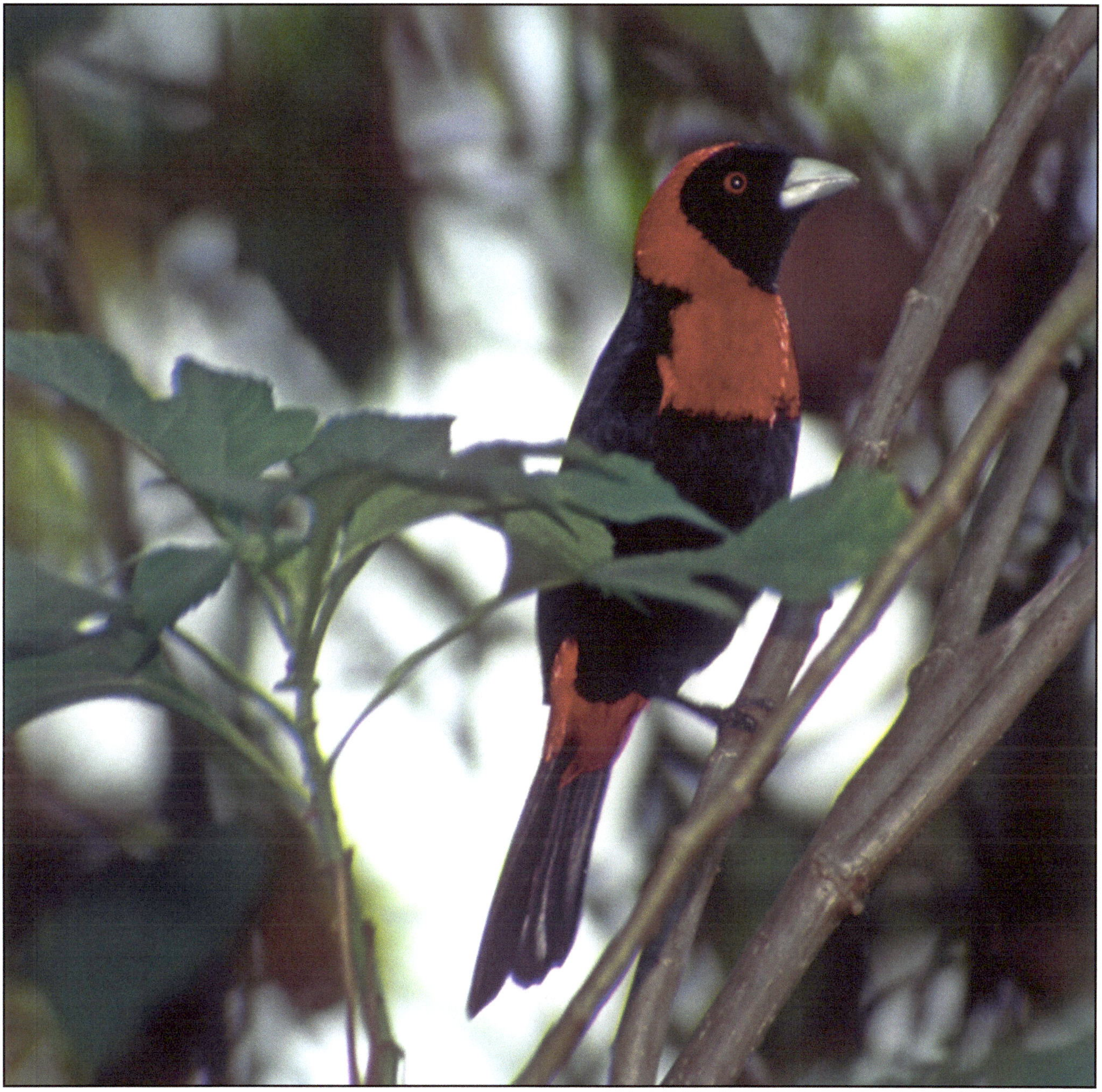

Crimson-collared Tanager (*Ramphocelus sanguinolentus*)

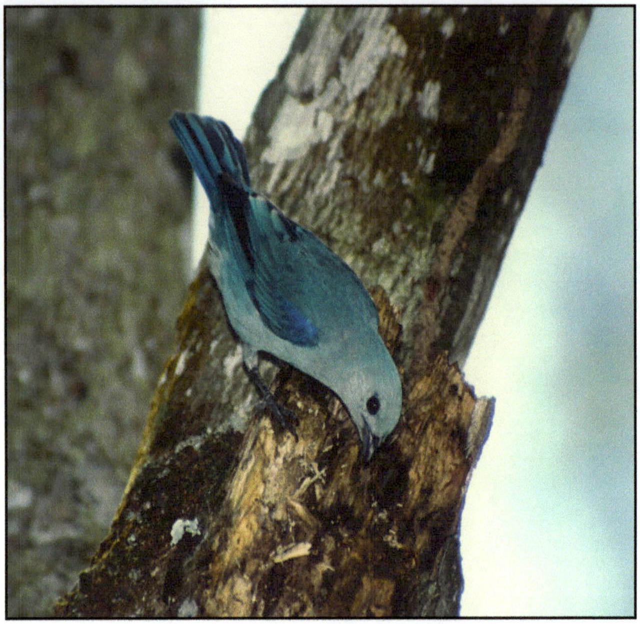

Blue-gray Tanager (*Thraupis episcopus*)

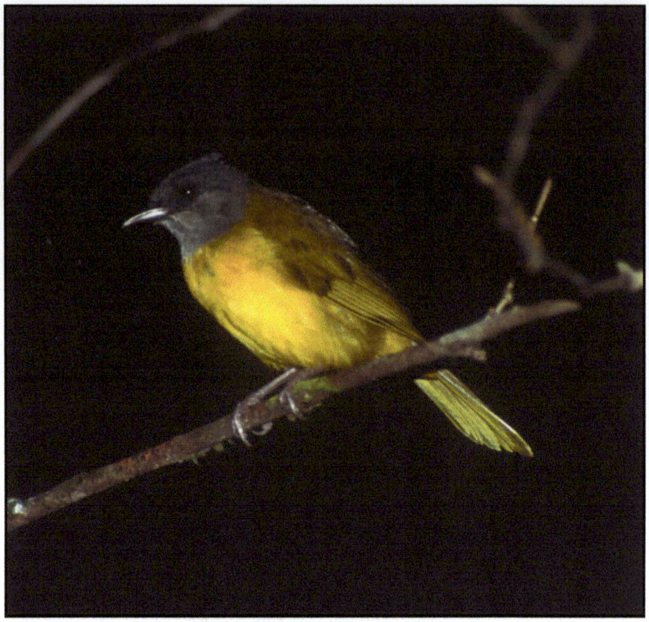

Gray-headed Tanager (*Eucometis penicillata*)

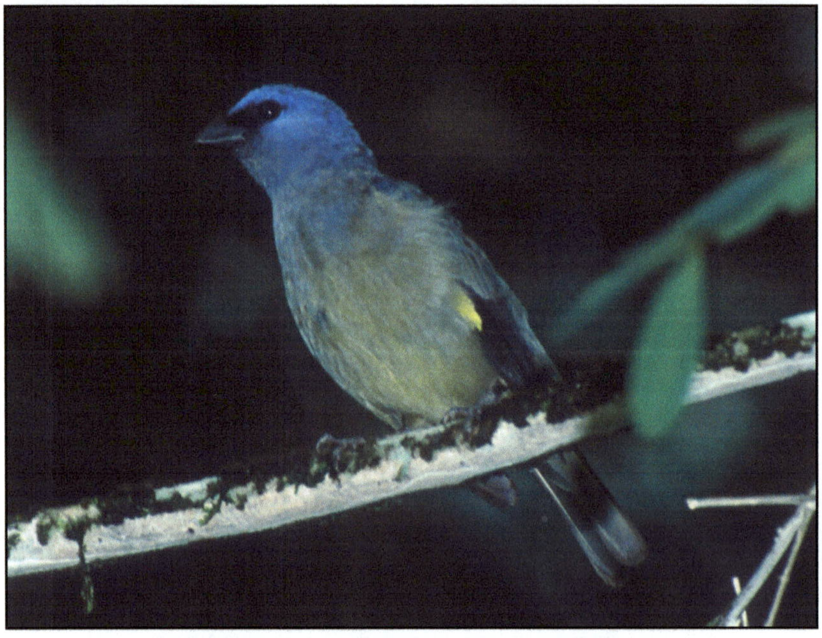

Yellow-winged Tanager (*Thraupis abbas*)

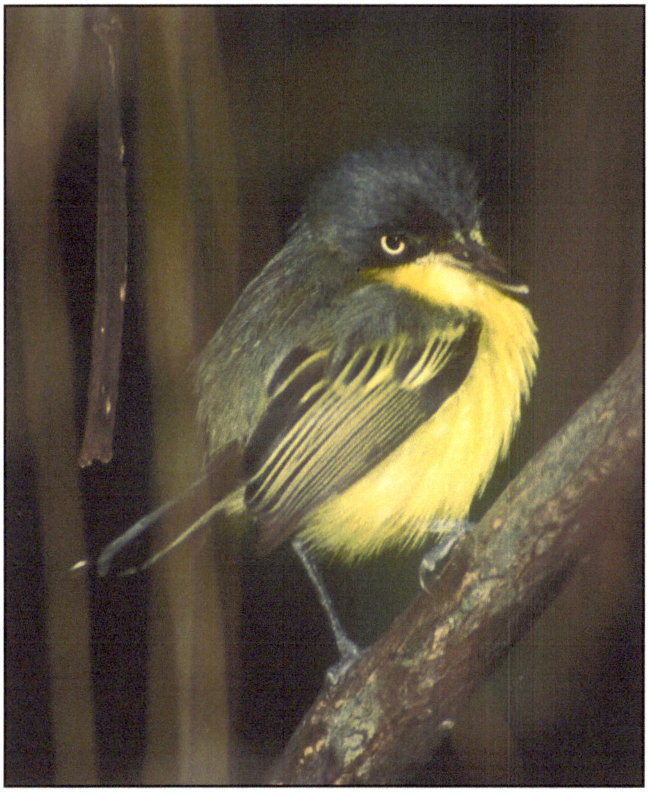

Common Tody-Flycatcher (*Todirostrum cinereum*)

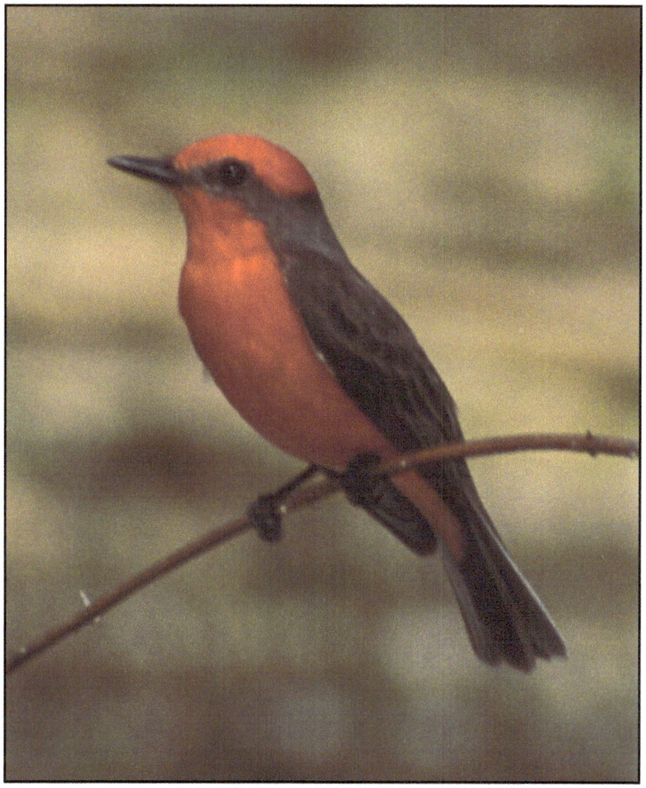

Vermilion Flycatcher (*Pyrocephalus rubinus*)

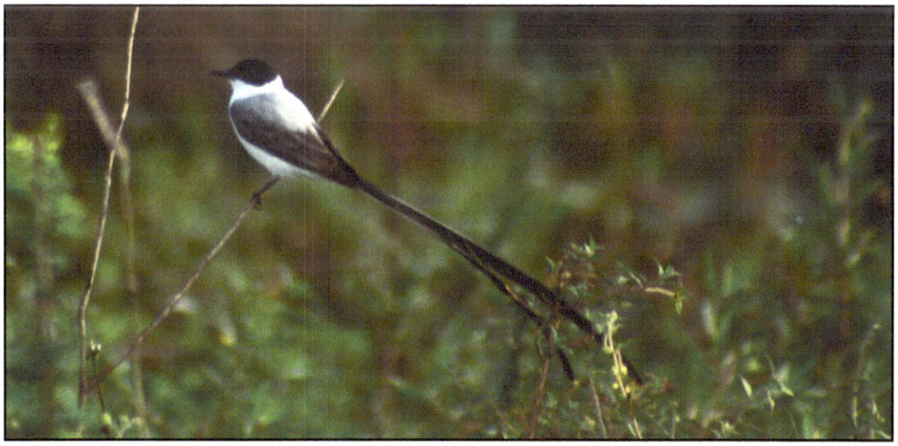

Fork-tailed Flycatcher (*Tyrannus savana*)

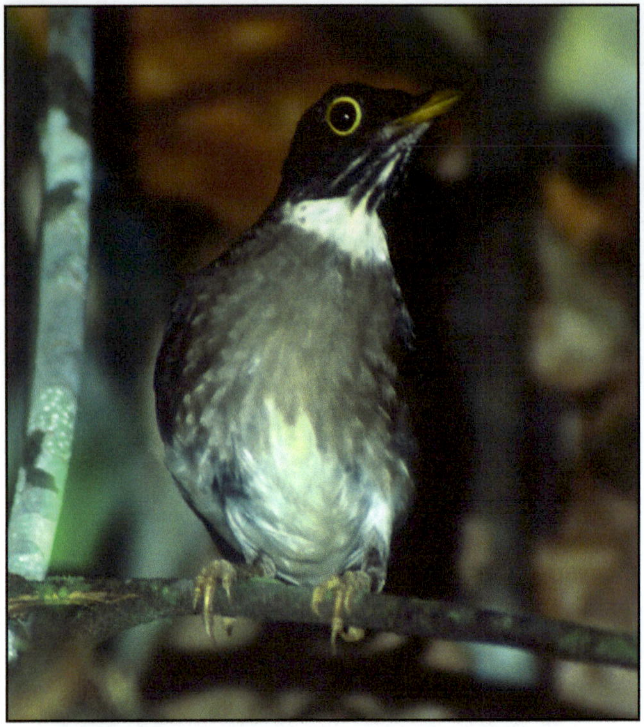

White-throated Thrush (*Turdus assimilis*)

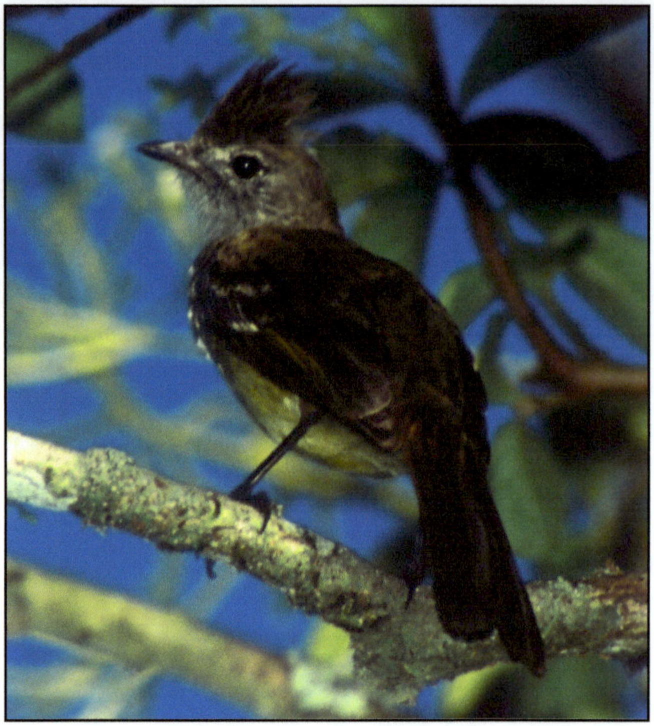

Yellow-bellied Elaenia (*Elaenia flavogaster*)

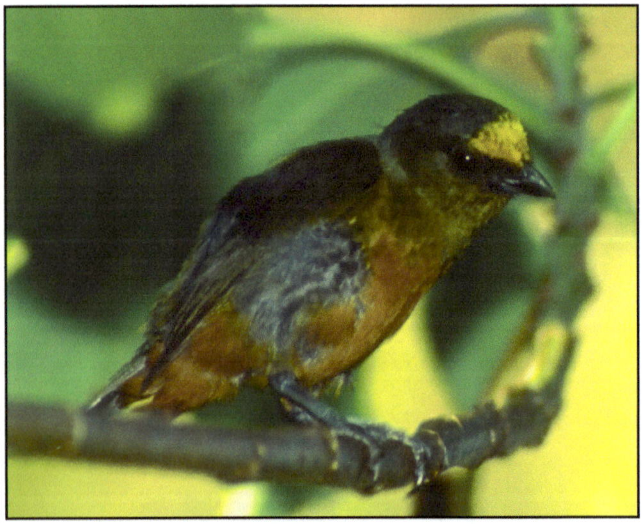

Olive-backed Euphonia (*Euphonia gouldi*)

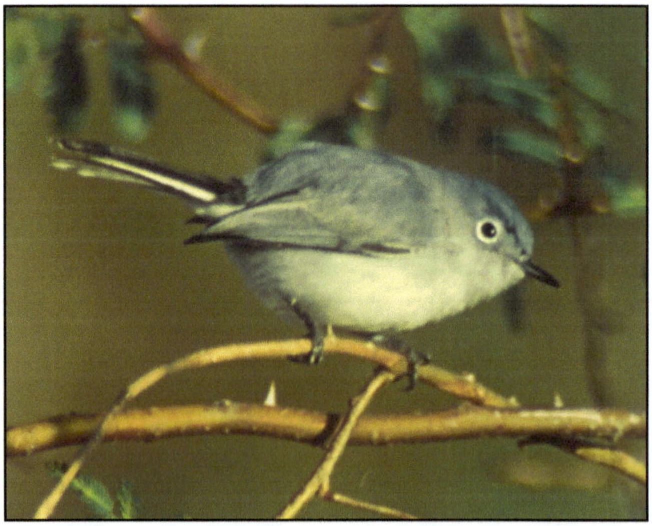

Blue-gray Gnatcatcher (*Polioptila caerulea*)

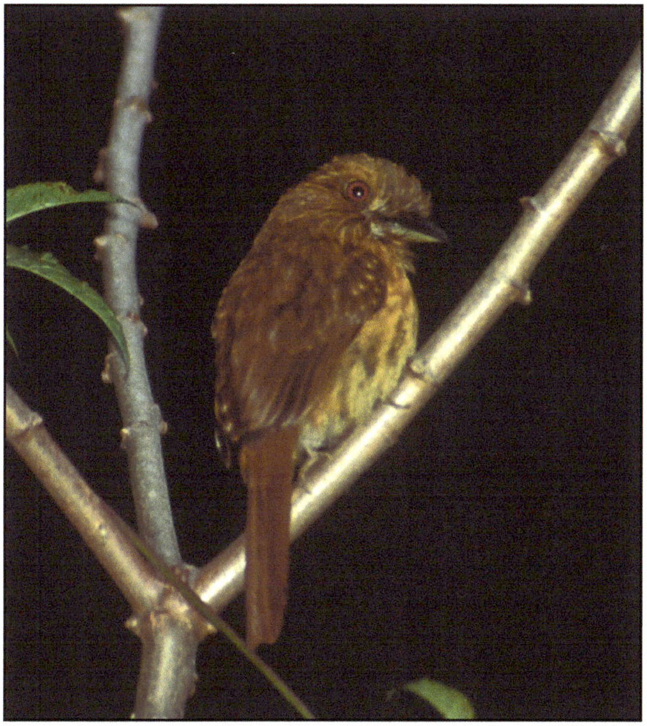

White-whiskered Puffbird (*Malacoptila panamensis*)

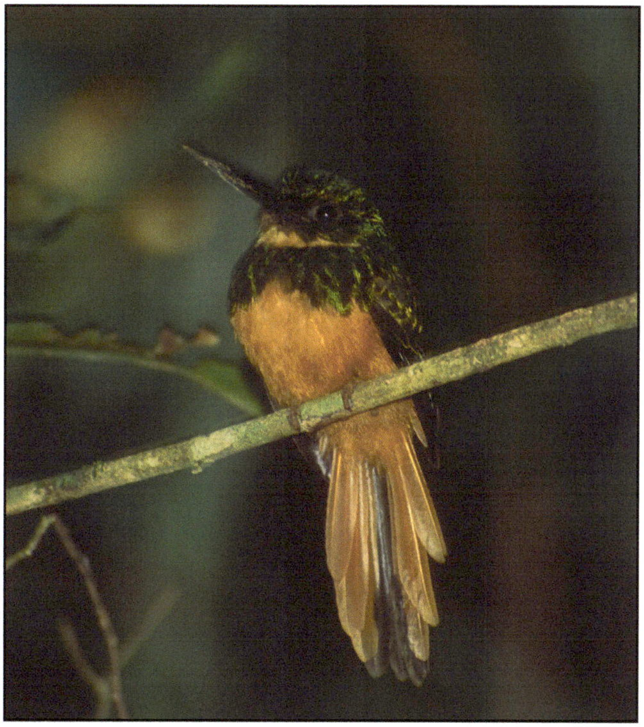

Rufous-tailed Jacamar (*Galbula ruficauda*)

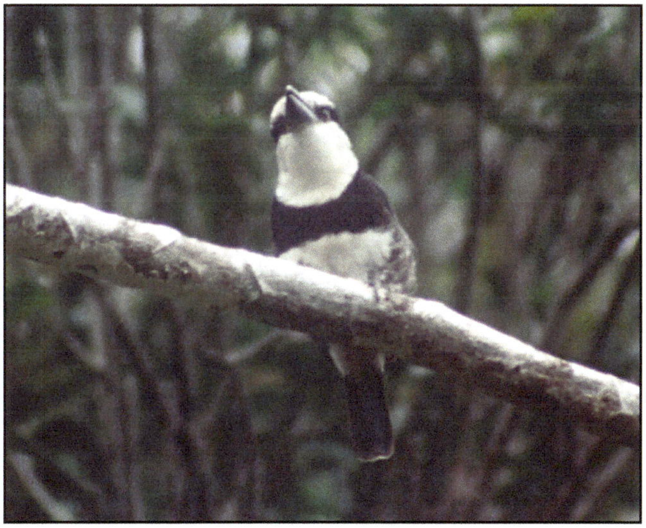

White-necked Puffbird (*Notharchus hyperrhynchus*)

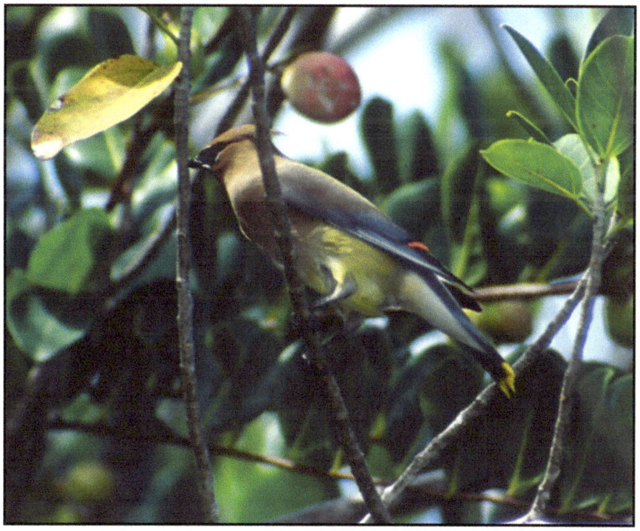

Cedar Waxwing (*Bombycilla cedrorum*)

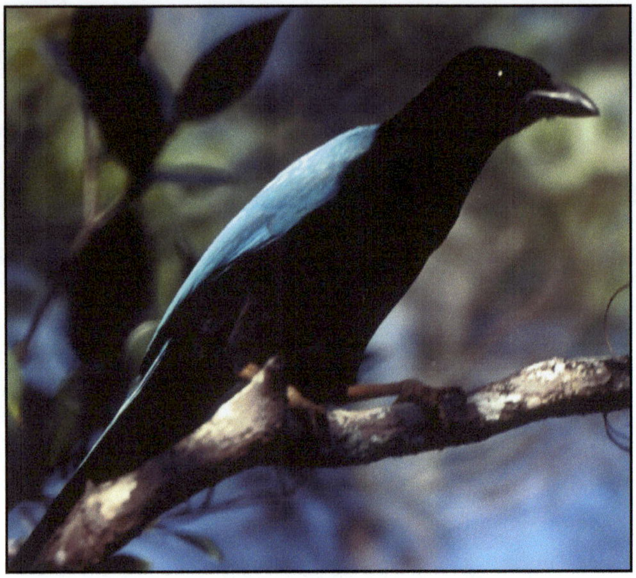

Yucatan Jay (*Cyanocorax yucatanicus*)

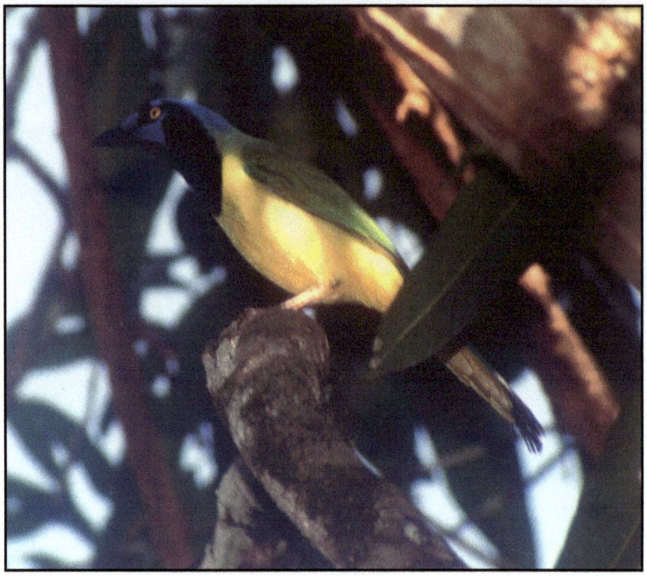

Green Jay (*Cyanocorax yncas*)

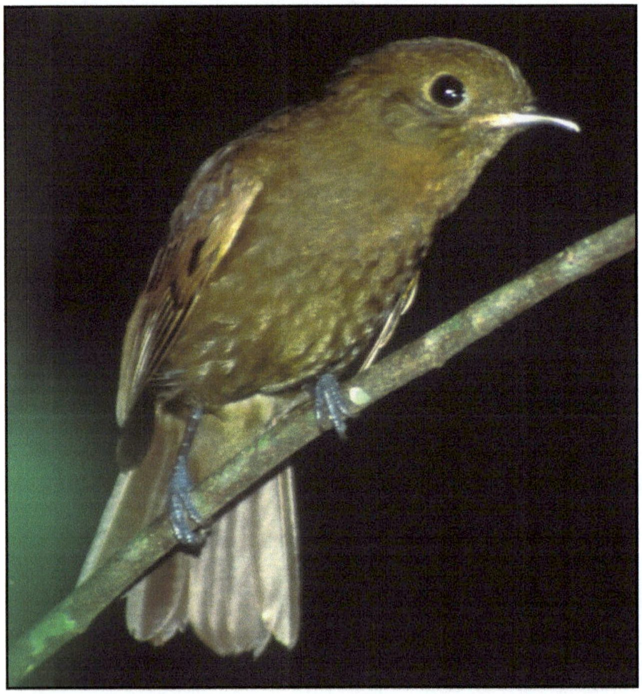

Thrush-like Schiffornis (*Schiffornis turdina*)

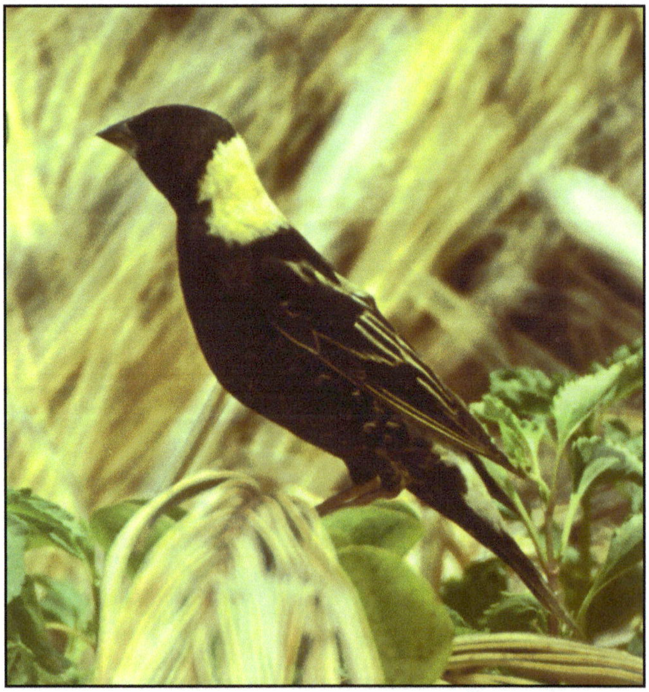

Bobolink (*Dolichonyx oryzivorus*)

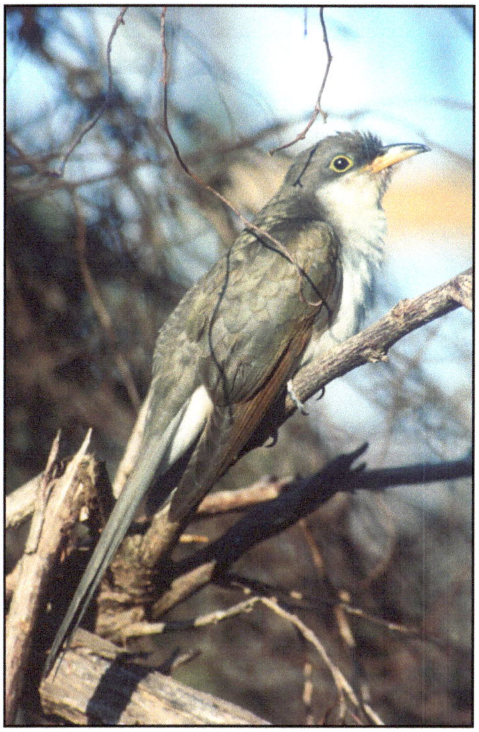

Yellow-billed Cuckoo
(*Coccyzus americanus*)

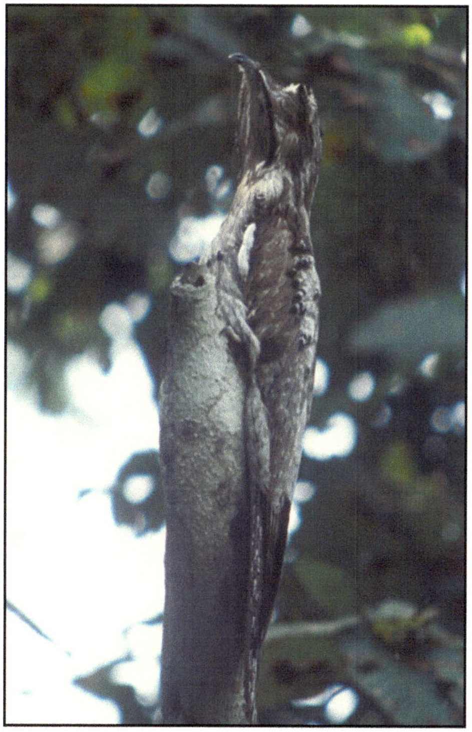

Northern Potoo
(*Nyctibius jamaicensis*)

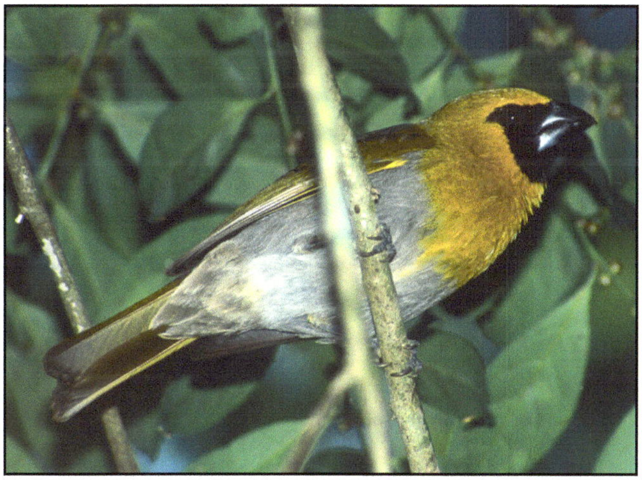

Black-faced Grosbeak (*Caryothraustes poliogaster*)

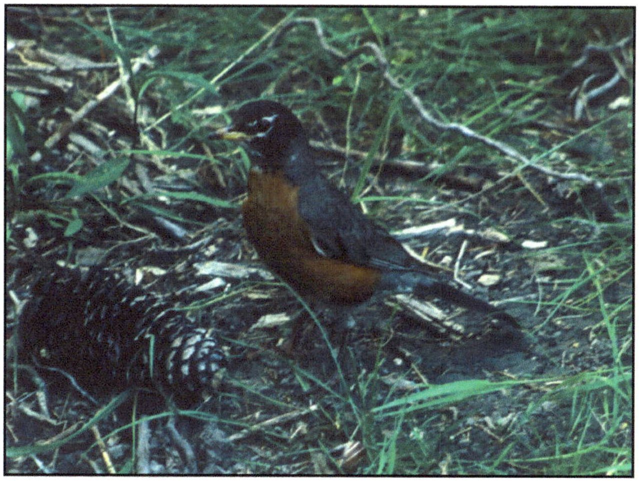

American Robin (*Turdus migratorius*)

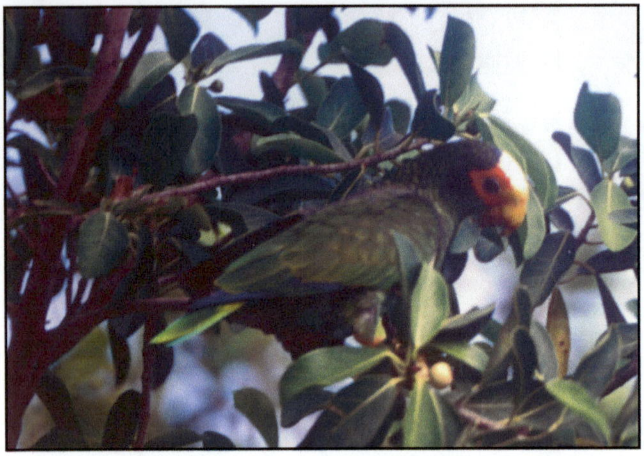

Yellow-lored Parrot (*Amazona xantholora*)

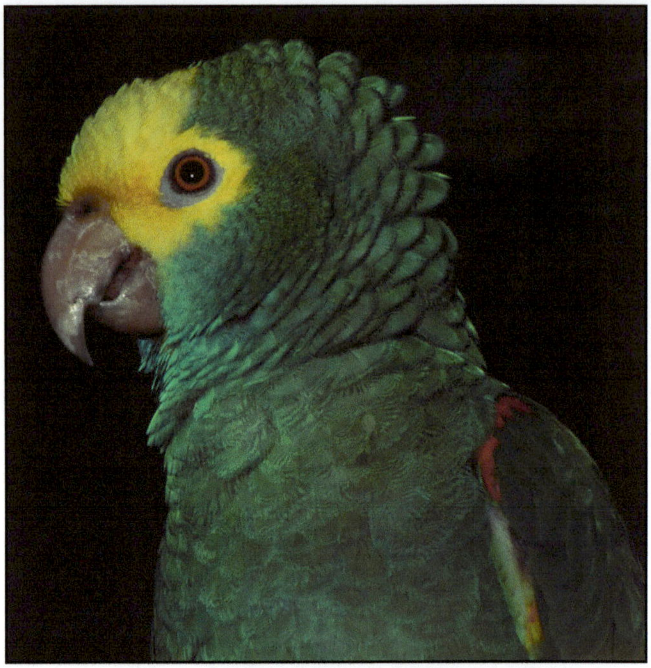

(*above and below*) Yellow-headed Parrot (*Amazona oratrix*)

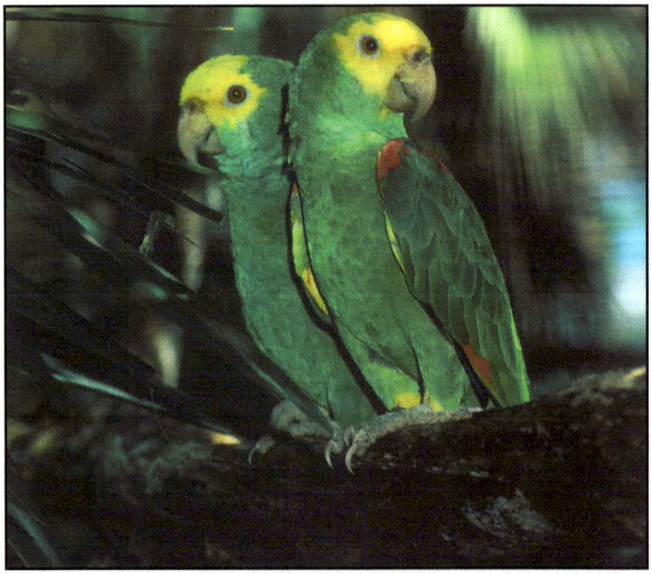

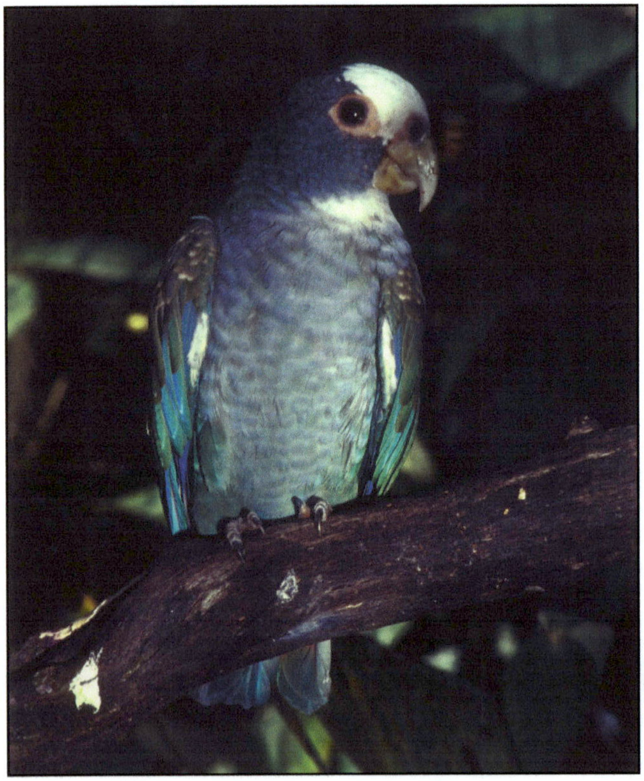

White-crowned Parrot (*Pionus senilis*)

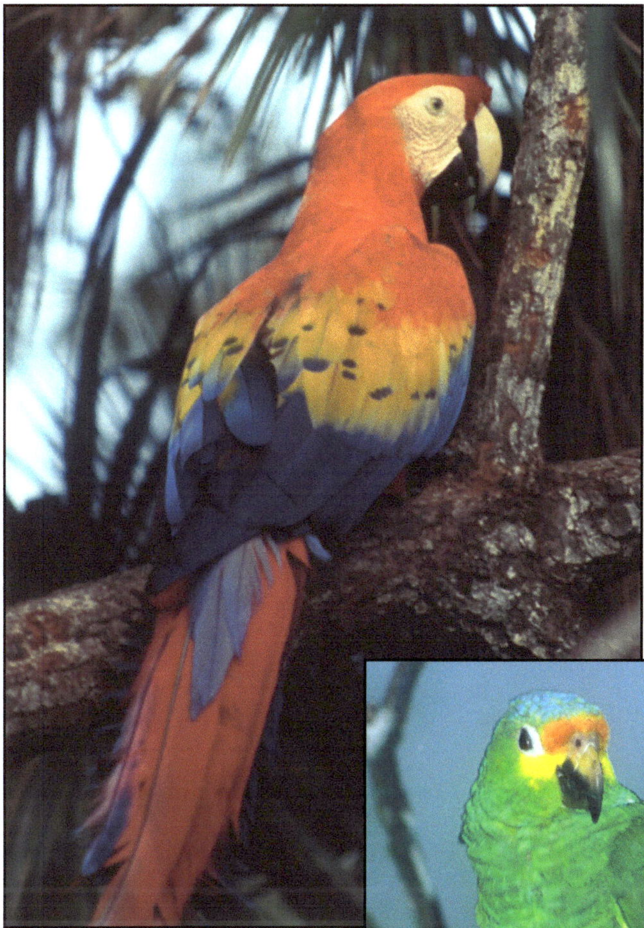

Scarlet Macaw
(*Ara macao*)

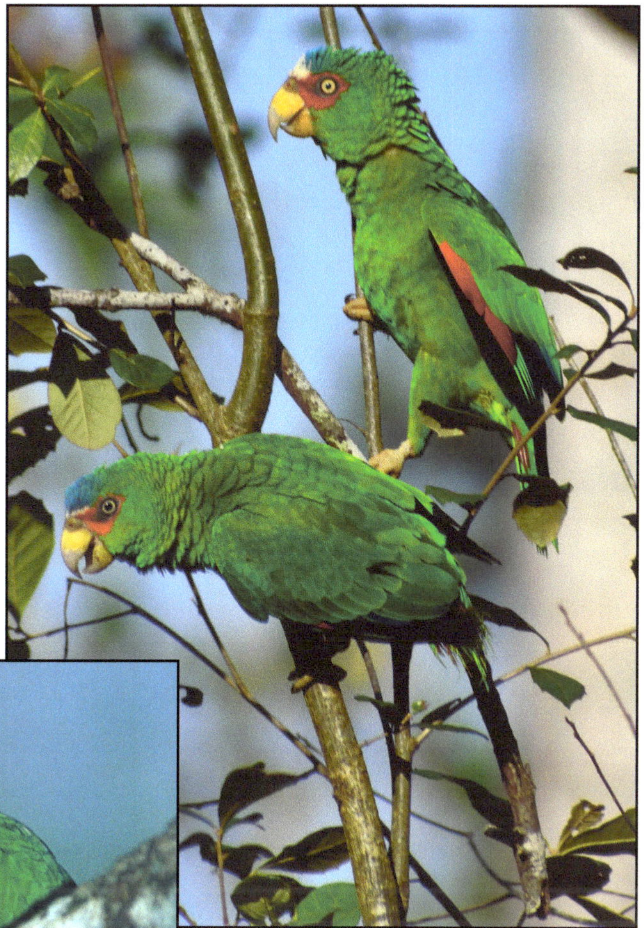

White-fronted Parrot
(*Amazona albifrons*)

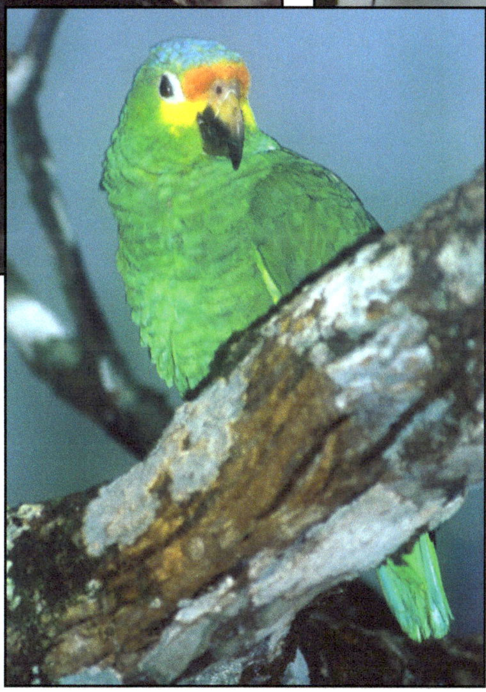

Red-lored Parrot (*Amazona autumnalis*)

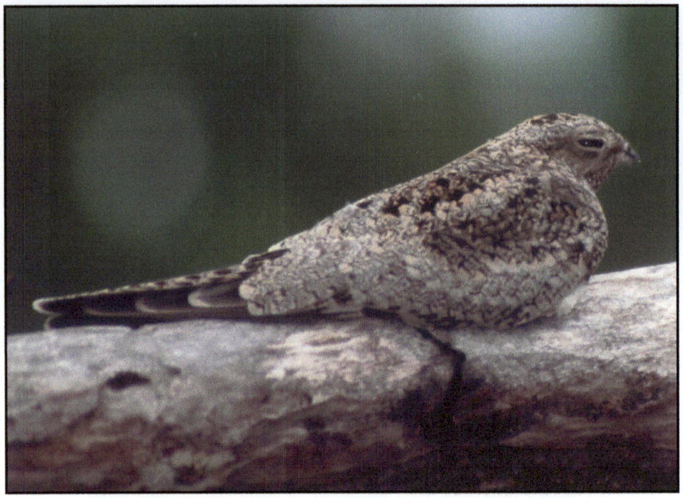
Common Nighthawk (*Chordeiles minor*)

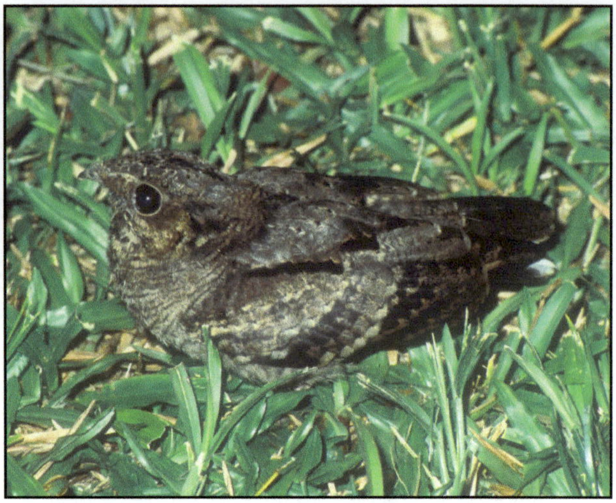
Yucatan Poorwill (*Caprimulgus badius*)

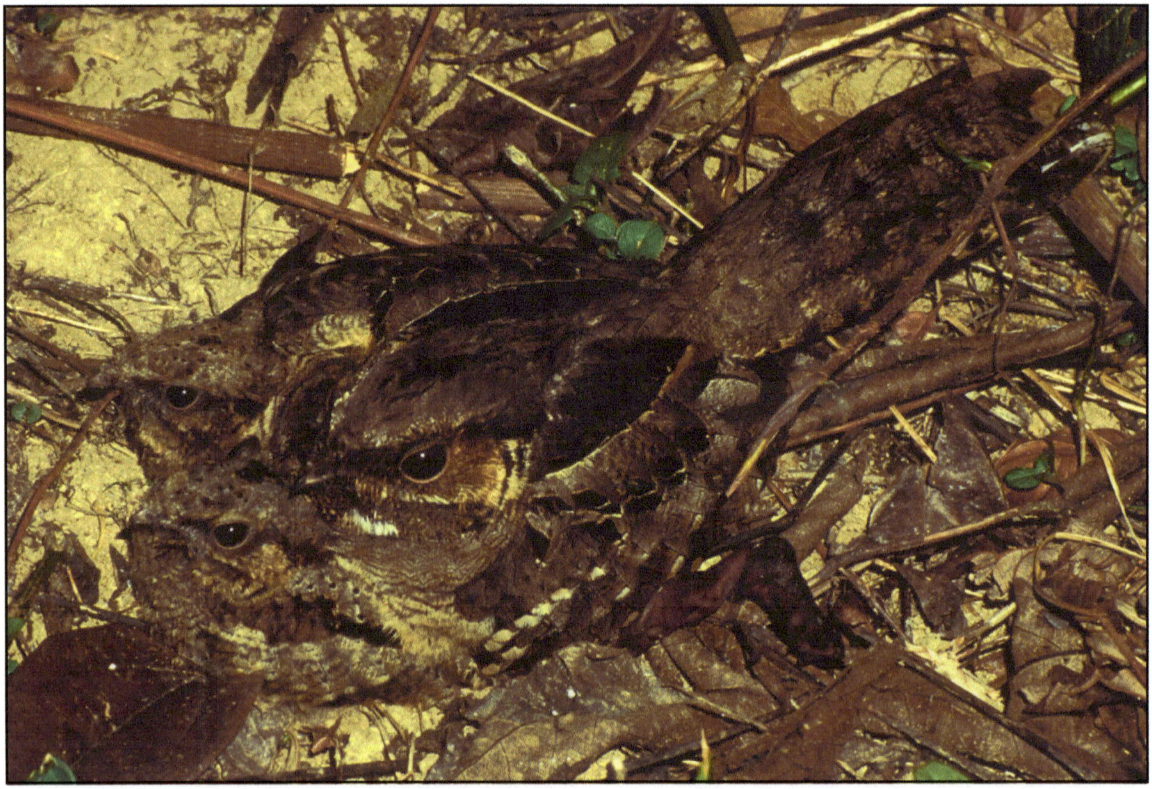
Common Pauraque (*Nyctidromus albicollis*) with two chicks

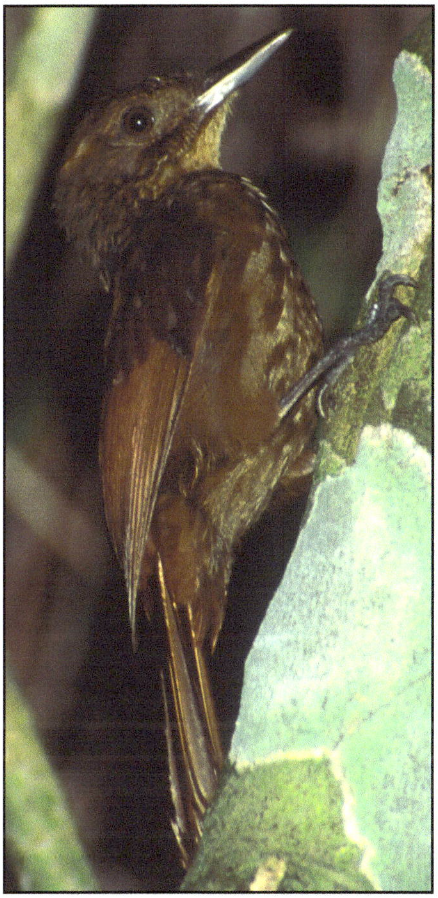

Tawny-winged Woodcreeper
(*Dendrocincla anabatina*)

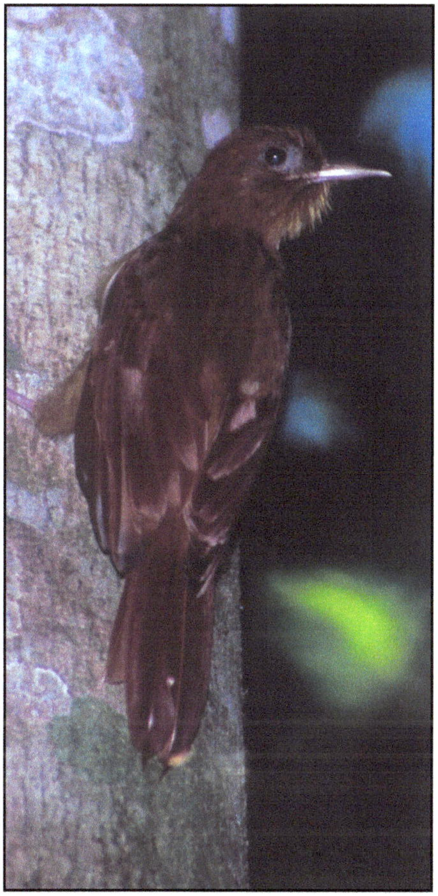

Ruddy Woodcreeper
(*Dendrocincla homochroa*)

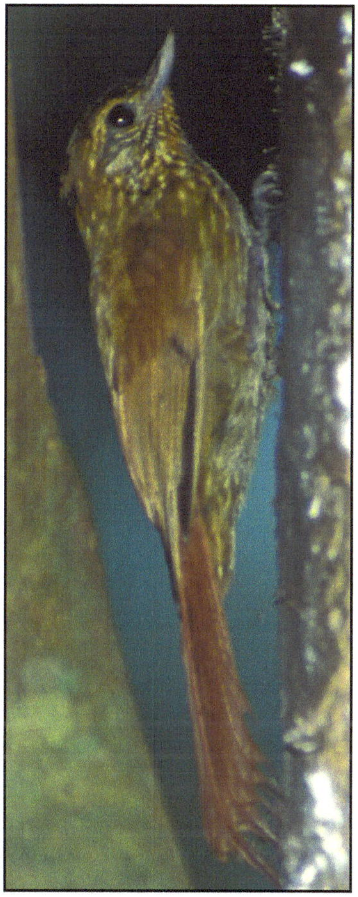

Wedge-billed Woodcreeper
(*Glyphorynchus spirurus*)

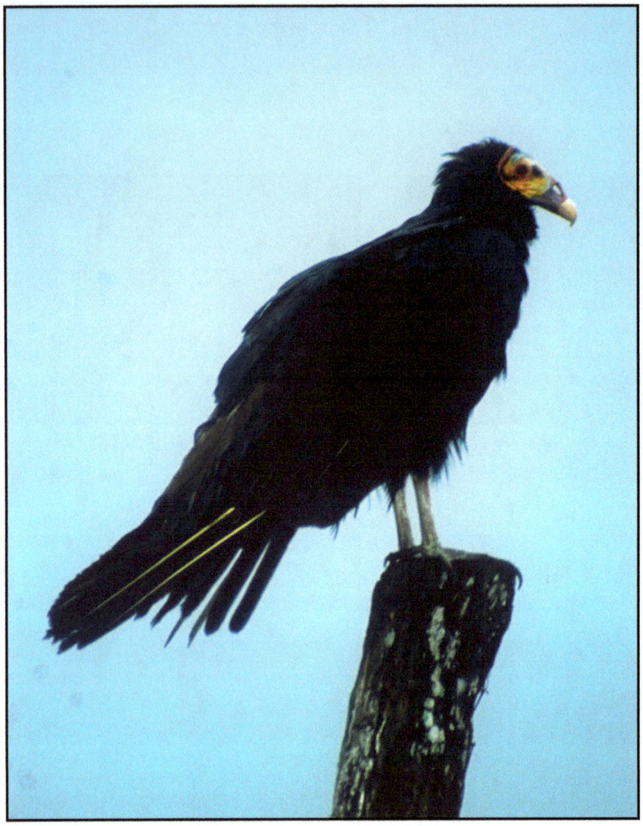

Lesser Yellow-headed Vulture (*Cathartes burrovianus*)

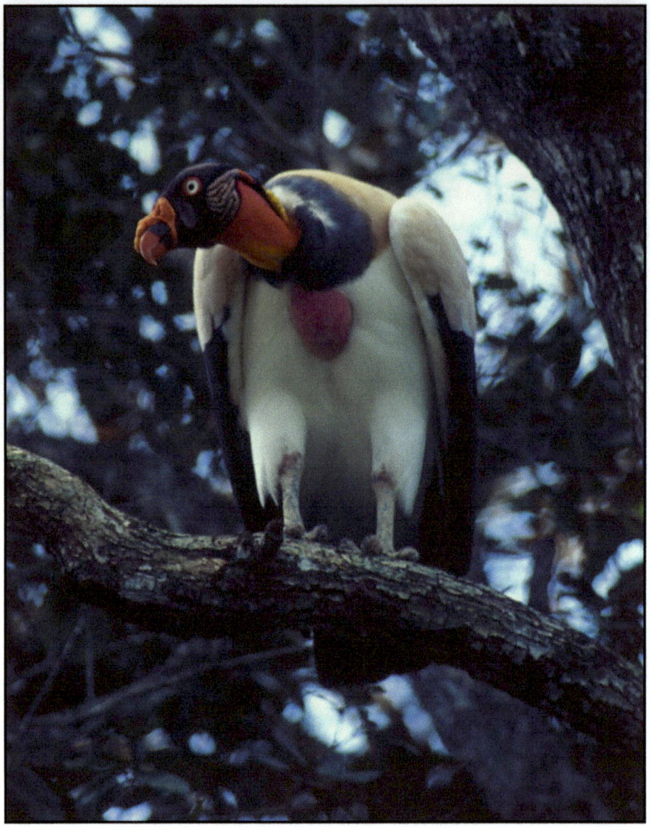

King Vulture (*Sarcoramphus papa*)

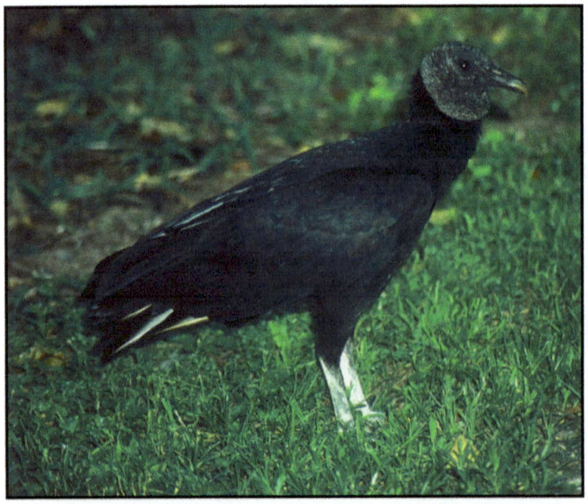

Black Vulture (*Coragyps atratus*)

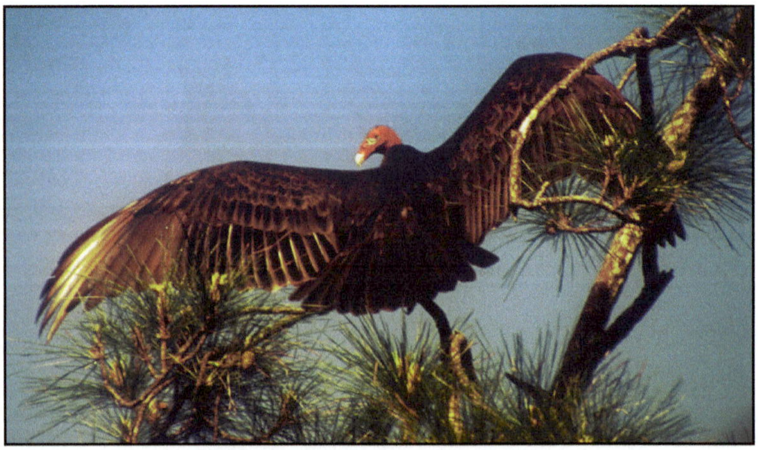

Turkey Vulture (*Cathartes aura*)

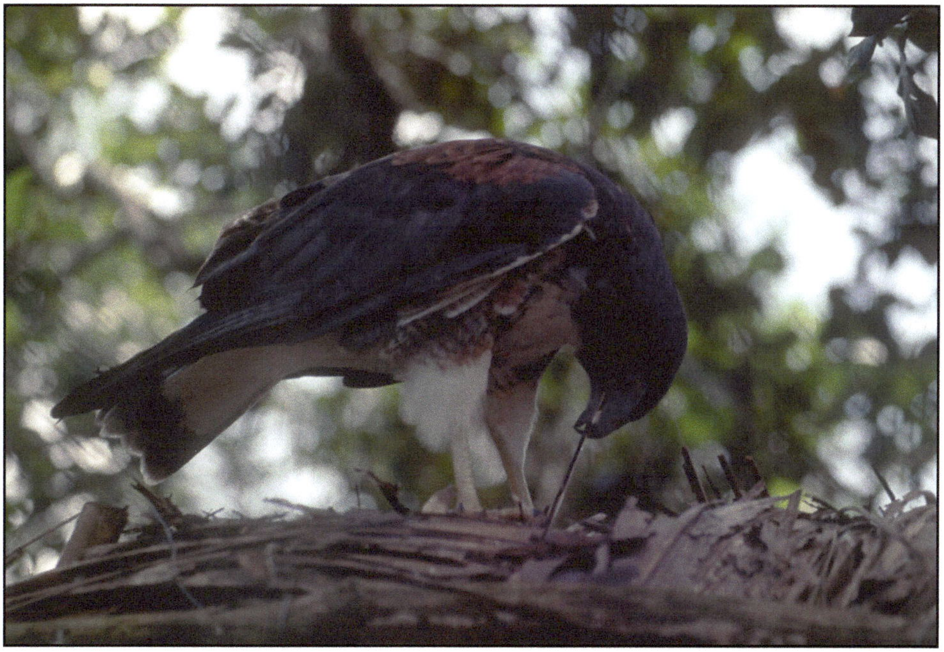

White-tailed Hawk (*Buteo albicaudatus*) using a tool

White-tailed Hawk

During the early 1990s I was walking along the Old Western Highway, a unused dirt road that parallels the Western Highway. After several miles I noticed a large dark coloured hawk on a thatched roof. As I got closer I recognized it as a White-tailed Hawk but it was its behaviour that caught my interest. It was tearing at the thatch with its beak. After several minutes, its demeanour changed. It stood back, then made a short ten-foot flight alighting again on the roof, and picked up a piece of broken branch about ten inches long and approximately three eighths of an inch in diameter. Holding the wood in its claws, it flew back to its original position where it tried to insert the wood in the thatch. It dropped the wood, and rested for a while. Several minutes later it again picked up the wood in its bill, but this time it used it as a probe to lever up a little of the thatch. Up until then I was too interested to lift my camera up, but then I snapped a photo. I watched for a few more minutes while the bird wrestled with the stick.

When it happened, it happened so fast, and I reacted too slowly to catch the climax on camera. A small brown rat leaped from the hawk's untidy excavation and swiftly ran down the roof. As fast as the rat was, the hawk was much faster. It took off and caught the rat before it had covered three feet. Then without stopping, it continued on, flying over the pine forest.

Again I was slow to react, and before I lifted the camera up, the hawk was almost out of sight, leaving me to wonder, did that really happen? At the time, this seemed to be important scientifically, because a few weeks prior, I had read an article in which scientists in Africa had discovered a chimpanzee that used a wooden twig to catch termites for food by pushing it into the nest, and then eating the insects which came out clinging to the rod. The heading was "Scientists discover a chimpanzee that uses tools!"

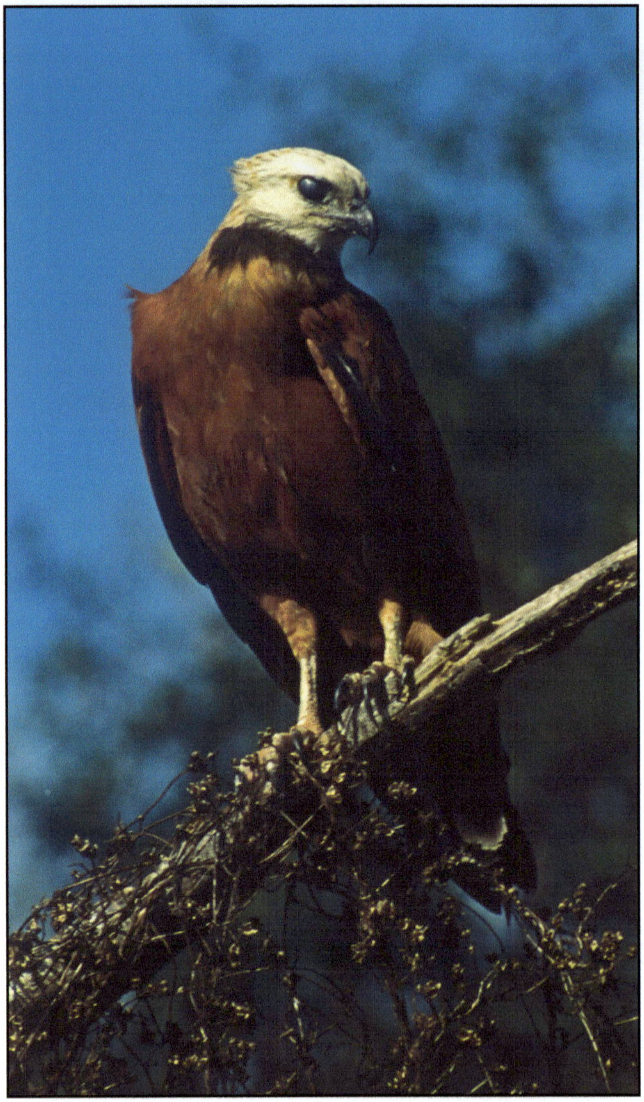

Black-collared Hawk (*Busarellus nigricollis*)

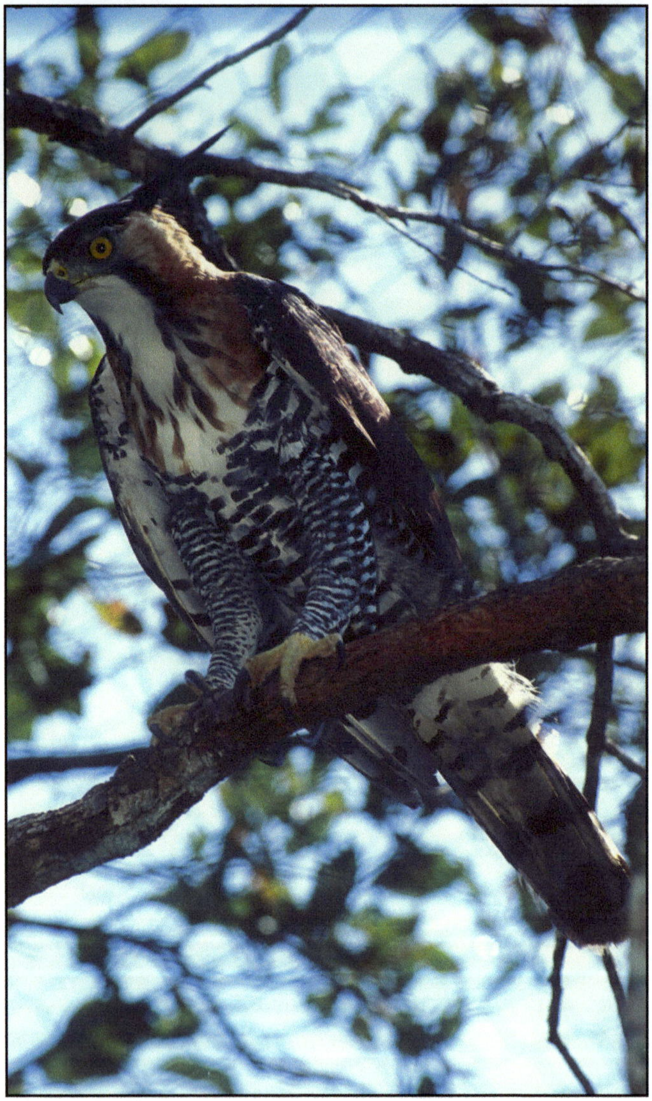

Ornate Hawk-Eagle (*Spizaetus ornatus*)

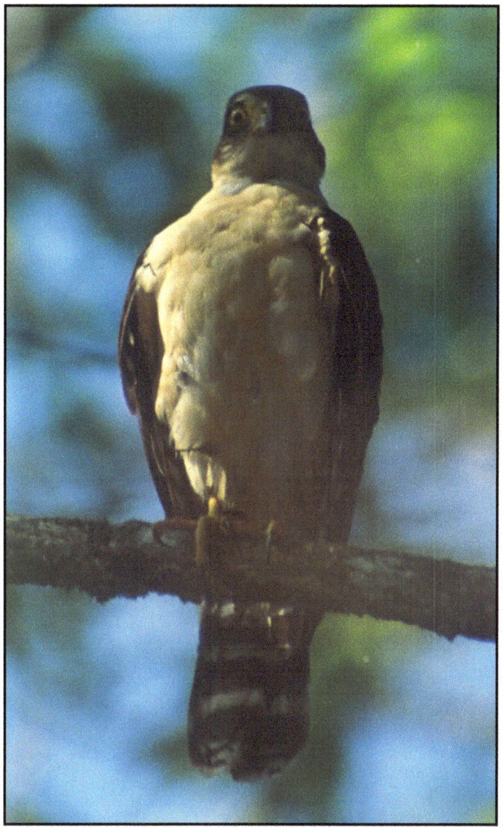

Bicolored Hawk (*Accipter bicolor*)

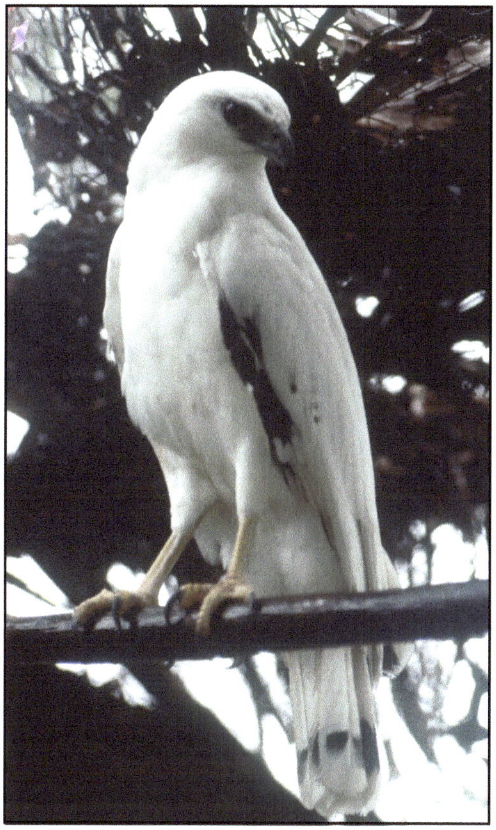

White Hawk (*Leucopternis albicollis*)
—*The Belize Zoo*

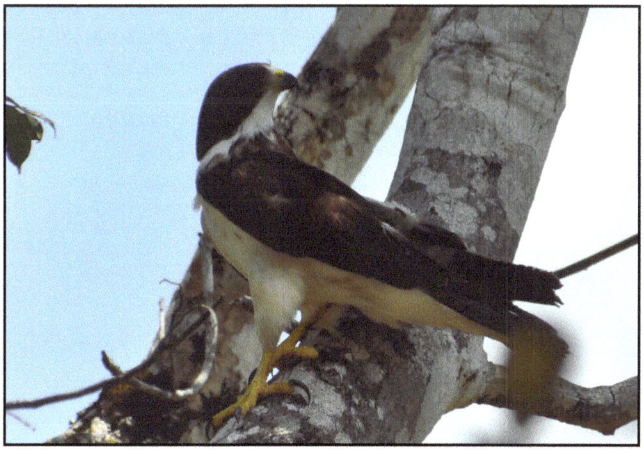

Short-tailed Hawk (*Buteo brachyurus*)

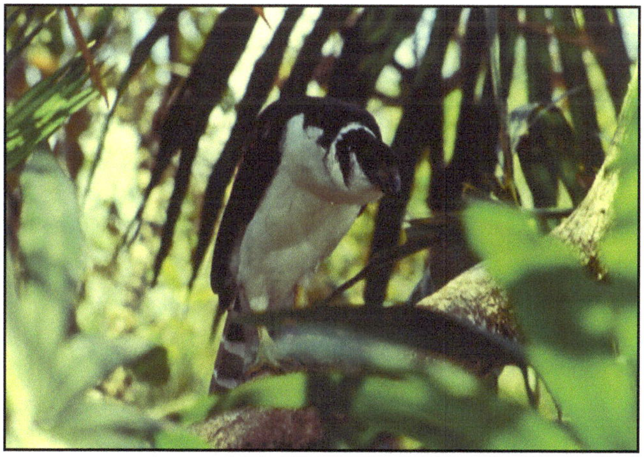

Collared Forest-Falcon (*Micrastur semitorquatus*)

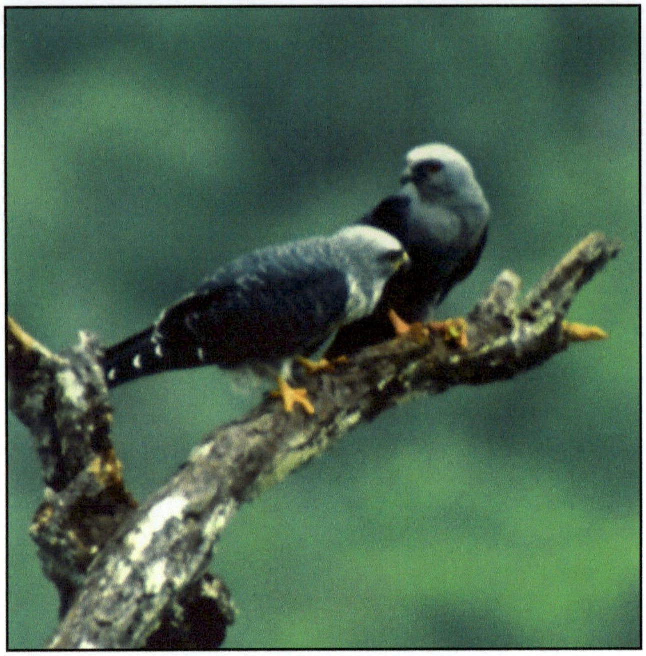

Plumbeous Kite (*Ictinia plumbea*)

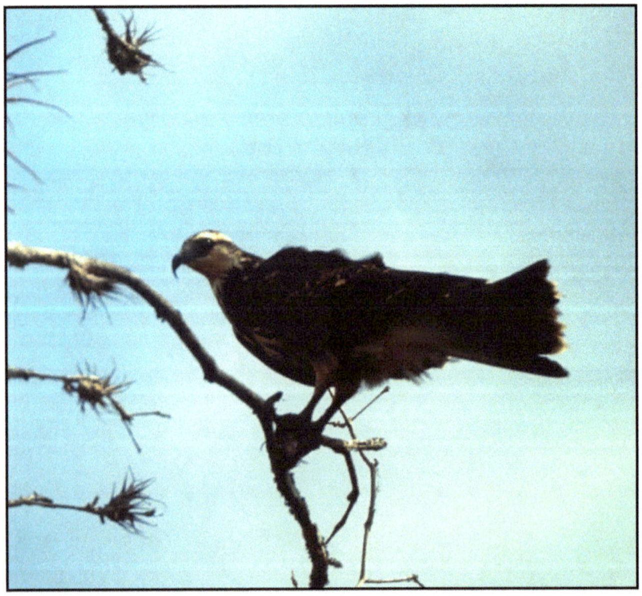

Snail Kite (*Rostrhamus sociabilis*)

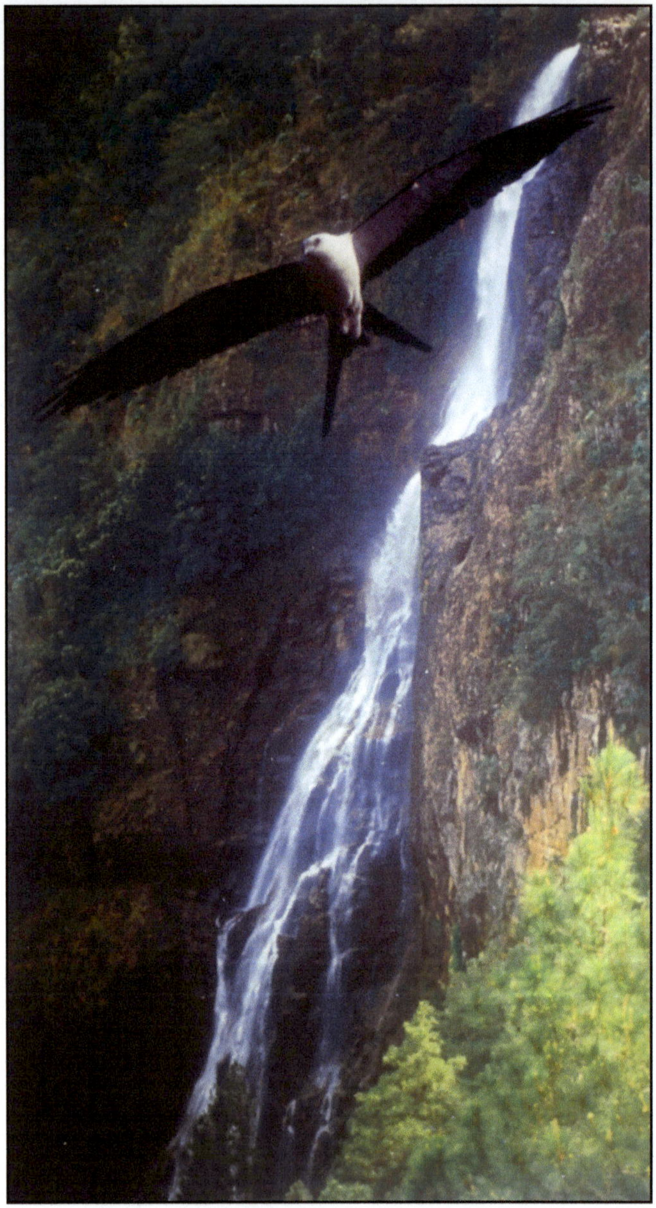

Swallow-tailed Kite (*Elanoides forficatus*) flying over Hidden Valley Falls

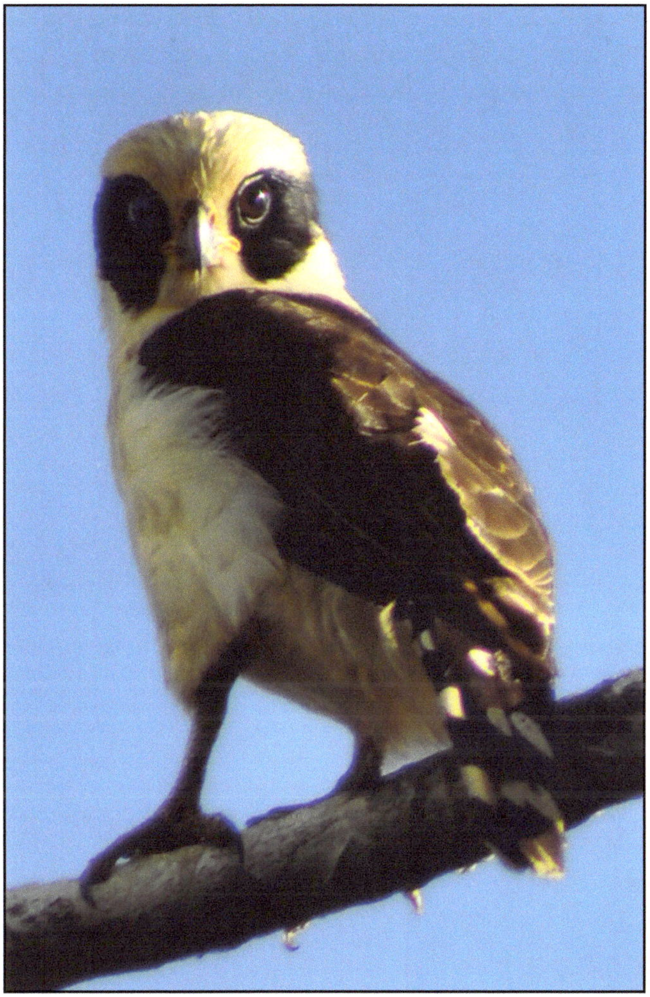

Laughing Falcon (*Herpetotheres cachinnans*)

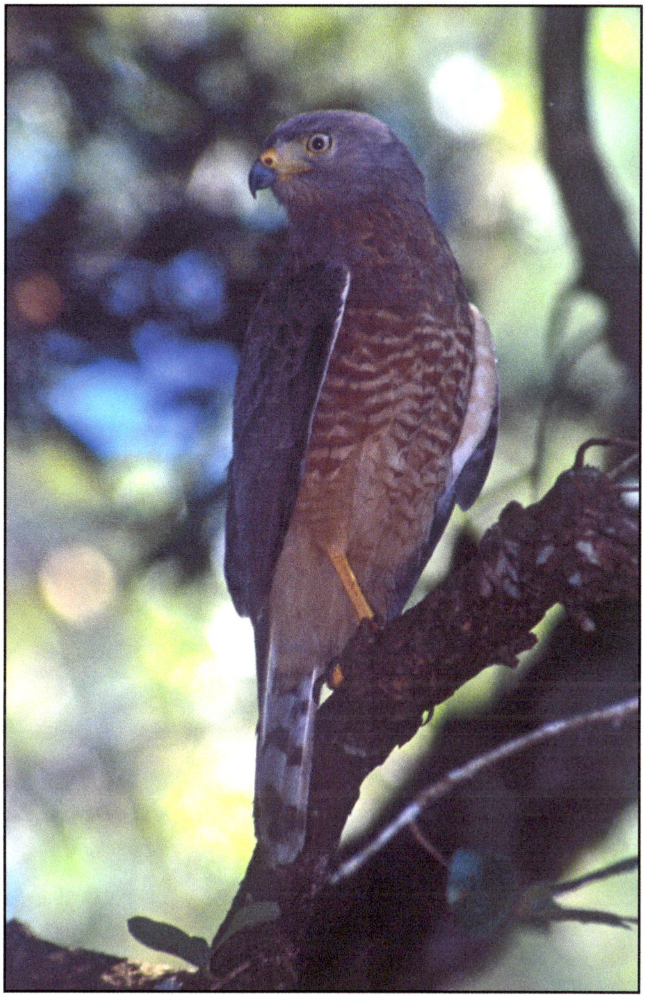

Roadside Hawk (*Buteo magnirostris*)

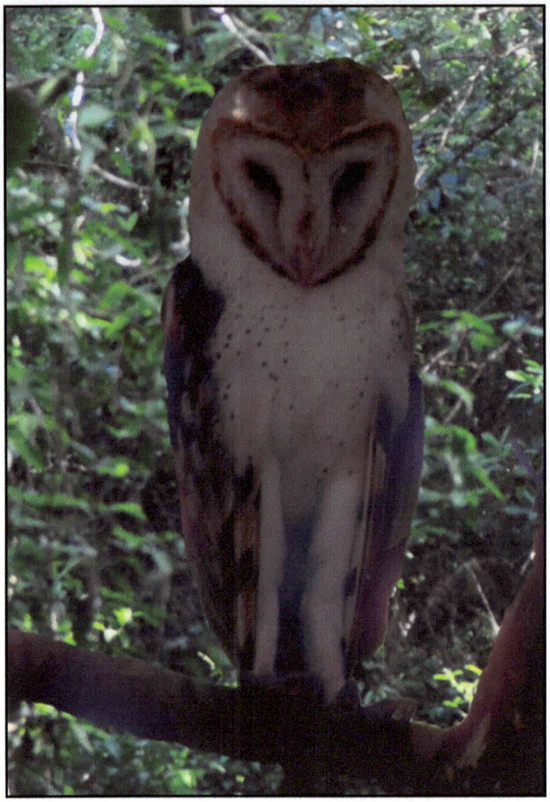

Barn Owl (*Tyto alba*) —*The Belize Zoo*

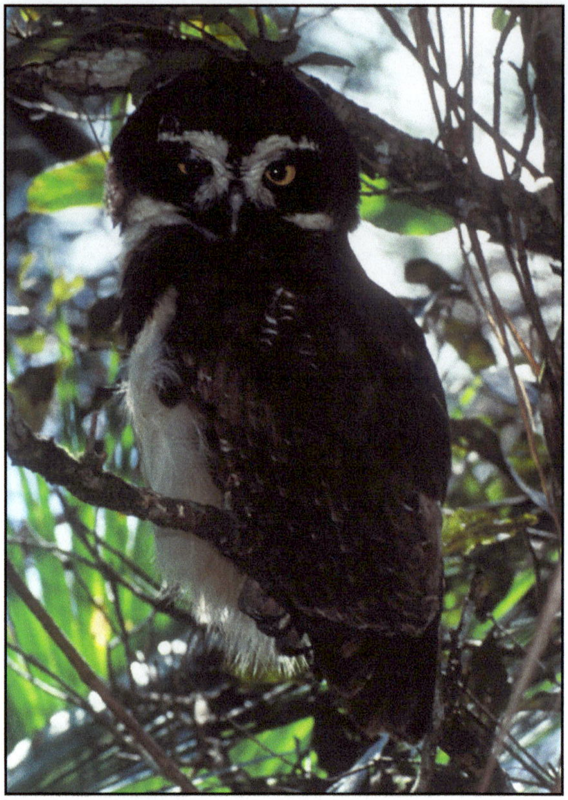

Spectacled Owl (*Pulsatrix perspicillata*)

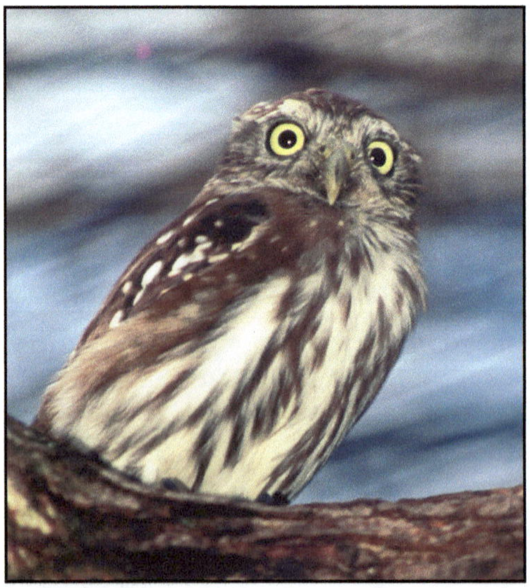

Ferruginous Pygmy-Owl
(*Glaucidium brasilianum*)

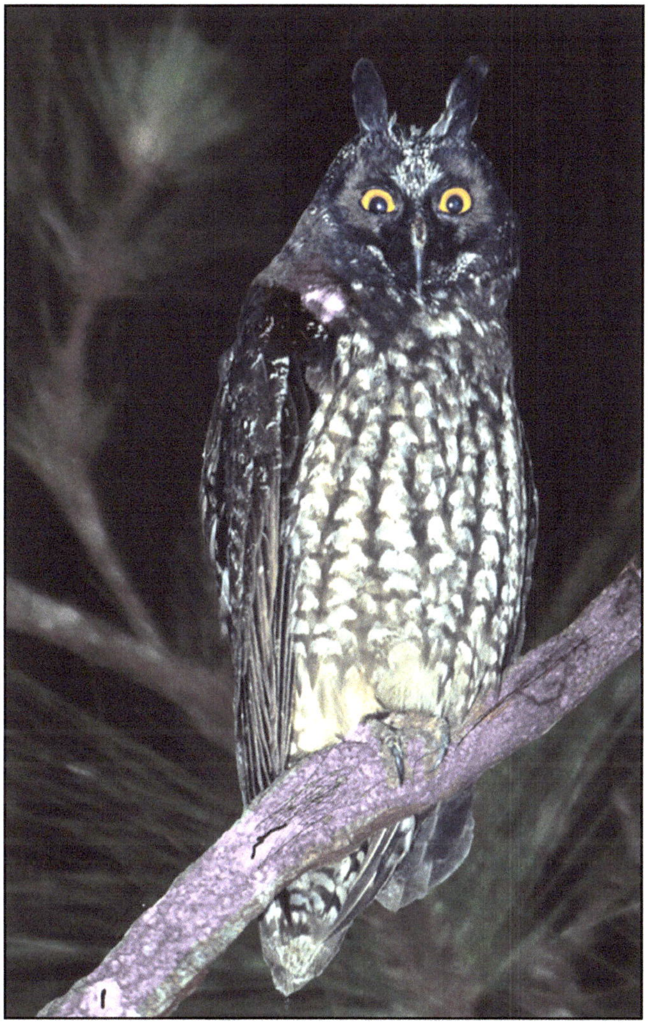

Stygian Owl (*Asio stygius*)

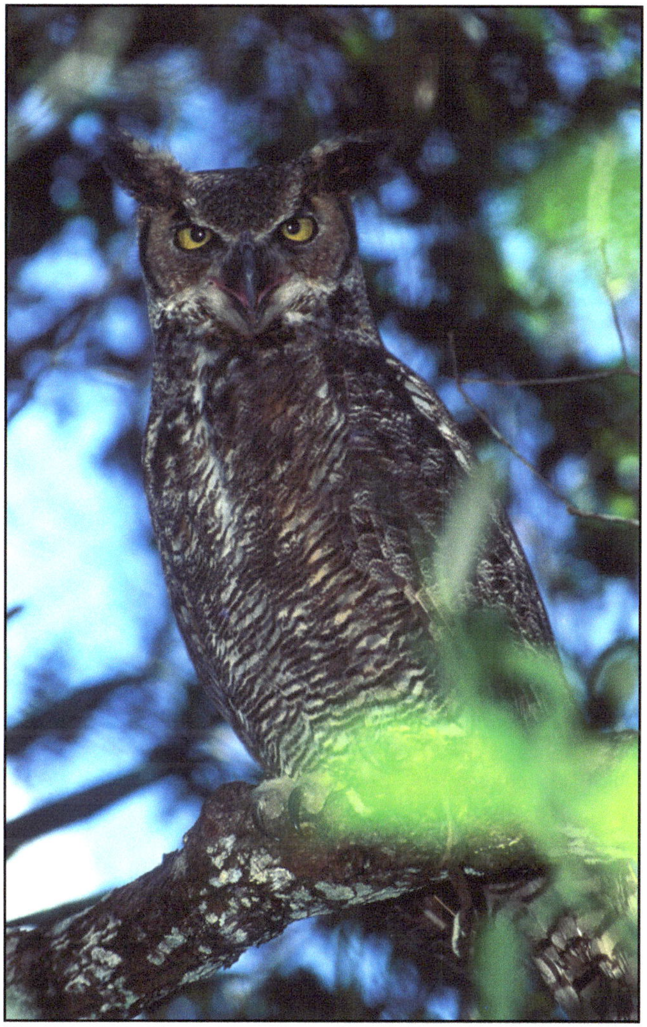

Great Horned Owl (*Bubo virginianus*)

Harpy Eagle

It was early in 1982 and I had been living in British Honduras for quite an extended period. But British Honduras was no more. Now it was Belize. As a British citizen and a resident in the former British Honduras, I was now a Belizean National. Being a Belizean was not a problem for me, in fact I loved Belize. My problem, at that time, was money.

For this reason I was leaving my new country for a while and heading north to Toronto, Canada. I had lived in Canada from 1965. Working in Toronto for one year I could make much more money than I could be working in the new Belize for ten years. I had worked as a plumbing contractor and my master's license was still valid, so, I opened my own company.

I placed an ad in several of the local newspapers, bought a Dodge window van, collected my tools from the friend who had obligingly stored them in his basement, and I was in business. It all worked out better than I could have guessed. There was an urgent need for a reliable tradesman who didn't mind coming at any hour of any day.

Accompanying me north to Toronto was Dorothy. Dorothy and I had been a pair for quite some time and she was very interested in visiting Toronto for a while. Dorothy also was busy. She found a photo developing shop, who taught the art of processing film and taking photographs.

As the Canadian winter approached again, I closed my business, stored my tools back in my friend's basement, and gave my clients the names of several of my old workmates who were reliable, actually more reliable than me, as they had no Belize to go to.

Once all my tools were removed, my plumbing truck was just an empty shell. I installed a double bunk, a small sink unit with pressurized hot water, a small 12-volt refridgerator, and a Coleman stove.

Unlike my Bell Telephone truck that took Reinhard, Maria and I through Baja in 1969, we carried no additional spare wheels on the roof. We did however carry a 14-foot canoe which we used along the way. We drove west then south; visiting Canada's Waterton National Park and crossed the border into the U.S. Glacier National Park. My favourite national park in the United States is the Grand Teton National Park and my second visit did not disappoint me.

Dorothy and I were following the trail to Lake Solitude and, as we approached a small gradient of the trail, an extremely large bull moose appeared over the crest of the hill we were approaching. At the time I was wearing a heavy red and black lumberjack shirt. I removed it and dropped it where I stood. Dorothy moved to the right of the trail and I moved to the left.

As the moose walked along the trail I observed that from my line of sight, the moose would make its way in front of Grand Teton Mountain. I had my camera in hand but I realized that from my perspective I was far too high to get the moose in the same picture as Grand Teton. I dropped to my knees as the moose approached. I didn't realize that the moose was going to be so close. I waited until its position was perfect and snapped a picture. The moose momentarily stopped; then moved on again. As this massive animal stopped, so did, I swear, my heart.

As the moose approached my lumberjack shirt, he stopped, backed off about five paces, then ran towards it and jumped over my shirt. I was extremely glad that I chose to remove my shirt!

When we returned to the campground, I told my story to the Ranger. He looked at me amazed. "Don't you even know how lucky you are?" he said. "The rutting season started a few days ago."

I always did have a way with animals!

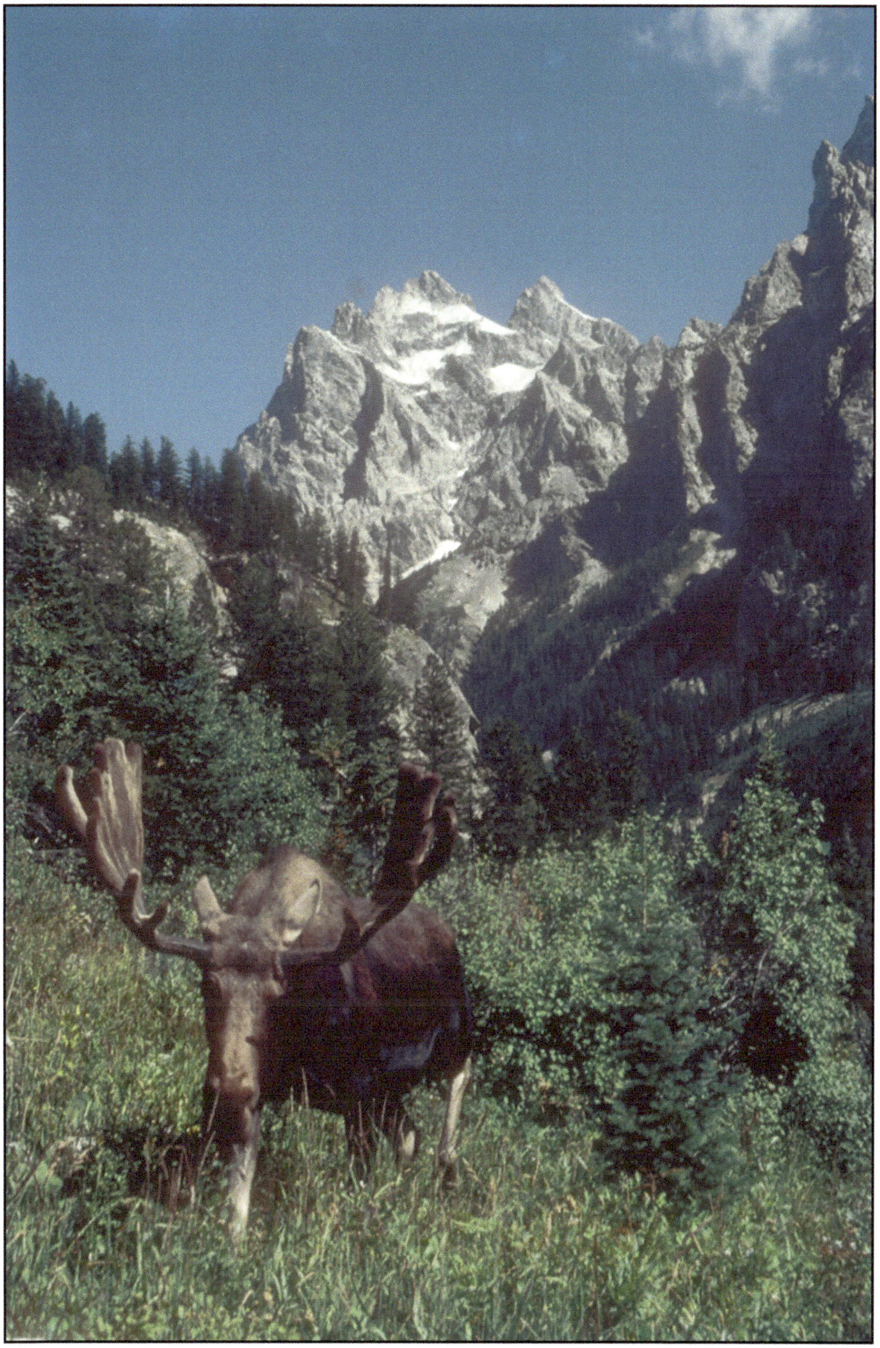

Bull Moose (*Alces alces*) at the Grand Tetons National Park, USA

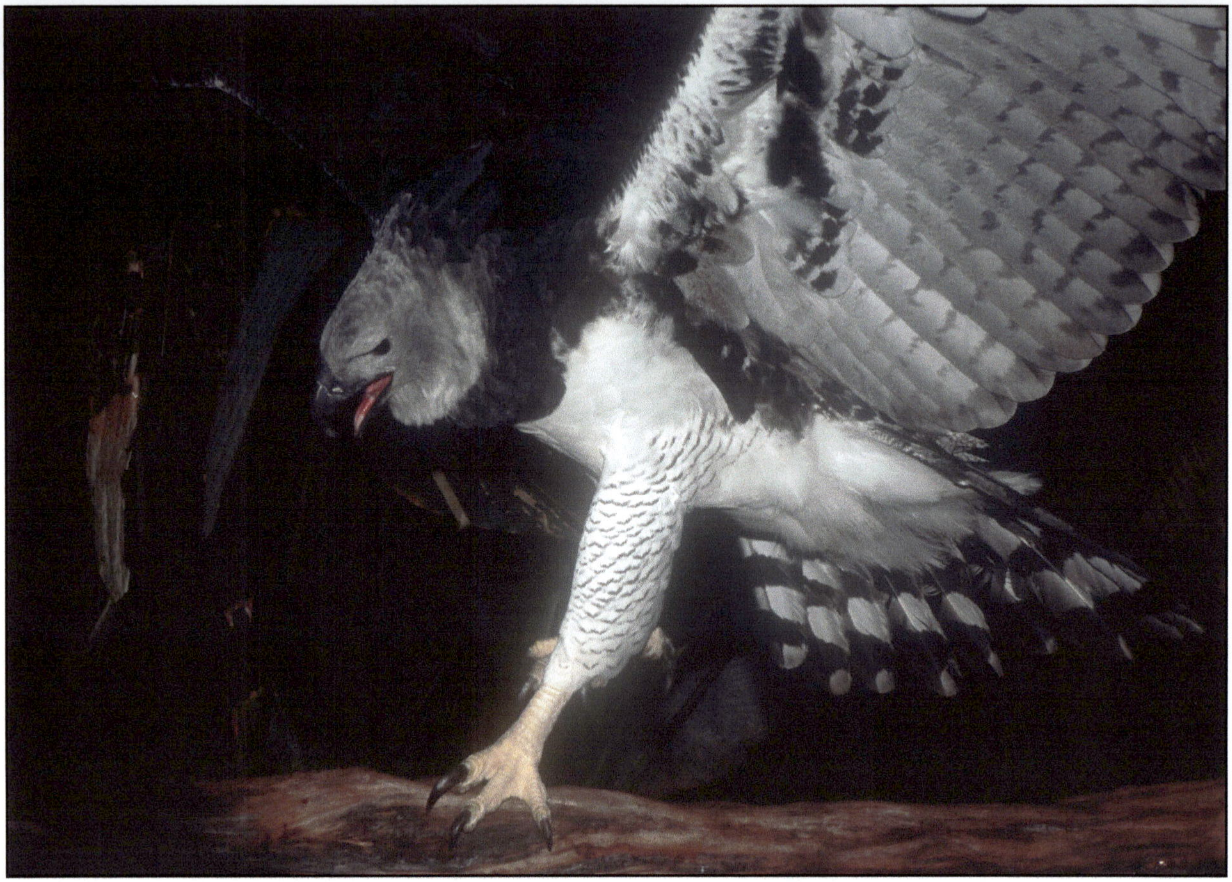

Harpy Eagle (*Harpia harpyja*) —The Peregrine Fund and The Belize Zoo

For those readers who are wondering why they are looking at a moose in Montana rather than a Harpy Eagle in Belize, my apologies. Please bear with me, the eagle cometh. The common denominator is the truck.

A week after we returned to Belize we used the truck to go on a trip to southern Belize, which was the only area I had yet to see. South of Belmopan the road became extremely rough with large potholes and large protruding rocks. We went only thirty miles and then our transmission died, leaving us stuck miles from anywhere. After half an hour a van driving south stopped for us.

The driver came out and checked out our van, found the problem, and introduced himself as Lloyd Pandi, the head mechanic for the orange processing factory about forty miles further on. He invited us to come in his truck to his workplace from where he would send the tow-truck to bring our vehicle in. We jumped at the opportunity.

On examination of our vehicle, Lloyd said he would need a day or two to replace the parts and invited us to stay until my van was repaired even offering us a bedroom. Concerned that we would be bored, Lloyd arranged a trip for us with a guide. Our guide was an older Garifuna man from Toledo.

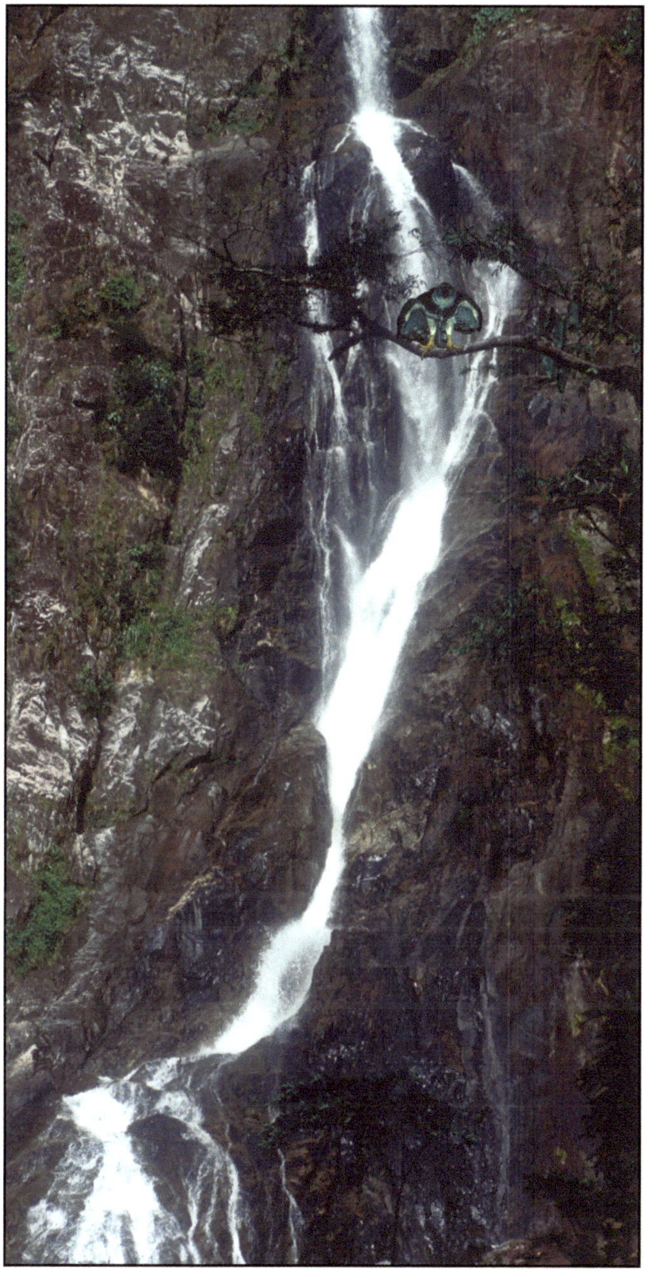

Davis Falls with Harpy Eagle

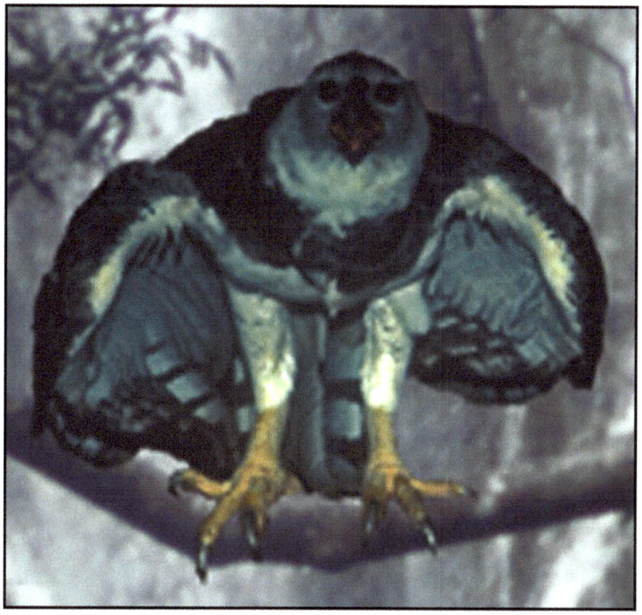

enlargment of Harpy Eagle (*Harpia harpyja*)

The trip was several miles each way over a rough and sometimes unmarked route. Our destination was a magnificent waterfall, several hundred feet high. I had kept careful note on our trail to the waterfall, which I found out was called Davis Falls. About eighteen months later I went back to Davis Falls and camped overnight. The next morning, I noticed that a large bird of prey was roosting on the branch of a large tree that hung over the falls near the top on the right side. I did not have the expertise to identify the bird, nor did I have a suitable lens to take its photograph. I did have a good short and a medium length lens which I used. The 50mm lens allowed me to take the entire water-fall, bird and all.

At that time, I was not really into birds, so I couldn't identify it. During 1980s there were no bird identification books for Belize. It was not until Lee Jones' book, *Birds of Belize* came out in 2003, that I was able to identify it as either a Crested Eagle or a young Harpy Eagle. It turned out to be a harpy, the female of which is the largest eagle in the world.

Solitary Eagle

Towards the end of 2009, I was invited to be on the Board of a new organization, the Belize Raptor Research Institute (BRRI). The object of BRRI is to study little known raptors in Belize, and most specifically the raptors of the Mountain Pine Ridge. There were two birds of special interest, the little known Stygian Owl and the very rare Solitary Eagle.

The Stygian Owl is one of the rarest owls, but lives in no less than twelve Central and South American countries. In Belize, it is regarded as a rare resident that is restricted to the highland pine forests.

The Solitary Eagle is an extremely rare eagle that favours highland pine and pine oak forests. Although much larger, it is quite similar to the Common Black Hawk in plumage and body shape, both having an extremely short tail with a narrow white edge and a wider white band. On occasions the bird may take on a reddish cast.

The Solitary Eagles' range is from highland Mexico south to Chile. They makes their nest high in Mountain Pine Ridge forests, where the female lays a single egg every second year. The adult birds feed the chick on a diet of snakes, which they capture and kill by crushing their heads. So rare is the Solitary Eagle, that only two nests had previously been found. Both nests were discovered in Mexico, one in 1947, and the second in 1958. Sadly the adult eagles were shot and collected as museum displays.

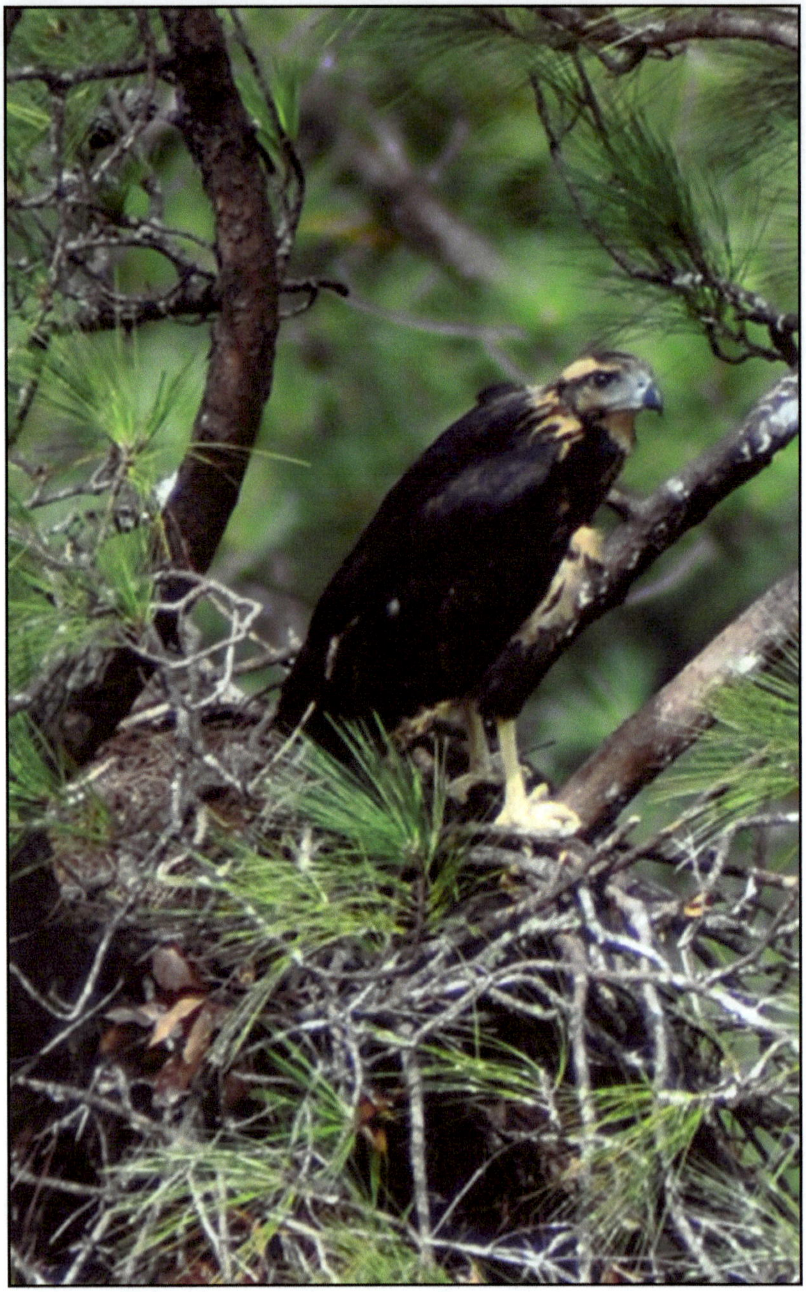

Solitary Eagle (*Harpyhaliaetus solitarius*) chick on nest

On 30th June 2011, Roni Martinez a BRRI member, assisted by several members of the Peregrine Fund, located a nest of a Solitary Eagle in Belize's Mountain Pine Ridge. He advised the other members of BRRI about the nest. One other trip was made with Jan Meerman, the vice president of BRRI, who made several suggestions regarding the threat of disturbing the chick and adult birds, by coming too close to the nest, which was only visible from one location, owing to the steepness and density of the pine forest.

BRRI Executive Director Ryan Phillips, who teaches in California, flew to Belize to see the nest for himself. On the 9th of July, I met Ryan at Belize International Airport and then drove to the Mountain Pine Ridge where we met Roni.

Two hours later, we had boarded BRRI's four-wheel-drive wagon which had been aptly named "The Raptor Mobil" and hurried towards the upper pine forest. The road was a narrow track with ditches on either side, and surfaced with red clay, which could be dangerously slick during rains.

After quite a few miles we found a spot to leave the vehicle, loaded our gear on our backs, and set out. The terrain was rugged and we made our way up and down multiple ridges, crossed one river, and started climbing up steep hillsides of loose pebbles which allowed two steps up for one we slid back.

I felt at a distinct disadvantage, for I was by far the oldest member of the group. Roni was a young 26 year old and Ryan was 34. I, on the other hand, was sneaking up on my seventieth birthday. Still I persevered! We stopped, set up the telescope, and found the nest at the top of the next hill. A half an hour later, we set off again, first down a hill and up the next hill. This last hill was steeper than the previous one with more loose pebbles. Spotted here and there were small charred pine trees from the many forest fires that had burned earlier that summer. In addition to my weary legs, my lower back was giving me problems. I recalled the three fractured vertebrae which I had received, courtesy of a Greyhound Bus Lines accident in 2002. I used the small charred pine trees to pull myself up the steep hill. While this eased the pain, it also blackened my hands, then in turn my face, when I wiped the sweat off my brow. Roni noted my tardiness and slowed down the last hundred yards or so.

At last we reached a somewhat level peak where I pitched my small pup tent. Further down a steep slope Roni and Ryan found a narrow level area to set up the telescope tripod, where they monitored the nest almost until dusk. After an hour of rest, I gingerly edged down the steep gradient so I could check out the chick.

The next morning a parent bird was at the nest giving the chick a snake for breakfast before taking off again. Ryan was glued to the telescope but both Roni and I were on our feet moving in the direction of the nest. The eagle moved off the mountain before disappearing behind the forest canopy. While I was waiting for it to emerge into the open, it became aware of us and flew directly above us. Roni was quick and snapped an excellent picture, as the eagle flew over him; unfortunately I was looking the wrong way.

In mid-afternoon, I announced that I was going to take a picture of the chick and Ryan volunteered to come with me. I knew that he wanted to be sure that I didn't get too close to the nest, for there had been much spoken regarding encroaching too near the nesting area. However, I had no intention of going too close, and welcomed his company. As we approached the area, Ryan said "There it is!" But I couldn't see it. Again he said it, I still couldn't see. The third time he said, "Don't you see that black spot in the tree?" I could indeed see the black spot, which was the chick! A couple more steps and he said, "you can take the photo from here."

I could see the black spot, but it was still pretty far away. The chick was standing upright at the edge of the nest. I crawled forward a yard or so and sat

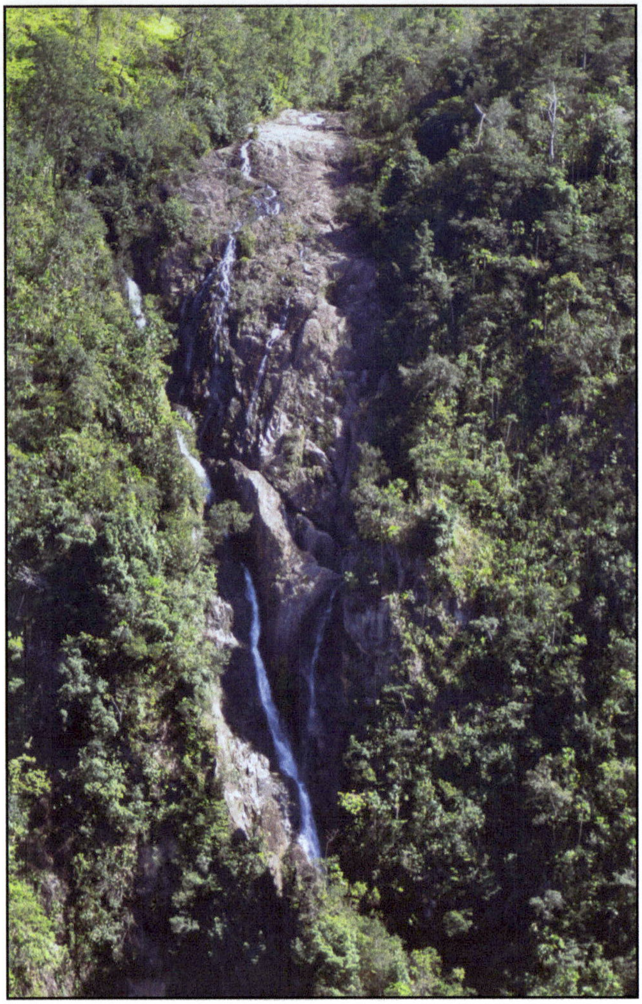

King Vulture Falls

directly behind a small dead tree with my legs and arms spreadeagle on either side of the tree trunk. I then lowered my telephoto lens so it rested on a five-inch branch which was approximately in the right area. Pressing my 400 mm lens against the tree trunk with it still resting on the branch, I had a reasonable facsimile of a tripod. I looked through the viewfinder and was glad to see that the eagle chick seemed to be unaware of my presence. I focused to the best of my ability, and gently pressed the shutter release. The chick moved! I waited until the chick stopped moving, and took ten more photos in a ten-minute period. Of my ten photos, over half of them could have been improved upon, three were reasonable and one was perfect. That was the only one I needed!

As dusk was approaching, we were surprised when three uniformed men, climbed up onto our little plateau. In the lead was a tall man who introduced himself as a British Army officer; the other two persons were guides. The British Officer explained that they had been looking for us for most of that afternoon. It seems that we had left the "Raptor Mobile" in the wrong spot.

The following day was the last day of a British Army training exercise. Unless it was moved, the Raptor Mobile may be blown up! He was politely asking Ryan if he could move our vehicle that night. It was already dusk, so for Ryan to move the truck himself meant a five-hour hike in the dark through some of the roughest and most rugged country in Belize. Ryan pulled out the keys for the truck and asked the officer if he could move it? The officier then radioed headquarters.

The reply was that the three of us, the officer, and his two guides, all had to get out that night and they would send a helicopter for us. In due course the Army helicopter landed and loaded the six of us and took off. I must say it got us back a little quicker than had we hiked out. That night we slept in the army camp, and the forecasted rains came in. The next day we drove the Raptor Mobile out. Ryan did an excellent job of driving on the treacherous red clay road. Even at 5 mph, the four-wheel drive truck constantly slid at right angles to the road and the eighteen-inch deep culverts. Finally we did make it to the hotel where Roni was employed as a naturalist guide.

I soon left to return home. A week later I found out Roni and Ryan had waited out the day of the mock war and returned to the nest site. On that day and the following one they had much better sightings of both adult eagles and at one point they watched as eagles carried snakes to the hungry chick. In spite of my aches and pains, it was a very interesting two days that I wouldn't have missed for anything.

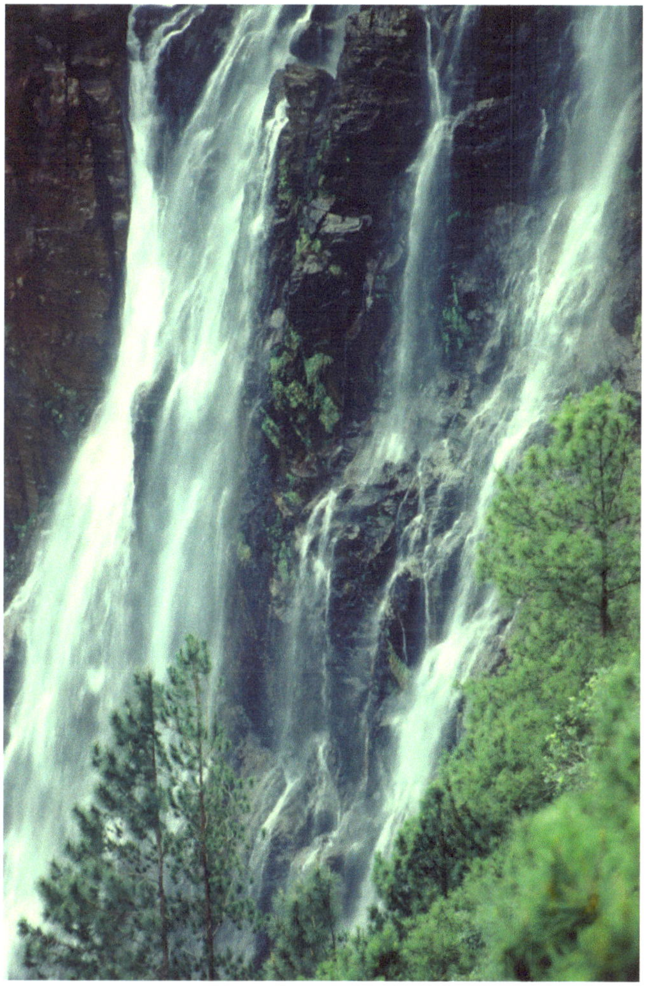

Lower Hidden Valley Falls

The Belize Underworld

For many years the main attractions in Belize have been fishing, diving, sailing, birding, hiking the rainforest areas, and exploring Maya ruins. But there is one other attraction – the cave systems. Two-thirds of Belize is karst, which is limestone that can be dissolved to form caves. Caves can be found in three forms: dry caves, wet caves with rivers, and submarine caves. Belize's cave systems have an additional attraction in that a large percentage of them were used by the ancient Maya, and relics of these times are still commonly found.

My first experiences with caves were mostly submarine. Belize's most well known submarine cave is the Blue Hole at Lighthouse Reef, where I guided for several years (*see Chapter* 2). A similar vertical cave is situated west of Columbus Caye, east-northeast of Dangriga. It is only 95 feet deep, with a much smaller diameter than the Blue Hole, but still interesting to divers.

In the north of Belize, there are two submarine caves, one near Caye Chapel and the other near Caye Caulker. Both are extremely dangerous for divers and the Caye Caulker cave has already claimed two lives through sheer ignorance and a total lack of essential equipment.

The Caye Caulker cave was partially explored by expert cave-diver Jim Coke and his dive buddy. They made several staged dives over the three-year period, 1986-1988. Part of the Caye Caulker Cave is under the island itself. This cave has been reported to be over a mile long, and one room about 1,800 feet in diameter with stalactites and stalagmites. The maximum depth of the cave exceeds 90 feet.

Since I moved to Caye Caulker, I have long heard stories of a house that suddenly sank and disappeared underground. In more recent times there was a rumor that in the construction of a large building, a steel pile actually disappeared through the ground.

Around 1974 Jim Smith, who operated a small dive shop on Caye Caulker, and I decided to explore the Caye Chapel cave, which was known to the locals as a good place to catch large groupers.We submerged through a vertical shaft until we hit the sloping horizontal tunnel, which we followed down to a point where it fell sharply away. Directly opposite, our dive lights illuminated a sloping crystalline wall which dropped quickly into the depth. Exceptionally beautiful, but because of the depth, very dangerous.

Jim and I looked at one another, then he raised his finger pointing back the way we came. I never went back down that hole and I'm pretty sure Jim didn't either. I don't have anything to do with either of these underwater caves.

Caving tourists are quite satisfied with the inland caves and Belize has spectacular examples of underground formations. The Actun Tunichil Muknal Cave (ATM Cave) was once part of the Tapir Mountain Nature Reserve but the cave became a popular tourist destination and the Nature Reserve designation does not allow tourism activites. So the cave was removed from the Nature Reserve to make the Actun Tunichil Muknal Natural Monument. The word "Actun" means "cave" in Yucatec Maya.

The entrance to the ATM Cave is found by following a river upstream into a cave system, and swimming across a small, but deep pool. Then there is a fairly long walk through water in a narrow canyon that eventually opens into a wider area. One climbs out of the water onto a limestone massive to gain entrance to a large open room with many artifacts, including human remains. The highlight of the cave trip is in a small elevated cave, set high in the wall and reached by climbing a non-Mayan aluminium ladder. Within this cave is the Crystal Maiden, a complete skeleton of an ancient, possibly sacrificial victim, preserved in calcite.

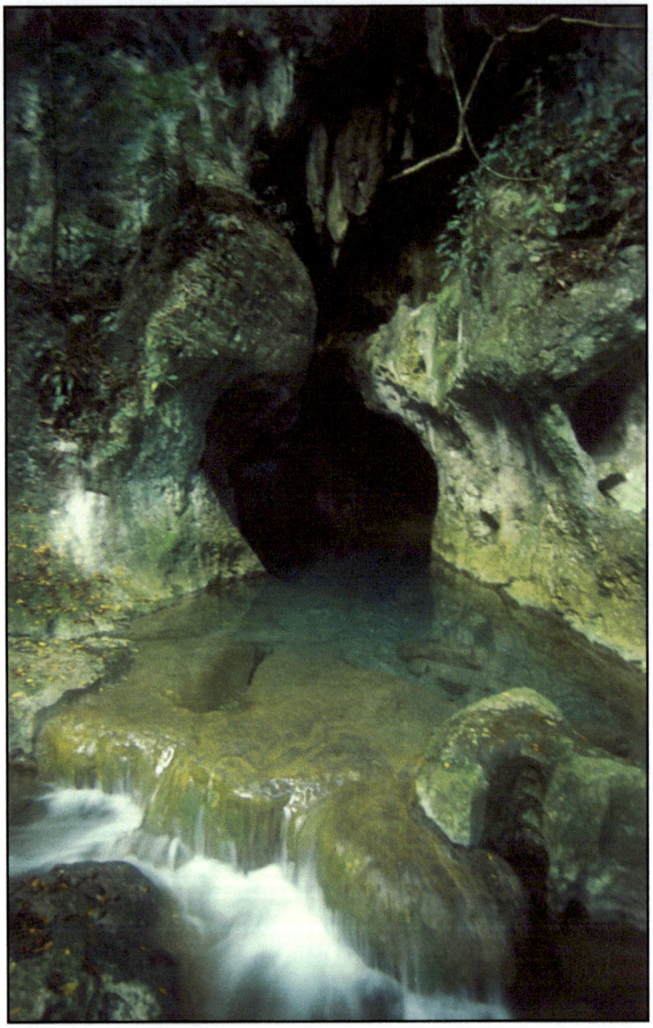

Actun Tunichil Muknal (ATM) Cave entrance, beyond which is a deep pool that one must swim across about 25 feet.

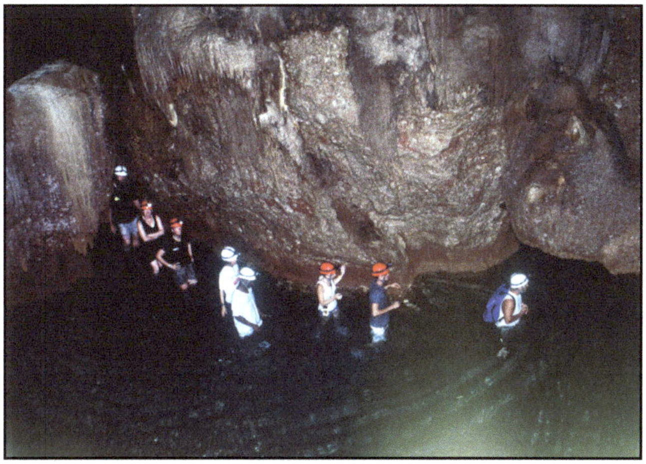

Visitors entering the ATM Cave walk in water through a canyon after swimming through the deep pool.

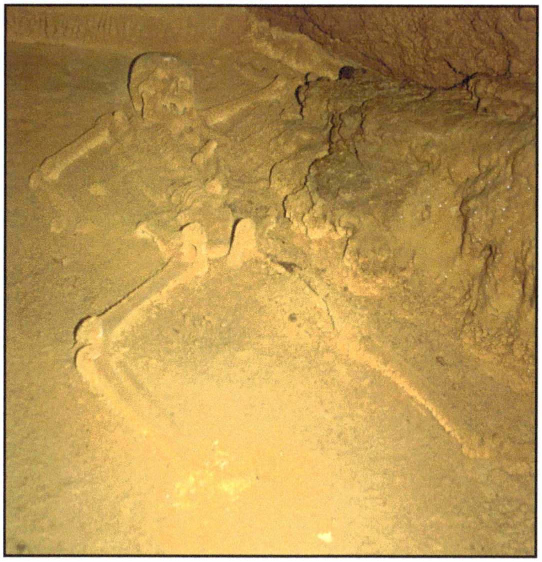

The Crystal Maiden in the ATM Cave

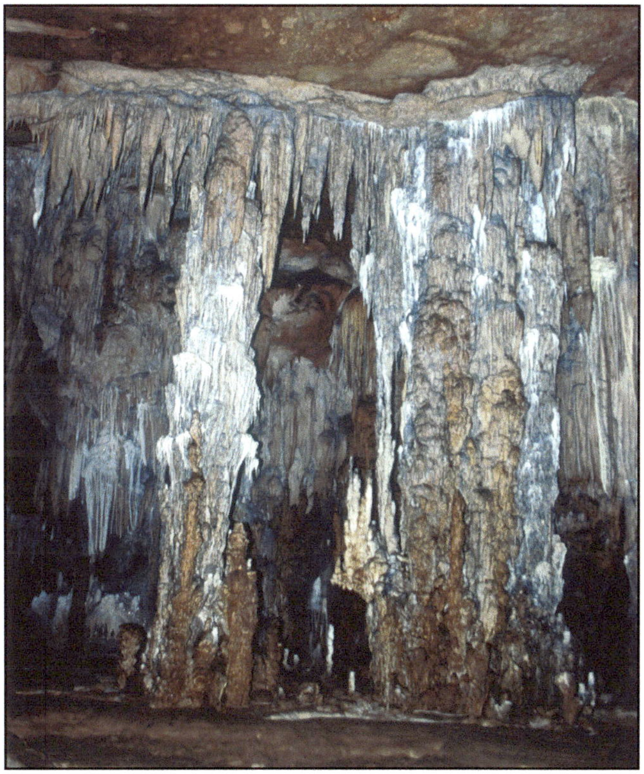

Columns in ATM Cave

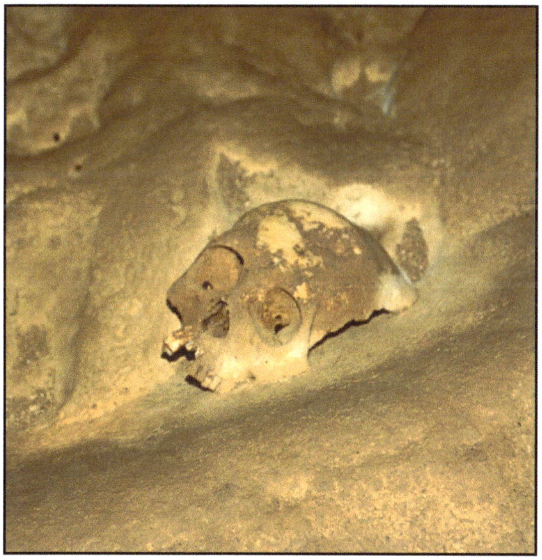

Skull in ATM Cave

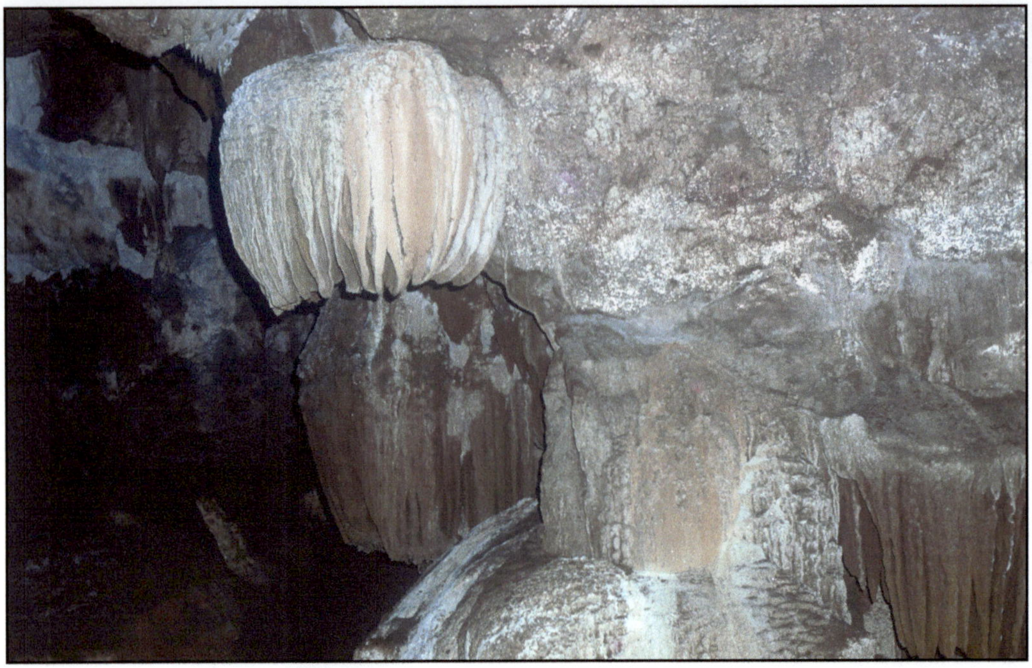

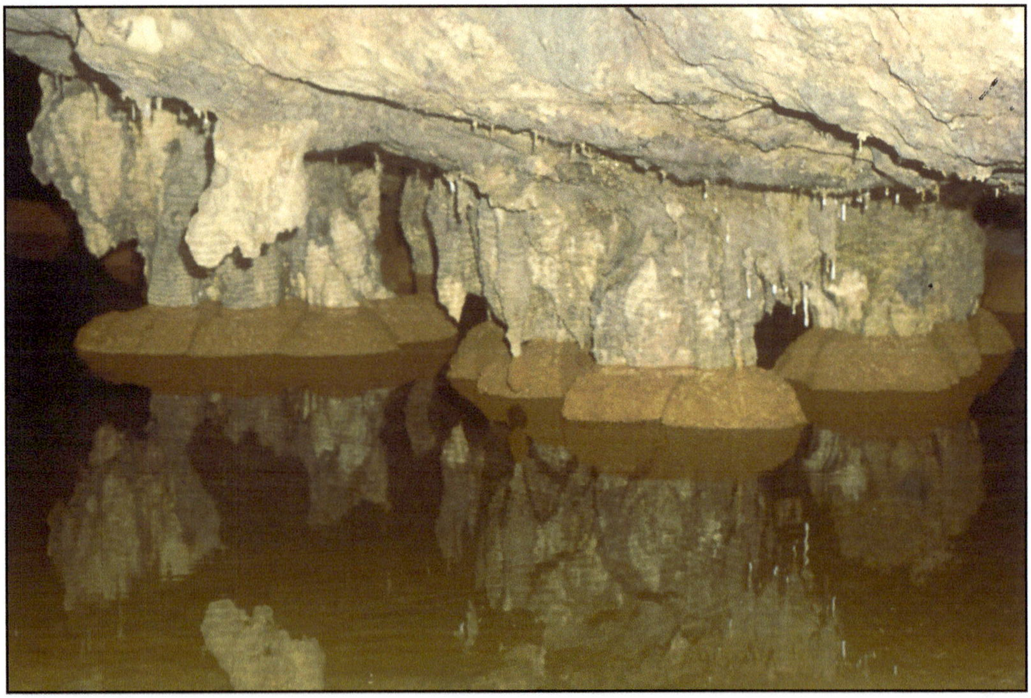

St. Herman's Cave
(above and right)

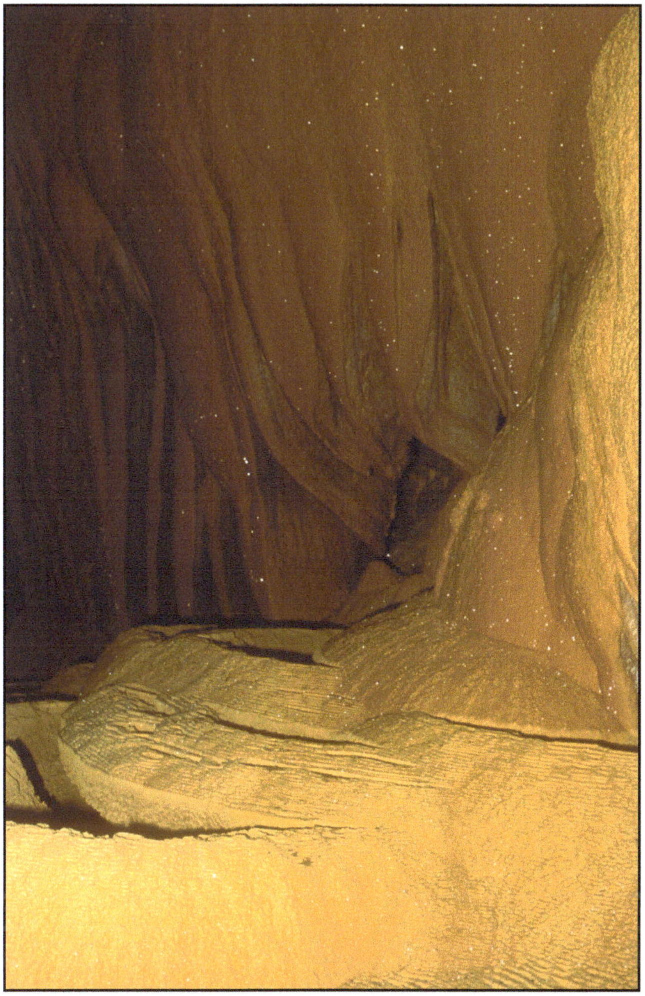

St. Herman's Cave

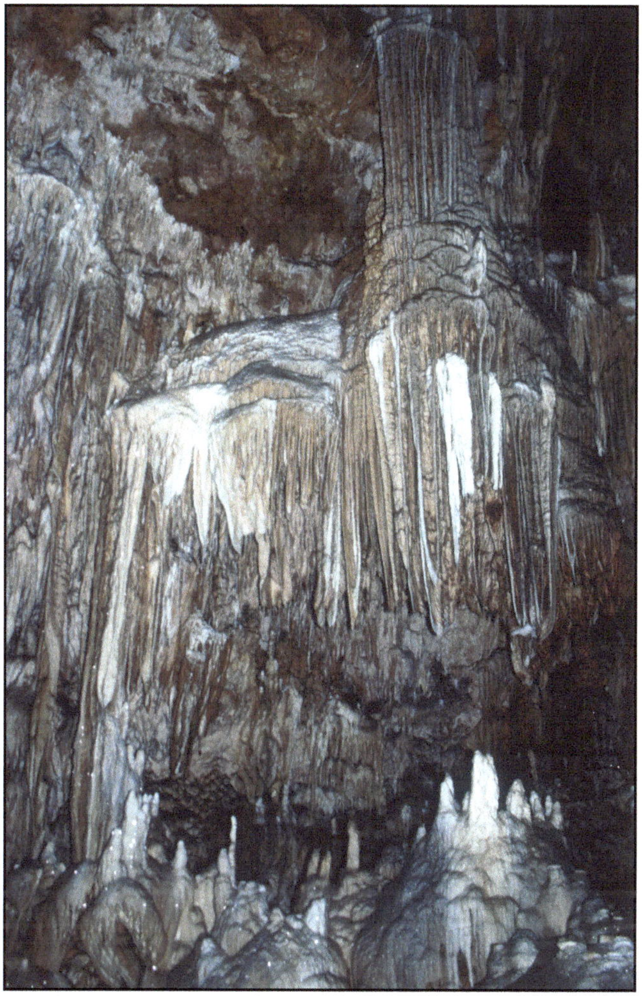

Crystal Cave

Other cave sites include Caves Branch, the name of a Sibun River tributary that runs underground. Tourists enjoy several hours floating downstream, with brief intervals through verdant forest pools before returning to the darkness of the cave.

The St. Herman's Blue Hole National Park, which is managed by the Belize Audubon Society, has two large cave systems, St. Herman's Cave that has a self-guiding trail, and Crystal Cave that requires a guide. Crystal Cave is one of my favourites. This cave was originally called "Mountain Cow Cave" because near the entrance there once was the skeleton of a tapir that had fallen into the cave and died!

Belize's longest cave system is actually four caves that combine to form the Chiquibul System, which is over 40 miles long with one entrance in Belize and another in Guatamala. These individual caves are Cebada, Xiblba, Actun Tun Kul and Actun Kabat.

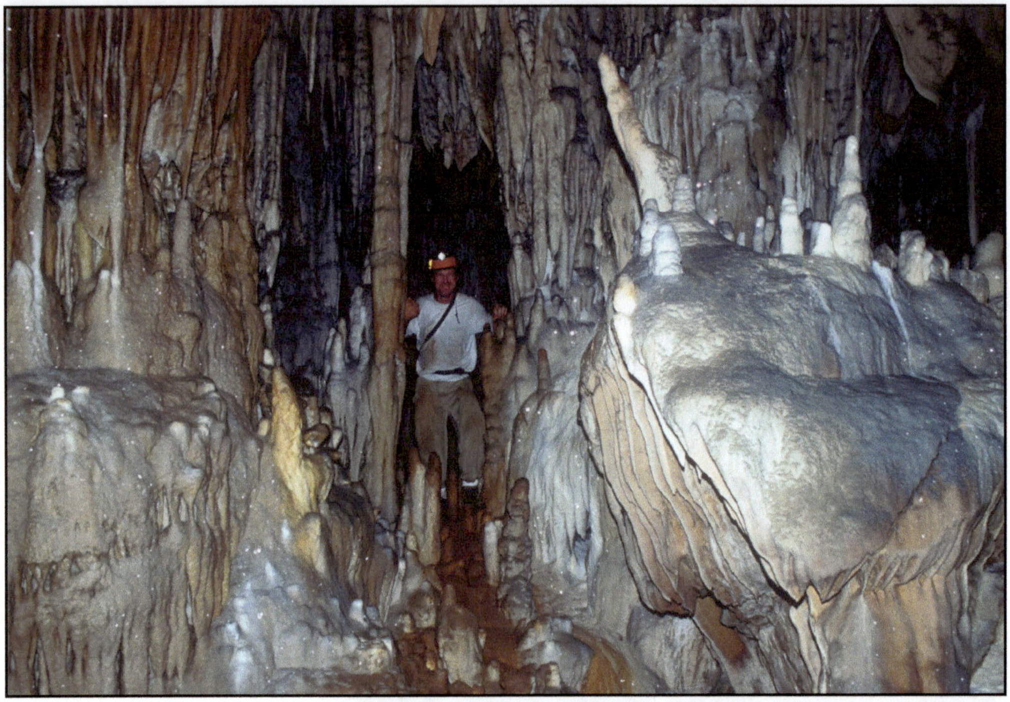

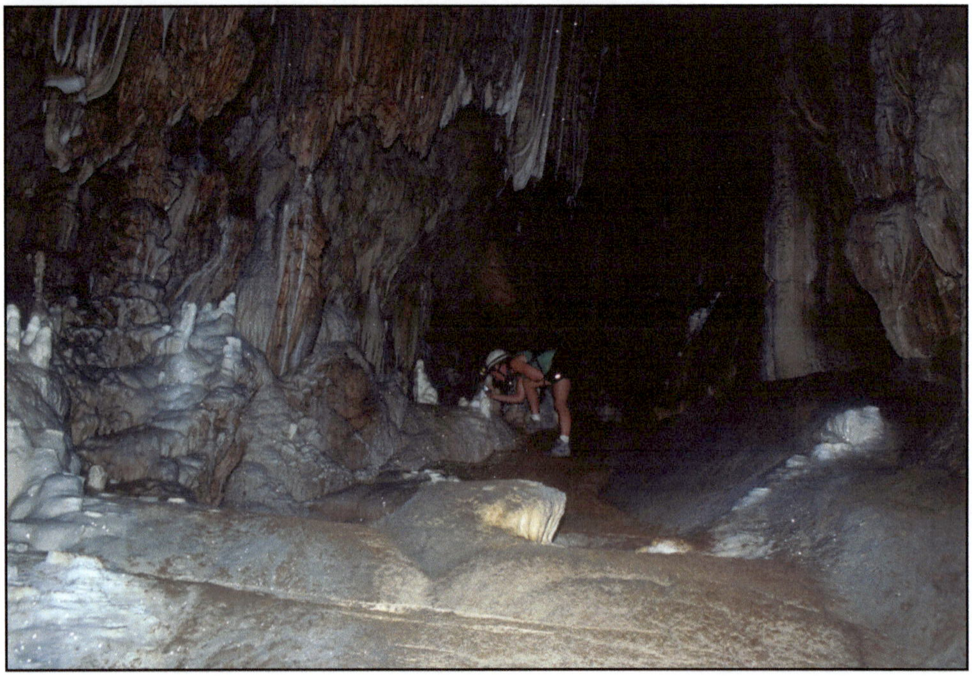

Crystal Cave
(*above and right*)

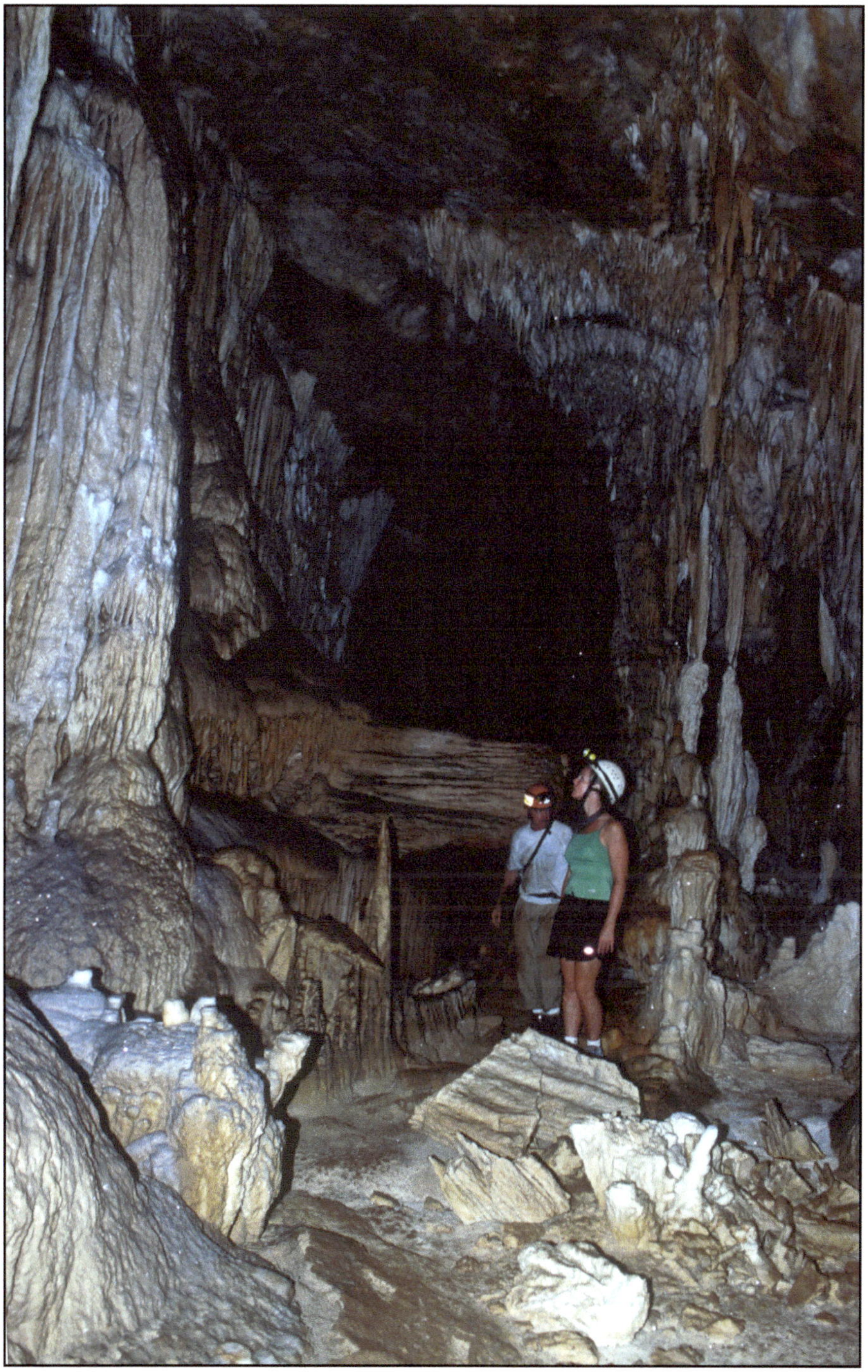

Crystal Cave

The Maya site of Chechem Ha in the Vaca Plateau was discovered by accident when a hunter, Antonio Morales shot an agouti. The agouti then ran through a crevice in a cliff face and was quickly followed by Antonio Morales's dog. When the dog didn't return after half an hour, Antonio started to pull out the loose rubble to look for his dog. Instead, he found a cave with large number of Maya artifacts including many large ollas, complete with lids. Antonio and William now serve as custodians to the cave. And close by, they offer meals, rooms and tours of the cave. Did he find his dog? I forgot to ask!

My first experience in a land cave was in the late 1980s with Greg Smith and Pat Barbutti, an ex-airline stewardess. Greg had already hired a guide named Moses from Gales Point, who knew the area quite well. The trail started at a point on the Coastal Highway, where the Gales Point bus dropped us off. The first day was spent hiking to our overnight camp, which was located on the bank of the Manatee River.

It was a pleasant evening that we spent watching spider monkeys while we ate our first meal of the trip. As evening darkened, we watched the large phosphorescent beetles, or lightning bugs flying around us on this moonless night. So radiant were their bright green headlights that one could easily let their thoughts wander and imagine that they were miniature extra-terrestrials checking us out, but then one made a crash-landing right among us, and it was, after all, just a large dull brown beetle, with pale green discs on the side of its head.

Nevertheless I found it quite interesting and decided to shoot a photograph for future identification. I set up my camera with a macro lens and put my flash at the end of a cable for remote hand-held and then took its picture. I was quite surprised when the beetle rolled over on its back, unconscious. Somehow I had caused a sensory overload. The lightning bug was insensible for over fifteen minutes, but slowly its legs

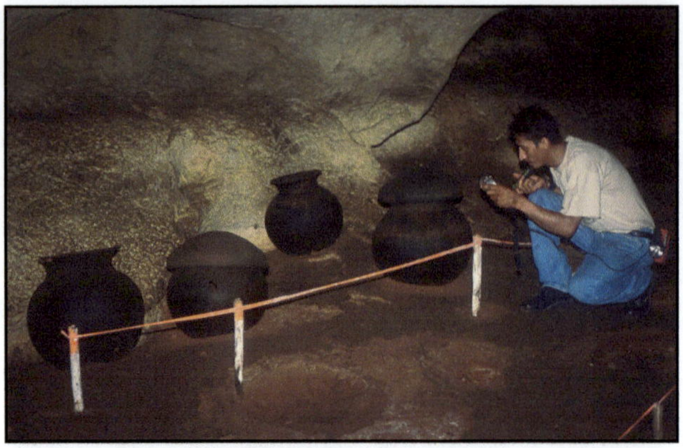

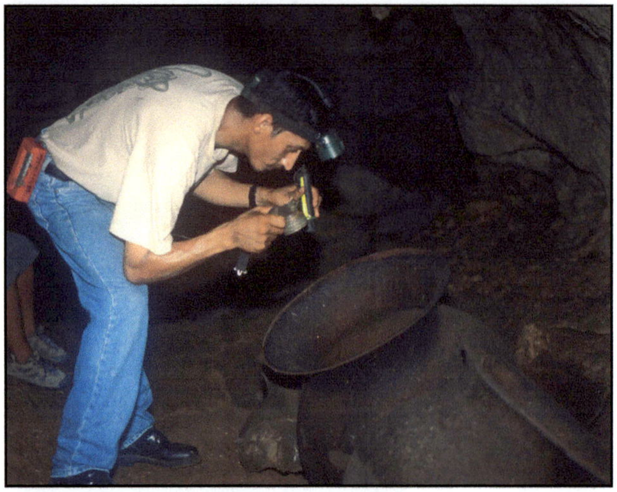

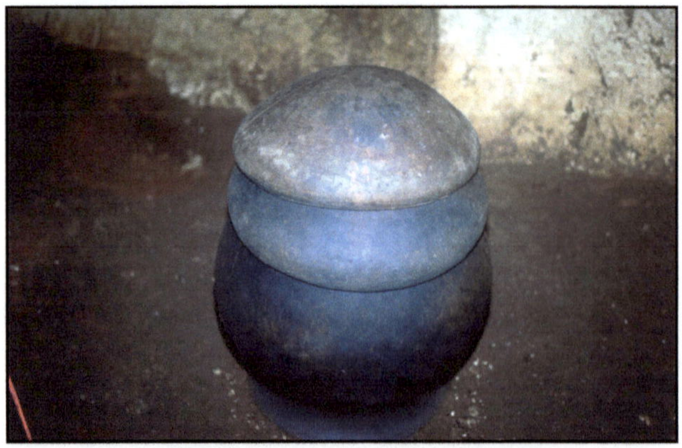

William Pleitez guiding in Chechem Ha Cave

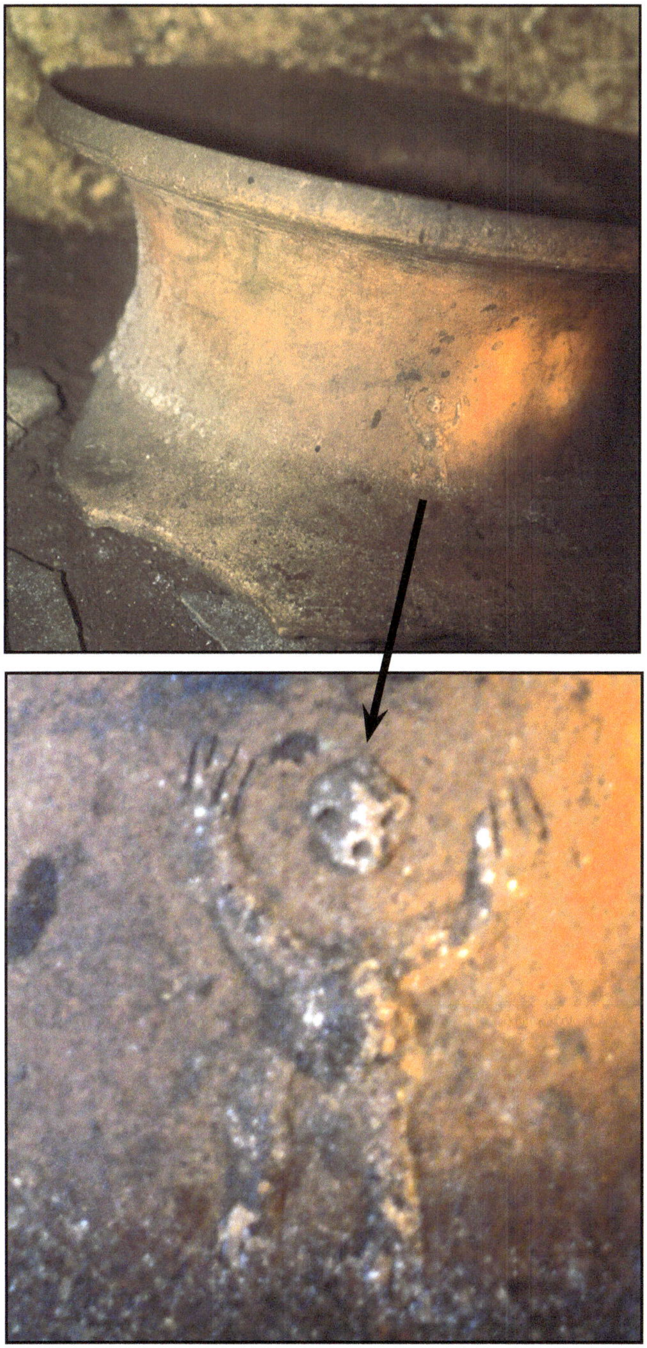

Monkey decoration on olla in Chechem Ha Cave

began to twitch and minutes later it was up on all six legs, preparing for take-off.

The following day, we arrived at the site of the cave by midday. The entrance was located about fifty feet up an overgrown karst hill and was vast, but not visible from ground level. If we didn't have Moses as a guide, we would never have found it.

We clambered down over large boulders until we hit the sandy floor of the cave. The ceiling was about 40 feet high. To the left was a river, knee deep and about twenty feet wide. At first we used the river as a road, and after about fifteen minutes, we came across a large white stalagmite in the centre of the river. The formation was shaped like a gigantic mushroom and water droplets from the cave ceiling landed exactly in the centre of the mushroom's dome.

When we climbed out of the river, we came across some beautiful formations. We stopped to take some photos, but time was pressing and we moved on, passing several more crystalline structures. Eventually we reached the far end of the cave. It exited into a large flat green area, with distant forest encircling it. Our trip was during the dry season, but my impression was that at wetter times this was probably a shallow lagoon. It was still wet underfoot, but there seemed to be no flowing water. I pushed through the five-foot high grasses. Jaguar pug marks were quite noticeable. On my way back I came across tapir tracks, which were still filling with oozing water.

This was my first trip to this cave which we incorrectly called Manatee Cave, not knowing at the time that there already was another Manatee Cave, which the Manatee River ran through. But for us it remained the Manatee Cave.

It was about six years later, that I mentioned this cave to my friend Rad. He was always ready for an adventure, so we decided that a cave trip was not a bad idea, especially as I had written out a detailed description of the route we took on that first trip,

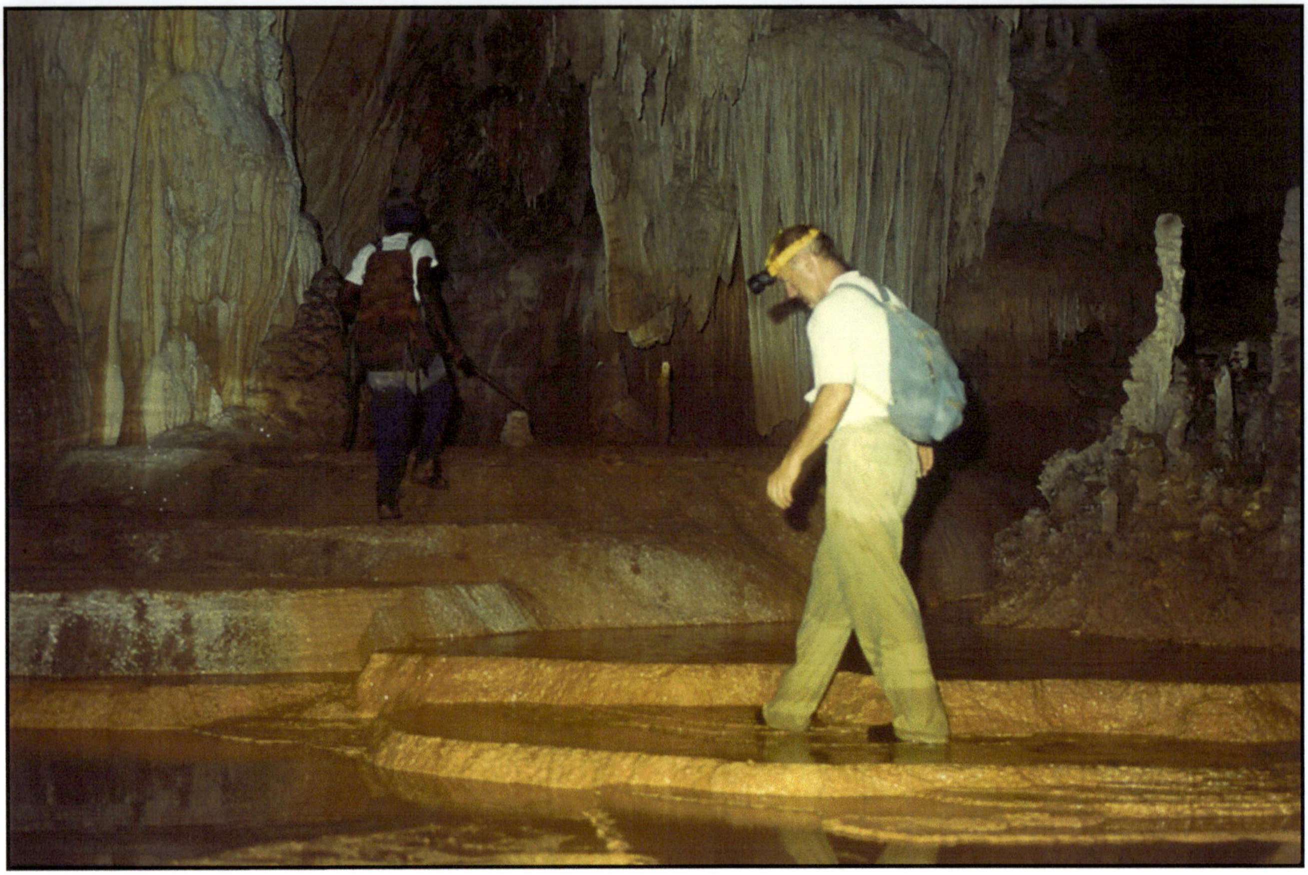

Moses leading Rad in Manatee Cave

noting times between points, plus other pertinent details. Just to be sure, I hired our old guide, Moses.

The trip started well, and at first the three of us made good time. It was after five pm when I realised we were climbing between two karst hills. I could not recall any karst hills on the first trip. I felt strongly that Moses was lost, so I suggested that we camp right where we were.

The next morning we retraced our path, found the proper trail, and after several more hiking hours, we made it to the cave. However, because of the lost time, we did not make it through to the far end of the cave. At the time Rad enjoyed the cave immensely, but at a later time I told him about the other side of the cave, and ever since he has wanted to do a return trip.

It was several years later when Rad and I tried once more. On the second day, our path led us to a group of men brandishing large chain saws and axes. Directly in our path was a maze of felled trees and more chain saws roared directly ahead of us.

What to do? Go back or go around? We decided to go around. We made a big loop around them, flagging our trail with red ribbons, as we went. Eventually we did reach the cave, but much later than we originally

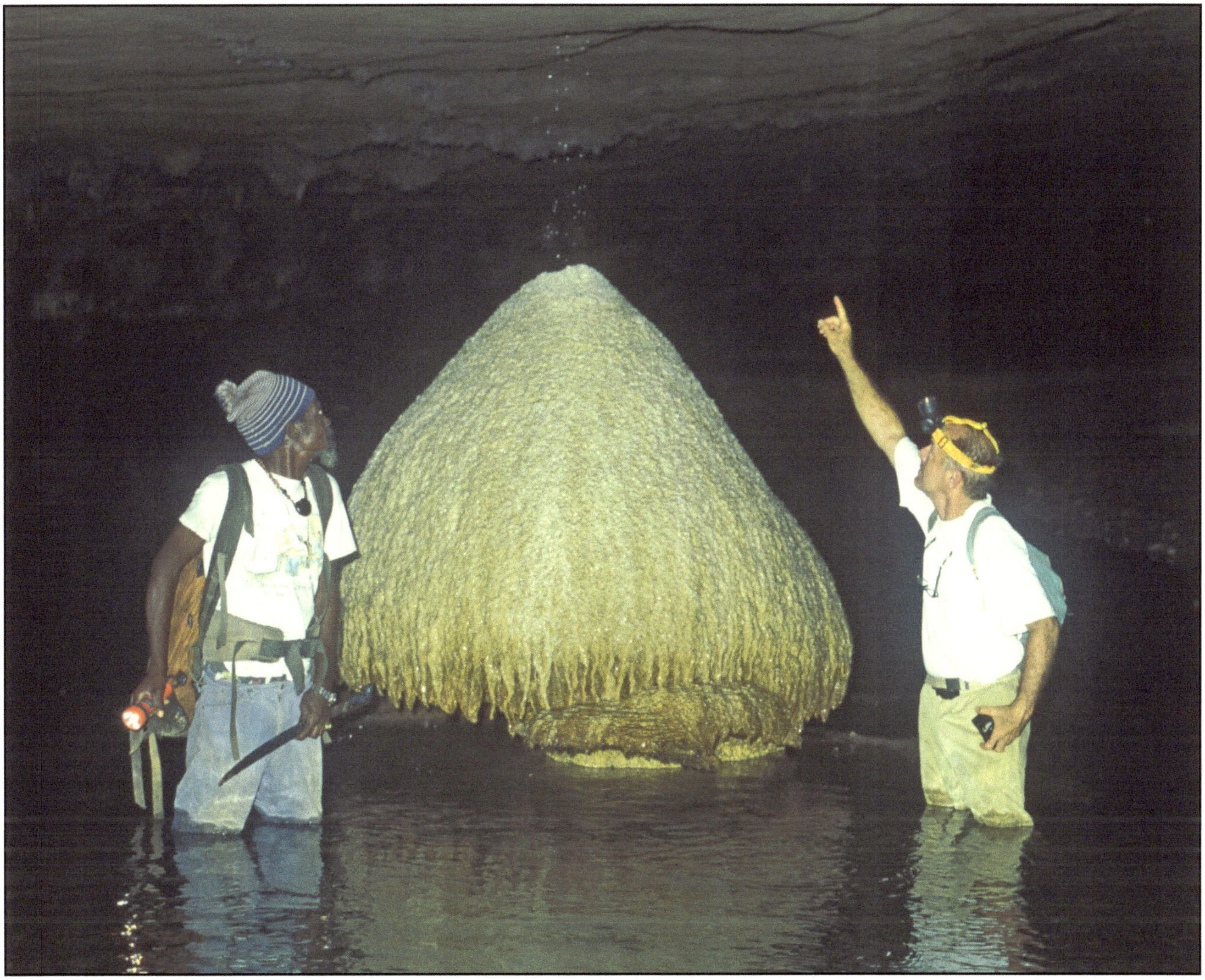

Water dripping on mushroom stalagmite in Manatee Cave

anticipated. Rad, normally a follower, lead the way. Suddenly I was alone. In the blackness of the cave he was out of sight. With the up and down elevations and horizontal turns, I could not catch a glimmer of his headlamp. I looked at my watch; it was creeping up on three-thirty. Normally it was a two- hour walk back to the river camp. I called repeatedly to Rad, at last he answered. I pointed to my watch. What to do? Go back to camp or sleep in the cave? It was decided to go back to camp. "But, I could see light at the end of the cave," said Rad. "Next time," I said. We made two more attempts, but never found the cave again.

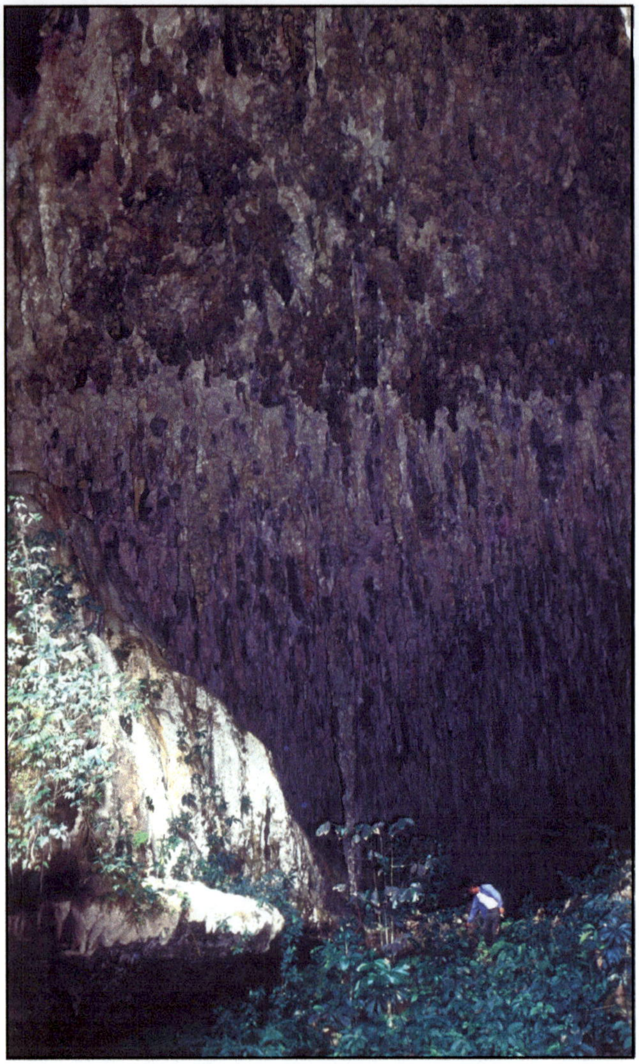

Karst rock wall at Manatee Cave

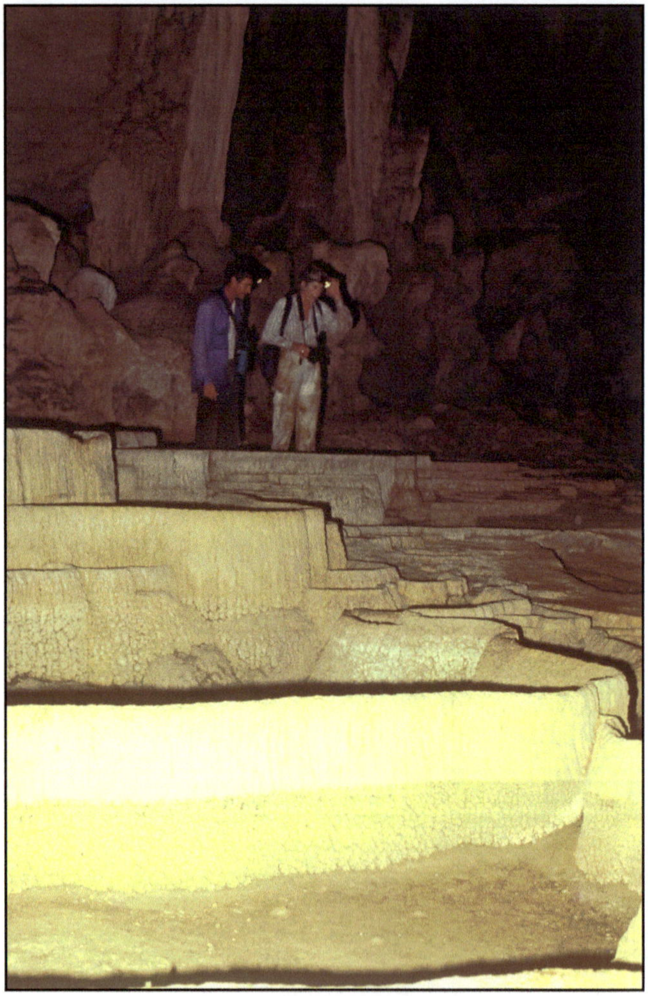

Greg and Pat in Manatee Cave

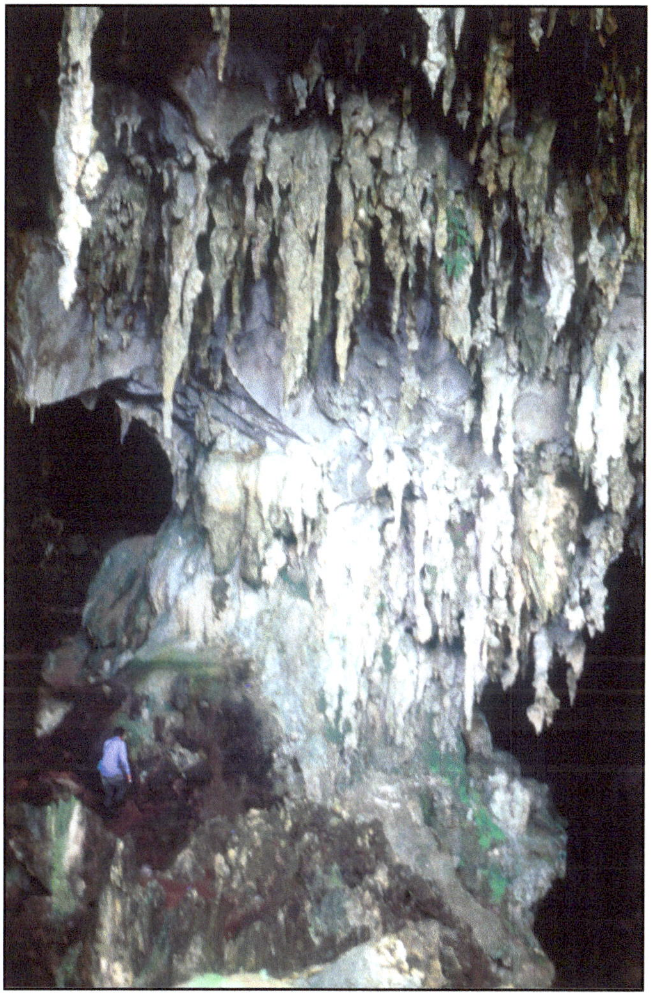

Greg near Manatee Cave exit

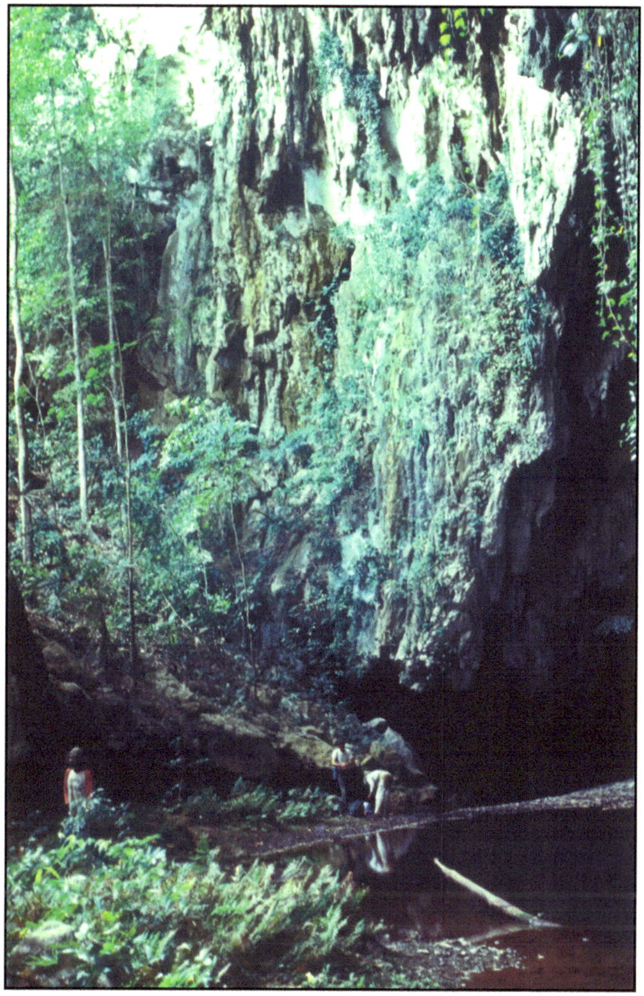

Manatee Cave exit

Beef Worm

No book about the Neotropics should ever be written without a mention of *Dermatobia hominis*. Lovers of nature, whether they are birders, orchid hunters, frog hunters, or just lovers of Belize's tropical forests, will probably never ever come across an adult botfly. Considering that it is a fly, it is a quite a handsome creature of bright green that blends in as part of the forest. However, for those who have come in contact with *Dermatobia hominis*, the memory of that encounter will quite probably stay with you until the day you die!

By evening, the botfly will take to the air to hunt mosquitoes. When it succeeds in capturing a mosquito it will lay approximately 20 fertilized eggs on the mosquito's belly. These eggs will lie dormant until the mosquito does what mosquitoes do best. Using their sensitive infra-red sensors they will hunt out a food source, normally a large mammal, or a bird.

On finding a source the mosquito will alight and feed. The body temperature of the targeted mammal also alerts the still microscopic botfly maggots, which will drop off the mosquito's belly and quickly penetrate the skin, either through the mosquito bite or a handy pore. Here the larva will remain to grow and develop. For the following twenty days, nothing will be noticed, except possibly a tiny red raised spot. As time passes, this spot will take on the appearance of a pimple.

Soon it may weep a little. At this stage if you are patient and have good eyesight, you may see a tiny white spot in the orifice, which will disappear while you are watching. As days pass, the pimple will grow into a boil that will weep continuously. You will also experience an occasional sharp stabbing pain which can make you jump. By this time you may walk around in public slapping or punching the offending area, causing passers-by to give you a wide berth! Shortly after the fourth week, the moment of truth is nigh! Abort it, or take it to term?

If you decide to take it to term, you will have more days of twitching, slapping and oozing a clear fluid, until its birth. This may happen at any time, possibly at 2 am when you are asleep in bed, or there is an equal chance that it will make its emergence in the middle of a presentation you are giving to twenty people. This can be particularly devastating to your presentation if your beefworm happens to reside in your forehead.

Most people decide to abort it. There are many ways to do this and none of them are particularly pleasant. Some methods have been carried down through generations. One such traditional method is to take a piece of raw meat, it has to be fresh and still oozing. This is then placed over the beef worm breathing hole and tied tightly. As the larva is an air breather, it will burrow through the beef in an effort to reach air. Unfortunately, this has never worked with me.

A second traditional method is to use a wad of chewing tobacco rather than meat. As I neither smoke nor chew tobacco, I cannot advise on this.

Another way of killing the larvae is by using Vaseline, butter, or clear nail polish, which is spread thickly over the breathing hole, which will either cause premature extrusion, or suffocate the botfly larva. However, it is best to pick a day when you have nothing else to do, as this treatment may take several hours.

In most cases, I have found it necessary to have a second person. Because I am devout coward I usually wait until the last minute and then get Dorothy to remove it for me. The method we use, I'd adopted approximately thirty years ago, after swimming in Five Blues Lake without wearing a hat! For my sins I got three botfly larvae in the crown of my head. Although I felt silly, we used both the beef and the tobacco method, but to no avail, the skull has insufficient flesh to squeeze. I was slowly going insane when I happened to meet my doctor in Romacs Grocery Store. I asked the doctor if he knew a way to get rid of botfly larvae. "Yes" he replies. "Use a household insecticide."

I replied that I'd heard of an incident where a mother had killed her child in an attempt to kill head lice using a household insecticide. So he explained that I was not to spray, but to use a Q-tip moistened with the poison, wait until the maggot's head appeared out the breathing hole, and dab the head with the insecticide-soaked Q-tip. After you have squeezed out the dead larva, have a hot shower to get rid of the poison. It worked!

Because I normally stand around for long periods, waiting for a photo subject I tend to get a lot of bot-fly larvae. At one time I managed to collect close to thirty bot-fly larvae in a one-week period while working in a rainforest habitat! I was extremely uncomfortable for several weeks, until at last I relented and had Dorothy do the insecticide treatment on me. As fully developed, the botfly maggot is more than an inch in length and club-shaped with five or six bands at the thick end, each band equipped with spines which hold it in place under the skin.

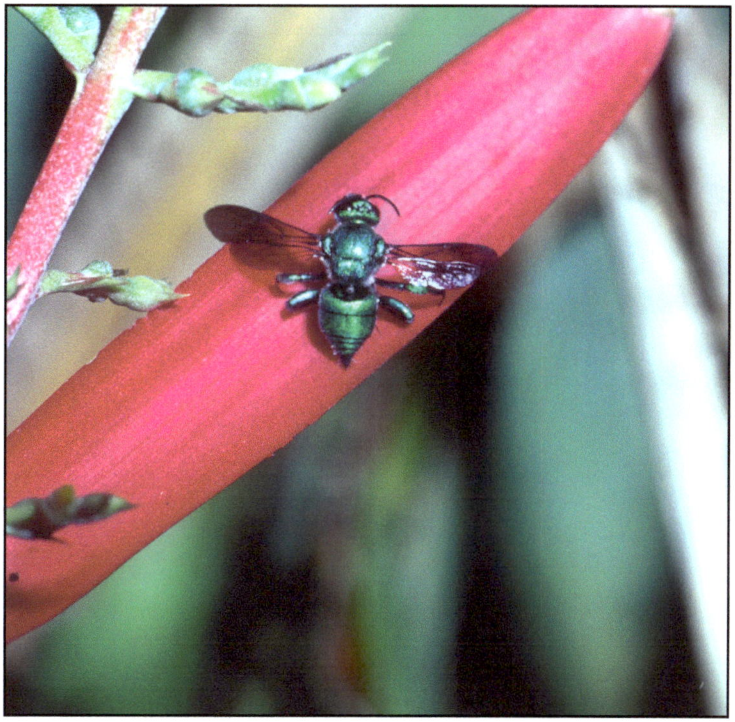

Botfly (*Dermatobia hominis*)

Esculent: delightful to eat, easy to digest

As a maggot is to a bird,
As a bird is to a snake,
As a snake is to a pig,
As a pig is to man,
As a man is to a maggot!

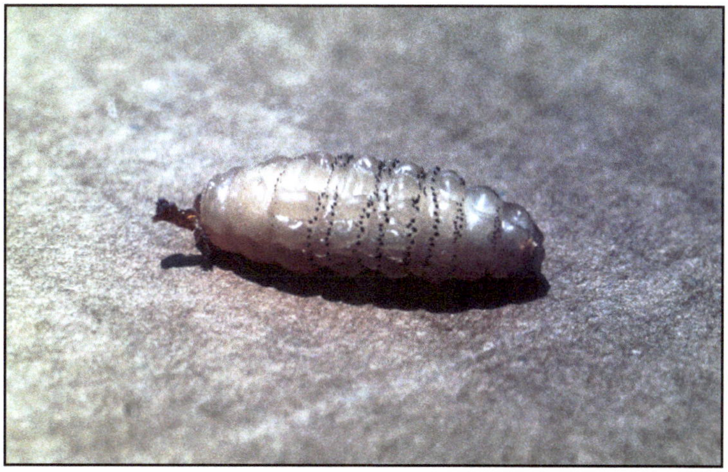

Botfly larvae

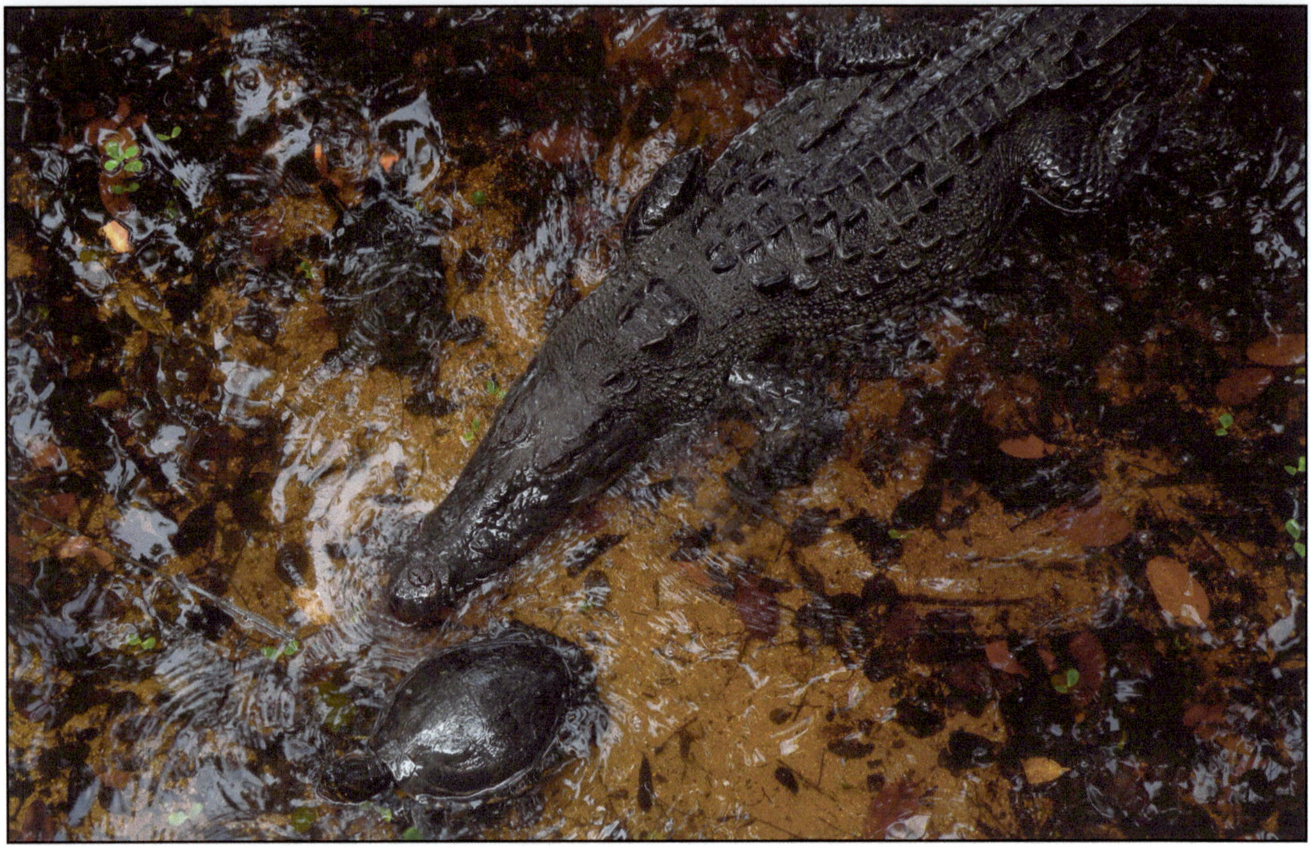

Common Slider (*Trachemys scripta*) and American Crocodile (*Crocodylus acutus*) taken from our back window

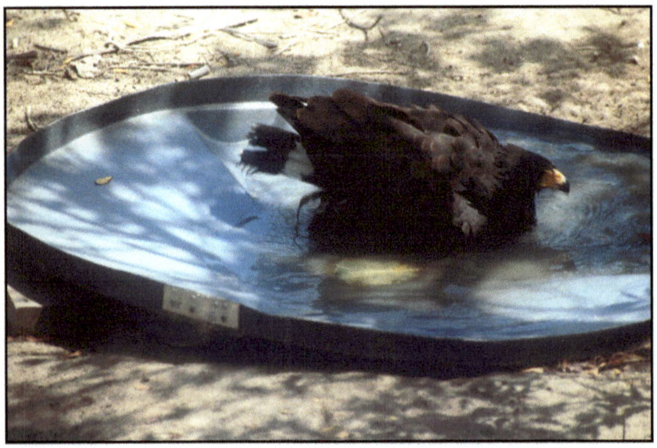

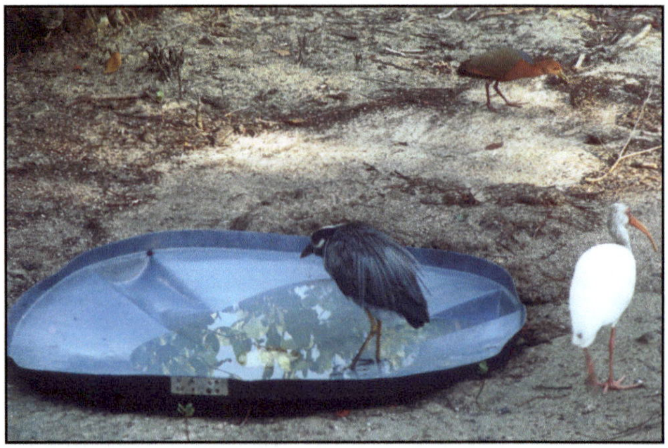

Birds bathing in mulltipurpose bath

(*left*) Common Black Hawk (*right*) Yellow-crowned Night Heron

Chapter IV

A Window to Wildlife

My interest in both crocodiles and birds pretty much appeared simultaneously on leaving Caye Caulker Village in late 1993, when Dorothy and I purchased a lot on the southern most point of Caye Caulker. I built our house on a strip of white mangrove forest, which is about 200 feet from the sea. Out our back window is a small lagoon and mangrove forest that is a six-acre protected area, which is seasonally flooded.

It wasn't long before we were aware of the rich abundance of wildlife on and around the property. Aquatic birds and crocodiles were frequent visitors to our yard, both of which shared a fondness of water and took turns bathing in the multi-purpose bath we provided.

When the first fall migration occurred in the following September, we were both surprised and amazed at the multitude of migrant warblers, vireos, and shore birds that shared our property and the surrounding area with us. At the time, I was totally ignorant of North American bird species but Dorothy was familiar with a few from her home state of North Carolina.

Sad to say, we owned neither a pair of binoculars, nor a bird identification book and at that time, nowhere in Belize could one purchase these items. This problem was solved several months later when Dorothy visited her family in North Carolina, where she purchased a pair of binoculars and a copy of the *Golden Guide to the Birds of North America.*

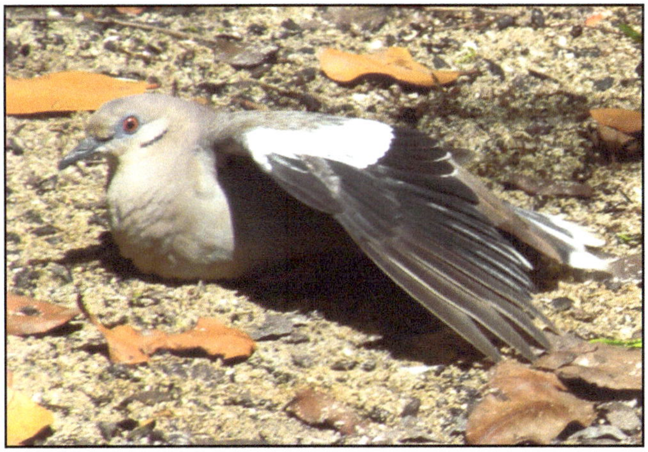

White-winged Dove basking in the sun
(*Zenaida asiatica*)

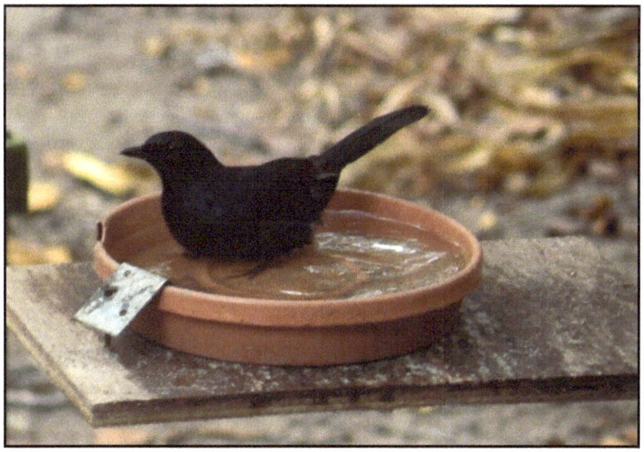

Black Catbird bathing
(*Melanoptila glabrirostris*)

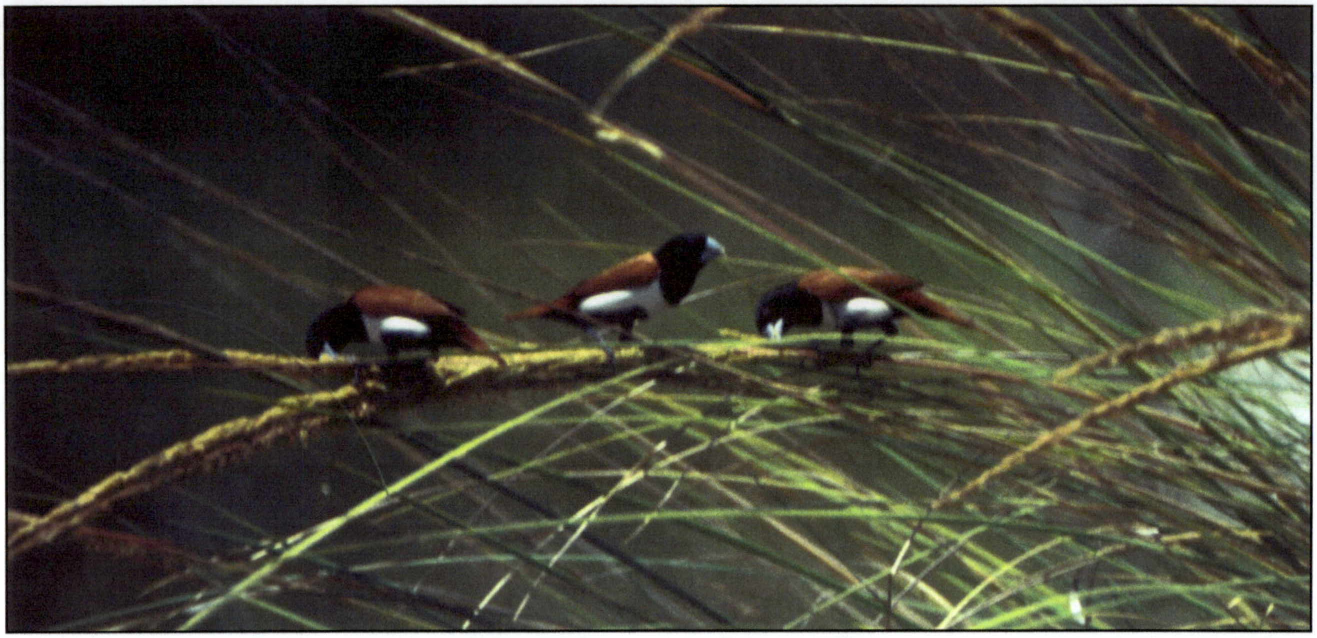

Tricoloured Munia (*Lonchura malacca*)

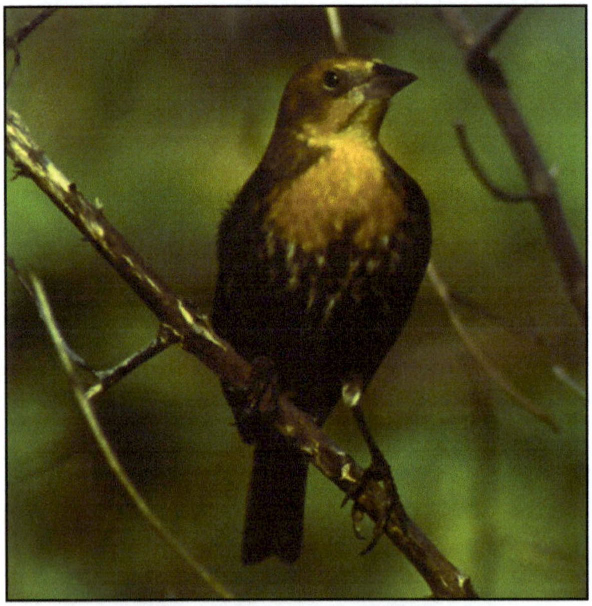

Female Yellow-headed Blackbird (*Xanthocephalus xanthocephalus*)

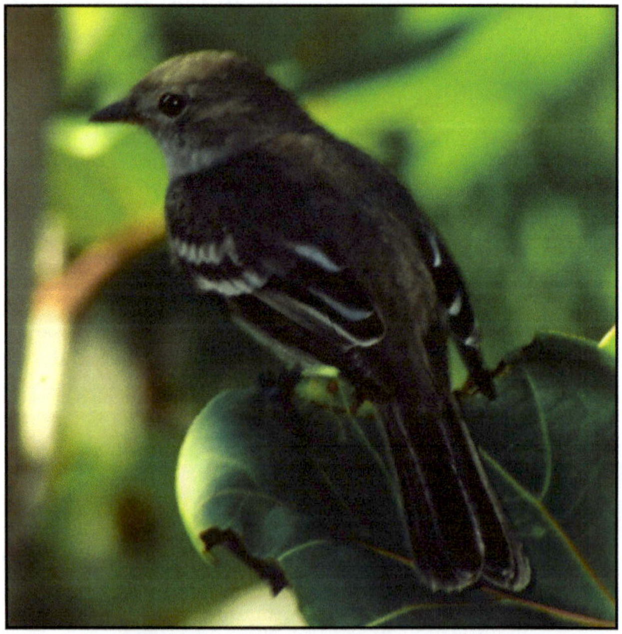

Caribbean Elaenia (*Elaenia martinica*)

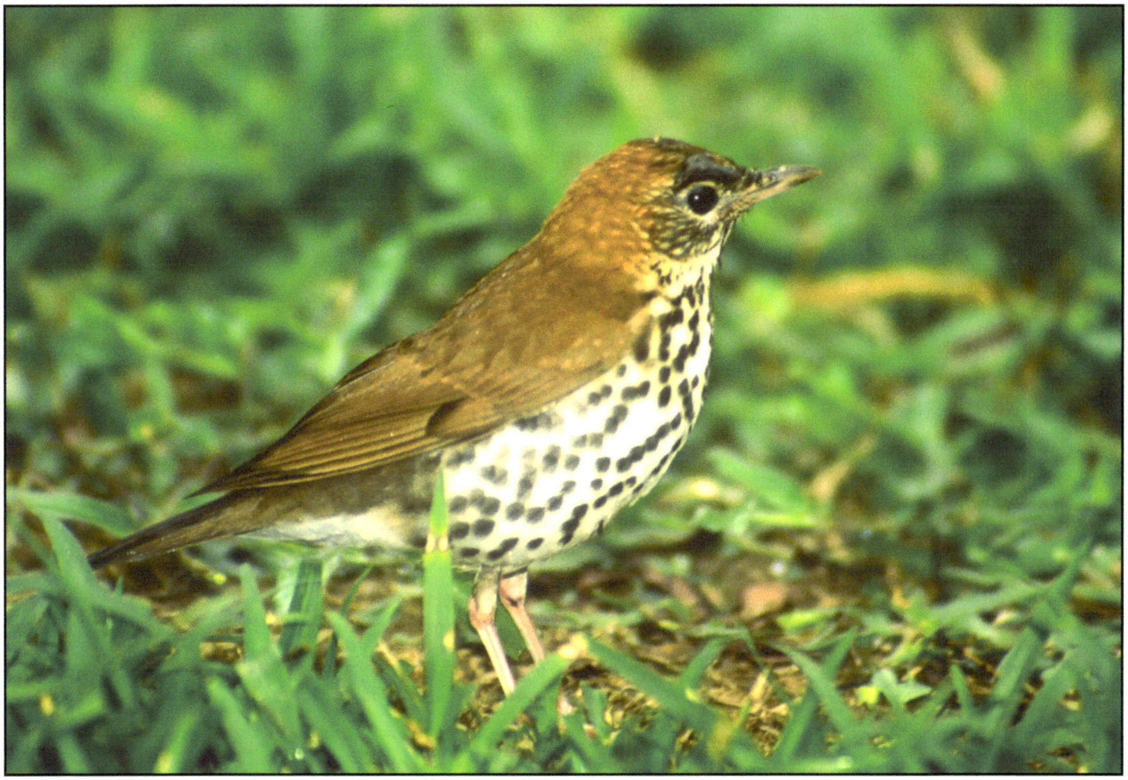

Wood Thrush (*Hylocichla mustelina*)

It was nice to be able to identify the migrant warblers, plus the many other species that visited our yard during the spring and fall migration seasons.

But the local species such as the Black Catbird, Mangrove Warbler, Yucatan Vireo and Cinnamon Hummingbird, all of which visited our yard daily all year round, had to wait until the latter part of the 1990s when Howel and Webb came out with their *Birds of Mexico and Central America*. At last we could name the birds that fed daily at our feeders.

There was also a multitude of waders and shorebirds that we could then name as we walked the mile and a half into the village. Caye Caulker has a total of 218 bird species as of September 21st 2010.

It wasn't until 2003, that Belize got its own *Birds of Belize* by Lee Jones, an American consulting biologist, who had moved to Belize in 1998. I was forever sending photographs of Belizean birds to Lee for identification.

I took pictures of two new species for Belize, which he identified as a Yellow-headed Blackbird (23 September 2001) and a small flock of an Asian bird, the Tri-coloured Munia (2 August 2003).

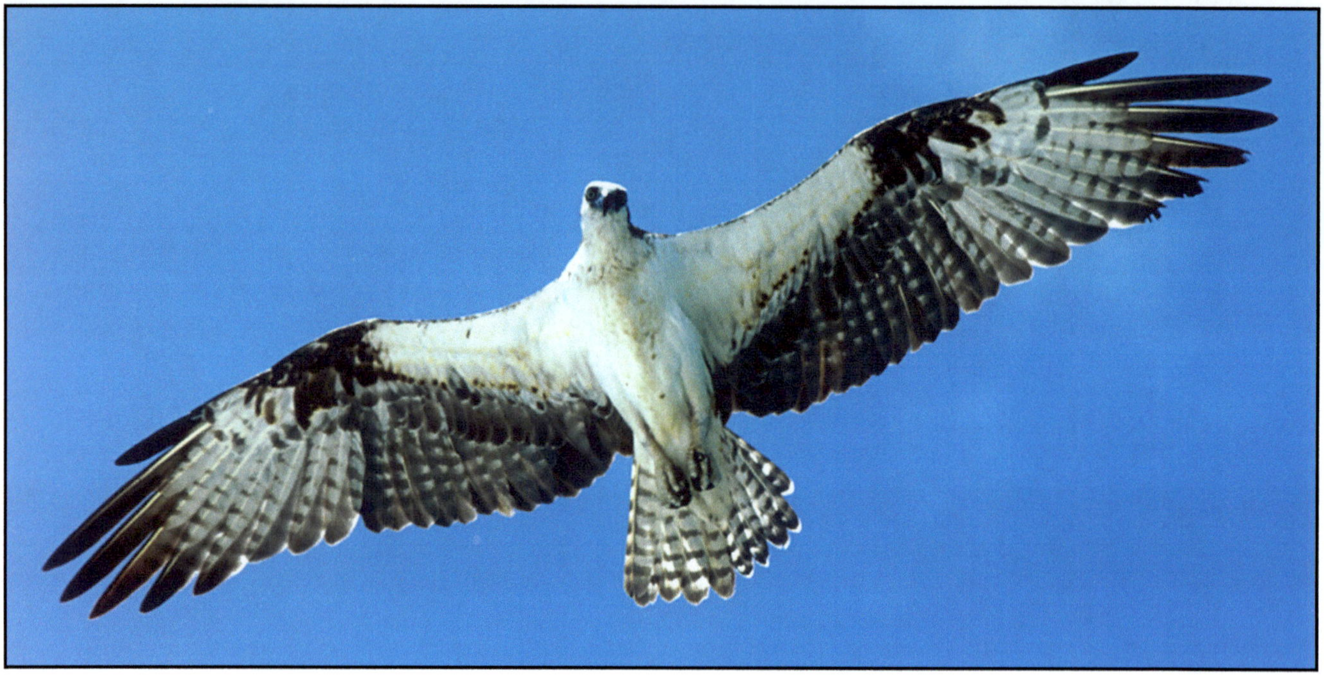

Osprey (*Pandion haliaetus*)

Osprey and Common Black Hawk

Caye Caulker has two resident raptor species, the Osprey and the Common Black Hawk. Normally, these two birds of prey live together on the island quite amicably. That is, unless the osprey has young or fledging chicks. Then the black hawk has to watch out. If it feels its young are in danger, the osprey will make every effort to drive the hawk away. The osprey is the larger of the two birds, and from the few occasions where I have witnessed this display of aerial combat, it also can out-fly the black hawk.

I was unaware of this drama in the sky until approximately twenty years ago an osprey actually drove a hawk into the sea. The hawk floundered, unable to take off as the osprey does so easily. If it wasn't for a group of scuba divers returning to the island, who knows what the hawk's future would have been, floating on the sea during an off-shore wind.

Since that occasion, I have watched their aerial dogfight several more times. Generally, the osprey is the aggressor, although the hawk may make one or two rallying attacks, but usually in vain.

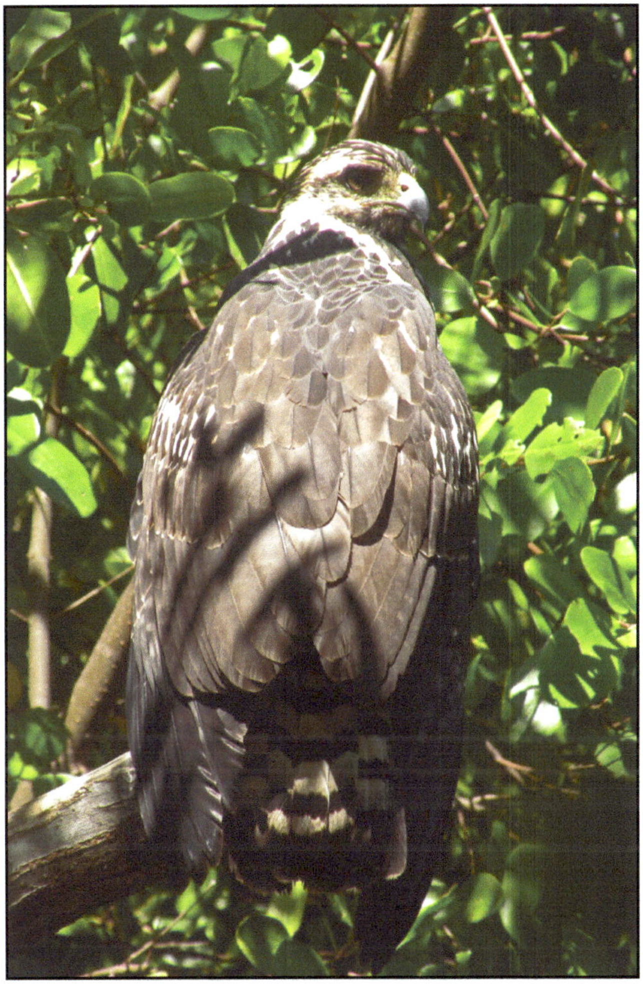

Juvenile Common Black Hawk
(*Buteogallus anthracinus*)

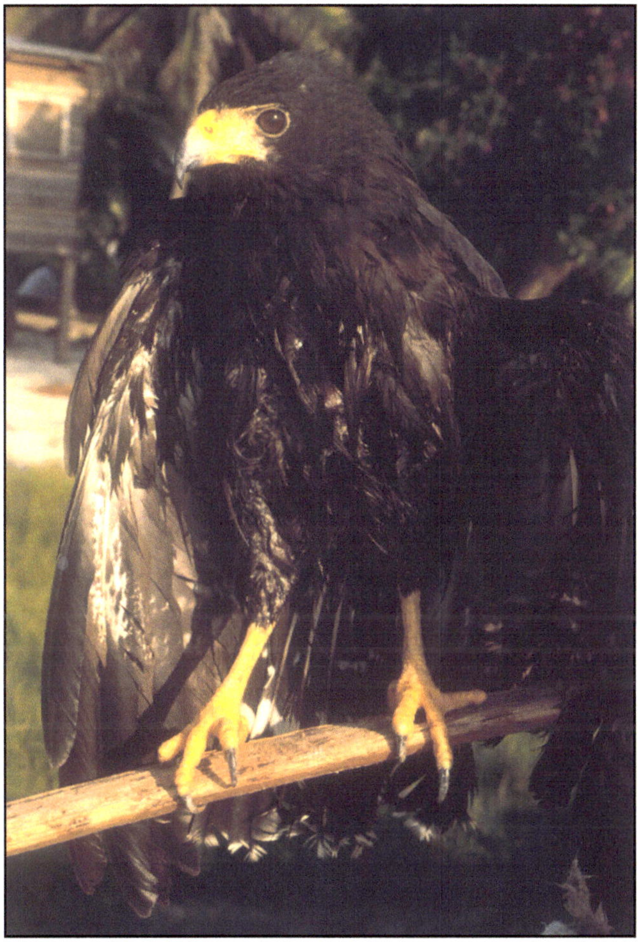

After being forced into the sea by an Osprey, this Common Black Hawk was rescued by divers.

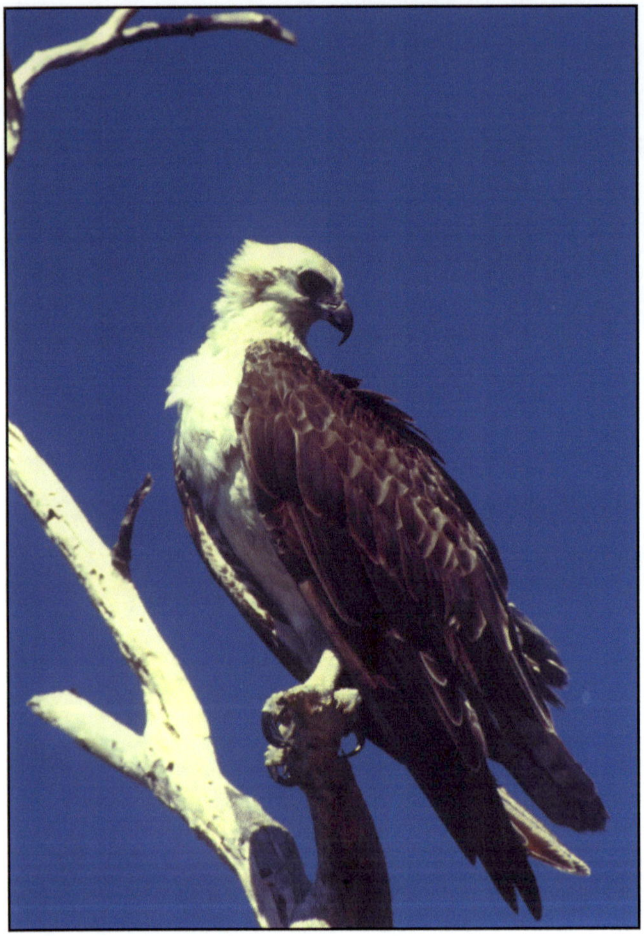

Osprey (*Pandion haliaetus*)

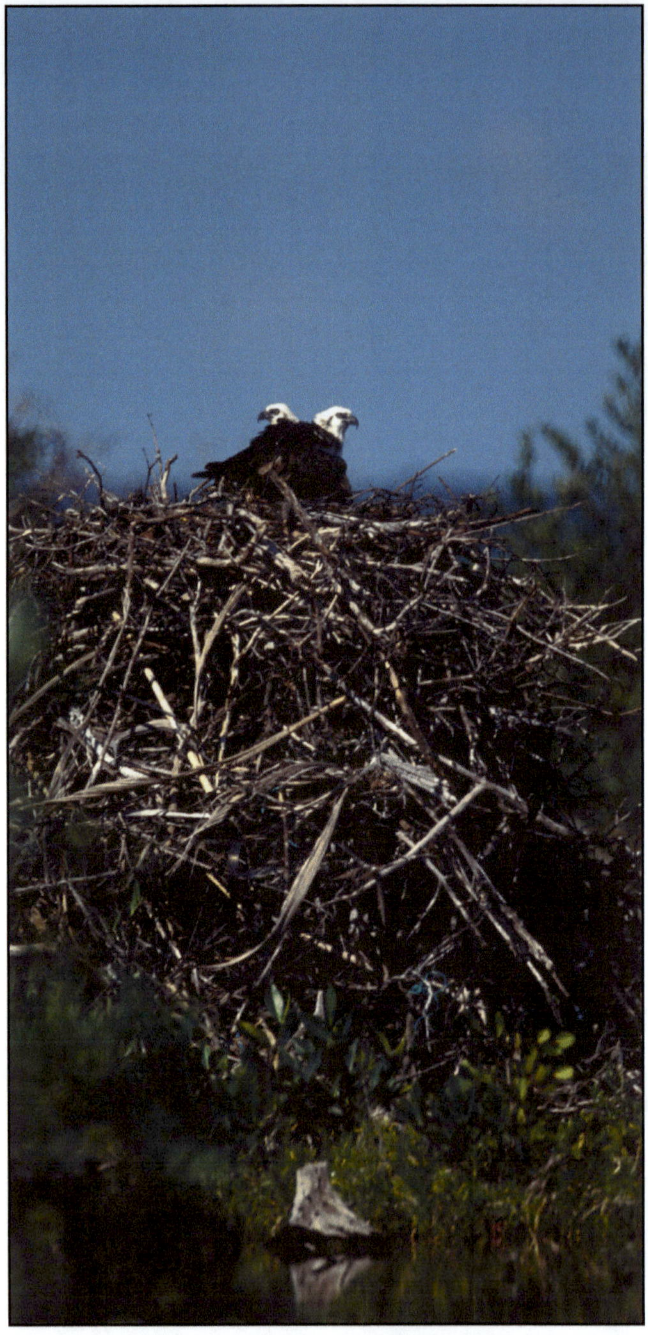

Unusual Osprey nest in lagoon

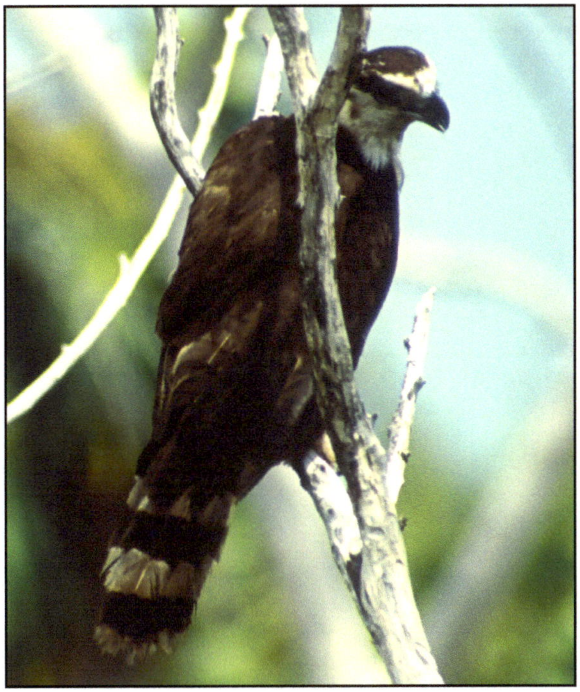

Juvenile Gray-headed Kite

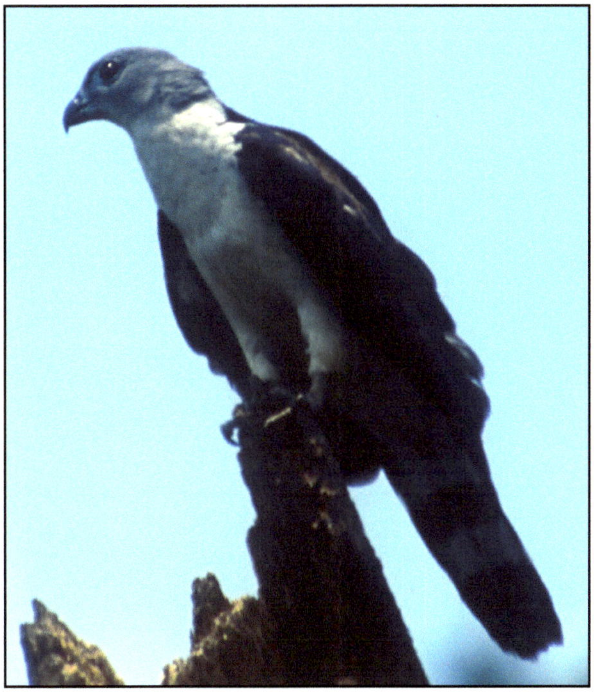

Gray-headed Kite (*Leptodon cayanensis*)

GRAY-HEADED KITE

A third raptor resided on the Caye for a little less than fourteen months. In February 2005 I noticed a large bird perched on a tree branch just south of the airstrip. It was neither a juvenile black hawk nor an osprey.

It had attracted two tourist birders. They asked me what kind of bird it was, but I had to confess my ignorance. Although they had a copy of Lee Jone's *Birds of Belize* with them, they could not identify it. I took several photos and looked in *Birds of Belize*, plus several more bird books with no success identifying it. I sent two photographs to Lee Jones who forwarded them to Bill Clark, a raptor specialist, who identified it as a juvenile Gray-headed Kite.

Dorothy and I monitored it for the remaining time it spent on the island which was relatively easy for it always seemed to be in the vicinity of our home, or the protected area directly behind us. It slowly took on its adult plumage, and after six month or so, it became identifiable.

Then it started to have problems with the local grackles. We began to see it more often, soaring over our property, closely followed by several male grackles which seemed to take turns harassing it.

Over the next week this continuous harassment grew more serious as the grackles continually pecked and pulled the kite's tail feathers. Around mid-February of 2006 we began to see less and less of this once handsome bird, whose tail was now very ragged and uneven. In March of 2006 we had one sighting only on the 26th. We were glad to see that his tail was starting to grow back. During April, we saw him twice, once on the 7th and on the 20th. After that, we never saw him again.

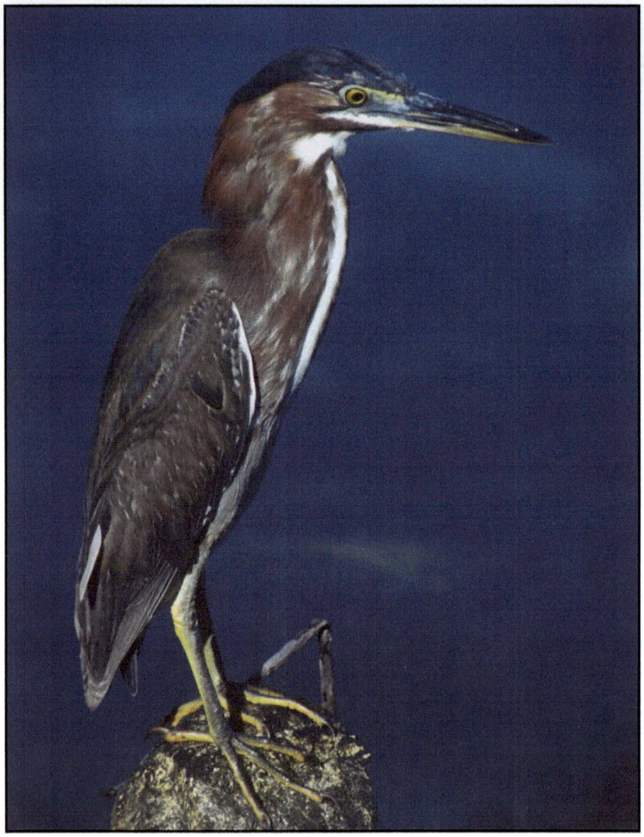

Green Heron (*Butorides virescens*)

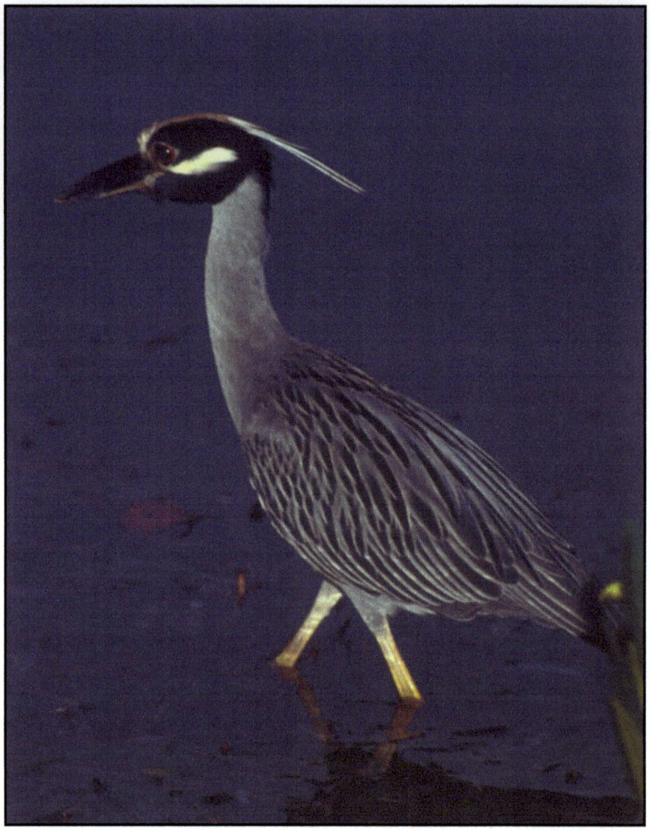

Yellow-crowned Night-Heron (*Nyctanassa violacea*)

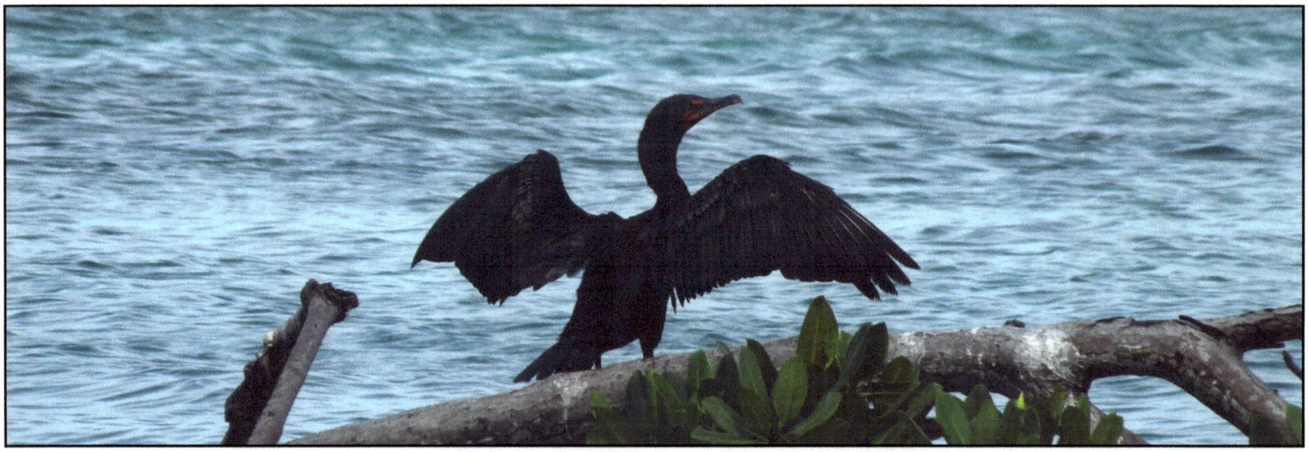

Double-crested Cormorant (*Phalacrocorax auritus*)

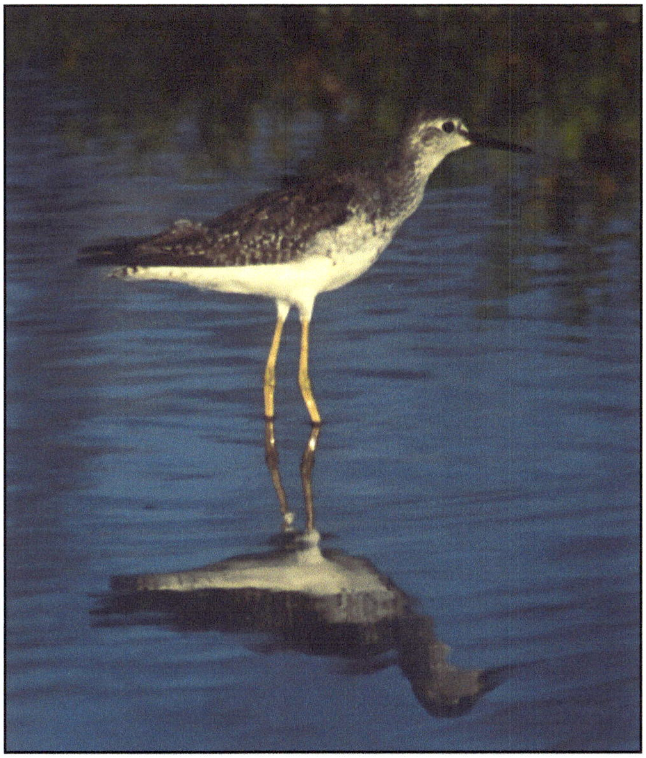

Lesser Yellowlegs (*Tringa flavipes*)

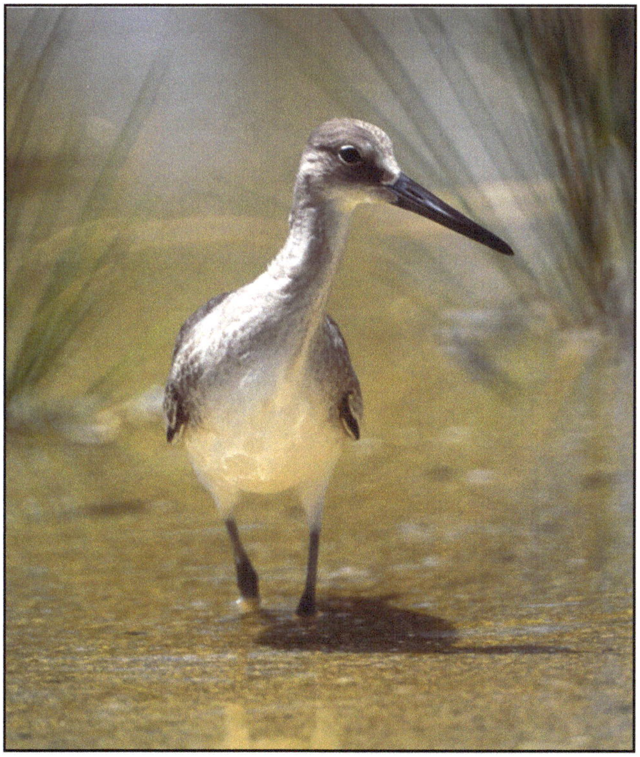

Willet (*Tringa semipalmata*)

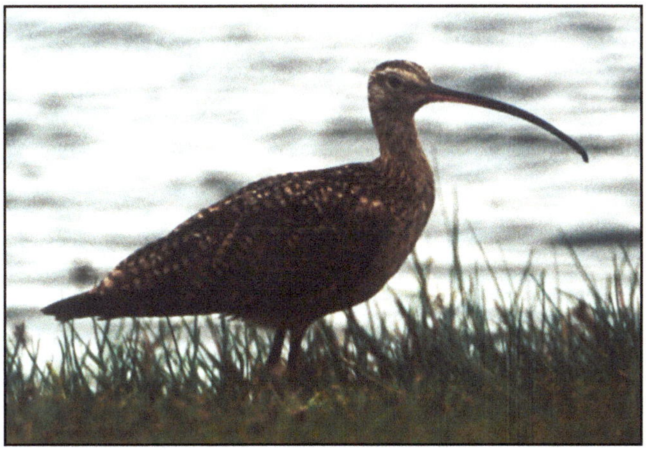

Long-billed Curlew (*Numenius americanus*)

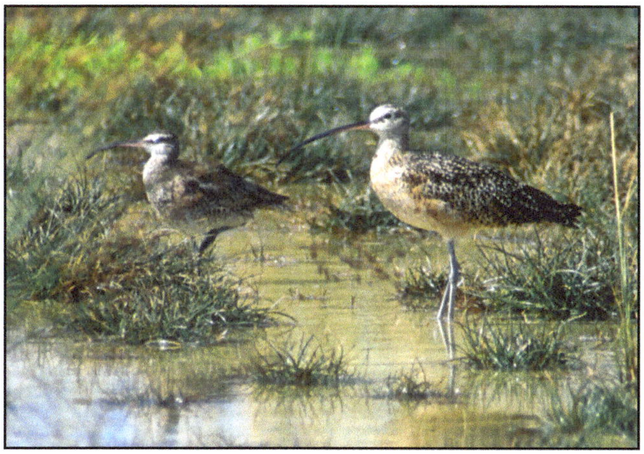

Whimbrel (*Numenius phaeopus*)
and Long-billed Curlew (*right*)

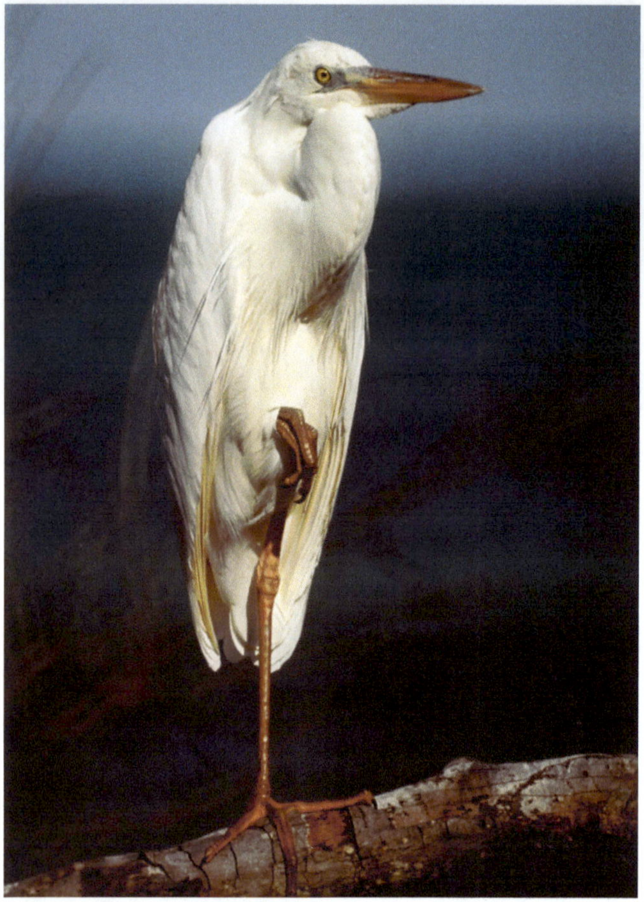

Great Blue Heron white morph

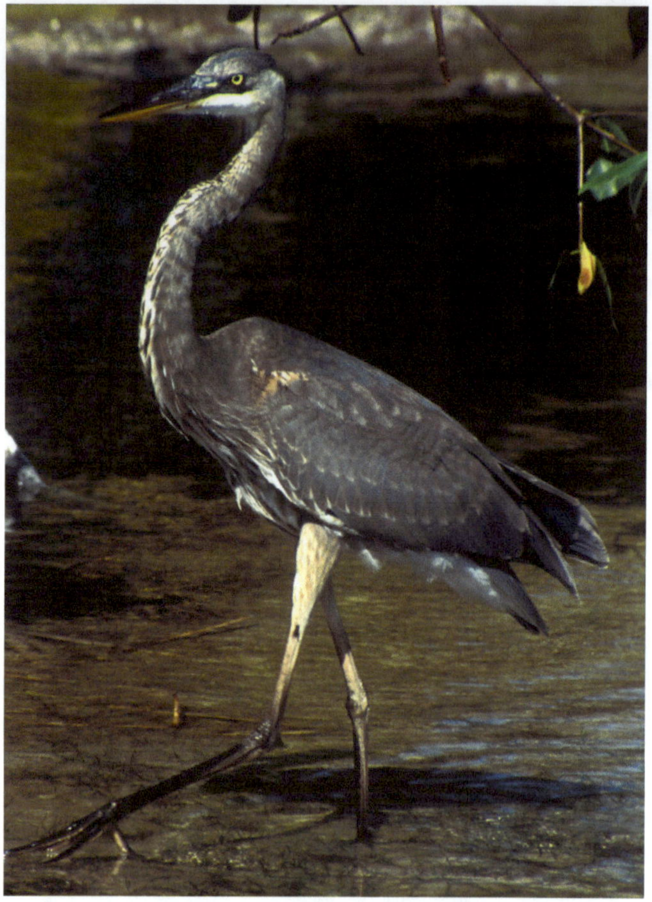

Great Blue Heron (*Ardea herodias*)

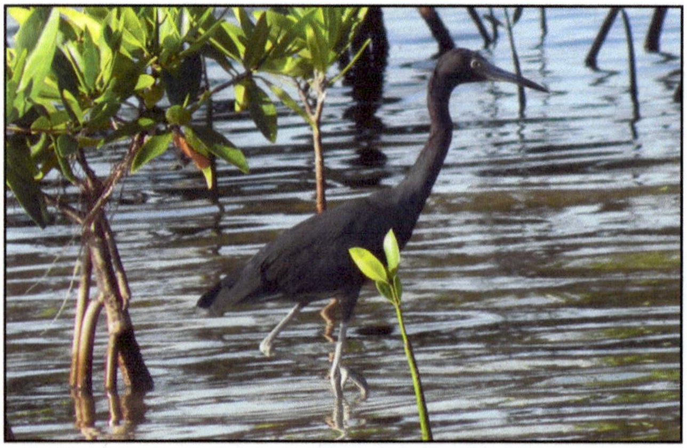

Little Blue Heron (*Egretta caerulea*)

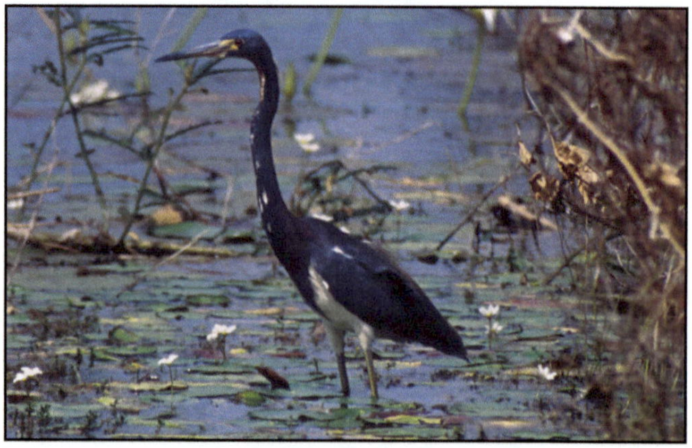

Tricolored Heron (*Egretta tricolor*)

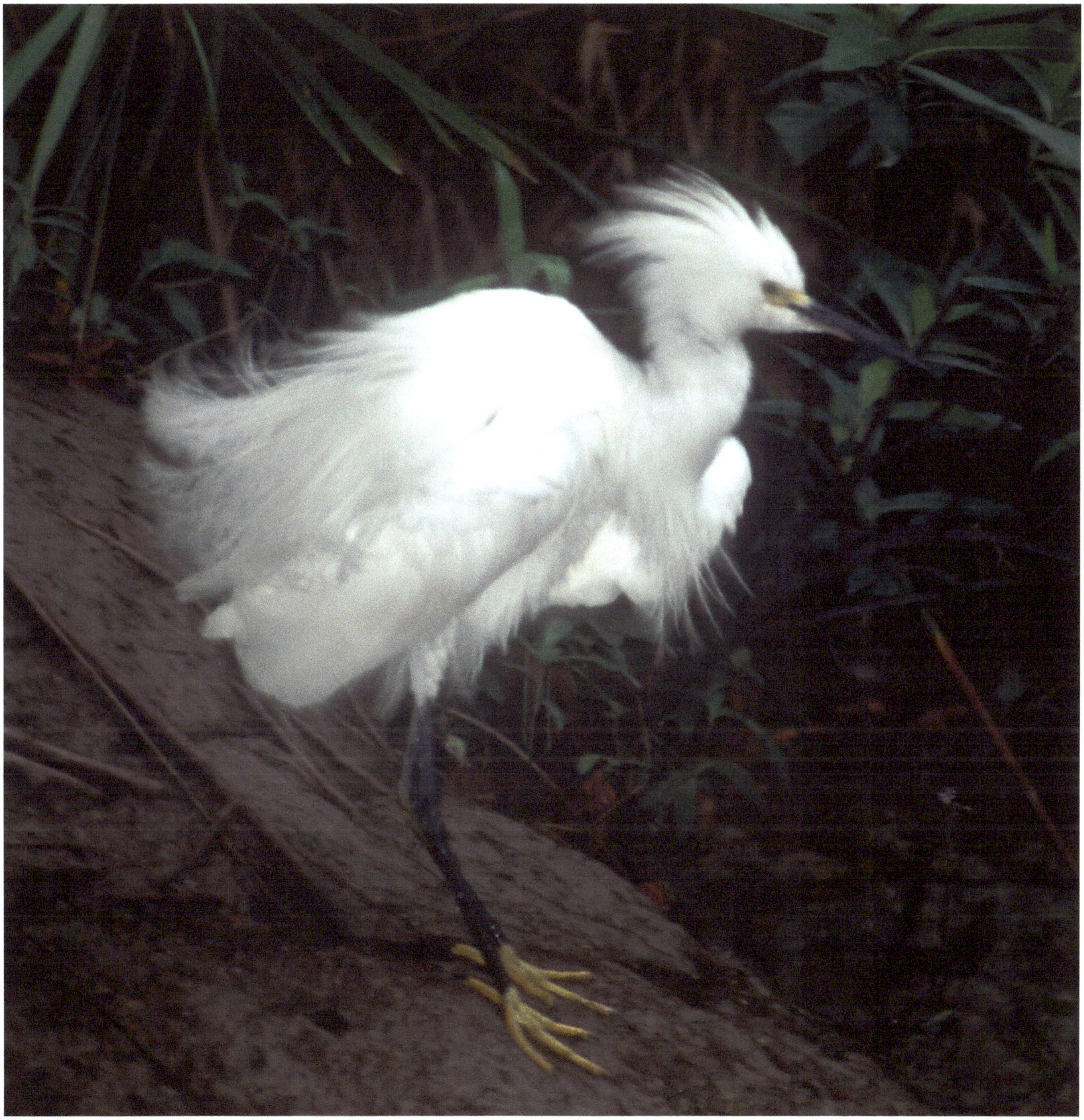

Snowy Egret (*Egretta thula*)

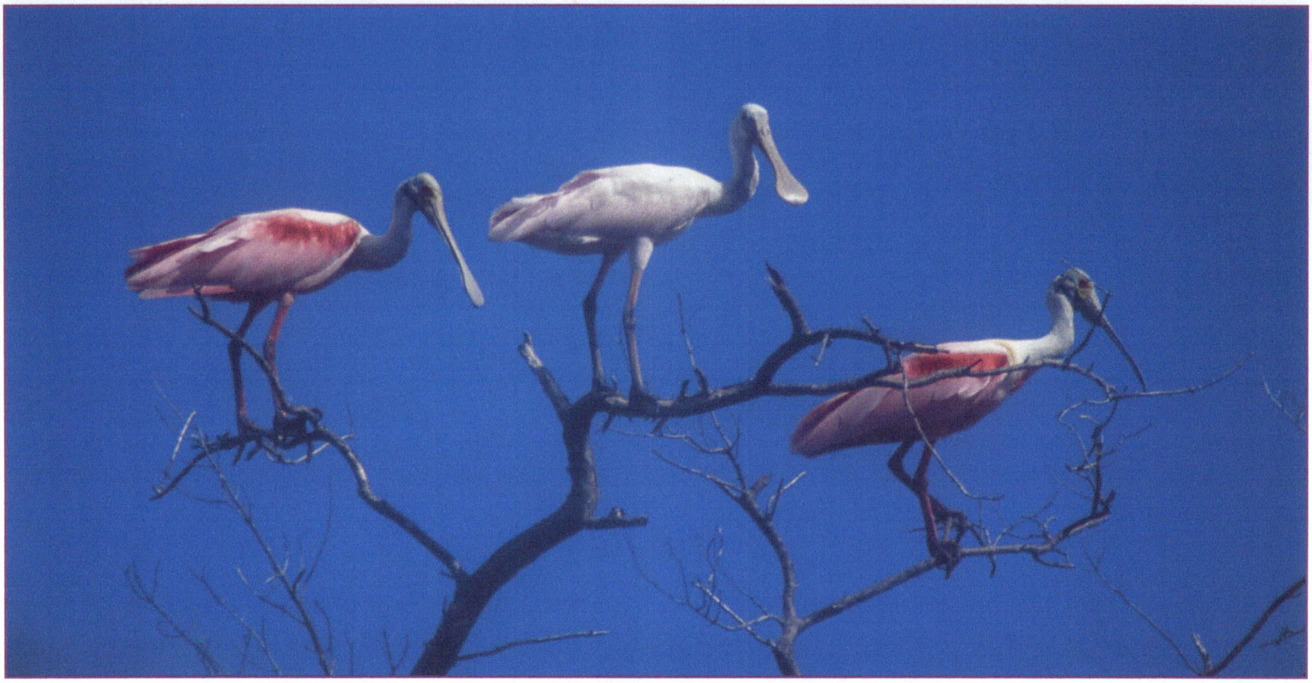

Roseate Spoonbills (*Platalea ajaja*)

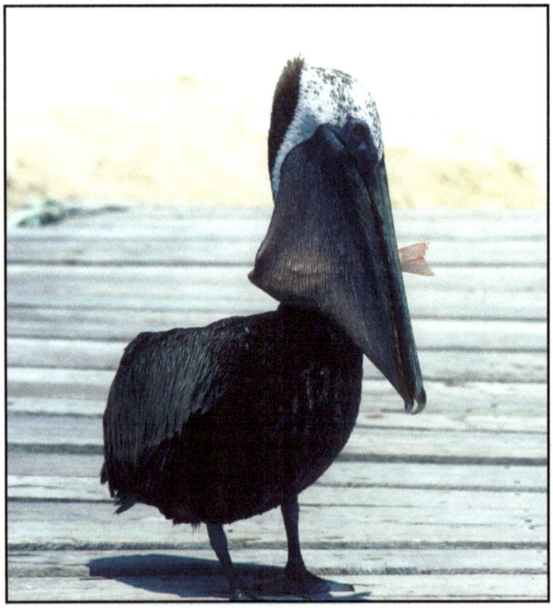

Brown Pelican (*Pelicanus occidentalis*)

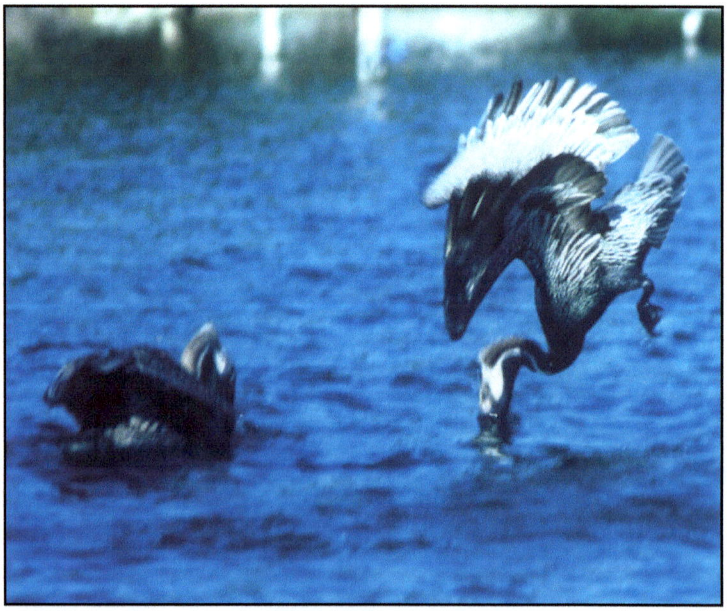

Brown Pelican diving

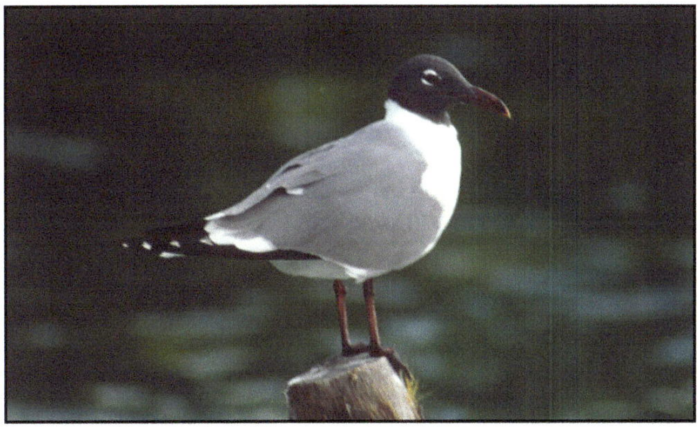

Laughing Gull (*Leucophaeus atricilla*)

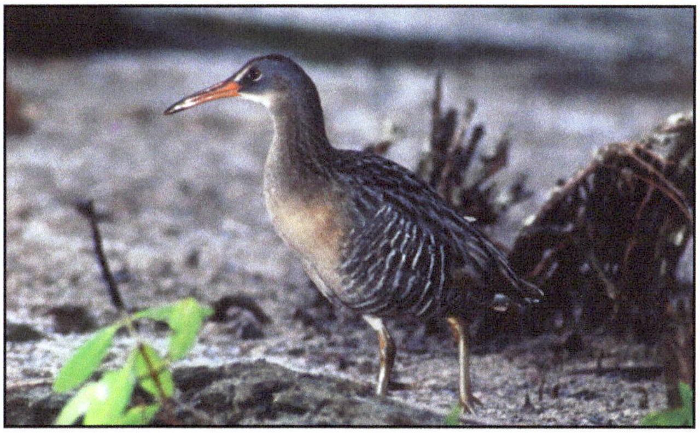

Clapper Rail (*Rallus longirostris*)

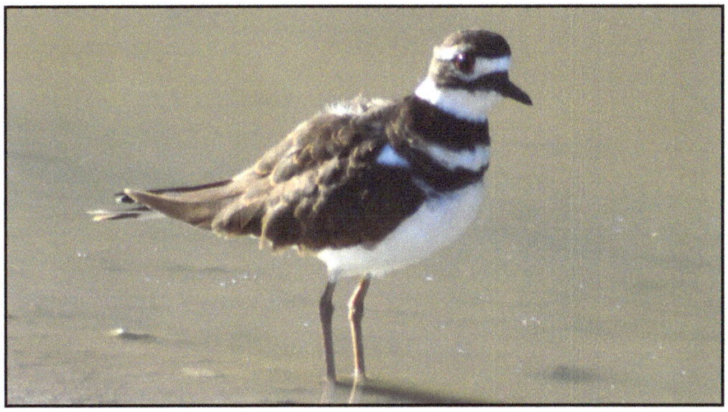

Killdeer (*Charadrius vociferus*)

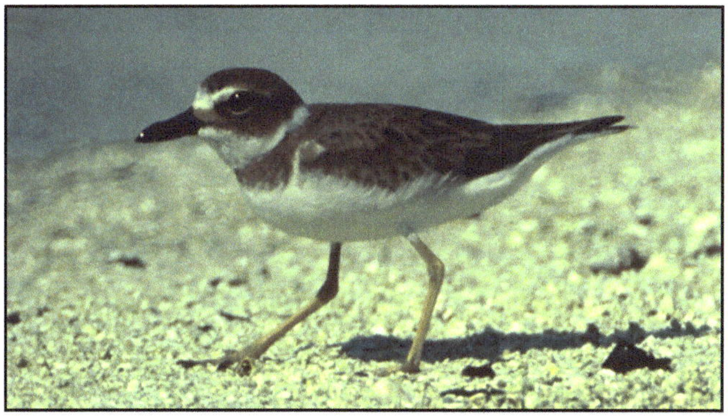

Wilson's Plover (*Charadrius wilsonia*)

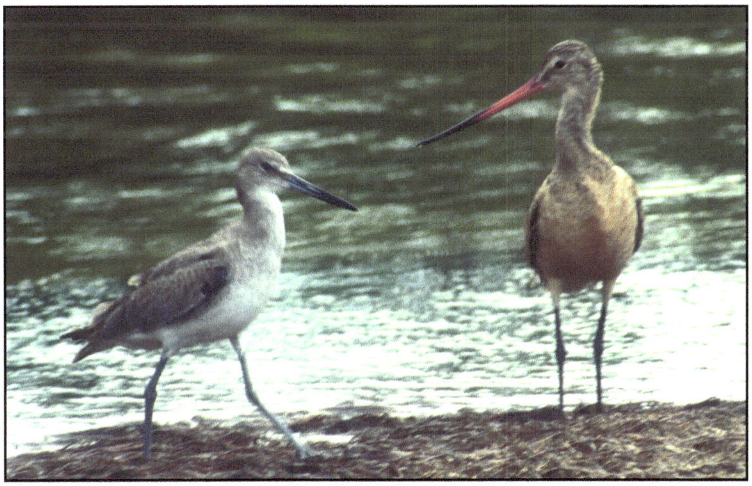

Willet (*Tringa semipalmata*) and Marbled Godwit (*Limosa fedoa*)

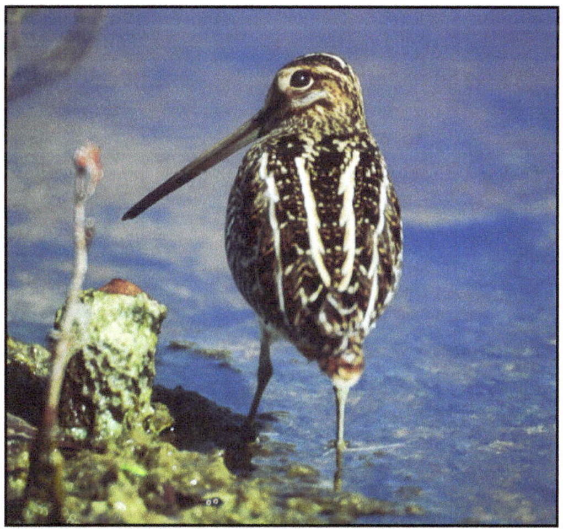

Wilson's Snipe (*Gallinago delicata*)

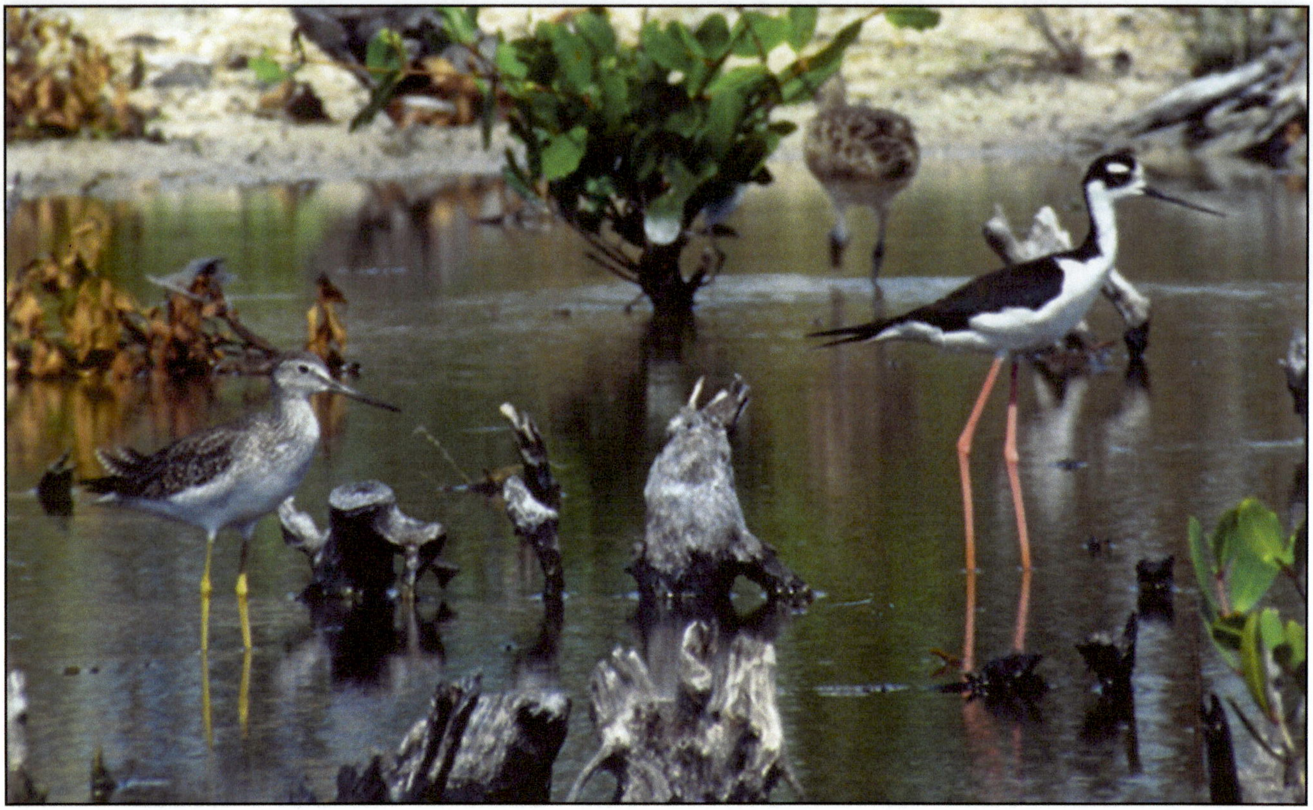

Greater Yellowlegs (*Tringa melanoleuca*) and Black-necked Stilt (*Himantopus mexicanus*)

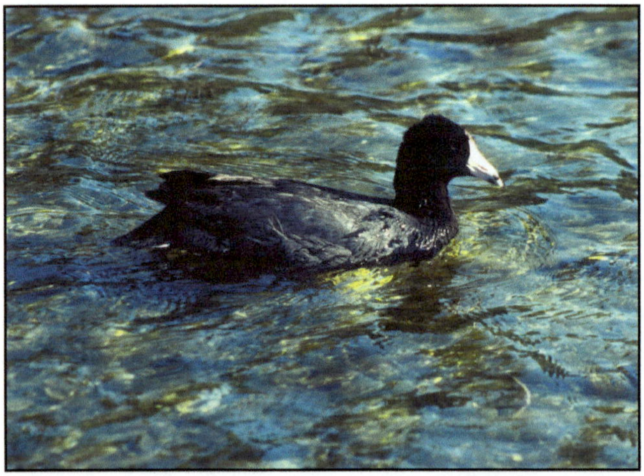

American Coot (*Fulica americana*)

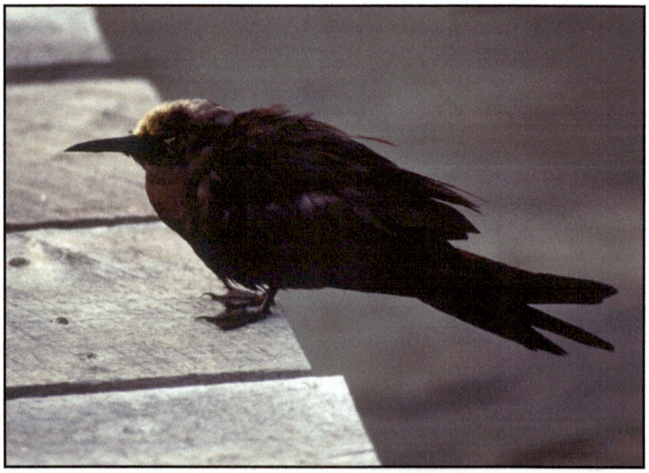

Brown Noddy (*Anous stolidus*)

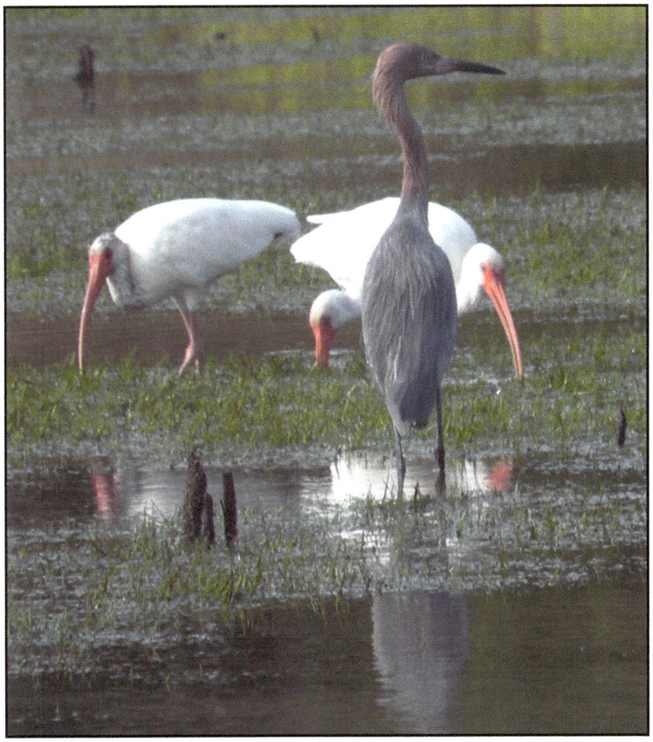

Reddish Egret with White Ibis

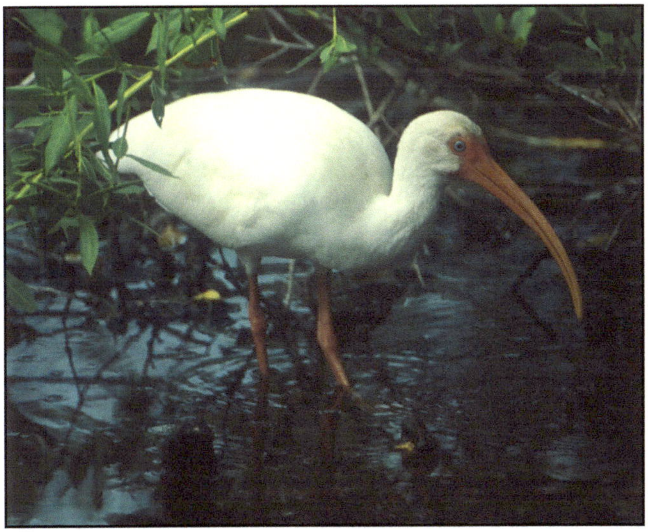

White Ibis (*Eudocimus albus*)

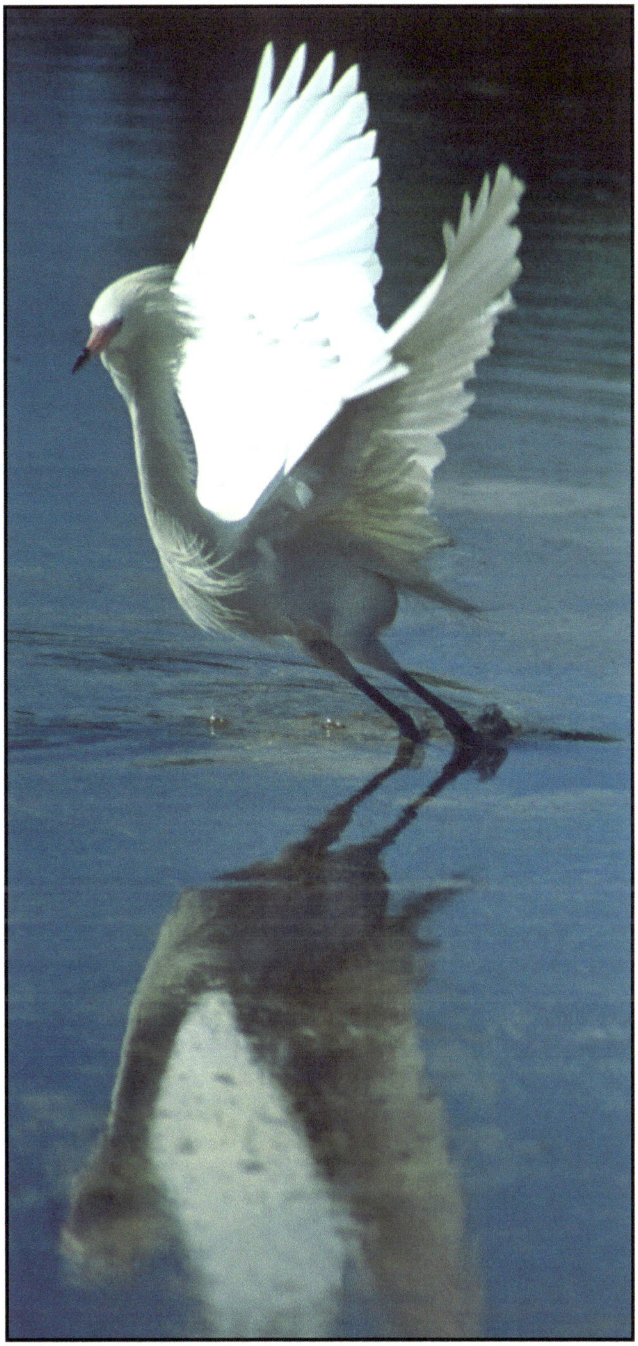

Reddish Egret (*Egretta rufescens*) white morph

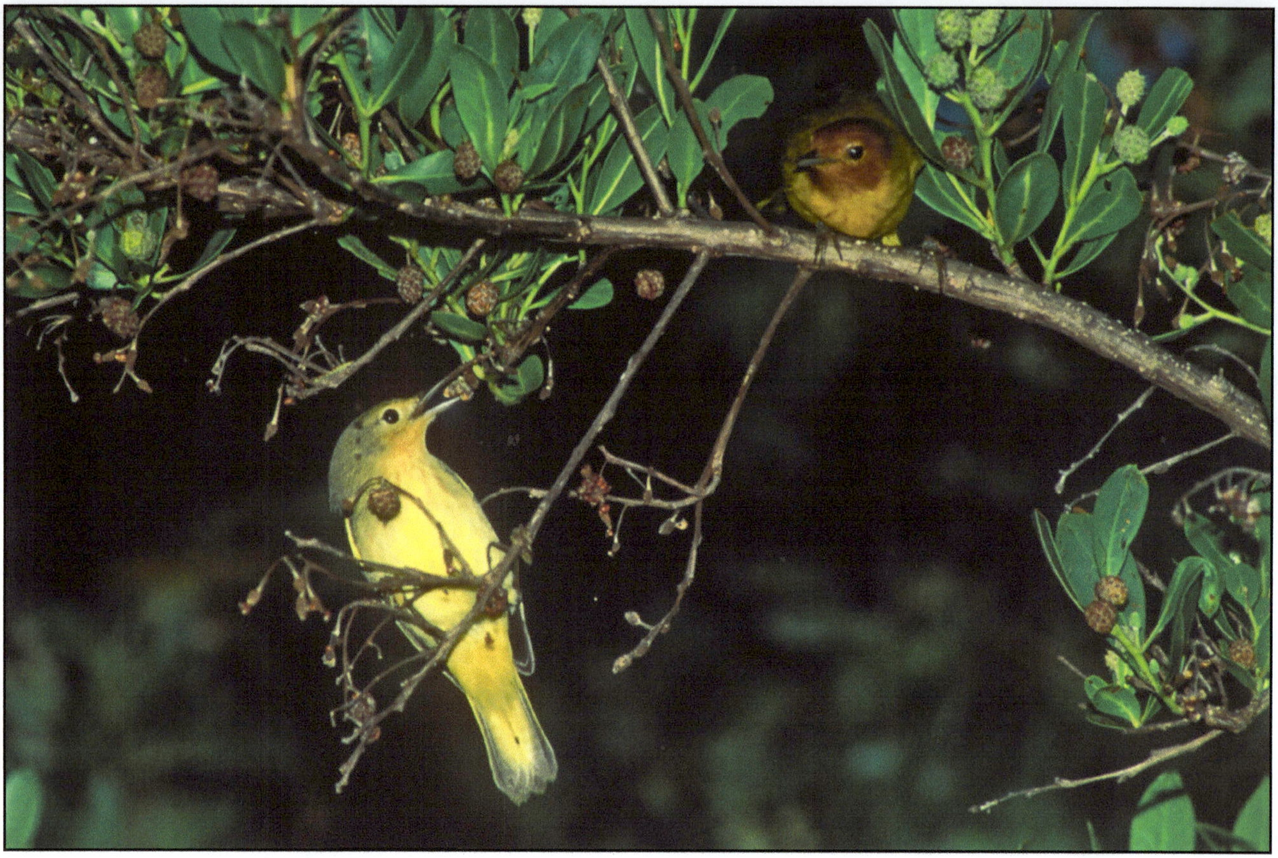

Mangrove Warbler (*Setophaga petechia bryanti*) female (*left*) and male (*upper right*)

Mangrove Warblers

Around our yard Dorothy and I have several bird feeders and we make sure that there is always food in the feeders. The most popular food for the birds is papaya.

For three or four years a pair of Mangrove Warblers have been the most frequent bird species to use the feeders.

The Mangrove Warbler is a subspecies of the Yellow Warbler of North America. It is a handsome bird, the female being similar to the Yellow Warbler of the U.S. but the male differs by having a mahogany red head.

For the last three years these warblers have nested very close to our home. I have to admit that I paid little attention to the pair during the nesting season, although Dorothy logged dates and other information.

However, during 2011 this situation changed. The reason being I took more notice when the adult male brought a single chick to the trees near our porch feeder and actively fed it insects from the foliage of the surrounding trees and papaya from our feeder. Both the male parent and chick were without fear, but in hindsight I shouldn't have been surprised for the male bird has been one of our most frequent visitors to our porch feeder. He was very tame and even used our cat's water dish to bathe while the cat watched!

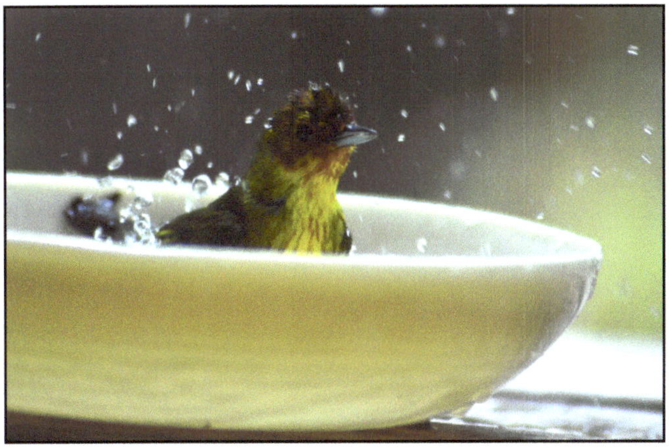

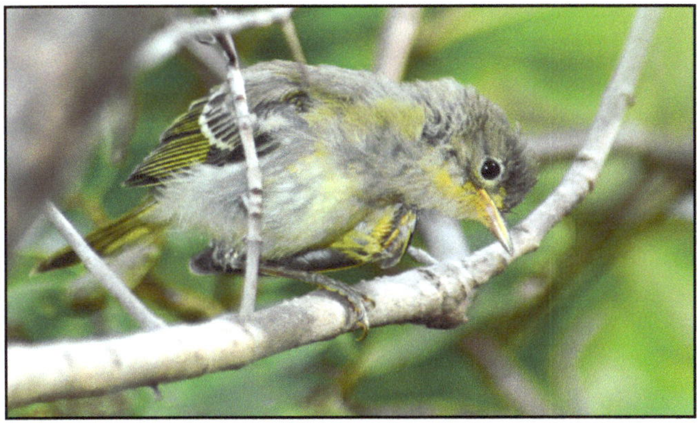

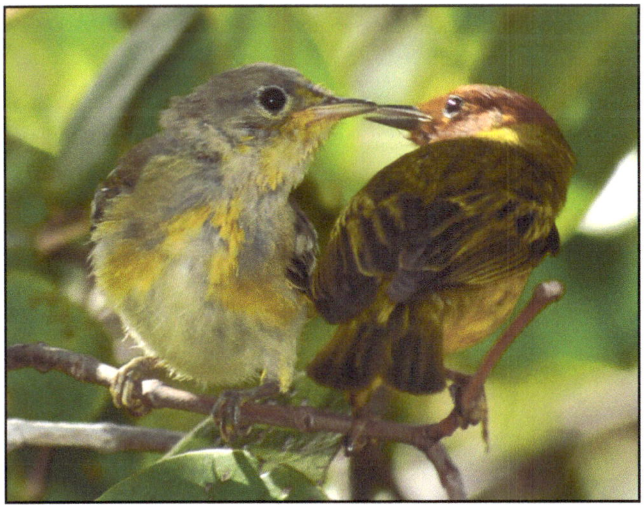

As the days passed, the male warbler was tireless but we saw no sign of the female warbler. The chick was pretty with a whitish belly and fluffy yellow and grey plumage to the upper body. It's breast very quickly became more and more yellow. The male was the perfect parent, always in sight of the chick, feeding it regularly with both bugs and papaya.

In the afternoon of June 25th, five days after the chick and male first showed up, we noticed a grey coloured warbler in the yard, it was quite small and we couldn't identify it. Four days later the grey warbler showed up again, then the following day the grey warbler returned once more and started to feed the chick, doing so for several hours. The male warbler then took over the feeding at mid-afternoon. Our mystery grey warbler was the mother!

Dorothy recalled that the previous year, she had seen a grey "chick" being fed by a yellow "adult" female warbler. Now in hindsight, it seems that the female Mangrove Warblers are grey when nesting.

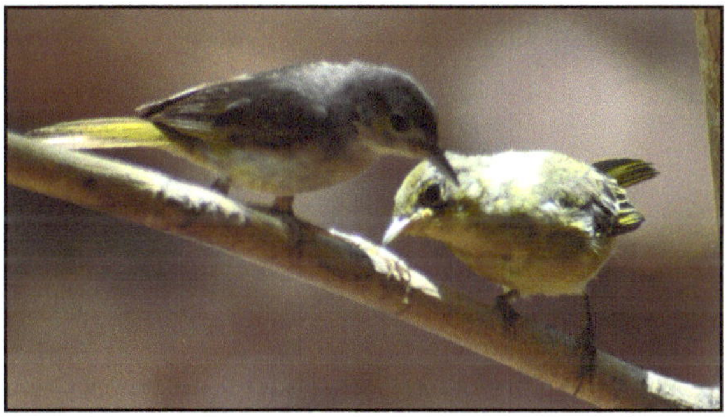

(*above*) Female Mangrove Warbler feeding chick

(*top left*) Male Mangrove Warbler bathing in the cat's water dish

(*center left*) Mangrove Warbler chick just after fledging

(bottom left) Male Mangrove Warbler feeding chick

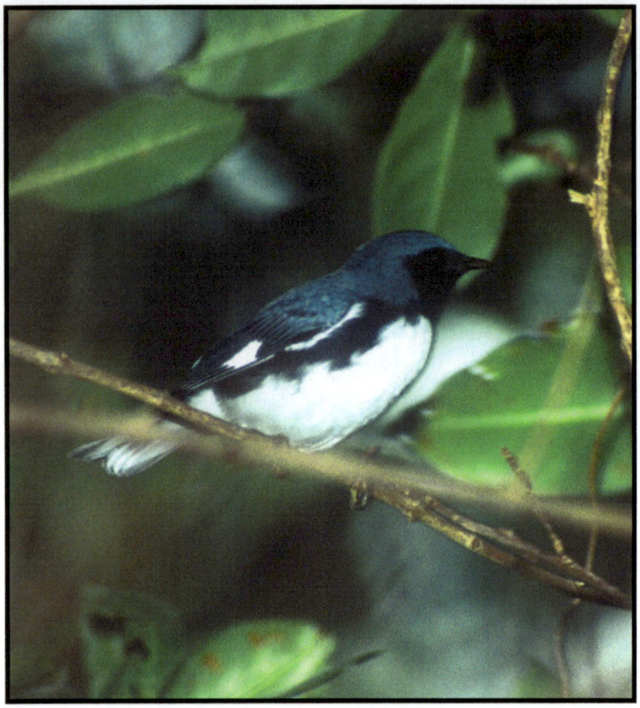

Black-throated Blue Warbler (*Setophaga caerulescens*)

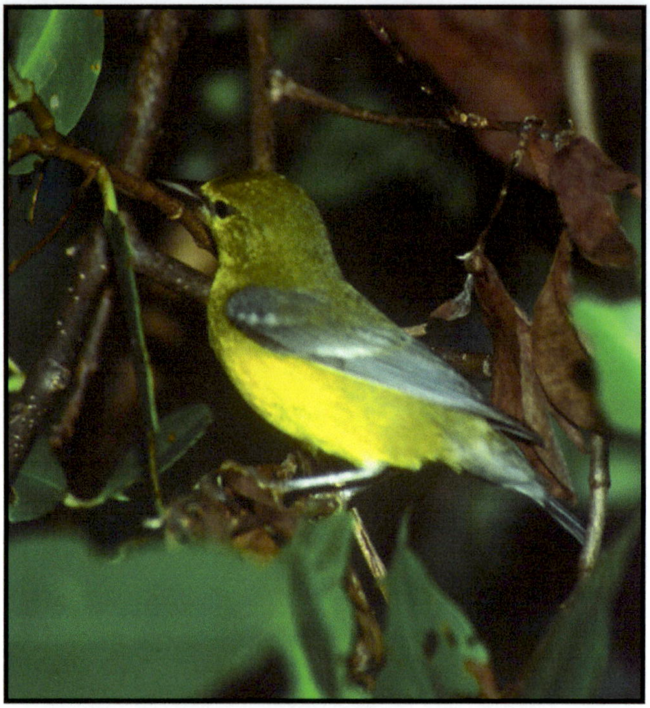

Blue-winged Warbler (*Vermivora cyanoptera*)

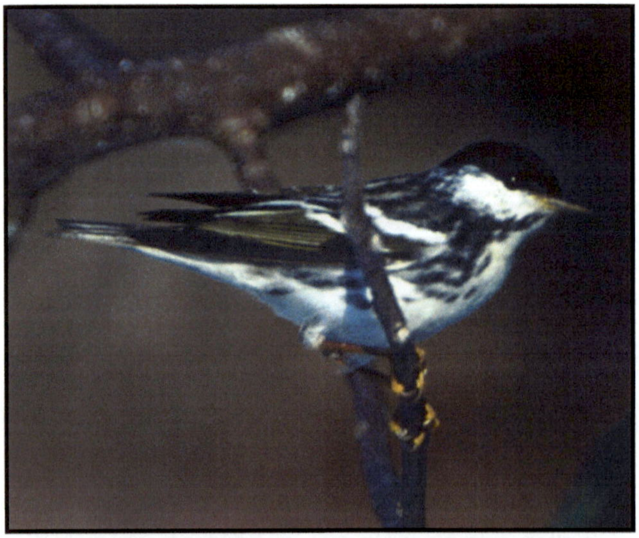

Blackpoll Warbler (*Setophaga striata*)

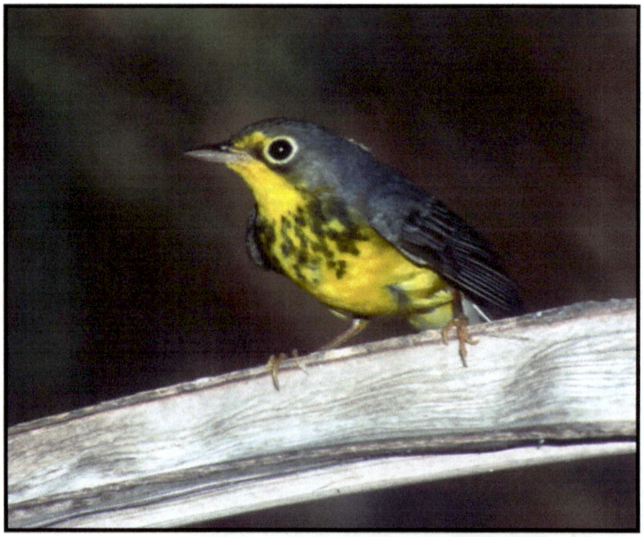

Canada Warbler (*Cardellina canadensis*)

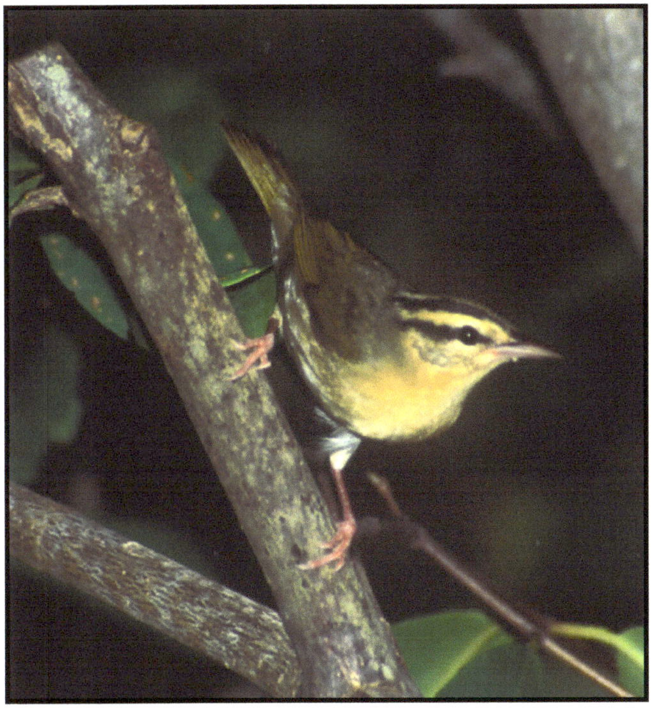

Worm-eating Warbler (*Helmitheros vermivorum*)

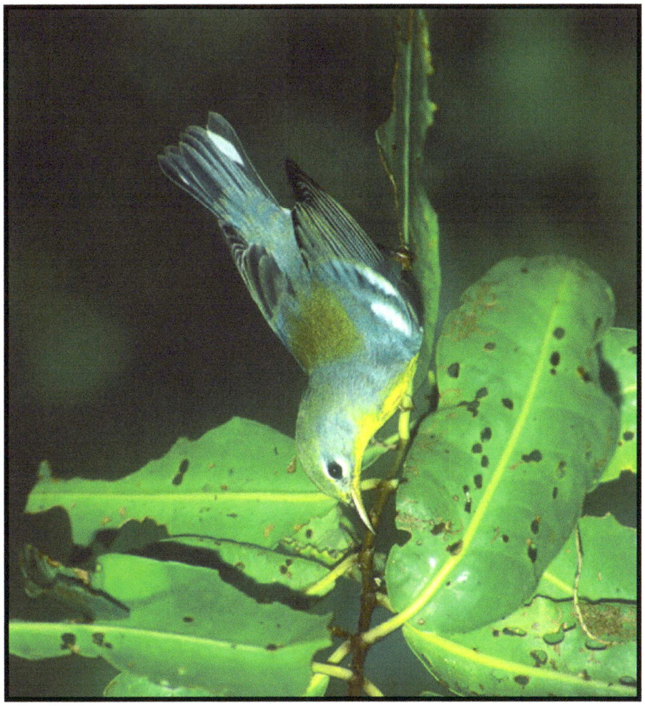

Northern Parula (*Setophaga americana*)

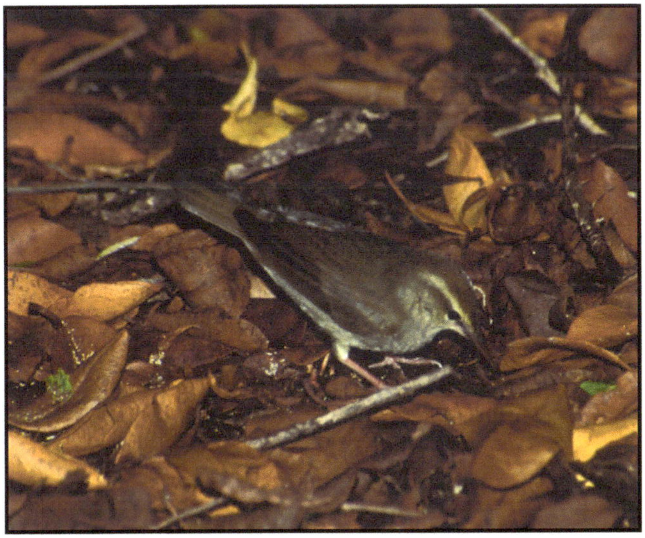

Swainson's Warbler (*Limnothlypis swainsonii*)

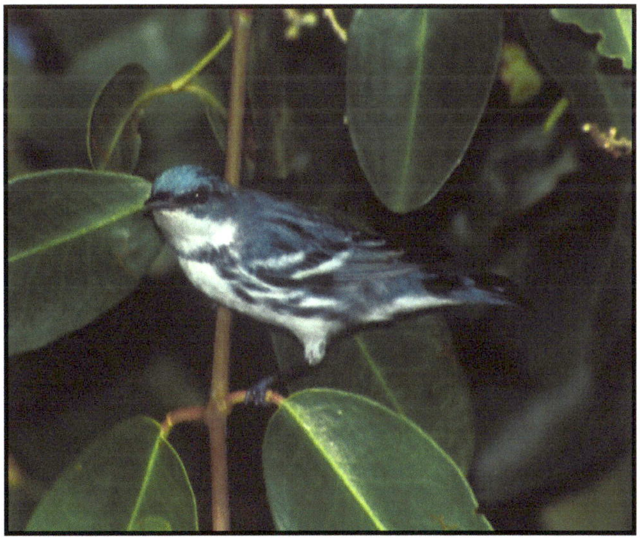

Cerulean Warbler (*Setophaga cerulea*)

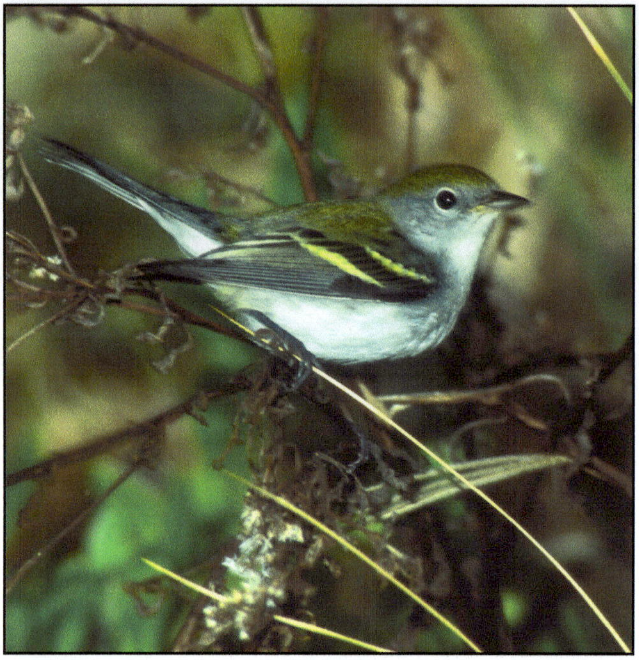

Chestnut-sided Warbler (*Setophaga pensylvanica*)

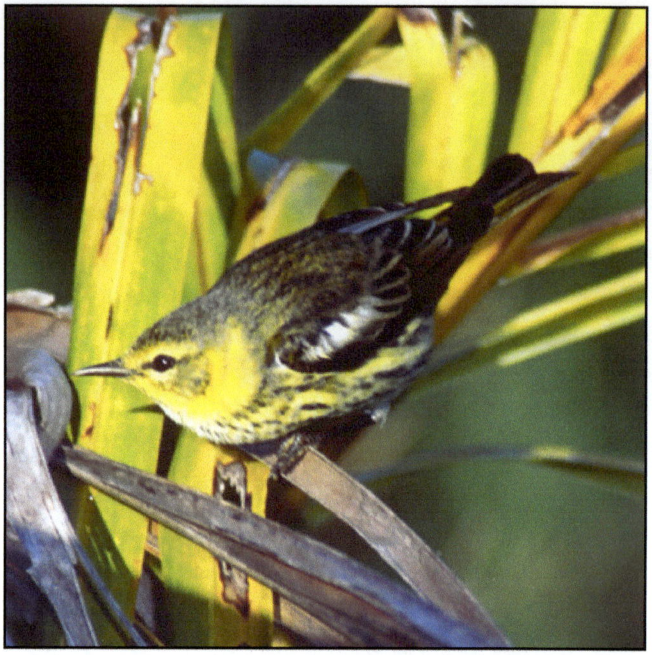

Cape May Warbler (*Setophaga tigrina*)

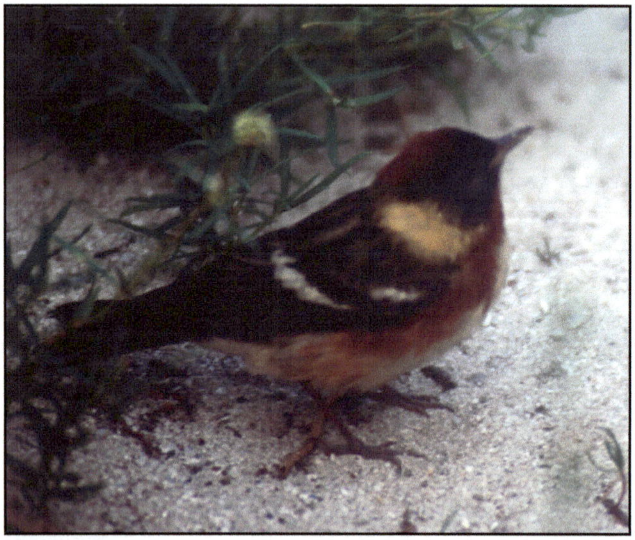

Bay-breasted Warbler *Setophaga castanea*)

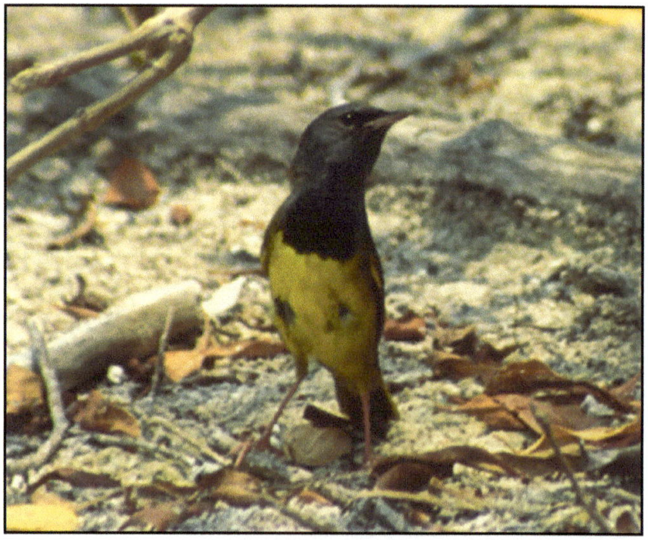

Mourning Warbler (*Geothylpis philadelphia*)

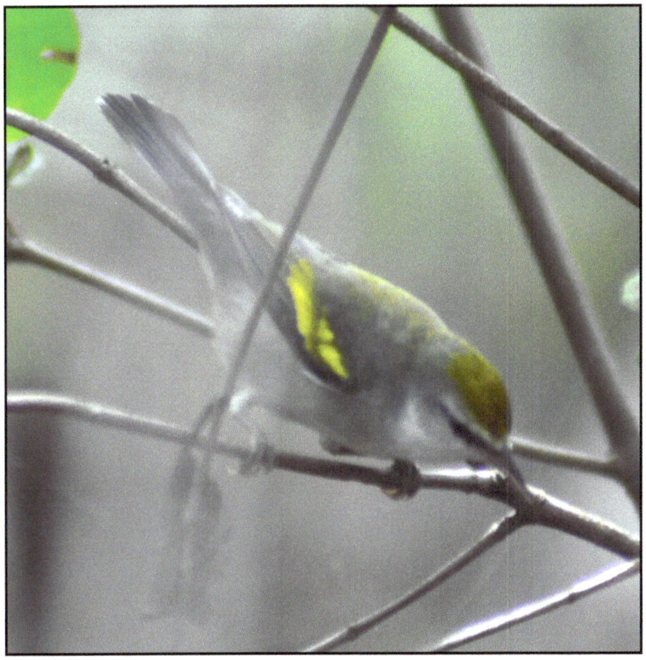

Female Golden-winged Warbler (*Vermivora chrysoptera*)

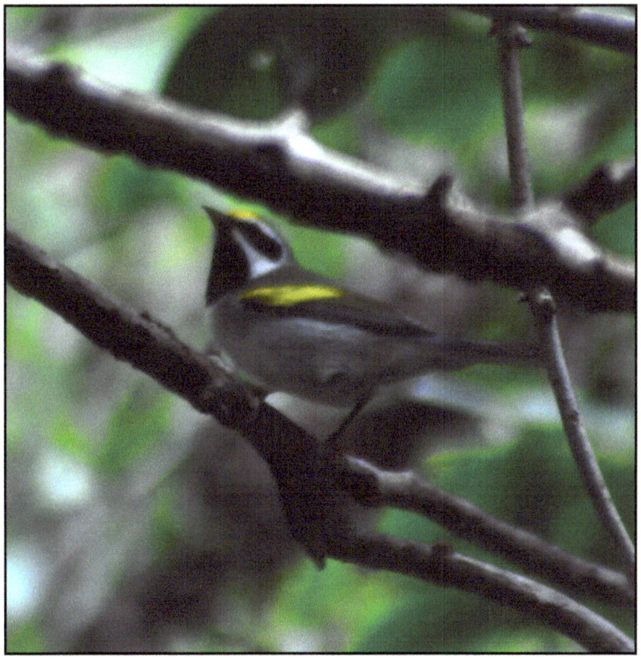

Male Golden-winged Warbler (*Vermivora chrysoptera*)

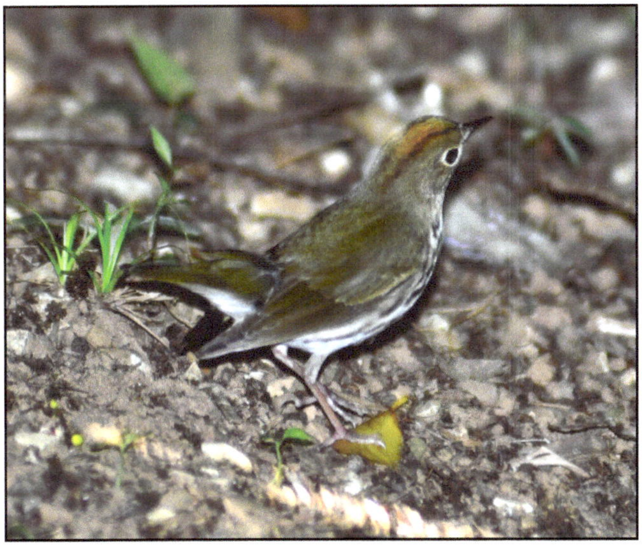

Ovenbird (*Seiurus aurocapilla*)

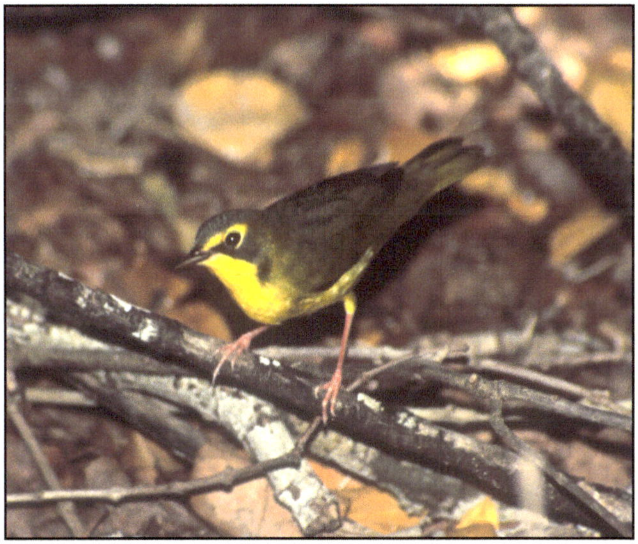

Kentucky Warbler (*Geothlypis formosa*)

Yucatan Vireo

One of our favourite yard birds is the Yucatan Vireo. We have a pair of vireos that are quite tame and visit our feeders quite frequently. One day Dorothy and I were out walking when we came across a vireo that was acting very strangely. The bird was hanging upside down in a leafy bush and making pecking motions at what looked like a green leaf. It wasn't a large leaf, but a green katydid, a large green plant-eating insect.

The vireo continued attacking the insect. Obviously it recognized Dorothy and me. We let nature take its course. No doubt the vireo intended to somehow get the katydid back to its nest, for it was nesting season and probably a pair of chicks were waiting for supper.

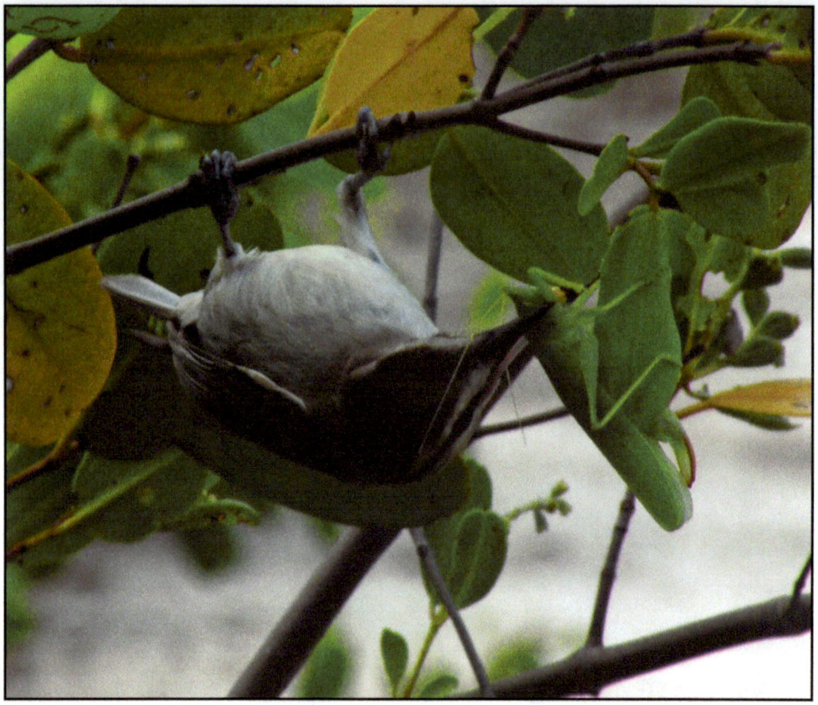

Yucatan Vireo eating katydid

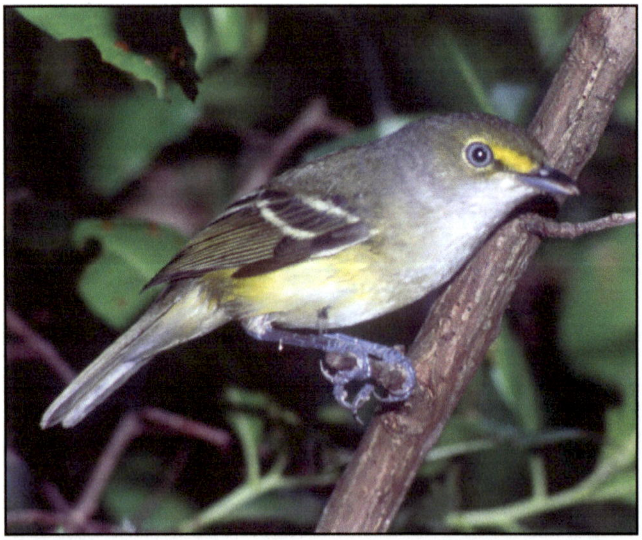

White-eyed Vireo (*Vireo griseus*)

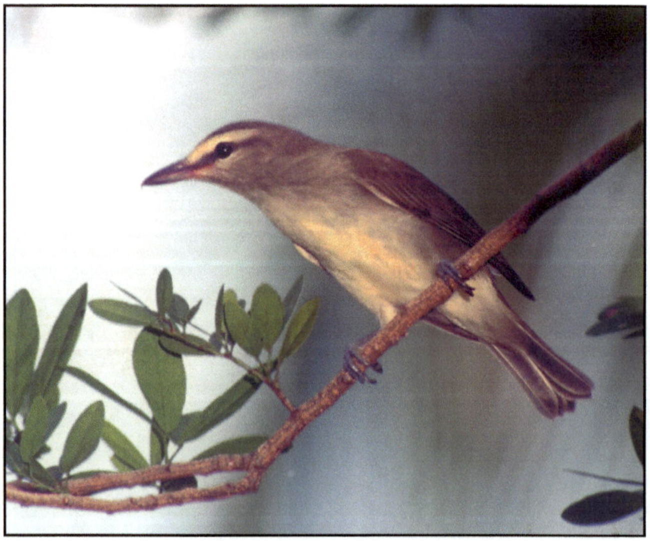

Yucatan Vireo (*Vireo magister*)

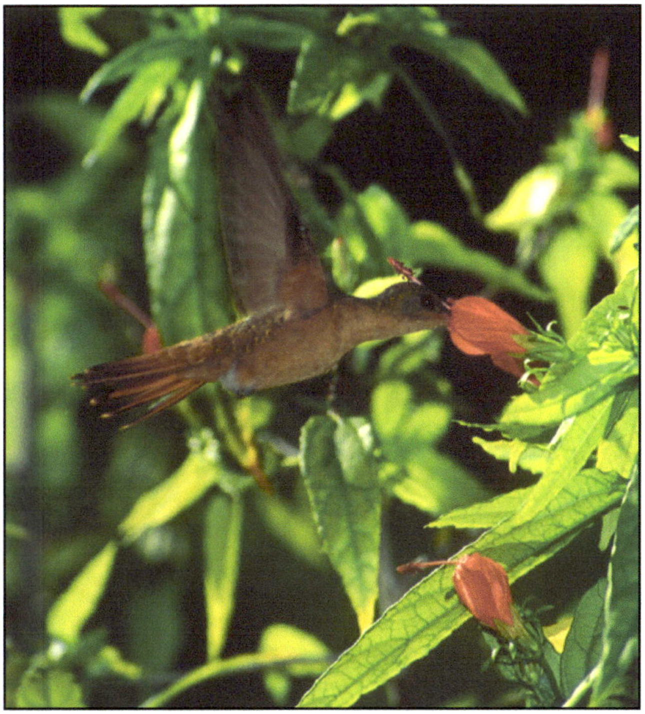

Cinnamon Hummingbird (*Amazilia rutila*)

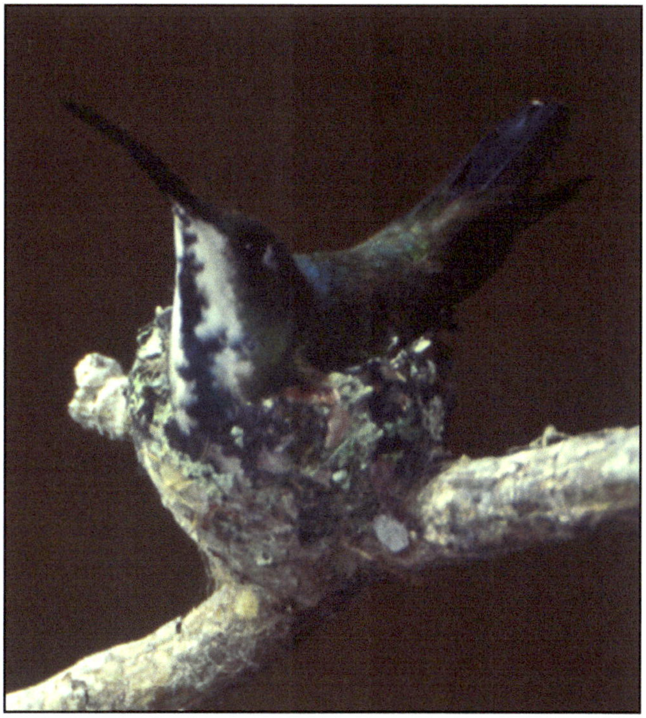

Female Green-breasted Mango (*Anthracothorax prevostii*)

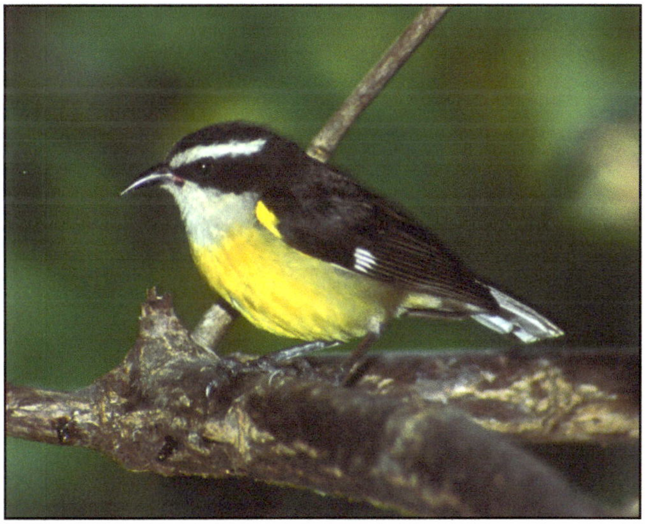

Bananaquit (*Coereba flaveola)*

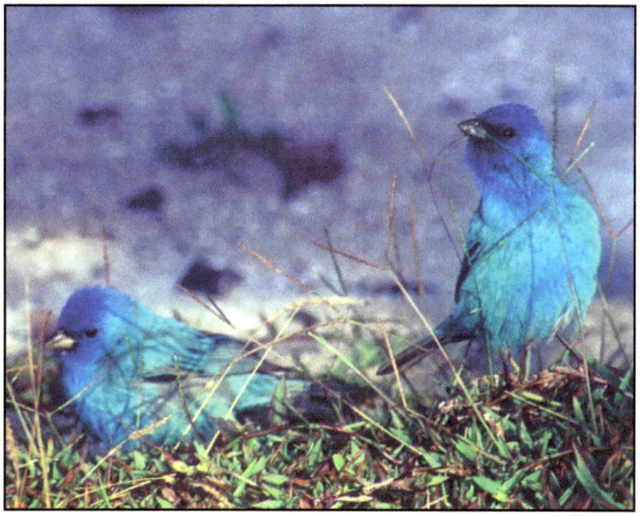

Indigo Bunting (*Passerina cyanea*)

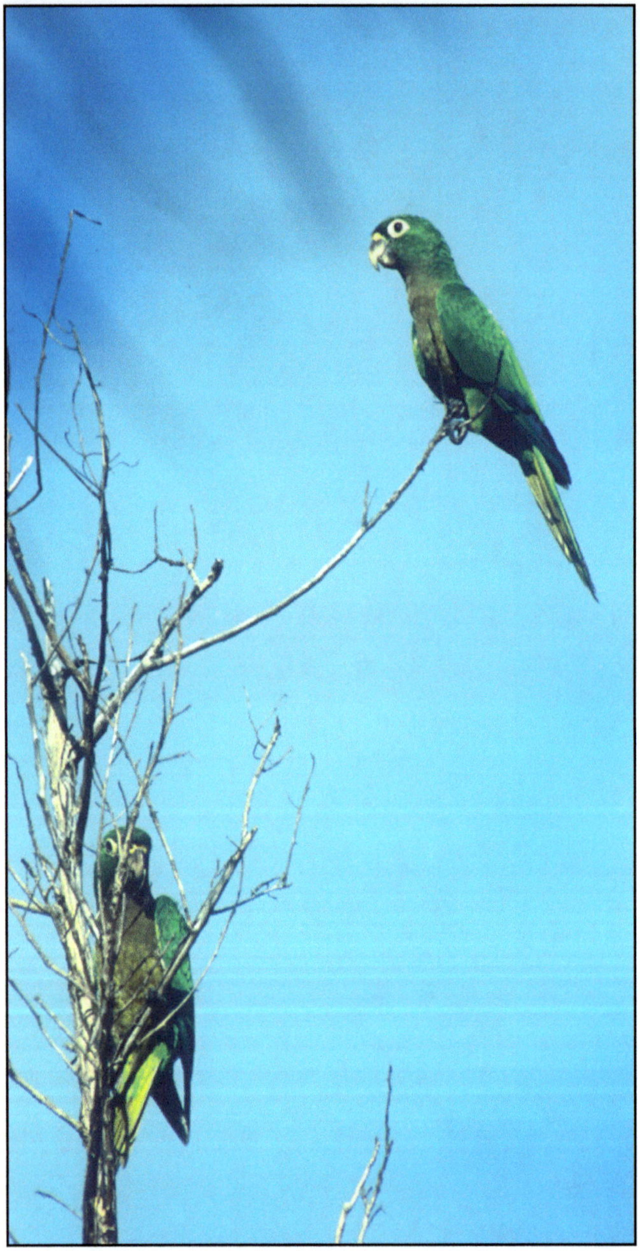

Olive-throated Parakeet (*Aratinga nana*)

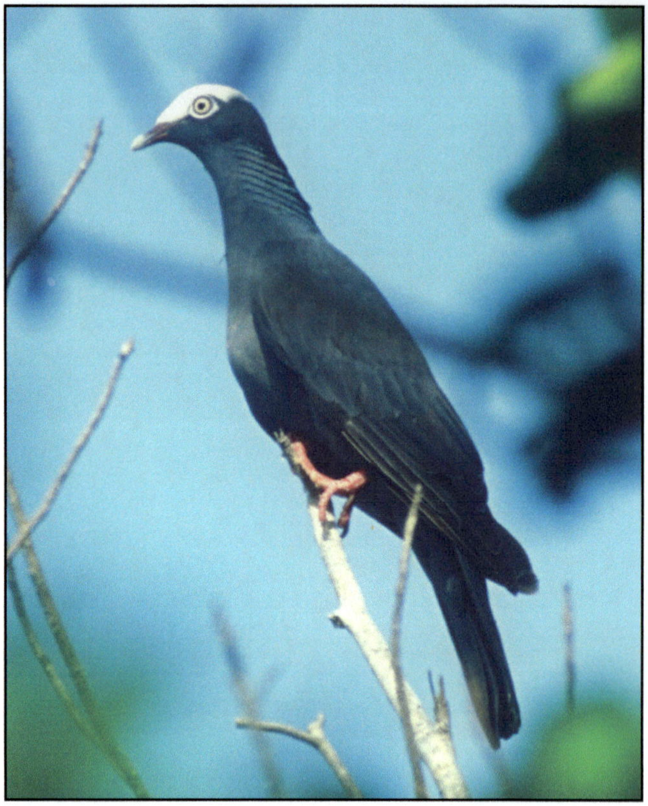

White-crowned Pigeon (*Patagioenas leucocephala*)

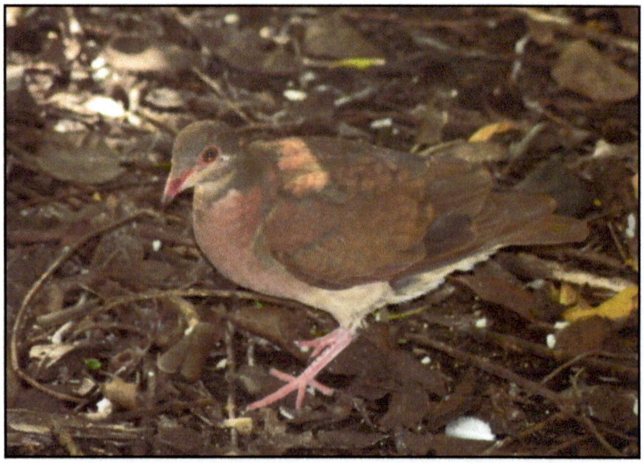

Ruddy Quail-Dove (*Geotrygon montana)*

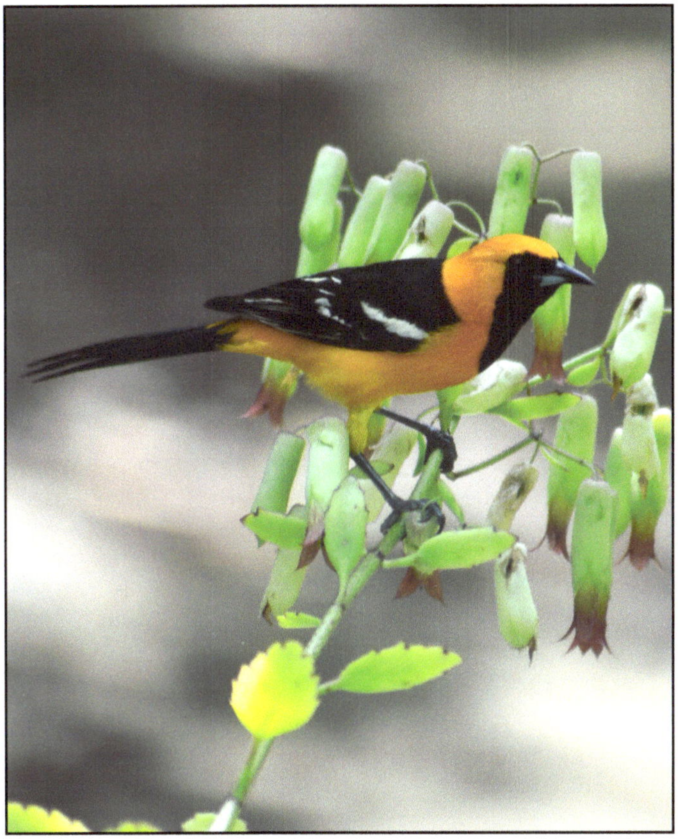

Hooded Oriole (*Icterus cucullatus*)

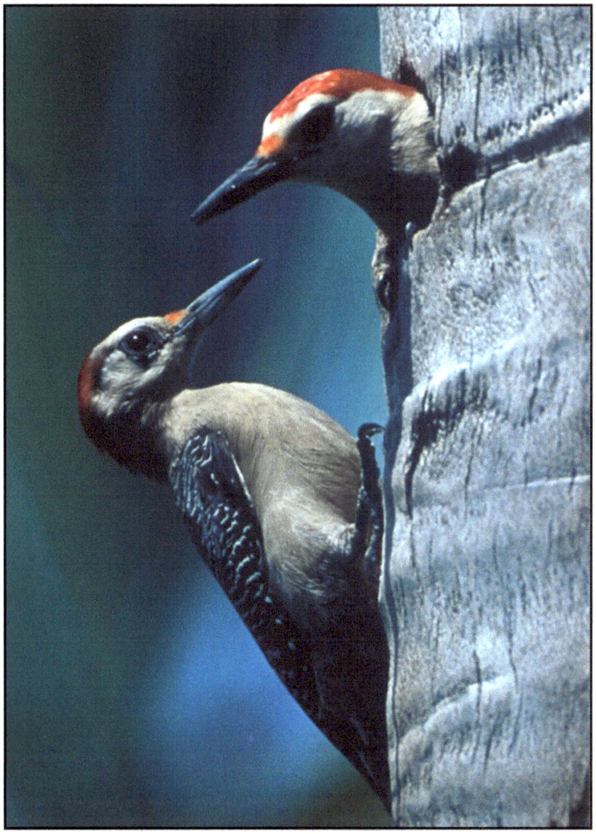

Golden-fronted Woodpecker (*Melanerpes aurifrons*)

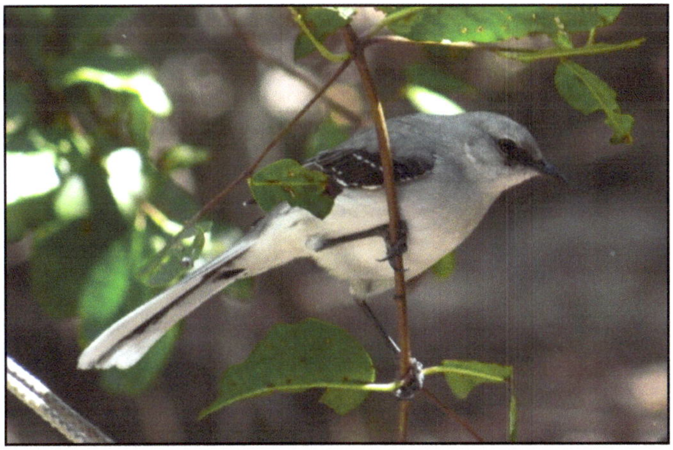

Tropical Mockingbird (*Mimus gilvus*)

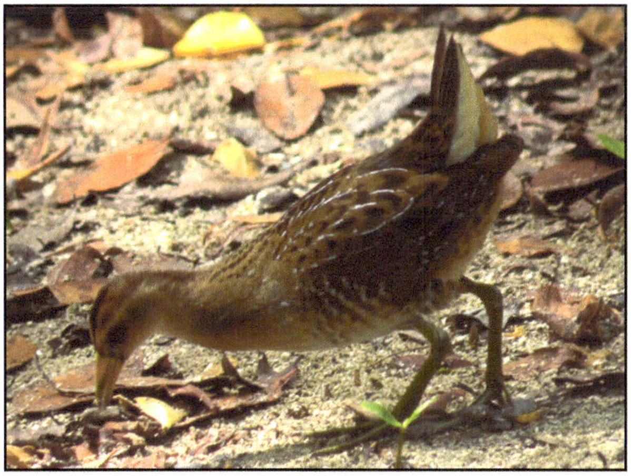

Sora (*Porzana carolina*)

DIFFICULT TIMES

Starting in October, 1998, and running until June, 2002, proved to be a difficult period for Dorothy and me. In 1998 Hurricane Mitch, the most deadly storm ever recorded in the new world, destroyed the *Lion*, the sailboat I had owned for over twenty-six years. From the *Lion*, I had done all my diving and underwater photography, so it was difficult living without it. When my friend Don decided to sell his 24-foot sailboat, I was definitely interested. He suggested that I do a trial period.

I checked it out in local waters. All went well, so I loaded the *Grunt* up with food and water, dive and photo equipment. I set sail for Turneffe Islands and stayed one week, diving the western reefs, shot a lot of photos, and altogether had a wonderful time. Every evening at six pm I turned on the radio to catch the local news and listen to the weather report. On the seventh evening, the 28th of September, 2000, they reported that a tropical disturbance was building up and already pretty close. I didn't hesitate, I packed away my gear and sailed west to English Caye Channel in the failing light. Once across, I anchored in the lee of the reef. At first light I headed north using both sail and motor, and reached the Caye by mid-afternoon.

Dorothy was waiting and extremely relieved that I was home. The tropical depression was now Tropical Storm Keith. Winds were already 50 mph, approximately 110 miles due east of San Pedro. Keith stalled over Caye Caulker around midnight of September 30th. There it remained stationary with winds building up to 135 mph. At noon on the third day, the centre moved to a point so that we were in the eye of the storm. Dorothy and I put on rubber boots and walked to the sea front.

The *Grunt* was still where I'd left it, except it was now sunk and a tree appeared to be growing from the mast. My other boat, had not fared as well. It was an eighteen-foot fibre glass cabin cruiser. It had been driven under the dock, and I could see more than several holes and fractures on the hull. It had been well and truly crushed!

As we walked back to the beach I noted that our neighbour's house was covered with swallows, resting where they could find a space in the lee. It was now flat calm, but they were too exhausted to leave. Many other avian species were sitting placidly on the bare branches of the trees. It took me a while to realize that the trees had no leaves!

I won't bore you with the details of the aftermath, except to say that I spent over four months repairing the house, boats, and dock. Don assisted me in salvaging *Grunt*, no easy task for the hatches had all been ripped off by Keith leaving it wide open, the seas then obliged by filling it with mud, grasses, sponges, and dead animals.

And I did purchase the *Grunt*!

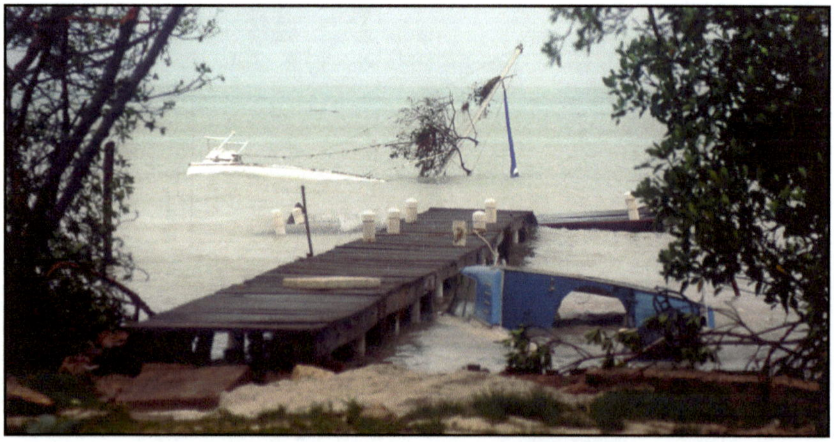

My two boats sunk in Hurricane Keith

I got into another adventurous misfortune in June of 2002 . Don now had his beautiful and roomy sailing catamaran, *Cats Meow*. He decided to sail to Florida. I had to visit Toronto, so I readily agreed to go with him. We set off on a leisurely cruise and stopped off at Isla Mujeres for a few days before continuing to Florida. On reaching Key West, I packed up and left ship.

That same evening I bought a ticket on a Greyhound bus that was heading north. I took the very back seat and pushed my camera bag under the seat in front of me. I was headed to Toronto via Chicago, but I never got to Chicago on Greyhound because the bus had an accident in the state of Kentucky. Close to midnight on the second day, the driver fell asleep at the wheel travelling at 80 mph. The bus left the interstate, drove along the central median for half a mile or so until it jumped up onto the southbound lanes at right angles, crossed over and finally stopped completely blocking a secondary road with its nose buried in a hillside.

Meanwhile in the back of the bus, I was bouncing between my very poorly cushioned seat and the metal ceiling. The toilet was making all sorts of aquatic sounds. For what seemed to be a long period, the bus alternated between running on either its left or right wheels. When it did stop, I was in considerable pain from my back and found it more comfortable hanging from the hand rails above. Eventually a Highway Patrol Officer arrived at the scene and boarded the bus. The first thing he did was to put his stern expression on me, and then ordered me to sit down. I cheated by supporting my weight on two rigid arms I kept my backside two inches above the thin cushion. I was the last person to be taken off the bus, and when they did so, it was on a back board!

At the hospital, I thought that I'd let them know immediately that I had limited money with me. The doctor said, "Don't worry, Greyhound will be paying for this!" After innumerable tests it was revealed that I had two damaged discs and one damaged vertebra. I was admitted to hospital. With the nurse's help, I called Greyhound to see if they would send my luggage to the hospital, after a wait I was told that my luggage had already been sent to Chicago! I asked why? "Because that was that bus' destination." he explained.

Great, now I had only the clothes I had been wearing since I'd stepped off the *CatsMeow*. Then I asked about the camera bag, and explained its location on the bus at the time of the accident. "There was no camera bag!" he said and abruptly hung up!

When this sunk in, I realized the magnitude of my dilemma! I was in a foreign country with only the clothes I was wearing, no travel documents and only eighty-five dollars in my pocket. On top of all that, I had three brand new fractures to my spine. Things didn't look too good! A gentleman brought me a large aluminium contraption, a back brace. The thing was a monster. He explained the workings to me, but didn't explain how I was supposed to sleep in it. The next two days were spent learning how to walk again, wearing the back brace.

I wanted to call Dorothy but a few days prior to my departure, the Belize telephone company had gone over to a new system and changed all the telephone numbers in Belize. I hadn't memorized my new number, but I did copy it down to my address book, which was in my camera bag. I asked a nurse if I could send an email to Belize. But no patients could use the computer. She said she would let the head nurse know.

Later that afternoon the head nurse, a very pleasant lady, came to visit. I explained my predicament and she said that she would send an email to Dorothy. After that, things started to move. Dorothy called me that same afternoon and then she called her brother Art, in Memphis.

On the fourth day, a large gentleman walked into my room and introduced himself as Kelly, my new lawyer. I was originally Scottish, so needing a lawyer had never entered my mind! Nevertheless, I explained my details of the accident and Kelly explained the American way to me. Art had called him regarding the accident. So I signed the necessary paperwork and said goodbye to Kelly.

As Kelly left, my sister-in-law Carol arrived to drive me to Memphis. When I checked out at the cashier's office, they already knew of me. Unfortunately, it was not the case that Greyhound would pay my hospital bill. Eighteen months later, I paid that hospital bill after the court case was settled.

I stayed with Art and Carol for several days, during which time Art managed to get my large bag sent from Chicago to Memphis. I flew to Buffalo, but flying was no small amount of hassle because of my very large and bulky aluminium back brace. However, I persevered and at last they allowed me to board the plane. My friend Helmut from Toronto drove to Buffalo to pick me up.

Luckily in my wallet I had my credit-card-sized Certificate of Canadian Nationalisation, so crossing into Canada was no problem. I stayed with Helmut and his wife Mei Lin. I applied for and received my new Canadian passport, driving license and money from my bank, where I still had an account.

After one month in Toronto I flew to Belize. It was a relief to be home to Belize, but I had to continue with doctor's visits and physical therapy for several more months.

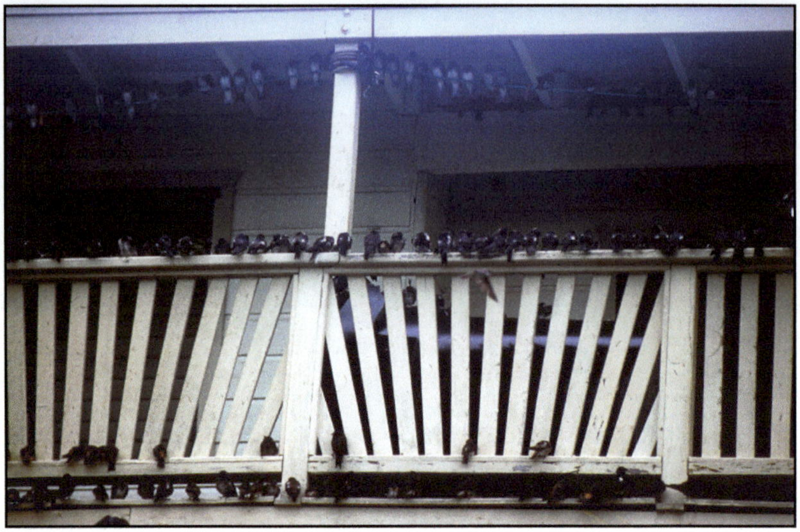

Swallows roosting in the eye of Keith

Feeding Birds after Hurricane Keith

Immediately after the departure of Hurricane Keith the forested areas on the island were completely destroyed and defoliated. At first Dorothy and I were blind to the starvation and distress the local and migrant birds were suffering.

It wasn't until we heard the noise of multiple birds outside our window. When I looked out, I was amazed to see a number of birds, squabbling over a piece of papaya rind that we had thrown out the window. Three of the seven birds were Black Catbirds, one was a migrant Gray Catbird, two were Tropical Mockingbirds and the remaining was a male Mangrove Warbler.

The Black Catbird is listed on the *IUCN Red List of Threatened Species*. This pretty black bird feeds almost exclusively on the small, white fruits of the palmetto (*Thrinax radiate*), a common but very slow growing plant on Caye Caulker littoral forest. Unfortunately, littoral forest is fast disappearing because it grows on high sand and well drained areas, the same environment that is ideally suited for building homes and invariably the first area to be clearcut for development.

Looking around, I could see no greenery, only bare branches, for Keith's 135-mph winds had stripped the trees of their leaves and berries. We found a large piece of bamboo that had washed ashore on the beach. Using my jigsaw and drill, I cut a large oval-shaped hole in each section of the bamboo to make a twenty-foot-long bird feeder. We purchased a large bunch of bananas and several large ripe papayas.

Our feeder was a big hit with the catbirds, but also every other bird on the Caye, or so it seemed. However, there were a few disappointed guests – the White Winged Doves. They stood on the ground, looking disappointed. So we cooked up ten pounds of rice and distributed it on the ground. The doves were well pleased.

The next morning we filled our feeding station once more, then relaxed at our own feeding station for breakfast. Around noon I found that we had two additional groups of guests. Seven Rufous-necked Wood-Rails were helping the doves clean up the rice and twenty-odd Great-tailed Grackles were cleaning up everything. These grackles were extremely thoughtful. I believe several were sent to the neighbouring Caye Chapel to bring back their starving relatives for a free meal. So began our dispute with the grackles.

Six months later the Black Catbirds were still dependent on our bamboo feeders. Following the aftermath of Keith the fruiting trees took a long time to bear. We were going through a lot of papayas and bananas. If it were just the Black Catbirds we were feeding, it would have been no problem, but for every catbird, there were three sneaky, thieving, conniving grackles. They didn't sit daintily on the feeder to eat their fill and leave. No, they snatched a whole slice of papaya and fled, and there were many grackles.

The Black Catbird's American cousins, the Gray Catbirds arrived in mid-October. Like the local Black Catbirds they subsisted on fruit and berries. After a long flight from the U.S, they arrived very hungry, and there was no food to eat. It didn't take them very long to find our feeders, but the feeders were empty except one or two Black Catbirds who were gulping down the last bits. The Gray Catbirds are the larger bird and quickly let out a stream of catbird curses, and then chased off their little cousins.

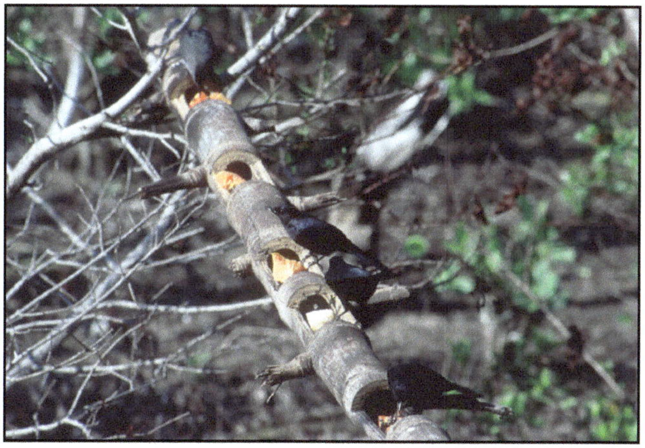

(*above and below*) Black Catbirds (*Melanoptila glabrirostris*) eating papaya at our feeding station

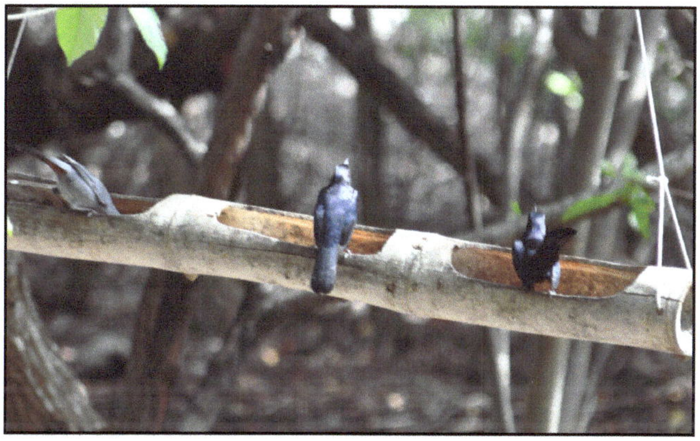

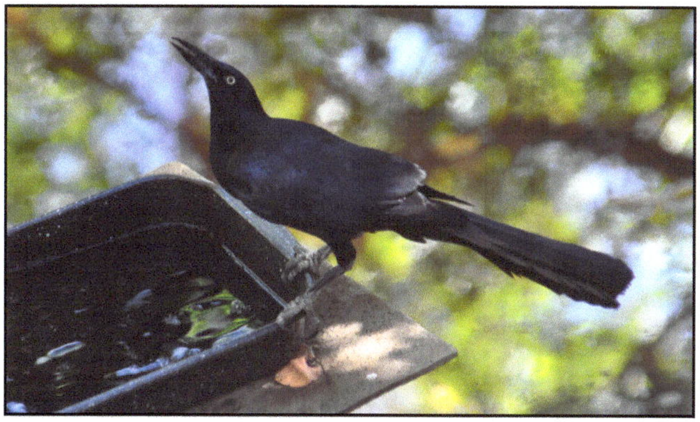

Great-tailed Grackle (*Quiscalus mexicanus*)

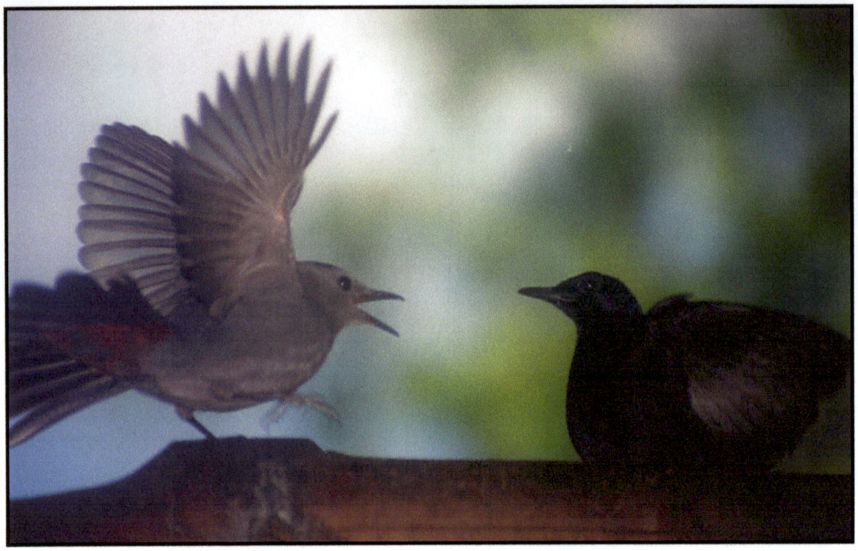

Gray Catbirds (*Dumetella carolinensis*) letting out a stream of catbird curses at his little cousin

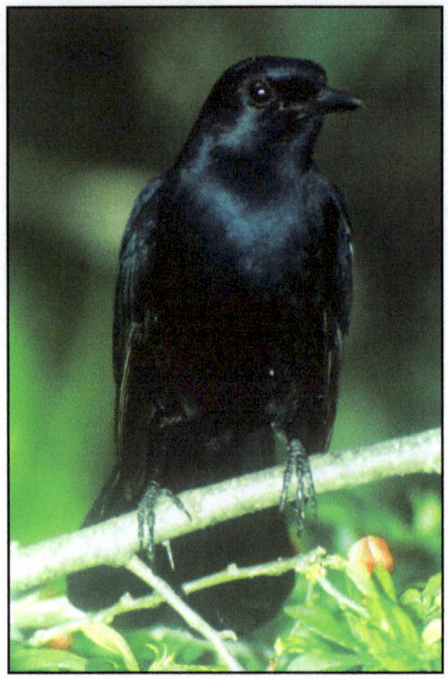

Black Catbird in tree waiting turn at feeding station

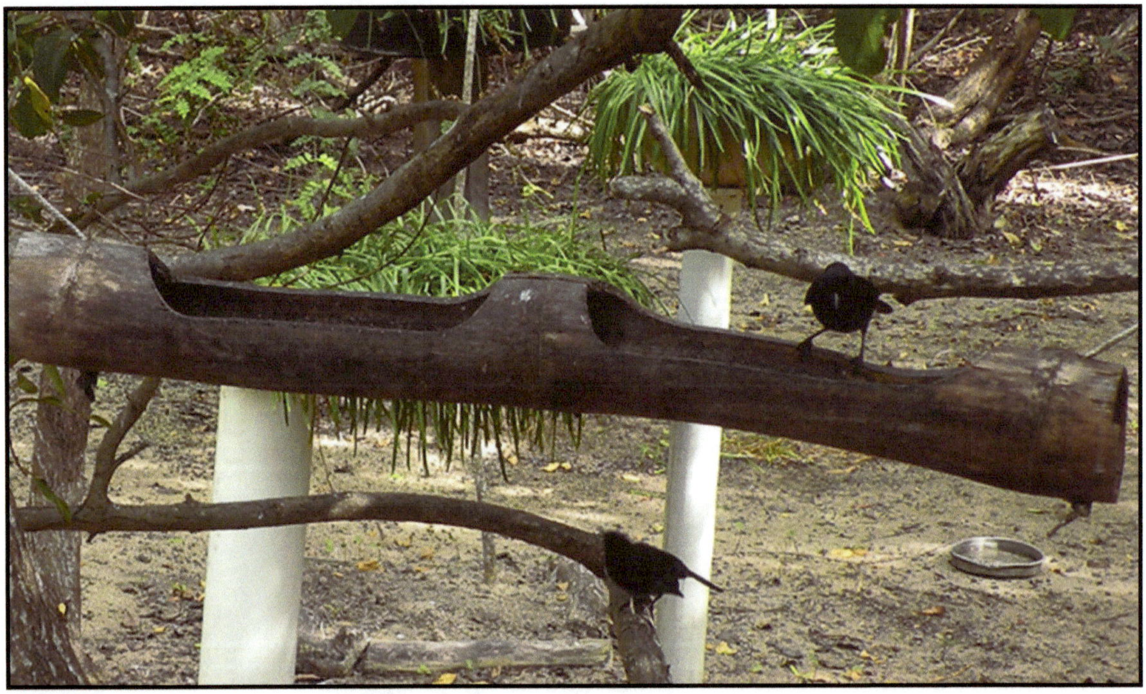

Black Catbirds at our feeding station

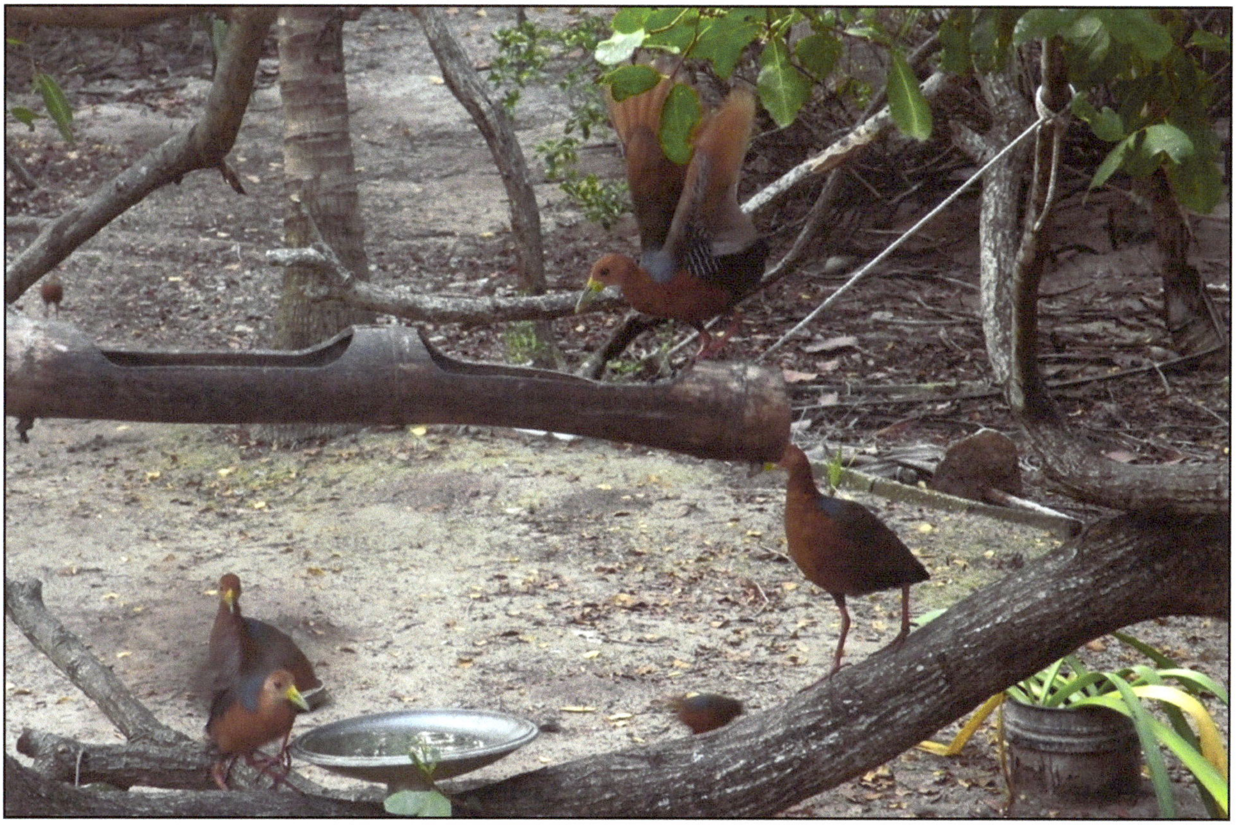

Rufous-necked Wood-Rail (*Aramides axillaris*) eating at feeding station in our yard

Rufous-necked Wood-Rail

We kept our feeding station going full time, for several months, and after a while our dinner guests became quite tame. The tamest of the birds proved to be the Rufous-necked Wood-Rail, which is quite surprising for such a shy and wary bird. Prior to the hurricane the rails were very secretive. One could only catch a glimpse of them now and again in the mangroves.

When they first visited our feeding station there were only seven that joined the White-winged Doves eating cooked rice. Pretty soon after their arrival, we began mixing cat food with the rice. The rails seemed to enjoy this more and the doves didn't seem to mind.

As the trees and shrubs recovered after the storm, the birds started to feed naturally on the local fruits and insects. Gradually the birds abandoned the feeders, except the rails and their buddies, the doves. As the years rolled on, the numbers increased. When the rails returned from nesting, the juveniles were introduced to the feeding station. By 2004, twenty-six rails were coming to our feeding station.

Prior to Hurricane Keith, and the introduction of our feeding station, it was thought that the Rufous-necked Wood-Rail was a permanent resident of Belize. Now, our observations pointed to the possibility that this is not the case. Our rails give up the feeding station in May, and don't return until late September, or early

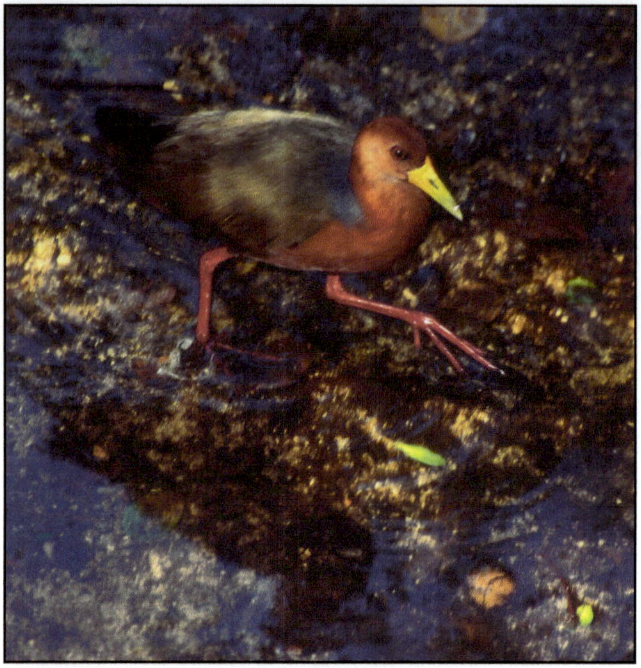

Rufous-necked Wood-Rail (*Aramides axillaris*)

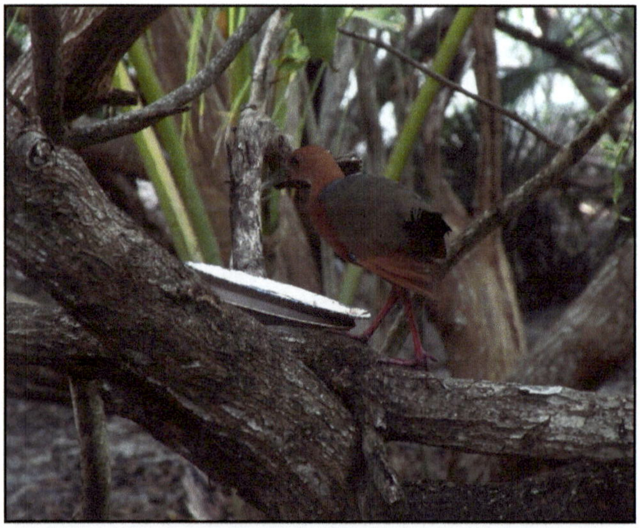

Droopy eating off a plate on a tree branch

October. There is a report that they are found in low montainous areas of El Salvador from May to October.

When they do return they seem to come right back to our yard. Occasionally the juvenile birds arrive with the adults, sometimes as many as four or five. Or the juveniles may come to the feeding station a week or two after the adults.

In May of 2010 we noticed that one of the rails had a droopy right wing. We gave her the name of "Droopy". As usual the rails were leaving in May. By the 26th of May we had but two rails left, Droopy and a male we called "Whitethroat". We gave him this name because he was in breeding plumage, a very noticeable white throat and bright red legs.

The two birds were inseparable, then on May 27th Droopy was gone. Whitethroat stayed a few more days and then on June 2nd, he took off also. The next day at 7:10am, a slightly dishevelled Droopy was standing on a branch in our back yard, waiting for us to notice her. Her wing was droopier than normal.

We surmised that Droopy had made an effort to go to the rails' breeding residence, but could not make it. Perhaps she stayed a day or two on one of the low mangrove cayes between us and the mainland and then on recovering her strength, she returned.

In the next months, Droopy almost became a family pet. Each morning she arrived at six am and gave a loud raucous cry from the tree outside our window to tell us that she was awake and hungry. We fed her twice a day with chicken fat cut into worm-like strips and rice mixed with cat food. On occasion, she was fed out of her own plate, which was fastened to a tree branch.

Now and again, she disappeared, a walk-about that lasted two to four days. She must have fed herself because she never appeared unduly hungry when she returned, at least until the next morning when she would make her breakfast order. On one such walk-about, I happened to met her a couple hundred yards from home. She actually recognised me and walked

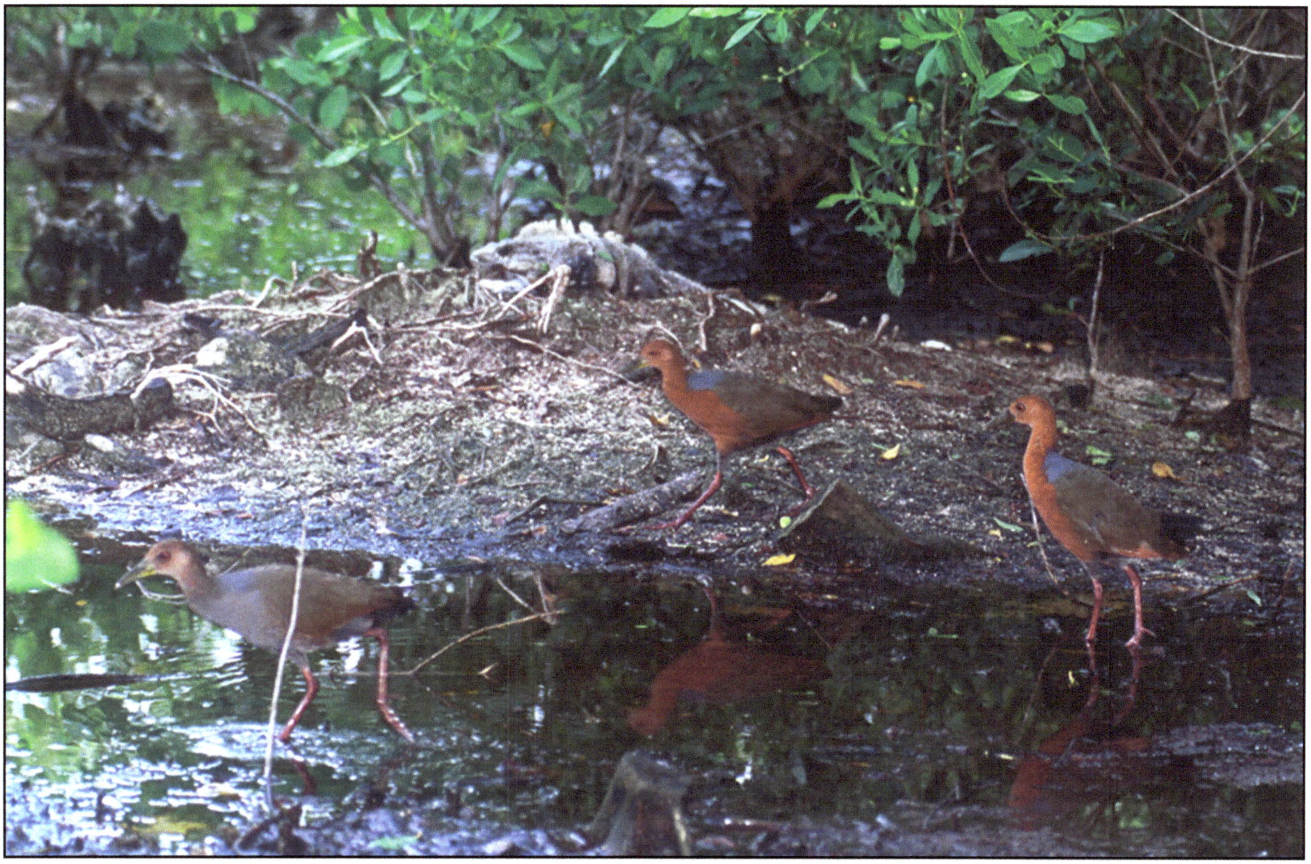

Rufous-necked Wood-Rail

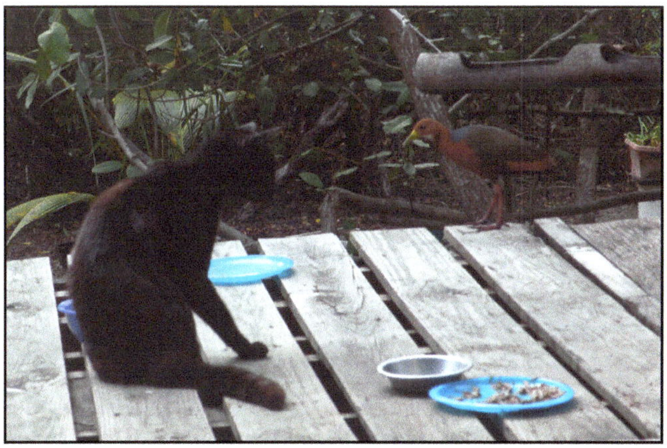

Rufous-necked Wood-Rail and Kitty

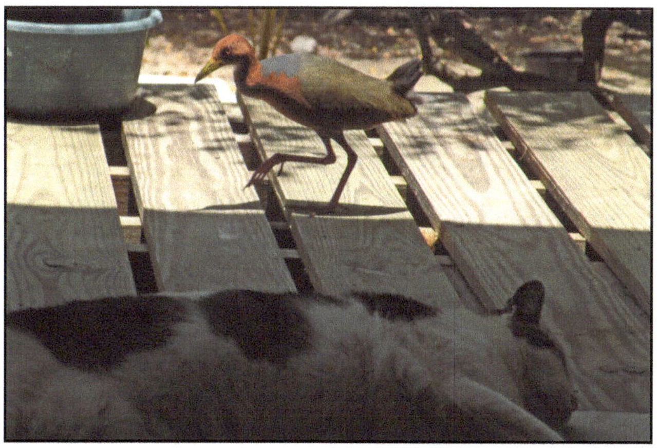

Rufous-necked Wood-Rail and Spot

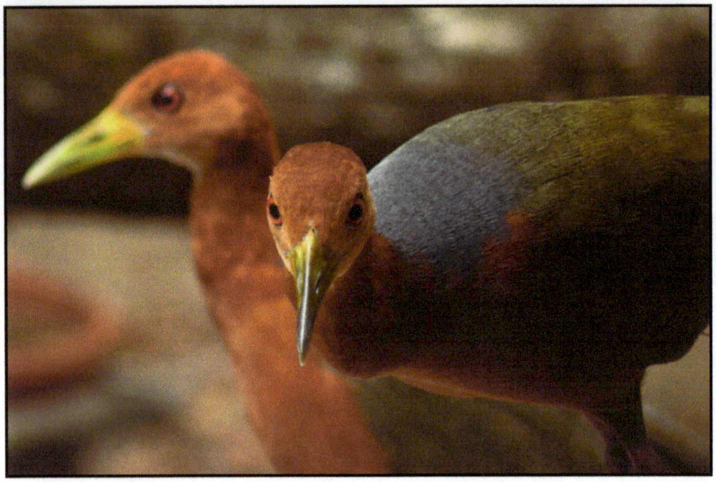

Rufous-necked Wood-Rail (*Aramides axillaris*)

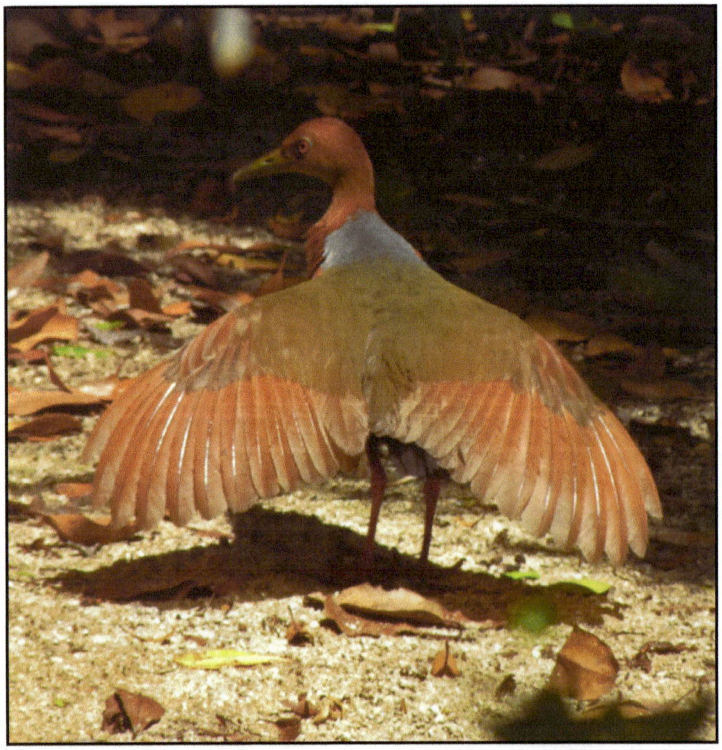

Rufous-necked Wood-Rail basking in the sun

with me for about fifty or sixty yards, until the path became a large curve. While I chose to follow the path, she decided on a shortcut, straight across dense, muddy mangroves. However, I still beat her by six minutes.

On September 15th, Whitethroat came back, but he no longer had a white throat, as courting season was over. However, he still had his red legs. So we changed his name to Redlegs. Droopy seemed glad to see him again, and seldom do they stray out of sight of each other. Four weeks after Redlegs returned, I could now no longer distinguish Redlegs, from the twenty odd new rails that have arrived in the last weeks. Come May, our rails once again left. Droopy's wing had healed and she left for her summer residence with the rest of them.

The Rufous-necked Wood-Rail does have at least, one problem on the caye, which they seem to handle extraordinarily well. Boa constrictors are extremely numerous on the south point of Caye Caulker. During dry weather, boas bury themselves under dry leaves and hide themselves so well that not a square inch of their hide shows. But the rails can sense the hidden snakes. One or two rails initially give a warning call; other rails then move in, as many as eight or ten, each crying a warning call. They keep it up until I move in the centre of the circle and look for the snake. At my arrival the rails disperse. They seem to have every confidence in me. Once I find the snake, I relocate it in another area, usually several hundred yards away.

During the rainy season, when the ground may be wet or flooded, the boa constrictors stay aloft in the white mangrove trees. Still the rails sense the snake and start their warning cry. However they can't pinpoint it, as they can at ground level and tend to wander. When I investigate, I cannot always move the boa owing to the branch density and the height of the tree. Then I leave it for it is unlikely it will catch a rail in the tree, but a grackle or a white-wing dove is also a possibility.

I had long wondered what sense the rails use to pinpoint hidden boas. I had suspected for several years that the rails scented the snake, for there seemed no other explanation. I consulted Lee Jones, who sent me a computer link, regarding "The Avian Sense of Smell". It seems that the sense of smell in birds is quite poorly documented, but physiologists do expect the sense of smell to be well developed in several species, including rails, cranes, grebes, and nightjars, based upon the relative size of the brain centre used to process information on odors. Although not 100 percent certain, this, plus my own observations, is good enough for me.

When the rails returned in October 2010, they brought with them five juveniles, which Droopy took under her wing, in a manner of speaking. I was looking out of our window when I noticed a group of six rails, which consisted of Droopy and the five juveniles. They were all leaning forward, staring at a slightly raised area of dead leaves and crying loudly. Sure enough under the leaves was a seven foot boa constrictor. I lifted it up and the rails quieted down and moved off in their own separate ways. Droopy chose to watch over these juvenile birds which were unaccustomed to the dangers in their new home.

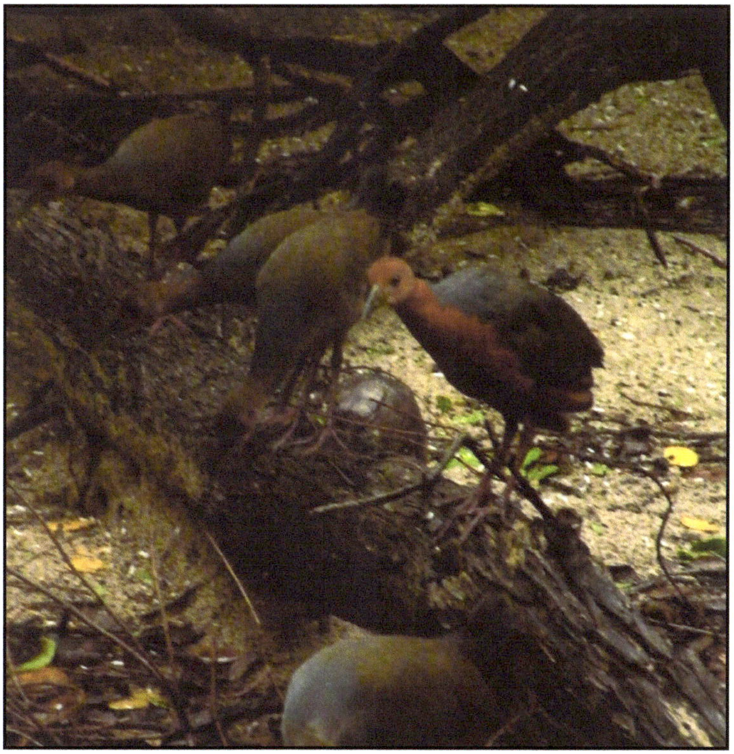

Droopy showing juvenile rails the boa

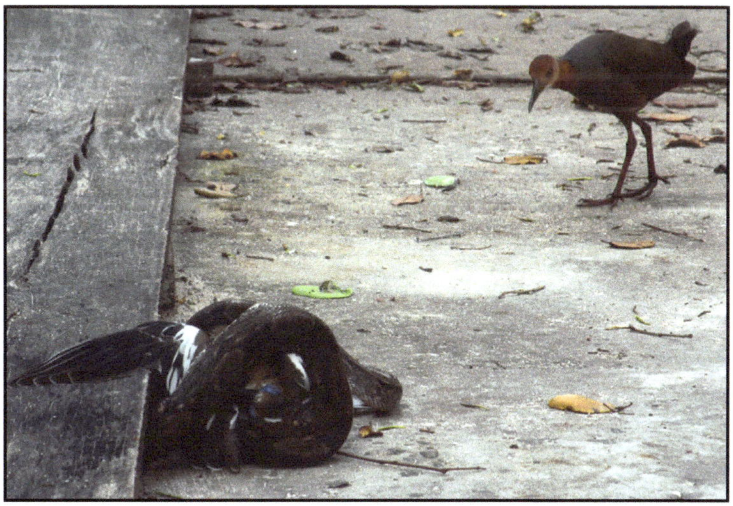

Boa killing a White-winged Dove
with a Rufous-necked Wood-Rail watching

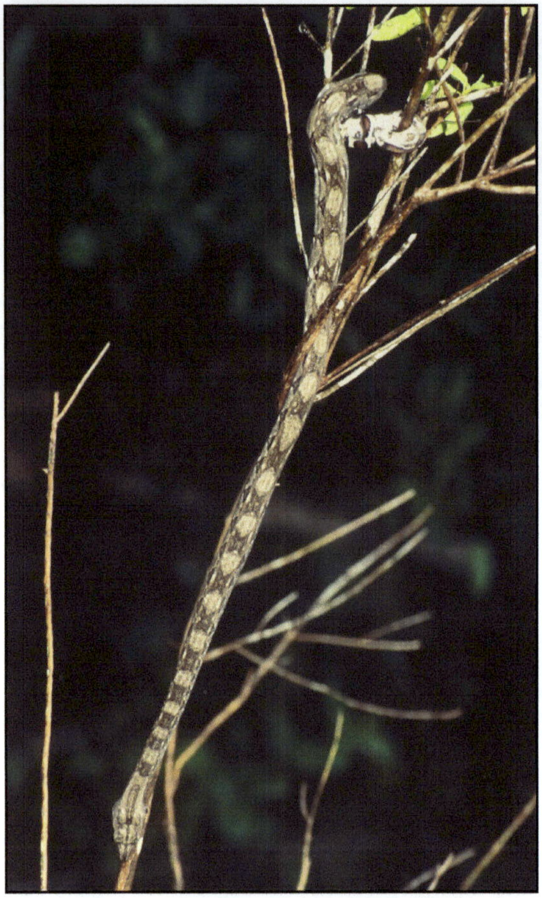

(*above and below*) Boa Constrictor
(*Boa constrictor*)

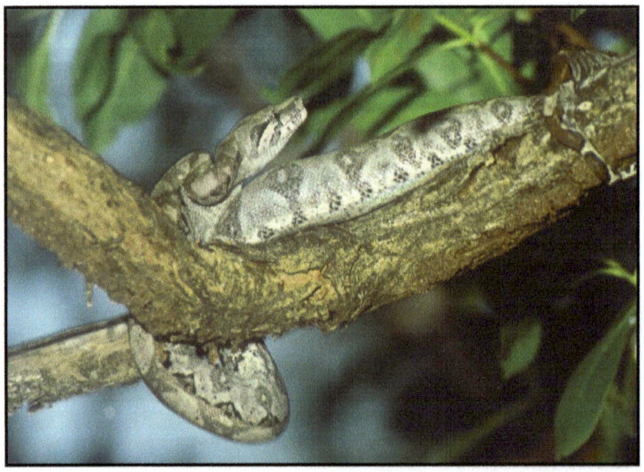

Boa Constrictors

When I arrived on Caye Caulker in 1972, I always seemed to be rescuing boa constrictors from the villagers, who believed they were venomous. I would then take them to the southernmost point and release them. This got me into no small amount of trouble on occasions. I recall one incident when I had come across a large eight foot boa surrounded by a number of children and villagers who were waiting for the arrival of a macho man with a machete. I picked up the snake and walked away with it. This act of bravado was not forgotten. When I returned, I was arrested by the local policeman who was waiting for me with handcuffs at the ready. I was arrested as an obeah man!

I spent the better part of that day in the police station arguing that boas were not poisonous, but to no avail. It didn't help that I was living with Margie at the time, along with her Irish wolfhound, which weighed as much as most of the local men, at slightly less than 140 lbs. To cut a long story short, we were given twenty-four hours to get out of British Honduras!

The following morning we visited the police station and I humbly asked if we might be allowed to stay another day, as the boat had already left. He hummed and hawed, but in the end, he allowed us to stay. Then I pulled out a large bottle of British Honduras's finest rum. I thanked him and asked if he would accept this small gift, as a token of our thanks. He did indeed accept our gift, and as we were leaving, he called us back and said we could stay another week.

One week passed and nothing was said, then another. From then on, I smuggled the boas to the south point, hidden in a backpack.

Boas are far from being animals that find their prey by chance. Our feeders are suspended from a tree branch by strings, two to three feet long. Several mornings we have found a 30-inch long boa curled up inside the feeder. Of course, we usually spot it when we put in the papaya or banana. The snake is then removed to another location. As far as we know this boa has not yet been successful in securing breakfast, but has tried three times.

One other small boa has been more ingenious. His method is to hide down a crab hole directly below a feeder which is hanging from a tree branch about six feet above. Birds are generally messy eaters and the area below quickly becomes covered with pieces of fruit, which in turn attract other birds. On their first nibble they are seized by a jack-in-the-box in a crab hole! So far this boa has caught two Gray Catbirds. At least two other Gray Catbirds have been caught by boas but none of the Black Catbirds.

To catch larger birds, predatory boas usually hide under dead leaves, and are totally invisible. Rails have never been caught this way, because they seem to scent them from a distance of five or six feet. Unfortunate birds that have been caught by the larger boas, six feet and over, include male Great-tailed Grackles, White-winged Doves and once a Yellow-crowned Night Heron.

One large boa tried a different approach. We have several baths. The one the rails favour is a bowl, about fourteen inches wide with about four or five inches of water in it. One day we were alerted by a rail screaming at his bath water. A seven-foot boa was totally submerged except for his eyes and nostrils.

Whether they are successful or otherwise, we move all the boas to another location, usually about ten to fifteen minutes walk away. But they generally find their way back to our yard. It took me quite a while to accept this homing instinct, even though we have an extraordinary number of boas that visit our yard. For example, in two months, December 2010 and January 2011, we removed 49 boas from our lot and the two adjacent lots on either side, both of which are vacant, and still dense with bush.

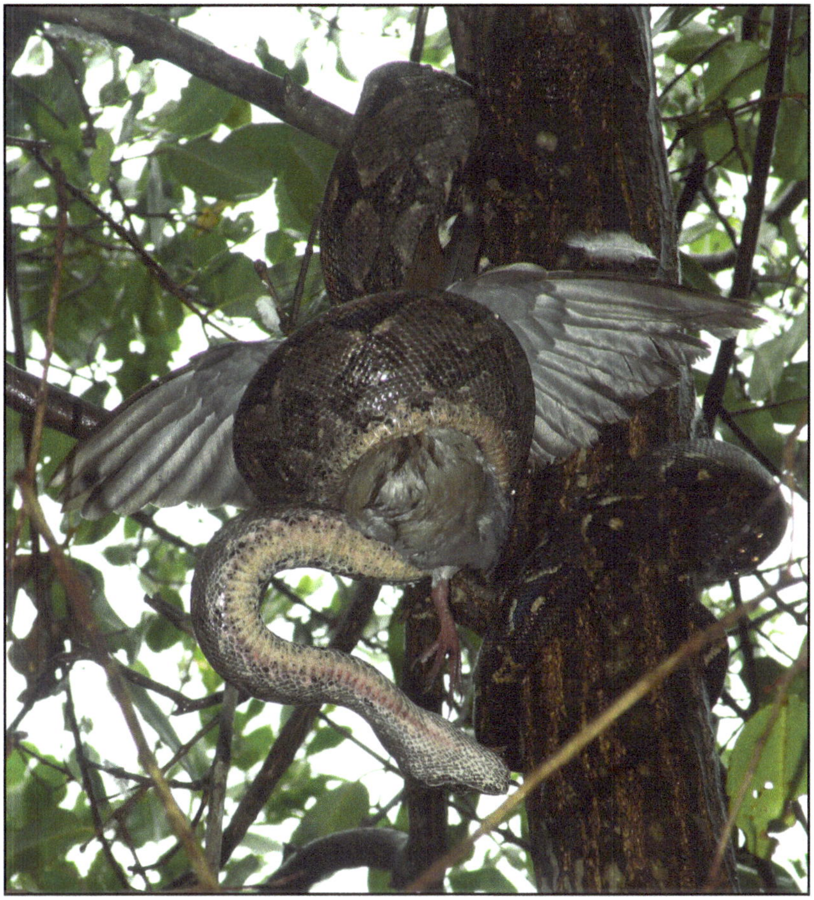

Boa Constrictor subduing its prey, a White-winged Dove

It wasn't until I captured a boa which I noticed had a white left eye that I had released about ten minutes walk away. He was a hefty six-foot-long individual that returned to our lot five days later. He was recaptured within a few inches of where he was previously caught, but I noted that he wasn't blind in his white eye as he reacted to a close hand movement.

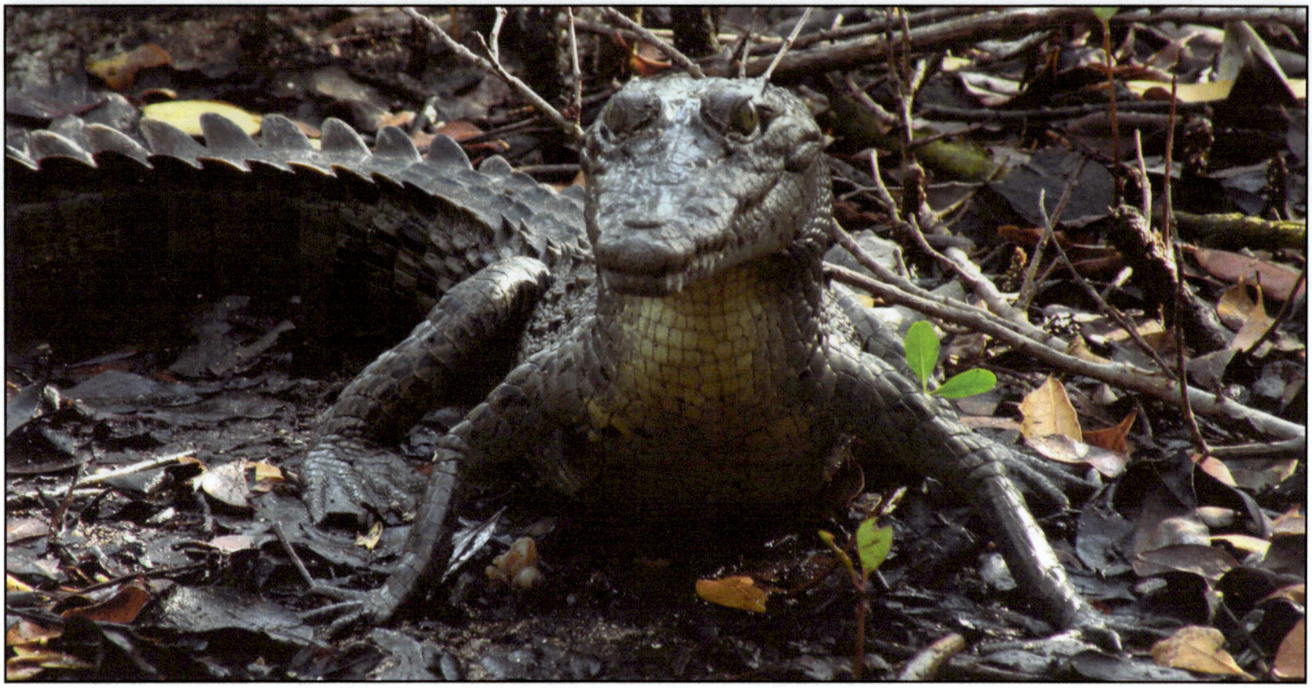

Smile for the camera, Charlie
American Crocodile (*Crocodylus acutus*)

American Crocodile

Although unrelated, crocodiles and sea turtles share several characteristics. They both are aquatic reptiles, lay eggs buried at water edge, and saw the dinosaurs evolve and go through extinction.

One hundred and fifty million years ago, sea turtles were land animals, not unlike today's tortoises, then around 110 million years ago, they moved into the sea.

Crocodilians have been around even longer. A 2010 BBC report indicated that footprints of a crocodile-like animal had been found in rock in Poland that was dated at 385 million years old. Today's crocodile species however, have only been around a mere 80 million years. To put this into perspective, recognisable man, *Homo sapiens*, made an appearance only 35,000 years ago.

It had long been known that around sixty-five million years ago, some unknown event caused the extinction of all the dinosaurs over a relatively short period, referred to as the K-T boundary. It was not until the 1990s that a vast crater was found in the sea east of the northern Yucatan Peninsula. This crater is 180 miles in diameter and it is thought that the cause was a meteor six miles in circumference.

The impact of this meteor caused the extinction of all the dinosaurs plus 75 percent of all living species on earth, but sea turtles and crocodile species survived the K-T boundary. However, now all sea turtle species are endangered, as well as most of the world's 23 species of crocodilians. What is endangering turtles and crocodiles? They are slowly being wiped out by a Johnny-come-lately species that is not so slowly taking over the earth!

In Belize, there are two crocodile species, the Morelet's crocodile and the American crocodile. Both species are protected in Belize. The Morelet's crocodile I know only from canoe trips and photo shoots, for it favours freshwater lagoons and rivers. On the other hand, I am well acquainted with the American crocodile. It frequents brackish and salt water habitats. Which coincidently, I also frequent.

American crocodiles, although they seldom show it, are remarkably intelligent. Dorothy and I have lived with American crocodiles as our neighbours for 17 years now. Our land is low and tends to flood during the rainy season. I filled in the front of the lot, but left the area adjoining the protected land as it was, as it attracted a lot of wading birds and rails. Our home has a large window which looks out on the back property, where we watch the wildlife.

Proper Charlie

The first crocodile to introduce himself lived in a shallow lagoon which was actually on an extension of our lot. I named this croc, Charlie. As more crocodiles showed up, I also named them Charlie, but gave them a Christian name also. The crocodile who lived directly behind us became Proper Charlie for he was a real gentle croc, who enjoyed my company, believe it or not!

Proper Charlie was a nine-foot-long male, whose neck and shoulders were stained red with tannin from the red mangrove lagoon where he lived most of the time. He loved fire. When I burned household garbage, he would always appear and run directly across the flames, then he would turn around and do it again on his way to the lagoon. At first I was quite upset at this behaviour, but I was hesitant to stop him from doing his fire act. It didn't seem too wise to stand between a running crocodile and a raging fire.

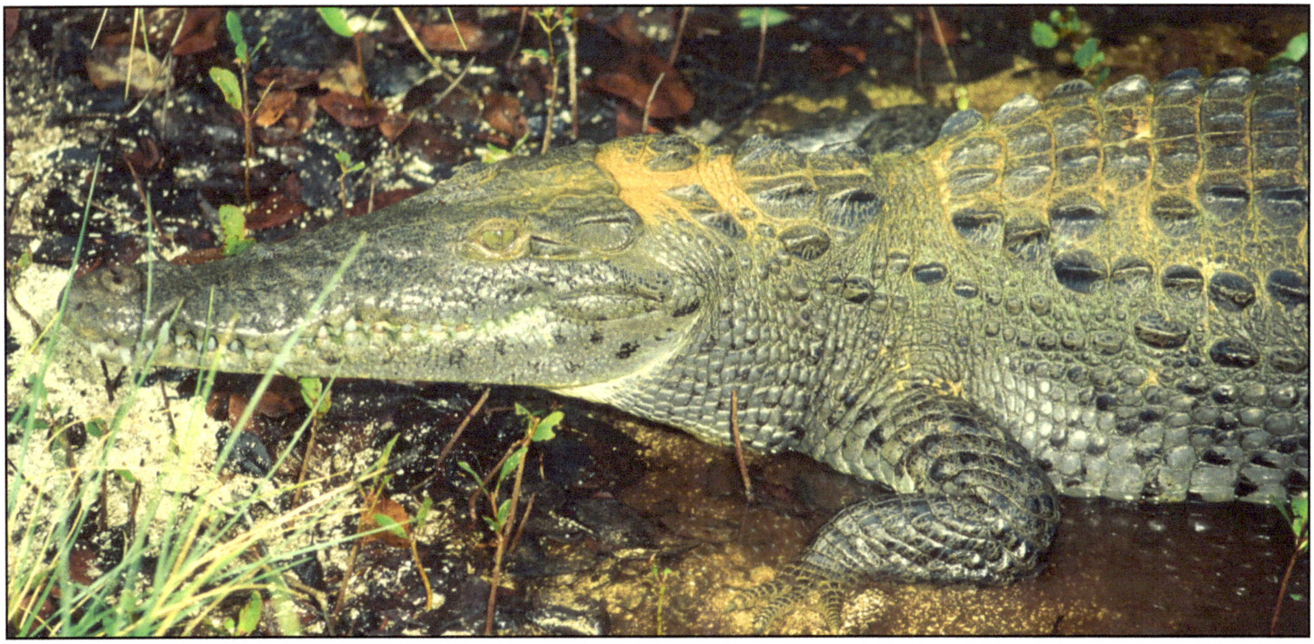

Proper Charlie

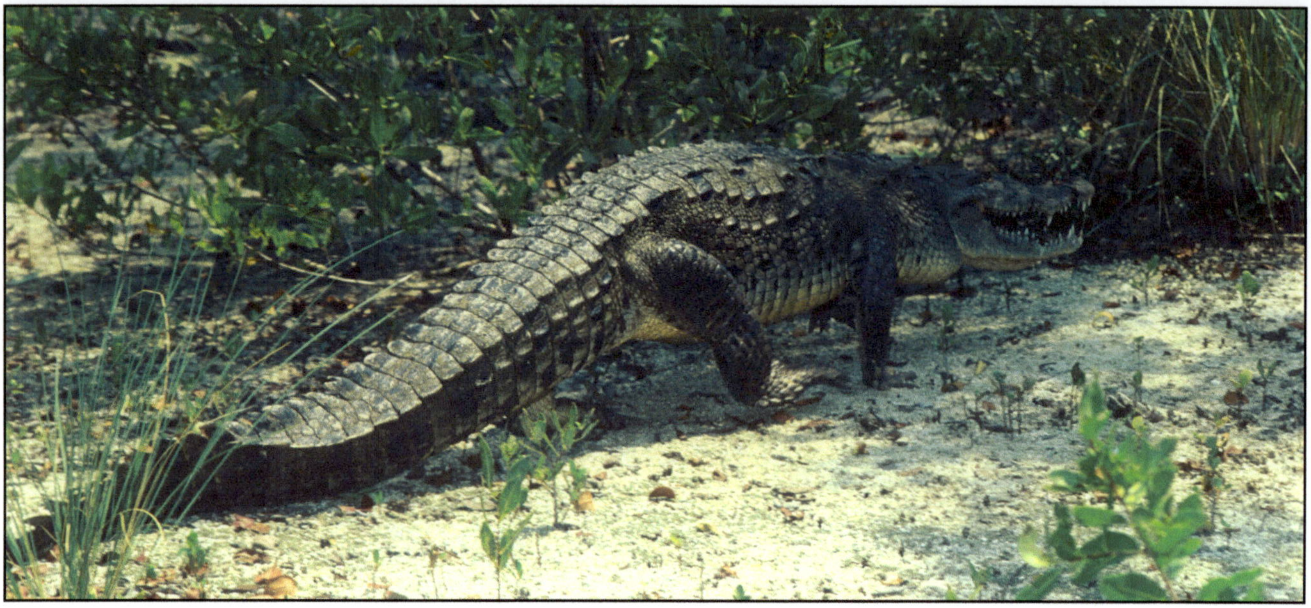

Big Charlie chasing Hard Luck Charlie

One day Dorothy was filling the bird feeding station. She was reaching up, standing barefoot in the wet yard, and felt something soft and gentle on her bare foot. She was horrified to see Charlie resting his lower jaw on her foot. I still kid Dorothy about the back flip she almost made through the window. She hurt Charlie's feelings by not accepting his gesture of friendship, for he never did it again.

Big Charlie

Another crocodile we called "Big Charlie" lived in a larger and deeper pool about 300 feet to the west. Big Charlie was also a male, about eleven feet long. He was extremely territorial and would not let any other male croc into his pool. This pool was also a favourite for wading birds, such as herons and egrets. I also visited Big Charlie's lagoon, or rather its surrounds. When I tried to photograph the birds he showed his displeasure at my closeness and defended his territory. He lifted himself high and ran at me.

A crocodile can run faster than a man and many dogs. I was aware of this, of course, and picked a well treed area when I took my pictures. If chased, I could run around a tree. A large croc has to stop to make a sharp turn. I took advantage of Big Charlie's absence one day to make a hide by the edge of the pond. The floor of the hide was three feet high, by three feet wide and six feet long.

One morning there were about ten snowy egrets, four spoonbills, and several white ibis feeding. Big Charlie was lying in the mud, pretending to be a log. While Big Charlie was looking the other way, I climbed into the hide and set up the tripod with my camera attached. I slowly pushed the long lens thru the palm fronds and focused on the spoonbills. While I was focusing, Big Charlie was already focused. He had moved up close to a point in front of the hide. Luckily the floor had gaps between the boards, so I caught sight of him thru the floor boards and slowly moved the tripod and me, back a yard.

In one second, I was watching Big Charlie through the crack in the floor boards, and in the next, Big Charlie had jumped up vertically and forced his head, shoulders, and forelegs through the palm fronds. The whole hide was shaking as I jumped back. Time to leave, I thought. The birds were now all gone anyway. I grabbed the tripod and exited quickly while Big Charlie was trying to free his forelegs from the fronds. I only used the hide that once. As I write, it is rotting by the water. Big Charlie was very aggressive with other crocodiles too, and was forever chasing them in the water and even on dry land.

A friend was returning from the village one evening after a party with a dim bicycle lamp on a moonless night. About 100 yards from home, he hit Big Charlie broadside as the crocodile was heading to the sea. The bicycle flew and cyclist flew further and rolled as he landed and kept rolling. When he looked back, his cycle was in Big Charlie's jaws and was being punished severely!

Around mid-1997, two other crocs came to stay behind our home. They were smaller than Proper Charlie, and although I never actually sexed them, I felt the smaller of the two was a female. This I deduced from Big Charlie's behaviour. He would allow the smaller of the two to use his pool, but would chase the other if it attempted to join them. I christened the two new crocodiles, Little Charlie, (the female) and Hardluck Charlie. Hardluck Charlie was so called because at one point while being chased, Big Charlie got a little too close and amputated a toe or two from the smaller animal's foot. I was worried at this because the water they lived in was quite foul with both fowl and croc excrement, but I could see no way of taking a seven-foot crocodile to Belize City to see a veterinarian. Actually, the wound healed extremely fast, almost miraculously. In years to come, Hardluck Charlie proved to be an exceptionally special crocodile.

Several months later, I got a call from a film director who had been given my business card by a friend. The caller told me that he was interested in doing a documentary in Belize about jaguars. I told him that a jaguar documentary would take at least three years. The phone went silent and then he inquired if I could suggest another animal to film in Belize. I suggested tree frogs, then humming birds—very definite no! He wanted an animal with more oomph! I was beginning to wonder about this guy.

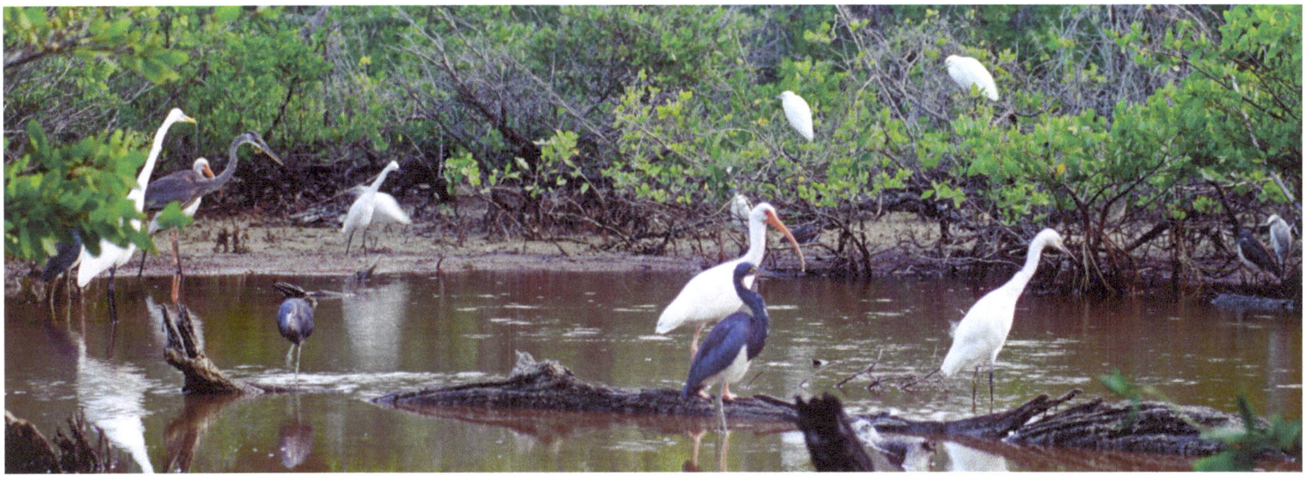

Wading birds inspecting Big Charlie's pond

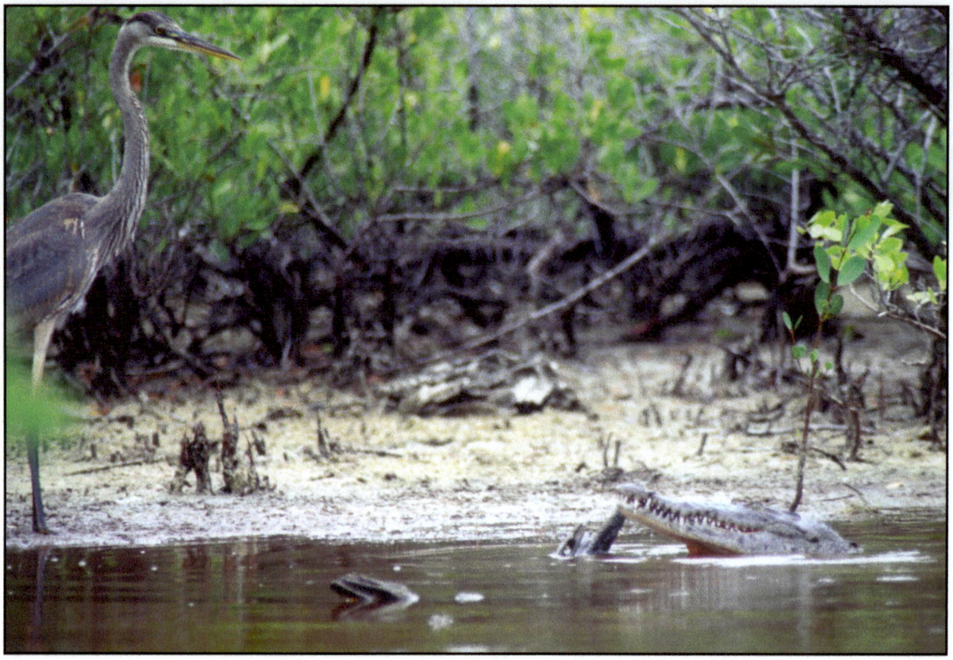

Great Blue Heron at the edge of Charlie's Pond

Then I heard Proper Charlie splashing in the background. How about crocodiles? I asked. "Great" he replied. He arrived on the Caye about ten days later, with two camera men and one gofer. Next day, the production crew came to our home. I spoke to the director at length, and told him what to expect. We all went down to our back yard where I advised the two camera men where to place the tripods. One tripod stood about five feet and the second, a small unit, held the camera only, about one foot from the ground.

Leaving the two camera men and the gofer in the yard, the director and I went upstairs to the back window where we could watch directly from above. When everyone was ready, I called the star with one word, "Charlie!" At the sound of my voice, Proper Charlie came out of the lagoon and started walking across the yard in the direction of my voice, but stopped about fifteen feet from the camera. He swung his head from side to side, lifted it high in the air; and then he spun around and ran back the way he came.

The director looked at me, "Can you get it back" he asked. I replied that I thought so, but then one of the camera men called. "Wait a minute, something is coming!" he said. "It's a gator, and a big one!" Then all three men below ran as if the devil himself was chasing them.

All my crocodiles answered to" Charlie." Although Big Charlie had been 100 yards away, he had made a beeline toward the sound of my voice. He had scented Proper Charlie, who had decided to be gone before Big Charlie arrived on the scene. The shoot was cancelled for the remainder of the day, but they did get some good footage of Big Charlie. When the camera men left suddenly, the little camera on its eight-inch tripod continued to run. It showed Big Charlie breaking thru the mangroves with jaws open, then running up to and over the camera! The crew left the next day.

CROCS IN HURRICANE KEITH

Hurricane Keith was an adventure for our crocodiles. At the peak, the water in our yard was over 5 ft. deep and just below the level of our floor. There were white caps all around our house. I noticed one of the crocs swimming in the yard. From his red-tinted neck, I was pretty positive it was Proper Charlie. He swam straight through and headed in the direction of the sea. I swear he had a grin on his face.

As I watched Charlie disappear, my attention was drawn to a large black saucer-shaped object which flew into my line of sight, travelling east, looking like a UFO (unidentified flying object). As I watched mesmerized, it sank below the tree line. I mentioned it to Dorothy, who looked at me rather strangely.

As we walked around in the eye of hurricane we saw over 20 birds, mostly swallows, but with a few night hawks, flying in circles, attempting to catch flying insects which were surprisingly numerous. As the birds were looking for a meal, so, too, was Charlie. He continually snapped at the low-flying birds as they flew by. That evening, Keith returned, but then it only had sustained winds of 100 mph. It almost felt balmy!

On the next day the storm had left. The sea was quite calm, and on the surface there was a surprisingly large amount of debris. I was watching when I saw some of the debris was moving, it was crocodiles, several of them feasting on dead animals drifting in the current. I went back for my camera, then walked west to the end of the Caye, where the current carried flotsom close to the island. Both Big Charlie and Proper Charlie had taken to the sea during and after the hurricane. No doubt they enjoyed a grand feast, for bodies, mostly birds, but also the occasional dog or cat was common flotsam for several days.

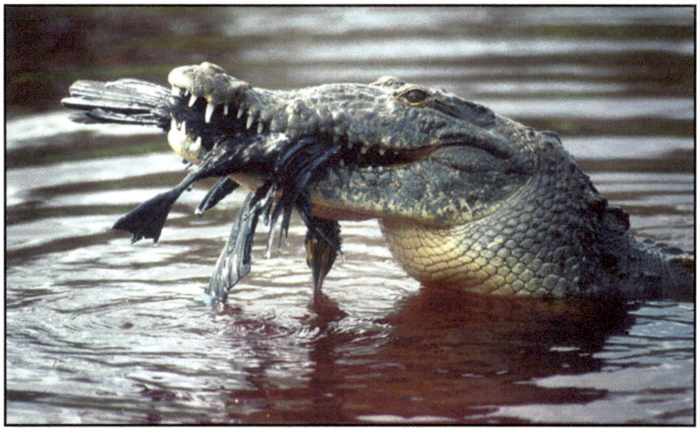

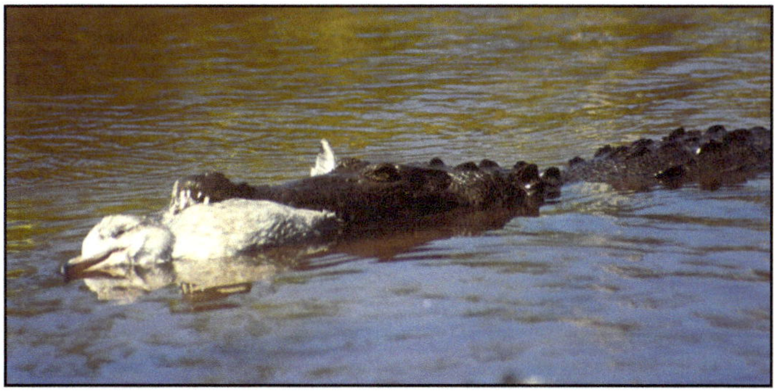

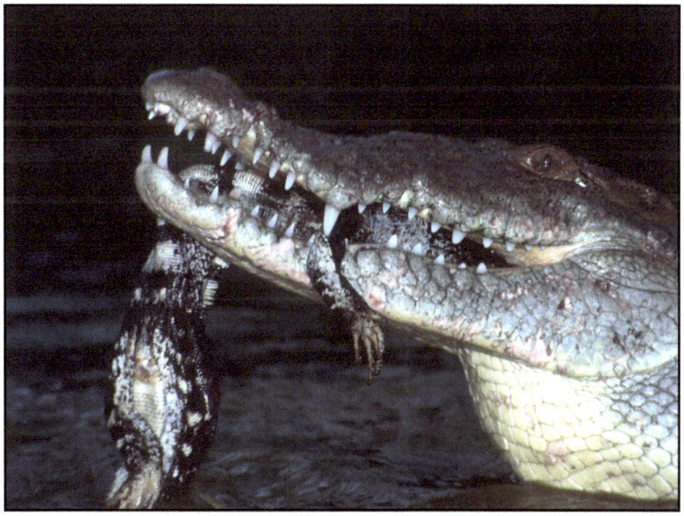

American crocodiles cleaning up dead birds (*top two photos*) and an iguana (*bottom*) after Hurricane Keith

Hard Luck Charlie and Little Charlie

But neither Proper Charlie nor Big Charlie came back after Hurricane Keith. I was on our roof, still replacing shingles, when I heard two gunshots over a period of three hours. I suspected what had happened, but so, too, did Little Charlie and Hard Luck Charlie. At each report of the rifle, they both let out a mournful wailing cry. Since then, I have heard this cry on four or more occasions, all of which were immediately after gunshots.

I didn't miss Big Charlie too much, but Proper Charlie was an amazing animal; he was sadly missed. We had two crocodiles left, Little Charlie and Hard Luck Charlie. They both moved into Big Charlie's Lagoon, but visited our lot frequently. And I visited them whenever I had some chicken fat or bones to throw away.

After Hurricane Keith I was very busy repairing roofs, salvaging boats and cleaning up the mess of flotsam that had invaded. So a month passed before I went looking for that UFO. Gaining access wasn't easy because of the large amount of fallen trees and broken branches, but it was quite obvious in the still naked trees. My black UFO was a lid from a large water tank lodged about fifteen feet high, in a naked white mangrove tree. Immediately I saw a use for it, a perfect watering hole for the birds. Later I learned that crocodiles also appreciate a clean bath now and again.

Near our house is a well that has not been used for years. One day on glancing down into the well I was surprised to see a crocodile looking up at me. What to do? I called Tony, a reptile enthusiast who worked at the Belize Zoo. He said he would be glad to help , but I would have to call the Conservation Unit of the Forest Department to get a permit.

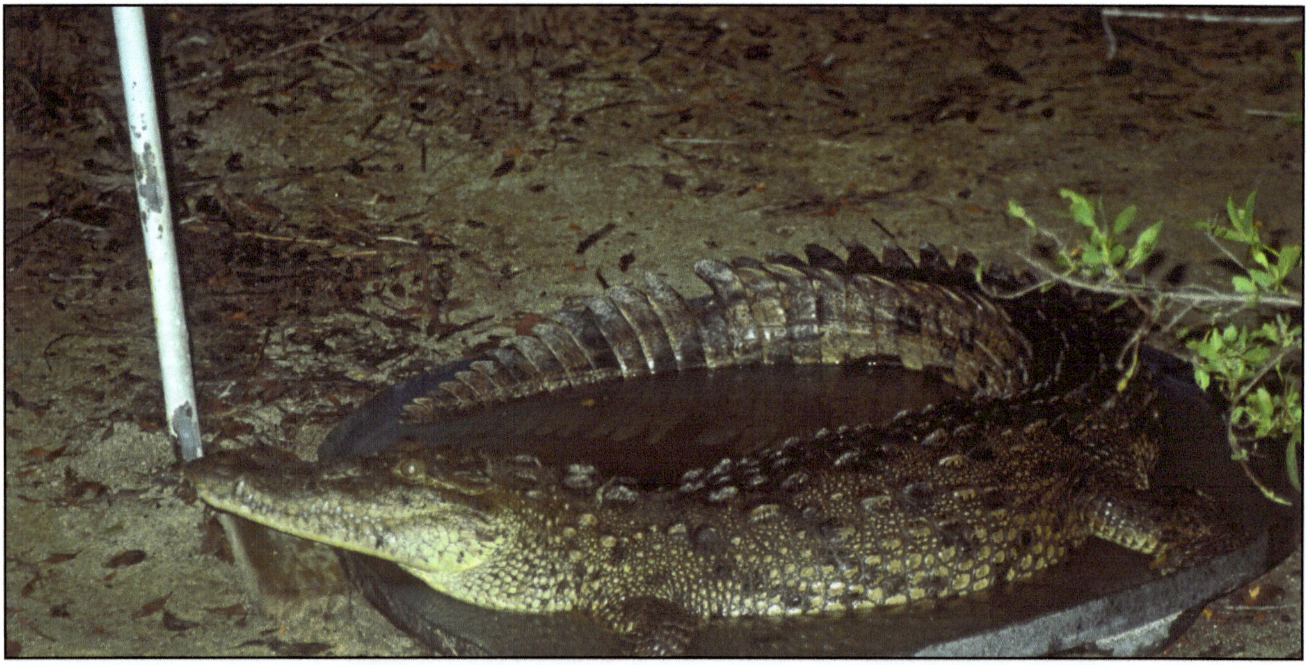

Hard Luck Charlie enjoying a bath in the UFO birdbath

I called the Forest Department. They asked what I was going to do with it. I was tempted to come out with a smart remark, but said only, "Return him to the lagoon where he has been living the last few years!" My permit was granted. Then I waited and so did the croc.

While waiting, I fed the crocodile. Every time Dorothy and I had chicken for supper, I took the fatty skin to the well. I didn't think Charlie would have the manoeuvrability to find the fat if I merely dropped it down the narrow shaft, so I tied the fat to a string and lowered it down to the crocodile. Ever so gently, he pulled the fat off the string and swallowed it.

One day just after I had fed the croc, I was startled by two men who had walked up behind me. "Are you Jim Beveridge?" the smaller of the two asked me. I replied that I was. "We've come to see what you've done to the crocodile," he told me. I replied "Nothing, he is still in the well!" He identified himself as Marcelo Windsor from the Forest Department, and his companion was Oliver, who was over six feet tall and very well built. Marcelo said "Ok, we have all the gear in the boat, let's do it now." The equipment was fetched and with no help Oliver snared the crocodile and hauled it up. It was then measured and weighed. For safety, his jaws were fastened closed with a half inch strip of rubber enclosing a slender stainless steel wire. By then I had recognised the crocodile as Hard Luck Charlie living up to his name. The three of us carried Hard Luck Charlie, who was a little over seven feet long, to the lagoon where Little Charlie waited.

As Hard Luck Charlie swam out, the long stainless steel line followed and Oliver held it. Soon the fine wire cut through the strip of rubber, which sprung loose from his jaws.

Hard Luck Charlie was free once more. That night, Dorothy and I were awoken by several crocodile roars. We thought the Charlies were celebrating. Three days later they came over to visit us in our backyard. We watched from our large back window as the two crocodiles grinned up at our two cats which were beside us, and making insulting remarks!

Both Charlies stayed for a couple of days in our yard, then they returned to the larger and deeper pool, formally occupied by Big Charlie. I continued visiting them after chicken dinners, when they received some chicken bones and fat.

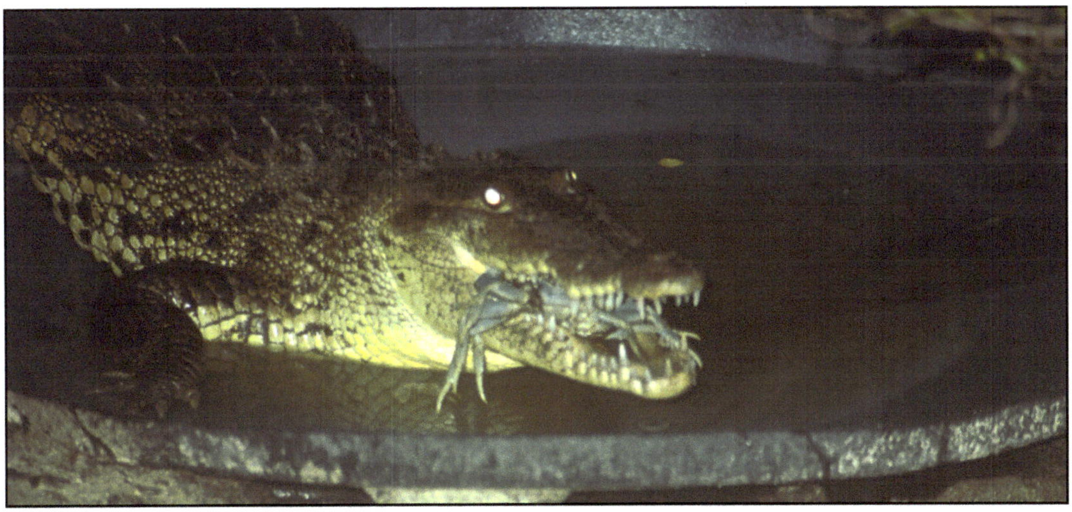

Little Charlie enjoying a midnight snack while bathing

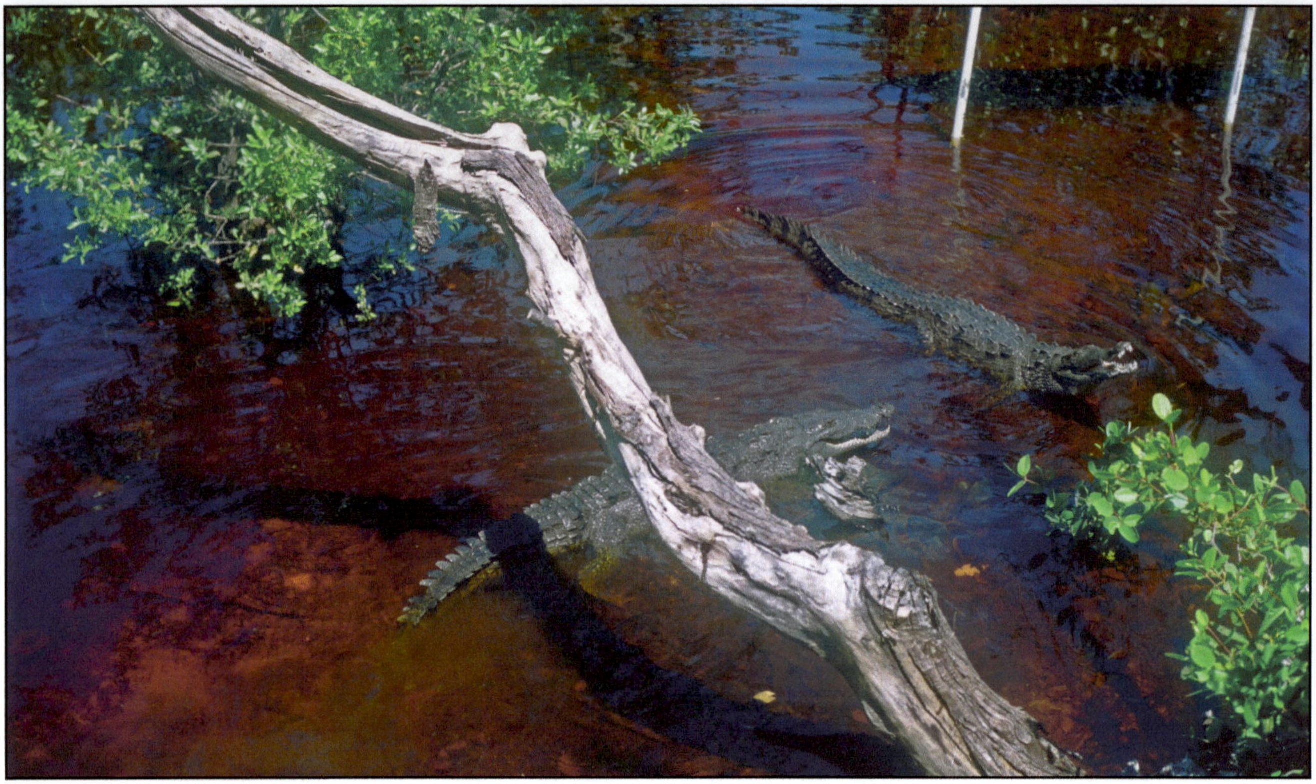

Hard Luck Charlie (*left*) and Little Charlie (*right*) laughing at the derogatory remarks from our cats

About a year later when I visited them, I threw chicken fat into the pool where it would float until it was swallowed. The bones as usual, I left at the edge of the water. I turned and walked away. I had made no path to the hole for I didn't want to leave an invitation to strangers or village kids. I had walked about twenty yards through the mangroves, when I heard the breaking of twigs behind me, I turned and there was a crocodile running towards me, and only fifteen feet away.

I turned to run, but my foot caught on a root. As I fell, I spun around to face the croc. There I was, my body weight resting on my heels and both my rigid upright arms. My legs were apart and bent at the knees.

In times of stress one thinks of weird things. I recall thinking that I shouldn't lower myself down, for there was a broken mangrove stump dangerously close to my rectum. Charlie slowed down and stopped, his head directly between my knees. A thousand thoughts raced through my mind. What do I do? Jumping on him seemed pretty much impossible given my position, so I thought the next move would have to be his.

We stared into each other's eyes for a while until I realised that Charlie was waiting for me to react. I pushed myself up with my arms so I was now vertical and I felt a little more in control. I lifted my left foot over his neck and placed it by my right. I stepped back, now recognizing him as Hard Luck Charlie, and said, "OK Charlie, let's go." Then I started to walk back to the lagoon. Charlie turned around and together, side by side, we both walked to where I'd left the bones. Once there, Charlie started eating. I stood by his side for ten seconds then walked home, all the time

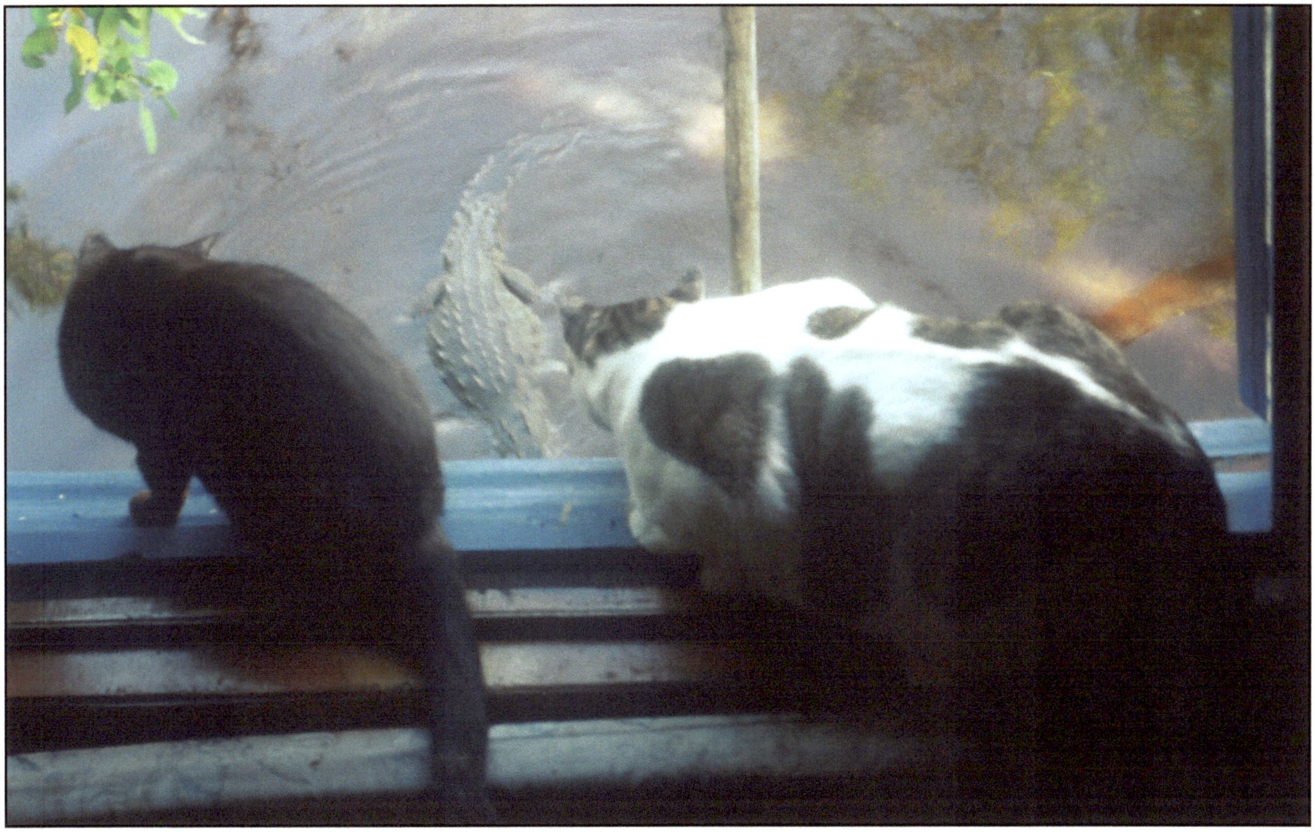

Spot and Kitty making derogatory remarks to the crocodiles

wondering, "Did that really happen?" When I told Dorothy, she recalled the time when Proper Charlie rested his head on her foot.

About two years later, Rad arrived in Belize with a friend who wanted to see a crocodile. I picked up a piece of chicken fat from the counter where I had just been preparing a chicken for dinner. At the edge of the lagoon, I called out one word, "Charlie!" Suddenly Hard Luck Charlie appeared in front of us with a dead white ibis in his mouth.

We watched as Charlie slowly swallowed the ibis. I wondered if he would have room for the chicken. I took the fat out of the zip lock bag, put it in the palm of my hand and said "Charlie". Immediately he started walking towards me, when he was about three feet away, I dropped the fat in front of its nose. He picked it up and swallowed, then looked up at me again. I showed him my open palms. He turned around, and went back to his hole in the mud and we went home to eat our chicken.

Eight months later and the seasonal rains were upon us. While working in my yard, I heard rifle fire. Almost immediately, Little Charlie screamed. I knew instantly Hard Luck Charlie was gone. I was devastated!

However, we again have two crocodiles, Little Charlie and Baby Charlie, who visit our yard when the lagoons are flooded.

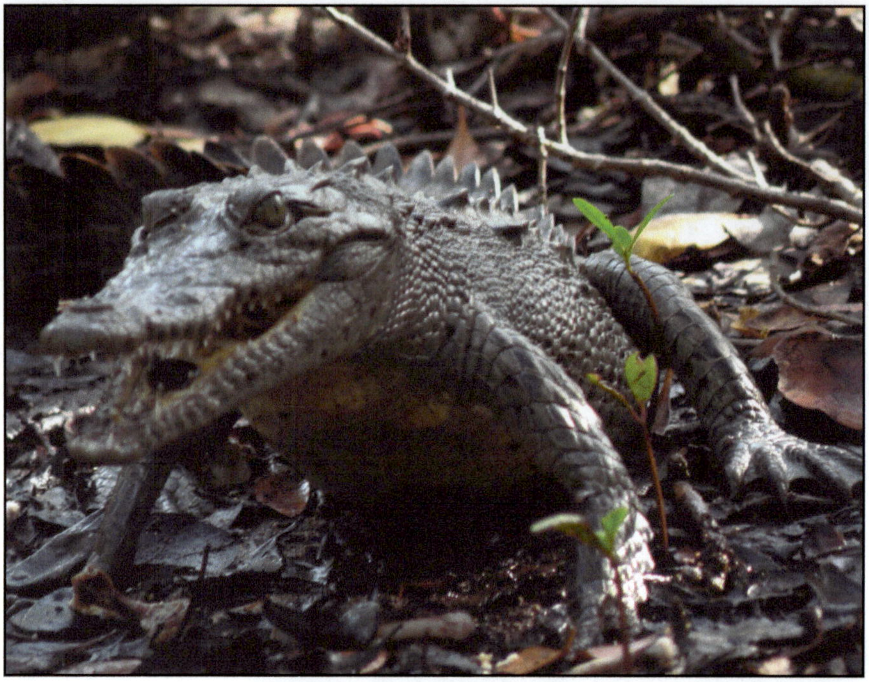

Baby Charlie and Wee Charlie

On February 14th, 2006, a new croc came into our lagoon that we called Baby Charlie; at that time it was approximately 28 inches long. By 2011 he had grown to four feet. He is very friendly and actually ran to meet me when he entered our yard a few weeks ago. Little Charlie still vists our yard, but she is no longer little, but over eight feet long. One morning in October 2011 we were out watching the sunrise and saw a 28-inch crocodile on a stump under a mangrove tree in the sea. We named him Wee Charlie.

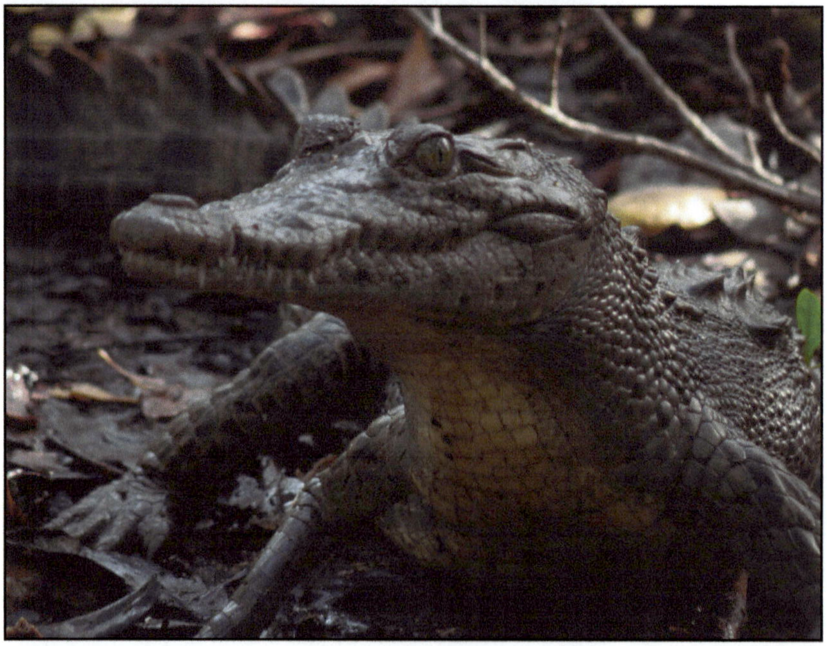

Two faces of Baby Charlie

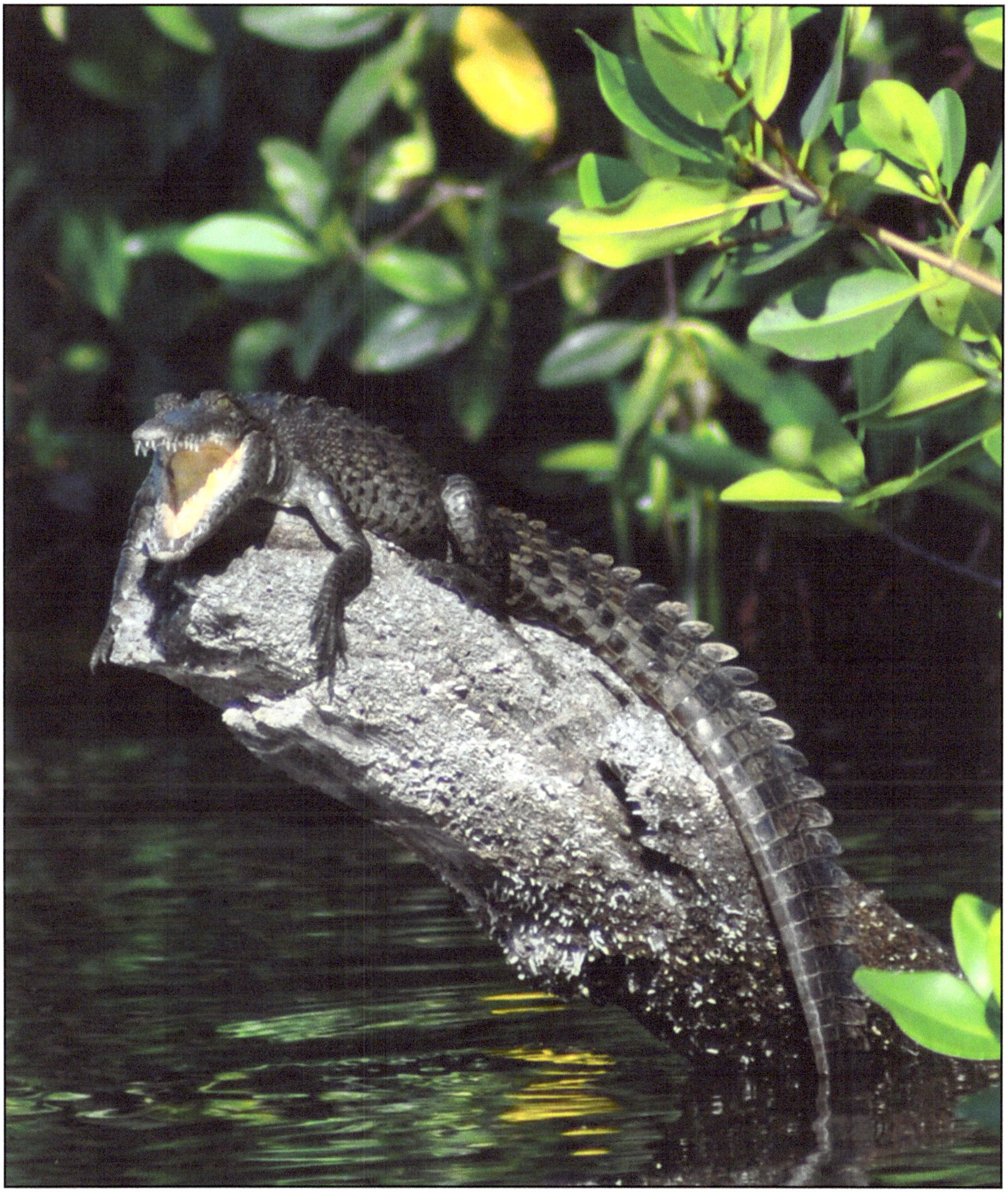

Wee Charlie

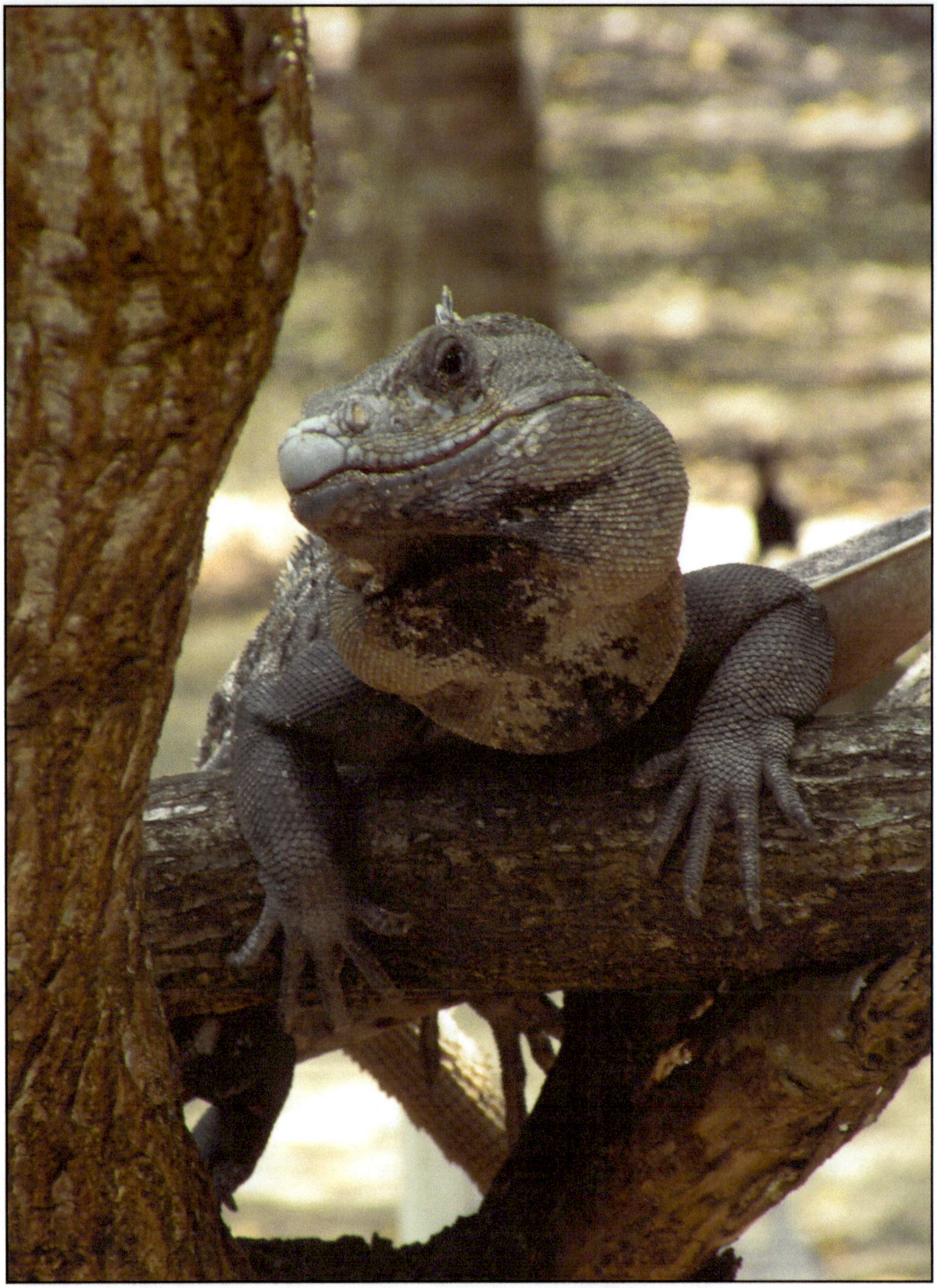

Black Iguana (*Ctenosaura similis*)

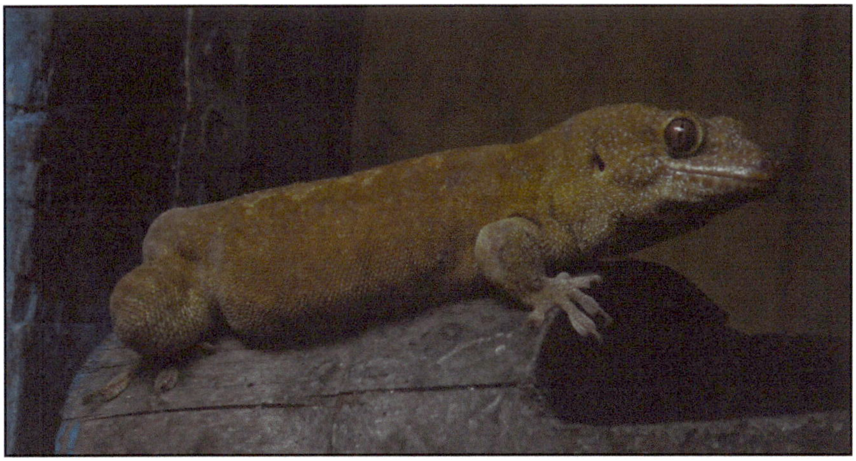

House Gecko (*Hemidactylus frenatus*)

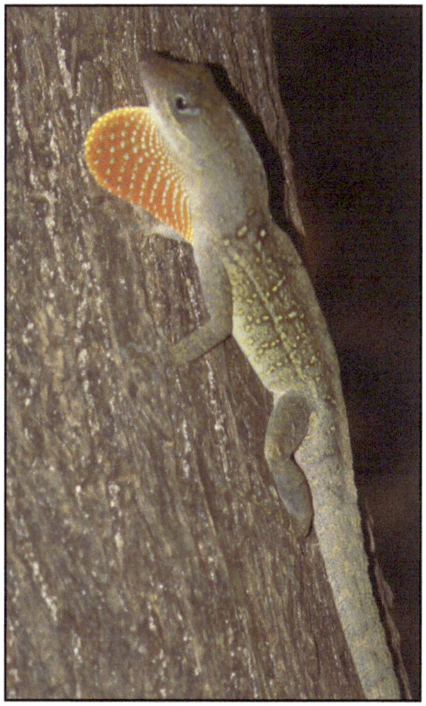

Brown Anole (*Norops sagrei*)

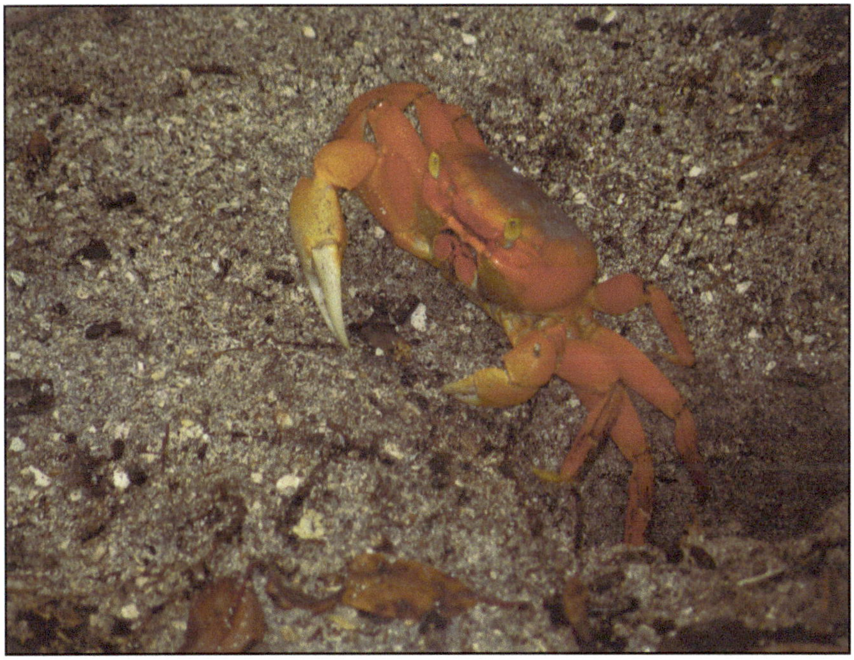

Land Crab

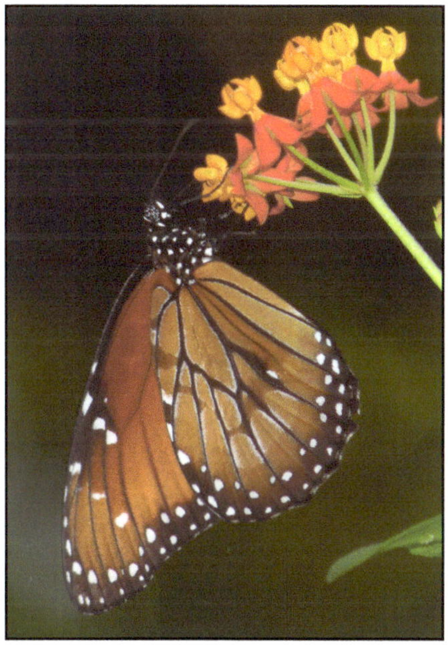

(*right*) Monarch Butterfly (*Danaus plexippus*)
on Tropical Milkweed (Asclepias curussavica)

Final Words

You are strolling along a beautiful beach
and suddenly you see a pretty seashell in the sand.

Let it be, there are many others
who are in a much greater need than you!

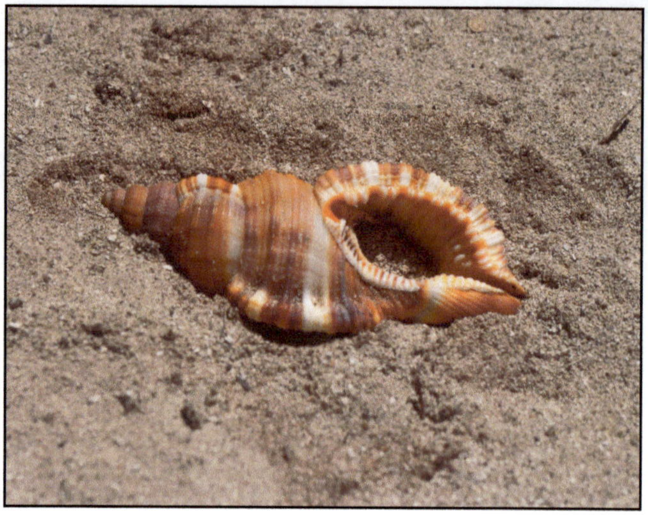

A tempting shell

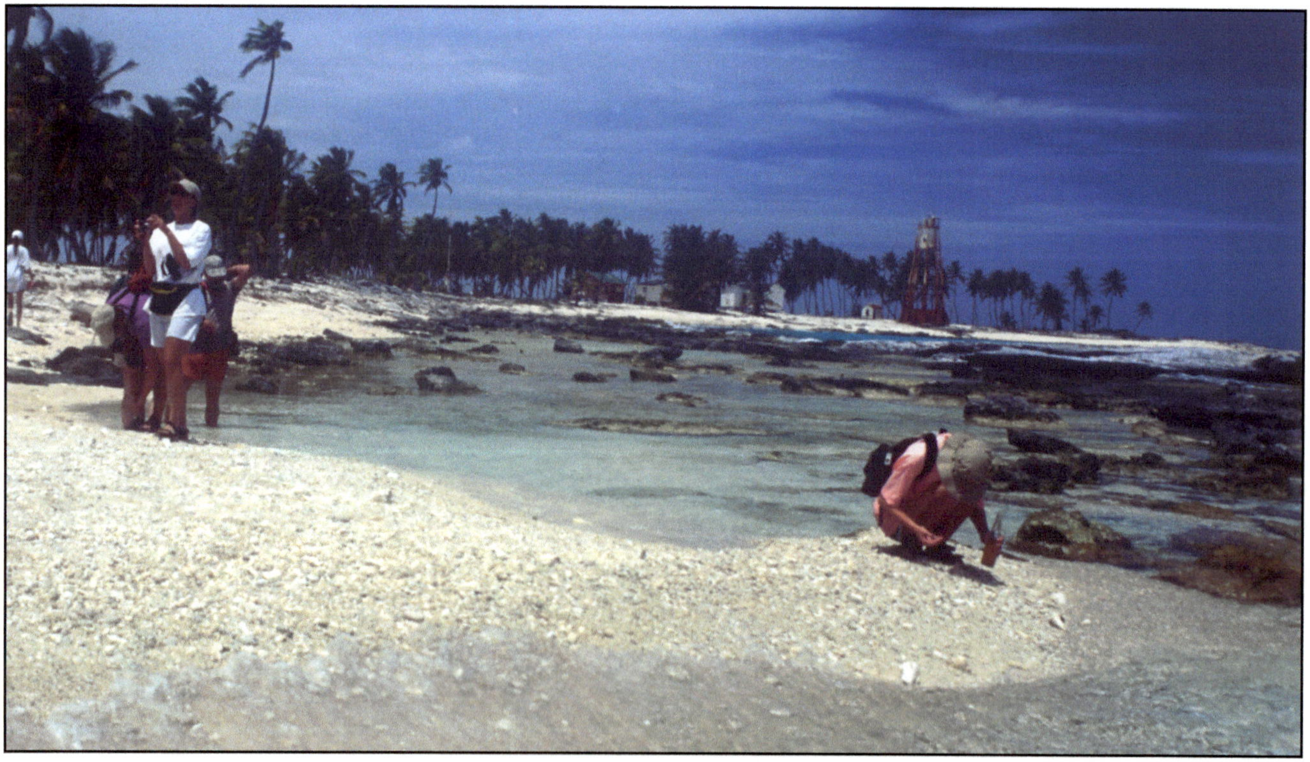

Looking for shells at Half Moon Caye

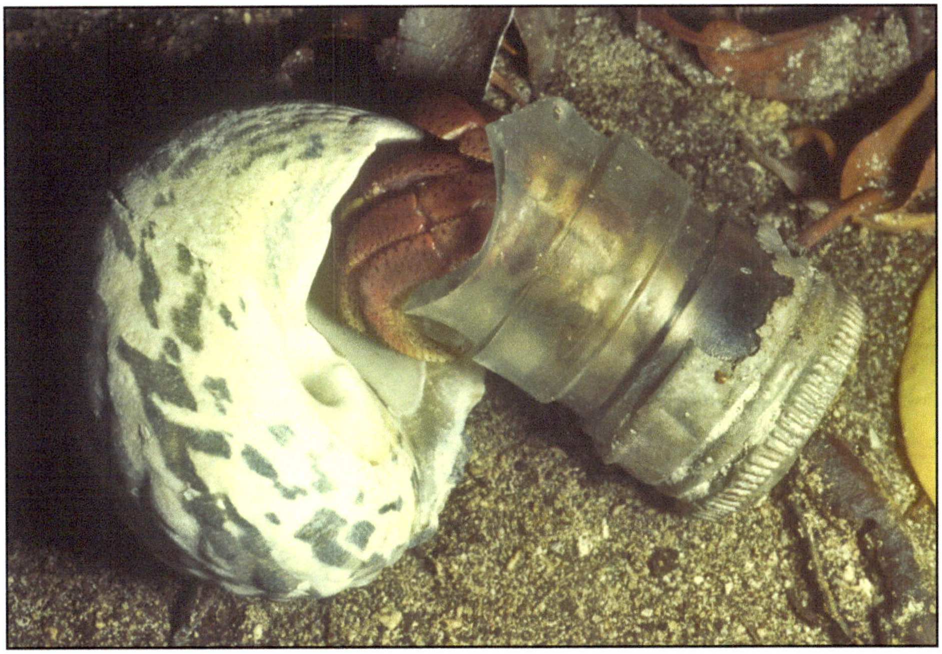

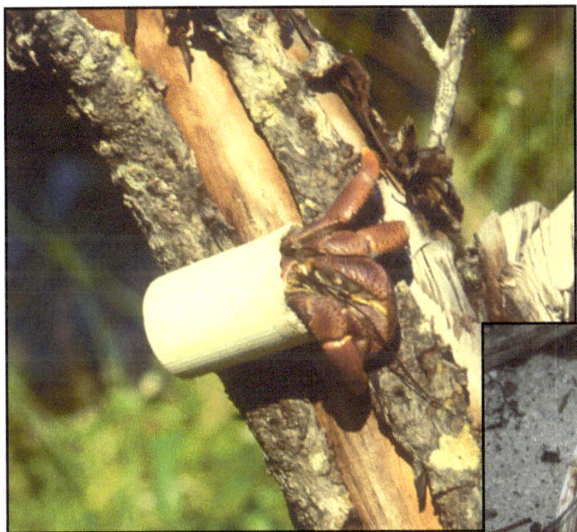

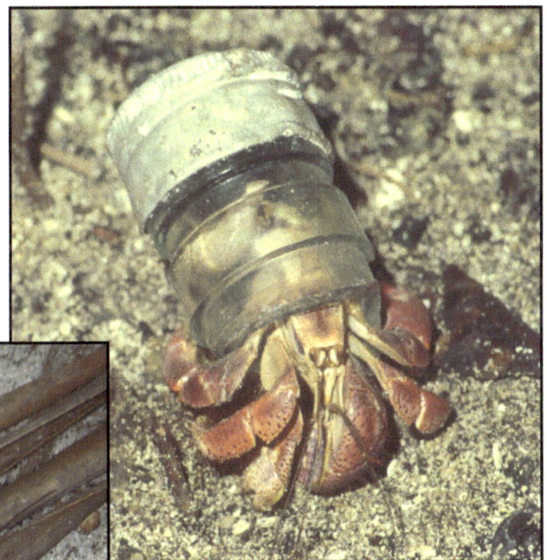

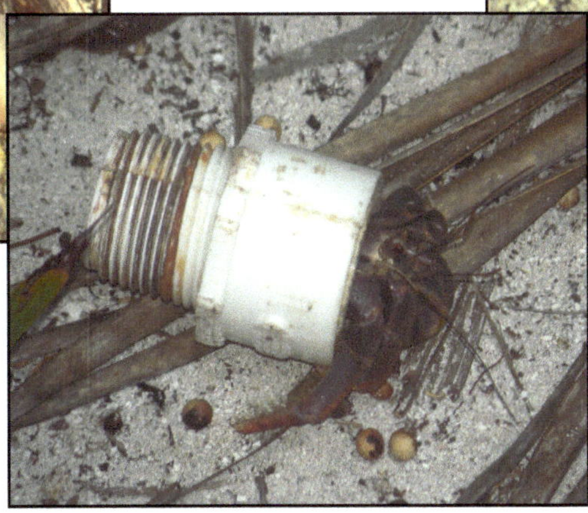

Caye Caulker hermit crabs use whatever they can find for their new home. The one at the top is moving from a bottle top to a shell.

Bibliography

Campbell, Jonathan A., 1999. *Amphibians and Reptiles of Northern Guatemala, the Yucatán, and Belize*. Norman, OK: University of Oklahoma Press.

Cousteau, Jacques-Yves and Phillippe Diolé, 1973. *Three Adventures: Galapágos, Titicaca, The Blue Holes (The Undersea Discoveries of Jacques-Yves Cousteau)*. New York: Doubleday.

Dunn, Jon L. and Kimball L. Garrett, illustrated by Thomas R. Schultz and Cindy House, maps by Sue A. Tackett and Larry O. Rosche, 1997. *A Field Guide to Warblers of North America*. New York: Houghton Mifflin Company.

Emmons, Katherine M., et al, 1996. *Cockscomb Basin Wildlife Sanctuary: Its History, Flora and Fauna for Visitors, Teachers, and Scientists*. Caye Caulker, Belize: *Produccciones de la Hamaca*.

Garel, Tony and Sharon Matola, 1996. *A Field Guide to the Snakes of Belize*. San Jose, Costa Rica: *Corporation Grafica*.

Howell, Steve N.G. and Sophie Webb, 1995. *A Guide to the Birds of Mexico and Northern Central America*. Oxford: Oxford University Press.

Humann, Paul and Ned Deloach, 2002. *Reef Coral Identification: Florida, Caribbean, Bahamas*, 2nd edition. Jacksonville, FL: New World Publications, Inc.

Humann, Paul and Ned Deloach, 2002. *Reef Creatures Identification: Florida, Caribbean, Bahamas*, 2nd edition. Jacksonville, FL: New World Publications, Inc.

Humann, Paul and Ned Deloach, 2002. Reef Fish Identification: Florida, Caribbean, Bahamas, 3rd edition. Jacksonville, FL: New World Publications, Inc.

Jones, H. Lee, illustrated by Dana Gardner, 2003. *Birds of Belize*. Austin: University of Texas.

Jones, H. Lee, 2011. *Belize Bird Checklist*. <belizebirds.com/bbrc/belize-bird-checklist>

Meyer, John R. and Carol Farneti Foster, 1996. *A Guide to the Frogs and Toads of Belize*. Malabar, FL: Krieger Publishing Company.

Reid, Fiona A., 1997. *A Field Guide to the Mammals oof Central America and Southeast Mexico*. New York: Oxford University Press.

Rietsema, Jacob and Dorothy Beveridge, 2009. *The Plants of Caye Caulker*. Caye Caulker, Belize: *Produccciones de la Hamaca*.

Stafford, Peter J. and John R. Meyer, 2000. *A Guide to the Reptiles of Belize*. London: Academic Press.

Index

D

E

F

G

H

I

J

K

L

Q

R

S

T

About the Author

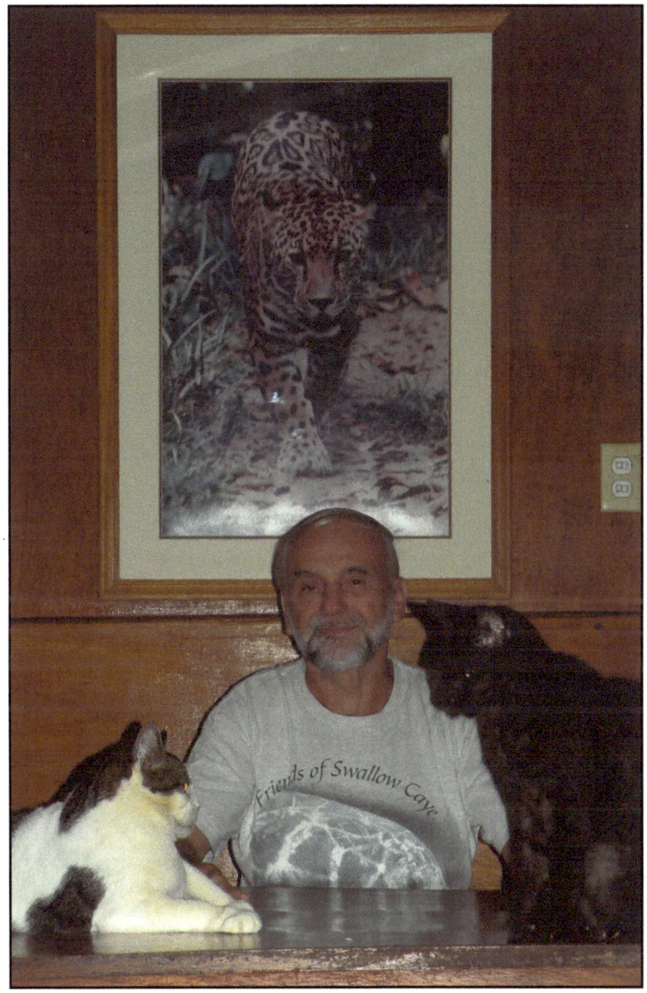

Jim Beveridge was born and raised in Scotland. He has traveled extensively in Europe and the Americas. He has worked in a variety of fields as a carpenter, salvage diver, master plumber, SCUBA dive master, and underwater and wildlife photographer. His photos have been published in international journals and books, such as, *Nature Conservancy Magazine, Watching Wildlife: Central America, The Rough Guide to Belize, Wanted Alive/Fragile Frogs, Cultures of the World: Belize, Tree Frog, Medicines from Nature*, as well as other books, textbooks, magazines, and calendars. Articles written and illustrated by Jim have been published in *Cockscomb Basin Wildlife Sanctuary, Aqua Geõgraphia, Belize Audubon Society Newsletter*, and other publications.

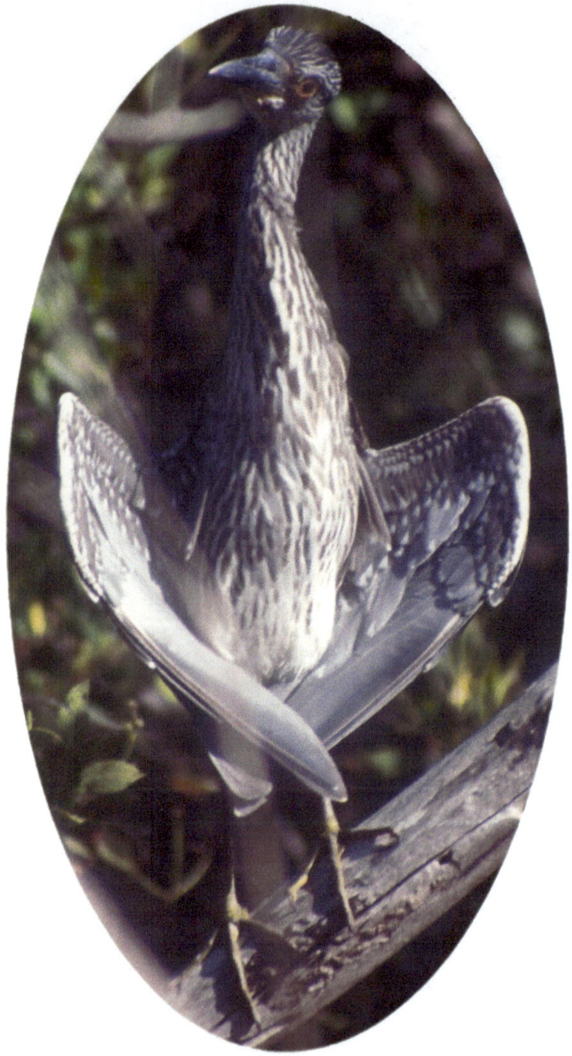

That's all folks!

www.ingramcontent.com/pod-product-compliance
Lightning Source LLC
Chambersburg PA
CBHW041320210926
51847CB00002B/10